T0245402

CAMBRIDGE LIBRARY COLLECTION

Books of enduring scholarly value

Technology

The focus of this series is engineering, broadly construed. It covers technological innovation from a range of periods and cultures, but centres on the technological achievements of the industrial era in the West, particularly in the nineteenth century, as understood by their contemporaries. Infrastructure is one major focus, covering the building of railways and canals, bridges and tunnels, land drainage, the laying of submarine cables, and the construction of docks and lighthouses. Other key topics include developments in industrial and manufacturing fields such as mining technology, the production of iron and steel, the use of steam power, and chemical processes such as photography and textile dyes.

Motor Vehicles and Motors

'Cheap or rapid or convenient road transport for man and goods is one of the most important of all contributions to national comfort and prosperity.' An early evangelist for the automobile, William Worby Beaumont (1848–1929) drew on his engineering background to produce the first volume of this work in 1900, when motor vehicles were still relatively new to British roads. Rapid developments in the automotive industry prompted the publication of a second volume in 1906. Replete with technical drawings and photographs, the work describes in great detail the design, construction and operation of the earliest motor vehicles, including those powered by steam, electricity and fuels derived from oil. Volume 1 traces the development of the automobile, from various attempts to produce steam vehicles light enough to run on roads through to the advances of Daimler and Benz. It also includes an overview of attempts to harness electrical power to propel road vehicles.

Cambridge University Press has long been a pioneer in the reissuing of out-of-print titles from its own backlist, producing digital reprints of books that are still sought after by scholars and students but could not be reprinted economically using traditional technology. The Cambridge Library Collection extends this activity to a wider range of books which are still of importance to researchers and professionals, either for the source material they contain, or as landmarks in the history of their academic discipline.

Drawing from the world-renowned collections in the Cambridge University Library and other partner libraries, and guided by the advice of experts in each subject area, Cambridge University Press is using state-of-the-art scanning machines in its own Printing House to capture the content of each book selected for inclusion. The files are processed to give a consistently clear, crisp image, and the books finished to the high quality standard for which the Press is recognised around the world. The latest print-on-demand technology ensures that the books will remain available indefinitely, and that orders for single or multiple copies can quickly be supplied.

The Cambridge Library Collection brings back to life books of enduring scholarly value (including out-of-copyright works originally issued by other publishers) across a wide range of disciplines in the humanities and social sciences and in science and technology.

Motor Vehicles and Motors

*Their Design, Construction
and Working by Steam, Oil and Electricity*

VOLUME 1

W. WORBY BEAUMONT

CAMBRIDGE
UNIVERSITY PRESS

CAMBRIDGE
UNIVERSITY PRESS

University Printing House, Cambridge, CB2 8BS, United Kingdom

Cambridge University Press is part of the University of Cambridge.
It furthers the University's mission by disseminating knowledge in the pursuit of
education, learning and research at the highest international levels of excellence.

www.cambridge.org
Information on this title: www.cambridge.org/9781108070607

© in this compilation Cambridge University Press 2014

This edition first published 1900
This digitally printed version 2014

ISBN 978-1-108-07060-7 Paperback

This book reproduces the text of the original edition. The content and language reflect
the beliefs, practices and terminology of their time, and have not been updated.

Cambridge University Press wishes to make clear that the book, unless originally published
by Cambridge, is not being republished by, in association or collaboration with,
or with the endorsement or approval of, the original publisher or its successors in title.

12 HP. Panhard & Levassor Daimler Motor Carriage. (*See pp.* 140, 380.)

MOTOR VEHICLES AND MOTORS

MOTOR VEHICLES
AND MOTORS

THEIR DESIGN CONSTRUCTION AND WORKING BY STEAM OIL AND ELECTRICITY

BY W. WORBY BEAUMONT

Mem. Inst. Civil Engineers, Member of the Inst.
Mechanical Engineers, and Member of
the Inst. Electrical Engineers

WESTMINSTER

ARCHIBALD CONSTABLE & COMPANY Ltd

PHILADELPHIA: J. B. LIPPINCOTT COMPANY

1900

BUTLER & TANNER,
THE SELWOOD PRINTING WORKS,
FROME, AND LONDON.

PREFACE

Objects and scope of book and illustrations

THE object of this book is to put its readers into possession of an accurate and critical account of modern forms of mechanically-propelled road vehicles of various kinds, and to give as many illustrations and descriptions as may be necessary to represent the chief types of motors and vehicles of the ninety-ninth year of the nineteenth century.

The history of efforts in the past to produce satisfactory self-moving vehicles is also so far treated as is necessary,—by drawings, descriptions and statistics,—to show what was accomplished or designed, and may be now usefully imitated or avoided.

It is intended, firstly, to be a book that shall be useful to engineers and motor vehicle constructors; and secondly, that those who take an intelligent interest in the construction and working of motor vehicles shall find in it more definite information, description, and explanation than has yet been placed before them on these subjects.

Nearly all the drawings have been made in the office of the author specially for this book, and now appear for the first time. Only two sets of the many herein published have been provided as they now appear, and it is claimed that the completeness of the drawings here given, showing not only disconnected details but everything in its place, makes them unique. Moreover, it may be mentioned that nearly all the line engravings herein appearing, having been produced from drawings specially made, will bear inspection with a magnifying glass; and in a few examples, such as Fig. 57, which show so much crowded into a small space, this may be necessary, but every detail and connection will, I believe, be found complete.

v

PREFACE

Rapid progress since 1896
Progress is now so rapid in the development of the various mechanical constituents of motor vehicles, more especially of those propelled by petroleum spirit motors, that even before the appearance of this book, some important details not included in the descriptions may appear. Sufficient experience has, however, already accumulated to show that the types of vehicles now most prominent are likely for a considerable time to be, in principle, and in the main features, those chiefly followed.

The motor vehicle a difficult mechanical problem
It may be safely said that the mechanically-propelled road carriage, which shall satisfy the requirements of a vehicle intended to take the place of a horse-hauled vehicle, offers to mechanical engineers and carriage builders one of the most difficult and comprehensive problems ever placed before them This is in great measure an explanation of the seeming slowness with which the motor carriage is taking the place it is destined to occupy.

Legal restriction and want of experimental experience in England
It must be remembered, however, that it was not until the end of 1896 that the designer and maker of a mechanically-propelled road carriage was permitted under the laws of the United Kingdom to make any road trials. Without such trials neither the experience nor the knowledge necessary to successful construction could be obtained.

Foreign freedom of use of roads and activity
Hence, although England is the true home of the motor vehicle, the constructors of other countries, particularly France and Germany, not fettered by such restrictions on the enterprise of the men who benefit their country by originating industrial enterprise, have had the honour of making their country the home of the modern motor vehicle.

In Great Britain are many thousands of miles of good high-roads, almost as good as an equal number of miles of the main roads of France, but our friends across the Channel are now making a dozen vehicles to England's one. In America, a great home of masters in practical mechanical engineering, there is not one mile of good road to our one hundred, and most of their street paving is so bad, or so spoilt by street railways, that trouble was the most certain reward of any attempts to use mechanical road carriages. Road construction is now, however, a subject of practical attention in the States, and this fact, and the potent fact of a probable market in Europe, is stirring

PREFACE

the severely practical American combination of mechanician and business man to that activity which will ere long put excellent American motor carriages of consolidated patterns on European roads.

British engineers and manufacturers are, however, now beginning to realise that, strange as it may seem, they are free to run motor carriages on British roads more or less free of Bumbledom, and the progress they have made in the three years of their emancipation has already shown itself in the practical results of many excellent vehicles and in the employment of some thousands of people.

Many, however, of those who are at work on the mechanical road vehicle are insufficiently acquainted with what has already been done by those who are far on the road to success, and they are in consequence repeating old designs and old experiments, only to arrive at similar conclusions and disappointments. To these, as well as to the users, or would-be users, of motor carriages, it is hoped that this book will be of service. Lessons from past achievements

Cheap or rapid or convenient road transport for man and goods is one of the most important of all contributions to national comfort and prosperity. For those who by ingenuity, diligence, much patience and much expense, provide this means of transport there should be no ungrudging reward. Unfortunately, there are no adequate means of securing this reward; for the end of years of work, though crowned with practical success, may leave the persevering worker in possession of a successful vehicle very little of which is protected under the Patent laws. That which is the most difficult to achieve may, after many trials, and many tons of scrap, which has cost much of several lives and almost its weight in gold, be achieved in the end by what may seem simple means. For this reason a knowledge of those things which have been the subject of past invention and of patents is of importance to those working in this field, and it is hoped that this book will provide much of that information. Importanc of quick and cheap transport Protection of invention in motor vehicle design

I wish here to acknowledge my indebtedness to several of those whose vehicles and motors are herein illustrated for permission to make sketches from their vehicles, and to others for tracings and photographs, from which suitable drawings have been made for reproduction here.

I would especially record the valuable aid I have received from my

PREFACE

assistant, Mr. C. R. D'Esterre, by whom many of the drawings herein given have been made entirely from measurement sketches of vehicles and their motors and gearing, and also for his assistance in the preparation of the chapter on electrical ignition apparatus. In respect of several vehicles drawings sufficient for the intended completeness of engravings for this book were not forthcoming from the makers. From sketches of the vehicles, their engines or motors and mechanism taken to pieces for the purpose, with permission of their owners, many of the working drawings herein have been made, and in this respect I have to thank Mr. F. H. Butler, the Hon. Mr. C. S. Rolls, Major Holden, Mr. H. Hewetson, Mr. A. F. Mulliner, Messrs. Brown Bros., Dr. E. E. Lewhess, Mr. S. F. Edge, Mr. H. A. House, Jun., Mr. W. C. Bersey, Mr. H. N. Searle, Mr. R. M. Ford, and Mr. F. R. Simms, and several whose names occur in connection with the vehicles illustrated.

The chapters originally intended to be written by Mr. Dugald Clerk, on the physics and economics of internal combustion motors, and a descriptive index to patents relating to motor vehicles, having been unavoidably delayed, will appear shortly as a separate volume.

W. W. B.

OUTER TEMPLE, LONDON,
  March, 1900.

Contents

ix

CONTENTS

List of Illustrations

LIST OF ILLUSTRATIONS

LIST OF ILLUSTRATIONS

xiii

LIST OF ILLUSTRATIONS

LIST OF ILLUSTRATIONS

LIST OF ILLUSTRATIONS

Introduction

HISTORICAL SKETCH OF THE MOTOR CARRIAGE

THE history of mechanical road locomotion offers very little encouragement to those who are entering upon it now. The conditions have, however, changed in three very important particulars.

Firstly, the railways, which absorbed all the attention and all the capital which the country could give to locomotion and to transport questions, and thus robbed the successes of Hancock, Gurney, Dance, Summers and Ogle, Maceroni and Squire, Hills, Russell, Church, Redmund, and others, of all the support which was required to establish steam locomotion on common roads, have now ceased to make material extensions either in length, means of distribution, or to be the most attractive means of investment. The railway trains, moreover, ran on private property, not the roads on which everybody was free to make demands and cause obstructions, or over which the baneful prejudice of bucolic magistracy had power, which the cyclist has lessened. This was a great advantage in favour of railway development.

British achievement sixty years ago

Past and present influences, conditions, and laws

Secondly, the far-reaching influences and beneficial results of mechanical engineering have again so far commanded a sympathetic ear in our House of Commons, that the laws for the suppression of mechanical road transport have been repealed. This, however, was not done until other countries, through their freedom from restriction in this respect, had begun the building of a great industry from which our manufacturers were excluded. The laws which covered us with ridicule, and which placed a heavy tax on one of the first necessities to national prosperity, namely, cheap transport and intercommunication, were of comparatively recent origin. In the days when Hancock ran his service of steam coaches between the City, Islington, Paddington, and between the City and Stratford, they did not exist, and were only partially represented by the arbitrarily applied turnpike laws, under which the toll farmers were permitted to multiply the ordinary charges for four-horse coaches by from five to twelve against the steam coach. Between Prescot and Liverpool the 4s. coach toll became £2 8s. 0d.; on the Bathgate road the 5s. toll became £1 7s. 1d.; between Ashburnham and Totnes the 3s. toll

Turnpike laws

1   1

INTRODUCTION

became £2. The opposition in some parts of the country to the introduction of the new mode of locomotion thus imposed a very heavy tax; but this would soon afterwards have been made impossible, for in 1831 a Select Committee reported to the Houses of Parliament completely in favour of steam coaches and the reduction in tolls. Moreover,

Enlightened government attitude in 1833

the Government, in 1833, introduced a clause into a bill then passing through Parliament, which exempted steam coaches from the operation of the Hackney Carriage Act. This was done to relieve steam carriage invention during its infancy, and like several other acts of the Government of the time, showed that there was as much, if not more, enlightened consideration and appreciation of the importance of the benefits derivable from mechanical road transport than there was in the twenty years preceding 1896.

Modern mechanical and material advantages

Thirdly, a most potent change has arisen from the all-pervading extension of mechanical appliances, namely, the general use of machine tools of variety and precision. These make accurate work not only possible, but cheap also, as compared with the means of 1830. From this has been derived the practical possibility and commercial advantages of the high-speed engines of modern times. These things have multiplied many times the power of a pound of metal in the form of a motor, for in prime movers, such as steam, gas, and oil engines, rotative speed is power, and at the stroke of a pen one horse has become two, three, four, or more. Hence the steam engine, which in 1830 required large and valuable space in the body of a coach, can now be put into a small box on the fore-carriage, or under a seat, or slung out of sight under the frame.

Present conditions and use of experience of the past

These three great changes have then placed the designers, makers and users of mechanical road vehicles under very different conditions from those under which Hancock worked during his twelve years of experiments from 1824 to 1836. The designer of to-day has the advantage of superior materials, superior appliances, and of considerable knowledge of the results of the costly experience of those who have gone before.

Against all this, however, has to be set the greater excellence which must now be attained and offered, as compared with that which would have commanded success in 1830, before the railways had accomplished so much in reducing the cost and increasing the facility of travel and transport. It must be remembered too that a knowledge of the results of the work of others is a small contribution to the means of successful achievement, as compared with the possession of the experience which led to those results. By experiment, by much trial and error, this experience has to be regained to reach and appreciate that which was achieved when those of bygone days left off, and to enable us to start on the improvements they would next have made.

To have before us, however, even the forms of the vehicles made nearly three-quarters of a century ago, is a great help, and these, with

2

INTRODUCTION

some knowledge of their reasons and results, will place those who patiently study them in a position of advantage, which may be easily under estimated.

From this point of view it is a matter for regret that the records of the work done then, are of so fragmentary a character, and that were it not for the full report of the Select Parliamentary Committee of 1831, many doubts might arise as to the reality of the actual achievements. Remarkable as these were, however, it is clear that at the end of all the enterprise of the time, none of the designers, even including Hancock, would have claimed to have reached more than those good results and practical experience, which were indispensable for their guidance in future, to produce commercially satisfactory self-propelled carriages. In this book it is not intended to deal at length with the history of the steam coaches, to which reference has been made. For much of this history reference may be made to the author's *Cantor Lectures*, published by the Society of Arts, in 1896; to his paper on "Developments of Motor Carriages," read before the Society of Arts, 27th November, 1896; to his papers on the same subject, read in March, 1896, before the Manchester Association of Engineers, and to the Cleveland Institution of Engineers, published in the *Automotor*, No. 7, p. 250, April, 1897; to his lectures to Liverpool branch of the Self-Propelled Traffic Association (now the Liverpool branch of the Automobile Club), published in the *Auto Car*, 12th September, 1896, p. 550, and *Automotor and Horseless Vehicle Journal*, No. 5, p. 196; No. 15, December, 1897, p. 105; No. 7. April, 1897, p. 250; and in the *Engineering Magazine* for September and for December, 1897, and January, 1900; in *Industries and Iron*, from December, 1896, to December, 1898. A description of Cugnot's locomotive will be found in the *Proceedings of the Inst. of Mechanical Engineers*, for 1853. Particular reference will only be made to the most instructive result; to those of vehicles which gave England the credit of originating practical road locomotion.

Select Committee Report of 1831

Chapter I

DESIGN, CONSTRUCTION AND WORKING OF EARLY STEAM ROAD VEHICLES

Hancock's early work

WALTER Hancock's work undoubtedly claims most attention. Hancock was born early in 1799, and died in May, 1852. He began by making a steam engine with two inflating and deflating india-rubber "cylinders," conceived the idea of applying it to a steam coach, found it would not do, and then, having had his thoughts turned to steam road locomotion showed that his knowledge of what others had done was not lost upon him, by first directing his efforts to the production of a suitable boiler.

Those who had preceded him, though only by a few years, and were cotemporary with him, were many in number,[1] and many companies were formed to make steam carriages, and run regular services with them. Only three, however, seem to have run carriages or coaches for hire, namely: Gurney, whose coaches were worked by Sir Charles Dance for four months between Gloucester and Cheltenham; Hancock, who established the services already mentioned; and Scott Russell, who established a service between Glasgow and Paisley.

Hancock's boilers, 1827-33

Hancock first made a boiler, consisting of sixteen horizontal tubes, $4\frac{1}{2}$ in. diam. and 4 ft. in length, connected together by small pipes, and a receiver, which acted also as a separator. The numerous pipe connections he found troublesome. He tried the same form of boiler with vertical main tubes, with a central smoke tube, 2 in. diam., through them, leaving a

---

[1] For an outline of the work done by many to whom reference cannot be made here, the reader may consult "Mechanical Road Carriages," *Cantor Lectures*, Society of Arts, 1896. *A Comprehensive Bibliography of Power Locomotion on Highways*, by Rhys Jenkins, should also be referred to for a complete index to the literature of this subject from the earliest times to 1896.

The first working road locomotive ever made, was undoubtedly that of a Frenchman, Nicholas Joseph Cugnot, in 1769. It was a three-wheeled vehicle, the front steering wheel carrying boiler and engine, the latter being a double cylinder high pressure, and actuating the driving wheel by two ratchet wheels and pawls with radius links, which were moved by direct connection with the piston rods. The vehicle was actually run on more than one occasion, and still exists in Paris; but it need hardly be said that it in no way led to either rail or road locomotion, any more than did the experimental carriage of Moore, of Leeds, in 1769, or so much as that of Murdoch in 1784.

smaller water capacity and increasing the heating surface. The numerous pipe connections were, however, equally undesirable, and hence his invention of the flat, narrow water-chamber boiler, illustrated by Fig. 1. This he patented first in 1837, and in the improved form shown, in 1833, the water-tube boiler being patented in 1825.

When first made, the flat water chambers of the 1827 boiler were separated by thin vertical iron bars, between which were numerous passages for the products of combustion. The steam pressure in the flat chambers or envelopes caused the thin sheet metal of which they were

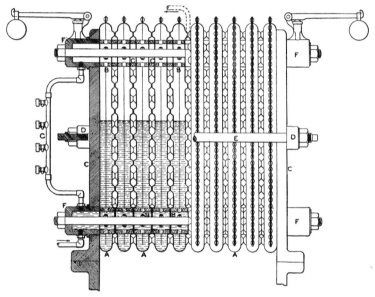

Fig. 1.—Hancock's Boiler, 1833.

made, to bulge out between each bar, thus converting them into corrugated side plates. The bars thus became less necessary, the summits of the corrugations when brought together made the plates mutually supporting, and left spaces for the passage of the hot products of combustion.

Hancock's most approved method of construction arose out of this result. It was that shown, in which the sheet forming the two side sheets of the chambers were embossed by hammering the metal into nearly hemispherical cavities made in a flat cast-iron mould for the purpose. One half the sheet was done at one fixing, and the sheet, when finished and bent round at A to the thin bag form, had similar bosses on either side to rest against the bosses of its neighbour.

The vertical edges of the bags were made up with a thin outwardly flanged rivetted-in plate, as seen at Fig. 3, or the side sheets were pinched together and rivetted, as shown at the right hand of Fig. 1. In the centre of the lower and upper part of these bags large holes were made for the passage through them of large stay bolts, perforated distance rings B, being placed inside the chambers at these holes and plain rings C, between the chambers. These chambers were placed between end plates C C, as seen at Fig. 1, which were pulled up together tightly by the upper and lower central tie bolts E, the pressure from which was carried by the distance rings, between which and the sheets were thin copper rings for making joint. The ring distance-pieces constituted a water communication between all the chambers, and to them, by the perforated rings and the upper series of rings, similarly formed a steam communication and receiver. In some cases the outside distance rings were dispensed with by outwardly embossing the sheets, as shown at the lower left-hand corner of Fig. 1. The end plates were also supported by two exterior bolts and buck stays. Safety valve fittings were fixed in the upper distance rings or on a continuation F, of the trunk pipe formed by them, as shown in Fig. 1. A pipe connection was made, as shown between the water trunk and the steam trunk, and on it were four well made test valves for ascertaining the water level in the boiler.

This boiler seems to have been an excellent steam maker and did not prime. The tubular boiler, like most of the boilers of the time, gave a good deal of trouble in this respect, even with the separator referred to. The large flat surfaces and freedom of the steam separation from the water in the chambers contributed to this. The use of a very small steam pipe connection at the boiler also helped to prevent priming.

Very few particulars as to the dimensions of these boilers can now be found. The steam pressure ordinarily carried by Hancock was 70 lb. per square inch, but he appears to have varied the pressure as circumstances, in the form of hills, loads, and bad roads, demanded, and used any pressure up to 200 lb., and over that.

As an example of these boilers, as used in the *Automaton* coach, the following approximate dimensions may be given:—

Eleven chambers of about 30 in. by 20 in. by 2 in. in thickness, and of $\frac{1}{8}$ in. charcoal plate. The heating surface of this would be about 85 sq. ft., and the grate surface about 6 sq. ft., but the heating surface of one of the boilers is given as 100 sq. ft. The engines used in the same coach had two cylinders of 9 in. diameter and 12 in. stroke, and ran at the same speed as the road wheels. These were 4 ft. in diameter, and at ten miles per hour would make about 70 revolutions per minute. The engines may therefore be taken as what they were stated to be nominally, namely, from 14 to 20 horse power, according to the pressure used and the speed.

This would for a mean power give about 5 sq. ft. of heating surface per average effective horse power, and about 0·3 sq. ft. of grate surface.

The numerous journeys on an 8-mile stage, which one of these coaches ran, gave opportunities for taking water and fuel measurements. By sorting down the evidence on this subject I find that the boiler evaporated as much as 10 lb. of water at ordinary temperatures per pound of

6

coke, and very little was done to heat the water on its way to the boiler. The boiler seems to have evaporated 8 lb. of water per square foot of heating surface per hour, and about 130 lb. per square foot of grate under forced draught of the fan.[1]

In some of Hancock's boilers the fire was directly under the boiler chambers, as in Fig. 1, but in others, which he preferred, it was placed to one side, as shown by Fig. 2. This arrangement constituted in a measure a gas furnace, and was probably an effective mode of converting a little of the exhaust steam into a water gas.

The whole of the exhaust steam seems to have been diverted into the case below the boiler at A, most of it passing upward through the space B

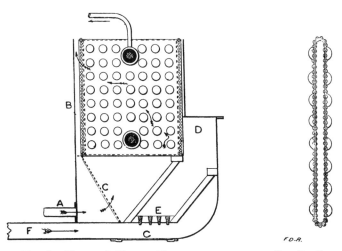

FIG. 2.—HANCOCK'S BOILER, FURNACE,
AND AIR ADMISSION.

FIG. 3.—END OF ONE
ELEMENT OF BOILER.

to the chimney, and being heated and dried on its way. Some of it passed through E and the finely perforated plate c, and was used as above mentioned.

The fuel was fed into the hopper D, and fell automatically into place as the coke was consumed below. The ashes were removed during the journey by shaking the bars E, which were cast in one piece. The two outer bars were provided with a rack underneath and a pinion geared into this for removing them end on through a hole in the side of the furnace case. As one set of bars was removed for cleaning, a second one was hauled in by a hooked end on the set hauled out. The hot set was left at a stopping-place on the out journey and taken up cold on the return. The main object of this arrangement was to reduce the difficulty in removing clinkers, a difficulty which used often to necessitate stops on the road.

[1] See papers by the author in *Proceedings of the Manchester Association of Engineers*, and in *Proceedings of the Cleveland Institute of Engineers*, March, 1897.

Forced draught

Hancock used forced draught in all his carriages, the air being sent by a fan through a flat air trunk into the space under the bars, as indicated in the engraving of the reconstructed *Infant*, Fig. 7, and in Fig. 2, where the air trunk is marked F. At the bottom of the latter was a sliding door G, through which the ashes and the whole of the fire could be dropped.

Hancock's first carriages 1828

Hancock made nine carriages, the first being the experimental carriage shown in outline by Fig. 5, on the top right hand of the page. This, like many of those by experimenters during the half-century following Hancock's endeavours, was a three-wheeled vehicle, the front wheel being the driver, driven by a pair of direct-acting oscillating cylinders, wheel and engine being carried in a frame jointed to part of the main frame for steering purposes. The outline engraving gives a sufficient idea of its arrangement. After running a few hundreds of miles its disadvantages led to a new design. Hancock has told very little of what these disadvantages were, but the springless foregear, the jointed steam connection, the heaviness of the mass to be moved in steering, and the small carrying capacity and incompleteness of arrangements for adjustment, were probably the chief. The new carriage which arose out of this was the *Infant*, which is shown in outline in Fig. 6. In this the engines were of the same type, but were fixed in a rear framing, and drove the wheels by means of an ordinary link chain. This vehicle, named the *Infant*, was a great advance, as it provided more seating accommodation, gave better steerage control, and inaugurated independent engines and chain gearing. The greatest difficulty is said to have arisen from the causes which are affecting many of our most recent carriages. He found that ashes, dust, and mud did not conduce to the longevity of piston rods, bearings and gearing, nor did these parts run quite so sweetly as when not buried in more or less effective abradents. What Hancock records is : " The difficulty of keeping the machinery clean, owing to its proximity to the fire place, as well as to the road, was found in practice to be so strong an objection that this form of carriage had also to be abandoned."

Hancock's second " Infant," 1830-31

This *Infant* was then rebuilt, or more correctly another *Infant* was built. As shown by Fig. 7, it was a much larger vehicle ; the fore-carriage was the same, but in all other respects it was a different carriage. Vertical engines were employed, placed in a space near the centre of the coach. Behind this space was room for the engine attendant, and at the rear was the boiler, which was stoked by a fireman standing on a rear platform. The main axle was driven, as in the first *Infant*, by ordinary cable chain, the chain wheels on crankshaft and driving axle, being apparently of the same size, but this does not seem to have been the case on all the coaches. A feed pump was worked from the crosshead of each engine, and an adjustable radius rod at each end of the crankshaft maintained the proper distance between that shaft and the main axle. The adjustment was probably used to take up wear of the chain. This carriage ran to Brighton and back twice.

8

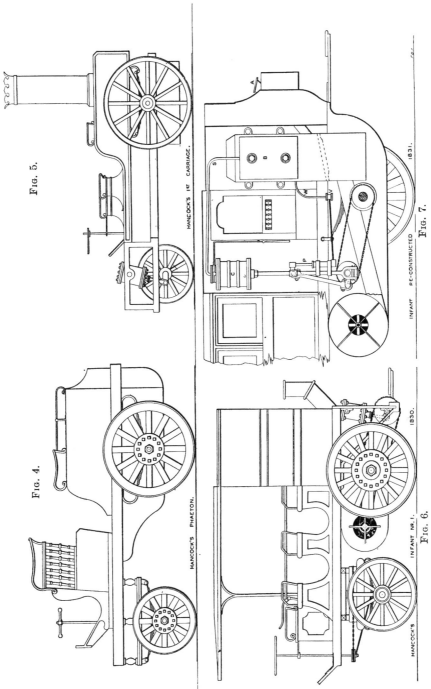

Fig. 5.

HANCOCK'S 1ST CARRIAGE.

Fig. 4.

HANCOCK'S PHAETON.

Fig. 7.

INFANT RE-CONSTRUCTED 1831.

Fig. 6.

INFANT Nº 1. 1830.

HANCOCK'S

Figs. 5, 6 and 7.—Hancock's First Carriage, the "Infants," and (Fig. 4) his Phaeton.

9

Hancock's "Era," 1832 Hancock's next vehicle was the *Era*, a still larger vehicle, with a double coach body in front, in place of the chars-a-banc of the *Infant*. The arrangement of the engine, gear and boiler seems to have been the same. This coach did a great deal of work. It was completed in 1832. Its boiler was made with chambers of about $\frac{1}{8}$ in. thick, instead of about $\frac{1}{24}$ in., which was tried and failed in the *Infant*. It was intended to run the coach between London and Greenwich, but the "London and Brighton Steam Carriage Company," for which it was built, never got into working order.

A London and Paddington Steam Carriage Company was started in 1832, and Hancock's next carriage, the *Enterprise*, was built to its order. The only noteworthy accident with any of these coaches occurred with this one. The engine attendant had fastened the safety valve lever down with copper wire, and had started the engines and blower while the coach was standing. The effect of the forced draught was that intended, but the steam not being used as fast as made, nor free to escape at the safety valve, it got out elsewhere, and at the inquest which followed, the evidence was more complimentary to the boiler than to its attendant. The real weight of Hancock's carriages may be judged from a statement in p. 83 of his narrative, to the effect that the *Era* and *Autopsy* together, in working order, weighed not less than 7 tons.

Hancock's later "Autopsy," 1833 Hancock next built the *Autopsy*, Fig. 8, which he completed in the autumn of 1833, and ran to Brighton with it, and afterwards ran it daily for twenty-four weeks for hire between Finsbury Square and Pentonville. In 1834 he built, in three months, a steam drag for a purchaser in Vienna.

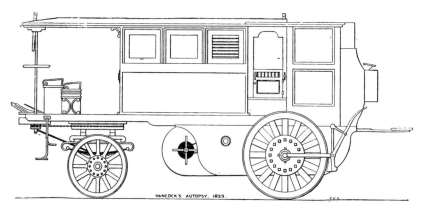

FIG. 8.—HANCOCK'S STEAM COACH "AUTOPSY," 1833.

This would carry six passengers, and driver and stoker, and haul a carriage carrying eight persons. This vehicle, as well as the *Autopsy*, had the same arrangement of machinery as that of the second *Infant*. It never went to Vienna, the purchaser first having ordered a carriage, then ordered its alteration to work as a drag, and then refused to pay the cost of the alterations.

In August, 1834, he completed another carriage called the *Era*, with a coach body and open front (it was afterwards named the *Erin*), and ran this and the *Autopsy*, from the middle of August to the end of November, for hire between the City, Moorgate and Paddington, and carried nearly 4,000 passengers. At the end of this time Hancock was induced to take the *Era*, re-named the *Erin*, to Dublin. It was shipped on board the steam vessel *Thames* on the 30th December, and arrived in Dublin on the 6th January, 1835. Here it ran for eight days, "sufficient to effect the purpose of its visit," and was re-shipped on the *Shannon* to Stratford. In August, 1835, the *Erin* ran to Marlborough and back, and afterwards to Birmingham and back. A London and Birmingham Steam Coach Company was formed, and more carriages ordered. The fares were not to exceed £1 inside and 10*s*. outside for the ride to Birmingham. Judging by a remark on p. 83 of Hancock's narrative, the *Era* was fitted with some kind of slow gear.

Hancock's "Era," 1834, and coach services

In the middle of 1836 the drag intended for Vienna, having then been altered and enlarged to a chars-a-banc coach to carry eighteen to twenty-two passengers, was named the *Automaton*, and put to work with others on a regular service between Stratford, Paddington and Islington. This service was continued for about twenty weeks; the carriages in that time ran 4,200 miles, carried 12,761 passengers, and the City was traversed over 200 times.

Fig. 4, the first of the four engravings on p. 9, is an outline of the last carriage Hancock built, namely, his phaeton. Of this he speaks with pride in his narrative (p. 98), and mentions it as intended for his private use. He had run it a good deal in and about London at from 10 to 12 miles an hour, but had reached 20 miles with it. There do not appear to be any existing drawings showing the arrangement of the engine and boiler, and no descriptive details.

I have given these particulars of the work done by Hancock with his carriages because they show that highly practical success was achieved, although it appears quite plain that commercial success was not very promising for town service at the time, and the railways proved much more promising for the long-distance services. Hancock published a narrative of his twelve years of experiments in 1838. From this and from cotemporary writings it is evident that extended practice and experience were required to make what, even with good roads, would have proved attractive and commercially successful vehicles. Somewhat frequent mishaps occurred, as might be expected, and it is questionable whether the comfort of the vehicles was even up to the standard of the time. True, the passengers were all in front of the machinery, but with powerful and unbalanced engines, and with the rough chain gear, the vibration must have been considerable. The *Automaton* had cylinders no less than 9 in. in diameter, and these engines had no fly-wheels. All these things were, however, matters for improvement, which would naturally have followed demand for the coaches and for improved tools and methods of construction.

Hancock's wheels

One more reference to the work of Hancock, namely, to his " wedge " wheels. These were the outcome of his first few years of experience, and are models of to-day. The great strength required for driving wheels soon forced him to special designs, and the wheel shown by Figs. 9 and 10 might be usefully copied, with slight improvements, to-day. The principle of their construction is indeed used in several well-known forms of wheels.

The essential features are metallic naves, forming what in the old language of the wheelwright would have been nave and box all in one. On the box A, Fig. 10, was cast a large flange, and fitting upon the box was a corresponding flange B, separated from the first by the thickness of the spokes F. There was thus formed between them a deep groove, which took

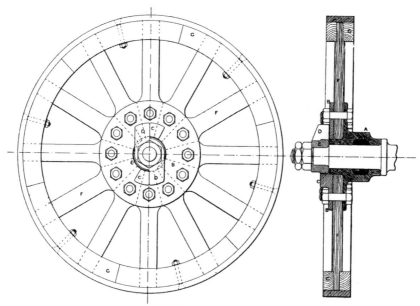

FIGS. 9 AND 10.—HANCOCK'S WHEELS AND DRIVER.

the place of the old nave mortices. This groove was completely filled by the spokes, which were made of a broad wedge shape, as indicated in dotted lines in Fig. 9, for the purpose. Bolts passed through the two flanges, one through each spoke, and thus firmly gripped the whole in one rigid connection. At first these flanges were made too thin; and on a trip to Brighton in October, 1832, the cast-on flange gave way all round, and was the cause of one of the many incidents of these early essays in a new field. Hancock's method of driving his wheels is also shown by Figs. 9 and 10. The wheels were loose on their axles, the ends of which carried drivers D. On the outside flange of the nave were cast two lugs C, with radial faces

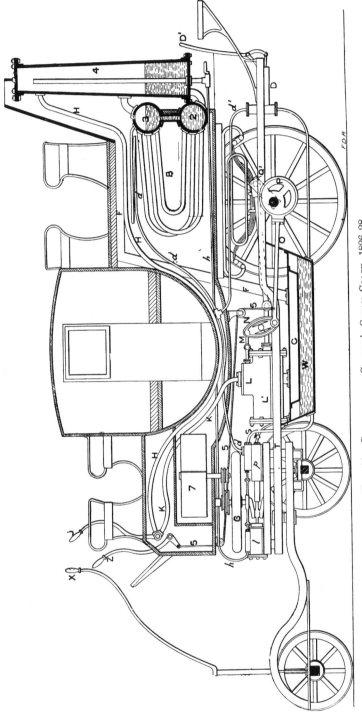

FIG. 11.—GOLDSWORTHY GURNEY'S STEAM COACH, 1826-28.

13

MOTOR VEHICLES AND MOTORS

like the driving faces of the pieces D. By this means there was nearly half a revolution of freedom for the wheels, with which one could overrun the other on curves.

Gurney's steam coaches, 1826

Goldsworthy Gurney made several coaches of different forms, such as that shown in longitudinal section by Fig. 11, and drags or tractors, like that shown by Figs. 12, 13, 15. As early as 1826 Gurney ascended Highgate Hill with one of his carriages, and the *London Courier* of the 10th September, 1827, describes one as working very satisfactorily. In the succeeding six years, he and Sir Charles Dance, for whom he appears to have made two or more coaches, experimented on a large practical scale. At an early stage in his experience he found the necessity for plenty of power, and used two horizontal cylinders, each of 9 in. diameter and 18 in. stroke, the driving wheels being 5 ft. in diameter.

Gurney used water-tube boilers of a novel construction. These are shown in two of the forms adopted by Figs. 11, 12, 13, 14. The pressures used varied from about 70 to 120 lb. or more, when circumstances demanded it, the boiler being tested cold to 800 lb. per square inch.

Gurney's oilers and steam separators

As used in the coach, Fig. 11, the boiler consisted of two horizontal cylindrical trunks, 2 and 3, placed one vertically over the other, and connected by nearly horizontal tubes B, of 1 in. internal diameter, bent into the form of a letter U placed horizontally. The two limbs of the C were placed vertically, one above the other, at right angles to the trunks, and the lower limbs formed the fire-grate (see Fig. 14). The upper trunk was connected to two vertical receivers 4, by a steam-pipe, at nearly mid-height, and the lower trunk by a water-pipe. From the top of these receivers, which acted as separators, a steam-pipe H, was taken to a valve under the driver's seat, and thence to the engines, the steam being dried, if not somewhat superheated, by passing it through the furnace casing above the top tubes of the boiler.

In some of his boilers Gurney used D-shaped trunk water and steam vessels, the flat side being a tube plate held on the other part by bolts through flanges, just as in some modern water-tube boilers. The water-tubes were fixed in the flat tube plate by nuts on either side.

The fire was supplied with air by a fan 7, on a vertical spindle, under the front part of the carriage. It was driven by gut or rope from a grooved pulley on a small engine, with a horizontal cylinder *l*, and vertical crankshaft, carrying a fly-wheel 6, as shown in Fig. 11, the same engine operating directly two feed-pumps *p*. The exhaust from this and the main engines passed into a space G, in the upper part of a flat water-tank W, below the coach, by which the noise was lessened and the feed-water heated. The vapour from this space was taken off by a pipe F, which passed upwards into and through the boiler casing into the funnel above the top tubes, where it was heated, and generally passed away without cause of annoyance The two cylinders L′ were fixed on a perch pole frame T, and were connected direct by rods to a bent crank-shaft G′, on which were

14

two excentrics P, for operating the valves in the steam-chests L, through rocking links N. According to the descriptions of cotemporary date, these excentrics were at 90° ahead of the crank, and the reversing of the engines was done by raising the ends of the excentric rods O, by means of a cord 5, and the lever Z, to the upper part of these links. This could, it would appear, only be done when the engines were stopped, the first of the rocking links being fixed; and it would further appear from this that the slide valves had no lead, a deduction which is supported by Gurney's frequent reference to his working by throttling or wire-drawing.

In giving evidence before the Select Committee, Gurney gave the weight of one of his coaches at 50 cwt., and another at about 60 cwt.; but the coach carrying about fourteen passengers seems to have weighed more like 65 cwt. With this he used about 10 gallons of water per mile, or 100 lb., and about 20 lb. of coke, giving an evaporation of 5 lb. per pound of coke; but, taking an average of the figures given, it seems sometimes to have been less than this. His engines were less economical than Hancock's. He used steam at from 70 lb. to 120 lb. per sq. in.; and taking the average pressure in the cylinder at 40 lb., his power was more nearly 28 than 20, and the steam consumption from 50 to 65 lb. per horse-power hour.

Gurney found that he used much more steam at or below 4 miles per hour than he did at 10 miles, and that a piston speed of 220 ft. per minute was the best. He found from experience with the several coaches that every 10 cwt. of steam engine was equal to four horses in draught power; and Hancock reckoned that two fourteen-place coaches, with engines of 10 nominal horse-power, would replace the 100 horses required for the two ordinary coaches on a hundred-mile stage, one coach up and one down.

Beside the two-power feed-pumps, there was a hand feed-pump D, at the rear of the coach. This took water from the hot feed-tank, and its delivery d' joined the delivery pipe d, from the main pumps, and passed upwards into the space over the boiler tubes, where it was coiled, and delivered the feed-water into the steam space of the upper trunk of the boiler. The pipe d', from the hand pump, was bent into one coil to give flexibility, and freedom from breakage by jolting.

The vertical separators 4, on the later carriages, were supplanted by a larger horizontal cylinder, as seen in Figs. 12, 13, both the other cylinders being drowned. Gurney had a good deal of trouble with his tubes, but on the whole the boiler was successful; but it seems to have been far inferior in efficiency to that of Hancock. This might be expected from the small amount of heating surface exposed to the hottest products of combustion, and the brief contact on the way to the uptake.

It may be here noted that Gurney used forced draft, water-tube boilers with water and steam trunks and separator, super-heated steam, heated feed-water, and separate direct-acting feed pumps. It may also

15

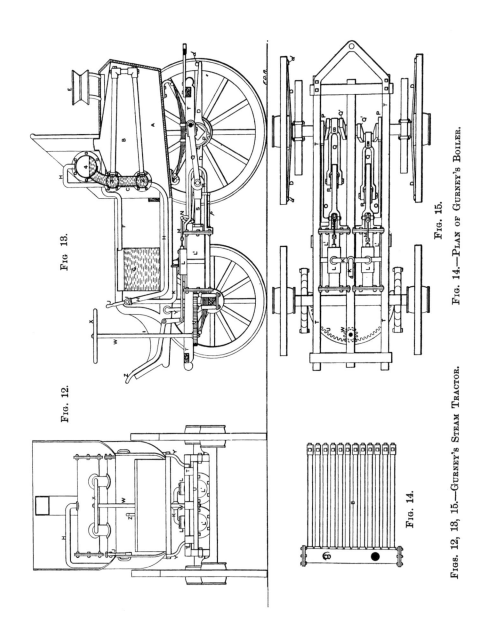

FIG. 12.

FIG. 13.

FIG. 14.

FIG. 14.—PLAN OF GURNEY'S BOILER.

FIG. 15.

FIGS. 12, 13, 15.—GURNEY'S STEAM TRACTOR.

be noted that he used water-gauge glasses; a steam pressure gauge, of the spring-loaded piston type, and fusible plugs.

In the steam drag or tractor, Gurney discarded the fan, and used the exhaust produce draught. As will be seen from Fig. 13, he used a wheel steering gear. The hand-wheel [1] was on a vertical shaft, at the bottom end of which was a pinion w, which geared into a strong semicircular rack u, the first instance of the use of this simple and effective gear. The whole of the machinery was carried by an under or perch-pole frame consisting of the longitudinal cills T, and two cross-pieces. This frame rested on the springs over the front axle through the intermediary of the locking plate formed by the steering rack u. At the rear the frame was carried by the bearings on the crank axle, and the springs were mounted above it, and carried the body in the usual way. At the front end the body was carried on vertical radius arm pivots, by which the proper relative positions of cylinders and steam-pipe were secured. The springing was thus in part to the under frame and complete for the body. In Figs. 13, 15, a double arm D D, will be seen extending from the crankshaft to a bolt d, in the felloe of the driving wheels. These arms were fixed to the axle at the back of the wheels, and drove by contact with the felloe bolts. They were free to turn half a revolution from one to the other direction of driving, Gurney having no compensating gear. Generally the driving was done by one wheel only, so as to secure facility for turning, and the bolts were only put in the other wheel when on a bad piece of road or very stiff hill.

The occasional insufficiency of Gurney's boiler led Sir Charles Dance, who had bought three of his coaches to run between Gloucester and Cheltenham, to design some modification of it. This is shown by Fig. 16. It formed the subject of a patent, and the main features were the larger heating surface within a given space; the method of connecting the lengths of straight or curved pipes by means of cast elbow-pieces formed with nipples and screw union connectors, and removable plugs for cleaning out; and the close arrangement of the tubes forming the furnace-bars. Dance ran his coaches about four months, from February to June, 1831, when he had to give up the project, beaten by the ignorant obstructiveness of the provincial horse and turnpike road interests. During that time he made 396 journeys, travelling 3,644 miles, and the total cost for coke on the whole was 4d. per mile. Cost of fuel is always a small item in the total.

William Alltoft Summers and Nathaniel Ogle had, when they were examined before the Select Committee of the House of Commons, in the autumn of 1831, made two steam carriages, with one of which Summers stated that they frequently ran at 30 miles per hour. It was a treble bodied phaeton running on three wheels, and weighing 3 tons, with water and fuel on board. It carried sixteen to eighteen passengers, and

Gurney's tractor

Dance's boiler, 1831

Summers and Ogle's carriage, 1831

---

[1] The method of carrying the motor on the driving axle at one end, and spring supported at the other was thus in principle and in effect the same as that patented in recent years for the support of electric motors in tramway vehicles.

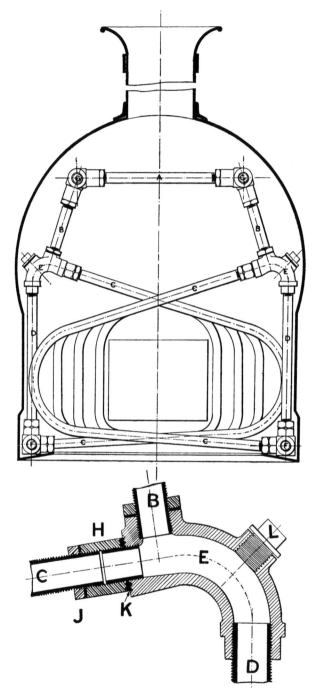

Fig. 16.—Dance's Boiler and Tube Connections, 1831.

had 5 ft. 6 in. driving wheels. These wheels had been 6 ft., but they gave way at the felloe end of the spokes, the failure being said to have been brought about by fine saw-cuts made by the hirelings of the then powerful enemies to the steam road-carriage. The spokes were newly tanged, and new felloes made for them, thus reducing the wheels to 5 ft. 6 in. in diameter.

The boiler of this carriage was the subject of Patent No. 5927 of 1830. It is shown by Fig. 17, except that there were six rows of tubes instead of five. It is a vertical water-tube boiler with internal smoke-tubes D, 3 ft. in length. It was 3 ft. 8 in. in height without the case,

Summer and Ogle's boiler, 1830

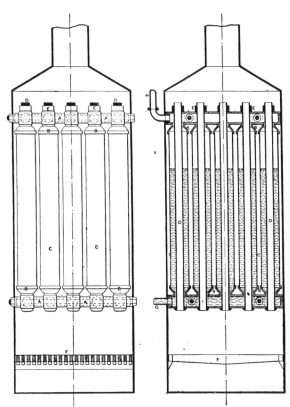

FIG. 17.—SUMMERS AND OGLE'S BOILER, 1830.

and 2 ft. 4 in. by 3 ft. in plan, and had 6 sq. ft. of grate surface. It contained thirty double tubes, the ends of which were let into cast iron cups B, cast with tubular heads of rectangular section at the cups, and round section between them, the tubes being 0·10 in. thick, of charcoal iron, and held in by double nuts E, on the ends of the smoke-tubes. It had a total heating surface of nearly 250 sq. ft., worked with a safety valve load of 247 lb. per square inch, and had been proved to 364 lb. The weight is stated before the Parliamentary Committee to have been 8 cwt. This is probably the weight without the case, which was of sheet iron 0·1 in. thick, but how lagged or lined is not stated. To keep it free from incrustation, it used to be blown out at a high pressure, said to be over 200 lb., but it must be remembered that the working history of these boilers was too short to have given much information as to their possible durability, although one was worked in a coach for more than a year.

Summer and Ogle's engines

The engines had cylinders 7½ in. diameter, and a stroke of 18 in. and they are said to have been 20 horse power. They were coupled direct to a crank driving axle. Judging by the stated consumption of coke and water the engines used about 56 lb. of steam per horse-power hour, and the boiler

Summer and Ogle's coach and boiler performance

evaporated about 7 lb. of water per pound of coke, weighing 40 lb. per bushel. In evidence before the Select Committee it was stated that 7 cwt. of water were evaporated in 40 minutes on a run of 8 miles, the fire being forced by fan draught. This gives an evaporation of 4·3 lb. per square foot of heating surface per hour, and 196 lb. per square foot of grate ; and 4·2 bushels were used per hour, giving a combustion of 28 lb. of coke per square foot of grate per hour.

Early use of high steam pressures

One cannot help expressing a feeling of admiration of these men of seventy years ago who dared and used steam pressures of over 200 lb. per square inch, especially when we remember the feebleness in this respect of the practice of our fathers from 1850 to 1880.

Maceroni and Squires' coach and boiler, 1834

The coach made and run by Maceroni and Squire in 1834–35 was arranged much like Gurney's as to the engine, and was fitted with a vertical water-tube boiler, having about eighty vertical tubes, connected at top and bottom by small stay tubes. It weighed 16 cwt. The top ends of eighteen of the water tubes were connected by small pipes to a central vertical steam drum, and the pressure was apparently from 100 to 150 lb. ; the coach carried eight passengers and driver in front, and a stoker at the rear.[1] It has often been stated that it ran about 1,700 miles without noteworthy repair, but this seems to be absolutely untrue. When running it cost 3d. to 4d. per mile for coke, the speed being about 10 miles per hour. At this time the horse coaches paid about 3s. per mile for four horses, for the fast coaches carrying eight passengers. The engines had cylinders 7½ in. in diameter and 15¾ in. stroke. The weight of the coach fully loaded would be about 3·5 tons.

Hills' coach, 1840, and Burstall and Hill, 1824

A coach was made by F. Hills, of Deptford, in 1840, not the Hill who in previous years, in partnership with Burstall, had made a cumbrous vehicle with beam engines and a flat chamber instantaneous steam generator. This was a failure, but was the first in which the front steering wheels as well as the rear wheels were driven by gear.[2]

Hills' differential gear, 1843

The coaches made by F. Hills should be noticed because of the departure he made in the construction of the under frame, which was similar to railway practice with respect to the spring hanging and carrying the driving axle between the pairs of horn plates. Also because he was the

---

[1] See *Cantor Lectures*, pp. 14–17, and for much information concerning the running of this coach see *Mechanics' Magazine*, vol. xx., 1834, and vol. xix., 1833.

[2] See Hebert's *Engineer's Encyclopedia*, vol. ii., p. 451, and Patent, February, 1824.

first to design for motor vehicles, if not to use the " jack in the box," or balance gear, to allow the wheels to rotate differentially, on turning corners. (Patent No. 9684 of 1843). His coaches, or one of them, were heavy and powerful. It would carry nine passengers and a driver, conductor and stoker, at considerable speeds up the hills on the road to Hastings and back, which it performed in one day, a journey of 128 miles, this road including Quarry Hill and River Hill, with gradients of 1 in 13, and 1 in 12, and short pieces of worse still. This coach appears to have had a pair of engines, with 10-in. cylinders and 18-in. stroke, coupled direct to the main crank axle. The driving wheels were 6 ft. 6 in. diameter, and were inside the main bearings. They were loose upon the crank, and either or both could be driven by the crank, through clutches operated by the driver; the clutch projections which engaged with the wheels permitting nearly half a revolution of freedom for turning ordinary road curves. For turning sharp corners one of the clutches had to be thrown out. The Patent No. 7958 of 1839, which describes this, also describes a two-speed gear of lathe back gear kind, but Hills does not seem to have used this. Neither did he use the differential gear on this or either of the coaches he made, for this was not patented until 1843 (No. 9684), about three years after the coach with which he did his famous running was made.

Hill's boiler was tubular, and weighed 28 cwt., without the 60 gallons of water it carried. His tanks carried 120 gallons, and at every 8 miles or thereabout he arranged for water supply, and took in from 80 to 100 gallons, according to the road travelled. His pressure was from 60 to 70 lb., according to the road. He is said to have reached 25 and even 30 miles per hour, and repeatedly made 3 miles in 7½ minutes, or 24 miles, but figures as to speed and weights were always generous. His differential gear does not seem to have been used by him. As described in Patent No. 9684 of 1843, it shows the gear with outside or overhung cranks, vertical cylinder engines, driving the cranks through bell cranks and levers. He shows also a condenser. Hills took out three patents: the two mentioned, and another 8495 of 1840, and all are of interest, especially with reference to his boilers, and numerous points in his 1843 coach design.

Having here arrived at this date, mention may be made of the first agricultural self-moving engine, made by William Worby, and exhibited at the Royal Agricultural Society's Show at Bristol, in 1842. He had the year before made the first portable engine for farm use. The self-mover was for hauling its own threshing machine. It was tested and awarded a prize,[1] and was apparently the first road engine driven by a pitch chain, and certainly the first by a rotary engine.

There were, as well as those now mentioned, several others who at this time turned their attention to steam road vehicles, and experimented

[1] *The Engineer*, 27th June, 1879; and *Journal Royal Agricultural Society*, vol. ii., p. 338.

with one or more. Among these were a Dr. Church, Henry James, Griffith, Scott Russell (of whose coaches several were built and ran between Glasgow and Paisley, until one burst and killed three people, in July, 1834).

Church's boilers
Church's boilers were really remarkable designs for the time. They were of two forms, and were firebox and water, or, alternatively, smoke tube boilers. The water space was too confined, and the firing more difficult than with modern boilers, but with slight modifications the originals might form excellent boilers.[1]

James's independent engines
James's arrangement should be mentioned, because he was the first to use four cylinders, two to each driving wheel—these wheels being on separate crank axles, so that no differential gear was necessary to enable the wheels to run at different speeds in turning corners.[2] He also showed tractors and trailers driven by vertical engines, with a fore and aft shaft and universal or Hook's joints (*cardans*, as the French have it), in 1824.

Gibbs' tractor and boilers
Joseph Gibbs took out a patent, in 1830 (No. 6302), for a boiler, a modification of which he afterwards used in a steam tractor, in which he was associated with Chaplin. The boiler consisted of a nearly rectangular firebox, delivering into two spiral tubes, down which the products of combustion passed, and escaped within about 2 ft. of the ground. The spirals were in two cylindrical shells, extending downwards from the box. The boiler was 5 ft. 6 in. in length from top to bottom, and each water cylinder 18 in. in diameter, the firebox being partly water-jacketed. The fire was forced by a fan blast. The engine was arranged on a frame, resting on the springs of the fore-carriage, as in Gurney's and others. The boiler evaporated 10 lb. of water per pound of coke, and the products of combustion left the spiral tube exit at a very low temperature.

Gibbs 'radiating axles
Gibbs used friction clutch bands, like excentric straps, for frictionally gripping either or both of his driving-wheel naves, so that either could be loosened for turning corners. The carriage hauled by this tractor is shown by Fig. 18, and is remarkable as having all four wheels steered, the wheels being all on radial stud axles, in accordance with the patent of Gibbs and Chaplin No. 6241, of 1832. It will be seen that the draw bar A is fixed to a vertical bar carrying a crosshead B. From this crosshead two rods, c and c', receive motion as the carriage is pulled in one or other direction. The rod c is coupled to the sliding bar D, to which the links F and F' are connected by means of two projecting lugs E and E', one pointing one way and the other pointing oppositely. The other ends of these rods are connected, one to the front-wheel axles G and G', and the other to the hind-wheel axle. Correspondingly the rod c', from the other end of the crosshead, is coupled to the slider D, which has similar lugs H and H', upon it, one to which the rods actuating the hind axles are jointed. By this arrangement the axles were made to radiate to the circles the wheels were describing, or nearly so. It will be seen that by means of a quadrant, instead of the rod A, the

[1] See *Cantor Lectures*, p. 15.
[2] Hebert's *Encyclopedia*, vol. ii., pp. 455, 456.

EARLY STEAM ROAD VEHICLES

arrangement described would make a steering gear to be directed by a person sitting on the car.

The whole of these steam coaches and carriages were one after another abandoned, until only those of Hancock were on the road in October, 1836, and soon afterwards not one was left on the road, and none of them are, as far as is known, preserved. Although many of them ran a good deal, the internal evidence of such records as exist, indicates that most, if not all, were incapable of continuous service without more incident than was agreeable to the passengers, or commercially satisfactory to their owners. Many of these incidents were, it is true, due to the machinations of friends of the old order of things, to which all were destined to revert; but there were, without doubt, many inconveniences attaching to the use of very

General abandonment of the steam road coaches in 1836

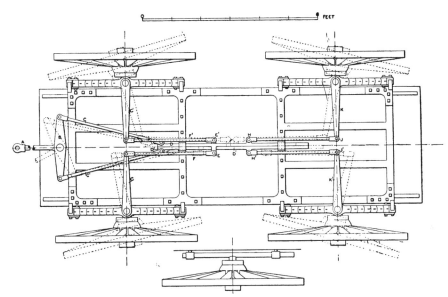

Fig. 18.—Gibbs' Radial Action All-Wheel Steering Gear.

heavy coaches, with heavy engines, requiring large quantities of steam and large water supply. The failure of the Tolls Bill recommended by the 1831 Select Committee was another blow to the steam carriage. Hence, with the exception of occasional spasmodic efforts, all practical enterprise ceased about 1835, and nearly all of the lessons of experience, and practical information of the men who had gained it, mental storage of inestimable value if those men could have continued in their work, have gone with them. All has to be not simply re-learned, but re-discovered, and most of it by the slow process of making elaborate scrap. This process may be said to have begun when Leon Serpollet made his first steam carriage, in 1893 or 1894; for although Gottlieb Daimler made, in 1884, those beginnings which have resulted in the new order of things in the hands of MM. Panhard and Levassor, these have not been re-discovery

Loss of the experience gained

23

MOTOR VEHICLES AND MOTORS

or re-learning. They have gone on continuously since Levassor began, and with immeasurable influence since the world awoke to the commercial and pleasure possibilities of the modern road motor vehicle, and made England ashamed of its repressive and narrow road locomotion laws of 1861 to 1878—laws which were the outcome of the notions of the traction engine makers of those dates, and which, through the restricted and inelastic nature of the Acts, were worse in their effect on steam carriage or any mechanical motor carriage, however refined, than the boorish treatment accorded to the men of 1830, whose work would have placed this country for many years as far ahead of all others in road locomotion as it was in railway locomotion.

Great Britain lost the lead

I have spoken of several of these coaches as having run successfully. It is as well to enquire, before leaving this historic point of the subject, what " successfully " means in this connection. There is no doubt that those coaches travelled from place to place, and carried from eight to eighteen passengers, and that some of them reached considerable speeds. They showed great ingenuity and originality and mechanical success in an entirely new field. They were creations which were much more than the mere roughing out of a good working arrangement. It is, however, equally certain that most of them were very heavy for the load they carried, that the wear and tear upon them was great, that the quantity of water used was a source of a good deal of inconvenience, that the heavy large cylinder engines they were worked by could not have worked without considerable vibration, that the quantity of exhaust steam to be got rid of must very frequently have been a nuisance, that the heat from the boiler must in several have caused much discomfort, that there was too much of engine-driver and stoker about them, and that every one of the designers, when they ceased to build, felt that their next coach would contain many improvements. The master hands had worked out all the main features, and in this they were successful; but it remained for future workers to apply day-by-day experience, and, as has been the case with the locomotive under its more favourable circumstances, to work out those improvements which would make the vehicles commend themselves and accommodate themselves to commercial requirements and public approbation. That has been left for this generation, led by the perception of our friends in France of the fitness of the internal combustion motor of Daimler, even more than by their work with the steam engine.

The outcome of the achievements of 1820-40

Before leaving this part of our subject some consideration should be given to the lessons derivable from the flood of inventions, practical experience, and results of the active men of the few years 1823 to 1840. Invention was prolific, although in these pages I have only given the results of some of the best of the practical efforts. From the ingenious but useless steam leg and foot propellers of Gurney, Gordon, and others, to the successful coaches and carriages of Gurney, Hancock, Hills, Squire, Gibbs, and Summers and Ogle there were but these few years; but in that time

Consideration of teachings of past achievements

24

these grandfathers of ours had left few of the main features of steam carriages to be discovered, and they had bequeathed to us, in Patent office and other records, the seeds and elements of so many of the then future requirements, that unless our Patent law is to be altered there remains, with respect to steam vehicles, little hope of protection for the inventor of anything but precise details. A designer may expend two or three years now in maturing the details and refining the arrangement, design, and combination of everything to meet the requirements of a steam carriage for the present time, and with all the advantages of modern materials and appliances, and although the outcome may be that which has not been before obtainable, he may find at the end of all his labours and expenditure that he has no more protection than has the architect who, after years of experience and study, plans the best-arranged house ever seen. Yet the writer of a shilling novel or the reporter of somebody else's speech have absolute property in the results of their more common-place and ephemeral work. So now it behoves designers of steam carriages to acquire some knowledge of what was done by the men I have mentioned. Firstly, that they may profit by suggestions of principle though not of dimension ; and secondly, that they may not found false hopes on that of which they are but second inventors.

It will have been noticed that all the successful vehicles of the time referred to, were, with the exception of Hancock's, driven by engines direct coupled to crankshaft axles. Gearing was tried and given up, and Hancock used common chain, said at the time to work well. Such a chain could however, not satisfy any but biassed witnesses for more than a very short time, as we know by experience, but it enabled Hancock to avoid the use of the cranked axle which had given other builders so much trouble, and at the same time to enclose the working parts of his engine from the dust. Against all this, however, was the fact that he used two shafts where the others used one, and introduced the further complication of a chain. This arrangement probably caused at least 7 to 15 per cent. loss by friction, but avoided the wear and tear of connecting rods, brasses, and other unenclosed working parts. The worst of these was the connecting-rod big ends, the wear of which would be avoided to-day by the use of dirt excluders, such as those used in the crank bearings of Cooper's steam diggers, as shown in Fig. 18. In this, A is a digger crank journal, with a connecting rod C and brasses D, the ends of the brasses and the connecting rod being covered by channel rings B, put on in halves, as shown by the separate view of the inside of one of them. This bearing proves successful under trying circumstances, and is better than others Mr. Cooper tried, with felt, which held grit and acted as grinding pads. Leather ring excluders were found better than felt, but the bearing shown proved best.

This being so, the simplicity of direct coupling, which the majority of the 1830 men found best, could be used without much disadvantage, more especially if outside connecting rods and Hills' differential gear were used.

Main elements of steam carriage design worked out

Property in modern invention

Main feature of successful early designs

Direct driving

Chain drive

Speed of engines, and size of drivers

Again, by the aid of the superior materials and tools of to-day, we could use lighter engines, and with only the disadvantage of smaller driving wheels, could use higher speed of rotations and lighter vehicles.

Proof of difficulties experienced

Clutches

Condenser

Speed gear

That the difficulties, which were made light of by several of the more enthusiastic builders, were very real, is shown by Hills' 1843 inventions, subsequent to all his experiments, which include means of surmounting three of the most important. Firstly, to avoid clutches for connecting one or both driving wheels to the shaft. Secondly, provision of a condenser so as to save time and extend the length of a stage to something beyond eight miles; and thirdly, an arrangement for running the crankshaft in a case flooded with oil. It is improbable that geared engines running four times the speed of the driving wheels could have been used in 1830, for it would have been almost impossible to have made good enough gear at reasonable cost. Even in the later period, with which we shall presently be con-

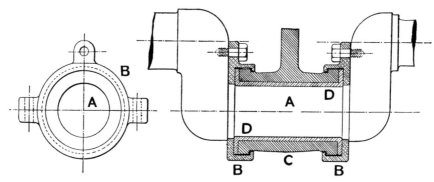

FIG. 18A.—COOPER'S DIRT-EXCLUDER BEARING.

Return to direct driving

cerned, gearing was displaced by direct driving, and, for steam, is in some recent examples being again used on account of its simplicity and high mechanical efficiency. It involves, however, either small wheels, which are bad on rough roads, or large cylinders which are heavy, or very high pressures, which only Jacob and Loftus Perkins have used with success.

High pressures and light boilers

To very high pressures, however, we must look for much advance in the future steam carriages. Light but strong boilers were made seventy years ago, and improvement in construction and materials is more required than new forms to fit them for the present time. Shell boilers, with water or with smoke tubes, or with both, were designed, to which the same remark applies, and the designs are preserved in the Patent records. The difficulties resulting from the use of hard water, were, to some extent combated by Dance in his water tube with removable plugs at every corner; but to prevent the formation of " fur," Gordon made a form of condenser, partly evaporative, in which two light steam trunks were connected by a number of them, ½ in. tubes, lightly clothed with hair-cloth, on which water dripped.

Early evaporative condenser

EARLY STEAM ROAD VEHICLES

The outlines of the best forms of wheels for heavy work were also made then; witness Hancock's, which only require improvement ; and the method of driving them at the rim instead of through spokes was invented and used by Gurney.

Steering gears of many forms were devised and used, and carriage bodies were spring-hung on springs separate from those which carried the motor parts. Radial axles were used, and stud or pivoted axles for steering, wheel and worm steering gear, quadrant-steering, single steering wheels in front or rear of the main steering axle were used, and many other inventions which only need the improvements now possible. **Steering gear and springs**

Too much space however could be occupied by a mere reference to the many things which the men of sixty to seventy-five years ago outlined for more service to the present generation than to their own.

I mention them because there is too much readiness in the present day to assume that road locomotion, methods and means are new problems concerning which nothing done three-quarters of a century ago can be any guide. This is a mistake, for things done then may be anticipations if not guides, and as such are often surprising in their completeness and prescience.

Chapter II

THE INTERMEDIATE PERIOD

TO what I have called spasmodic efforts some reference must be made. They have not been sufficiently connected to be of any use in maintaining effective continuity of knowledge and experience, but some of the results of the efforts may, as far as mere record will permit, be relearned with advantage.

Trend of design

Between the traction engine and several of the so-called steam carriages of from 1850 and 1894, the division line is not broad, because some of these steam carriages were really only self-moving boilers and engines with some seats attached. They were on the border-land of the traction engine in their design, with heavy thick boilers and large water contents, and with consequently heavy engine parts and heavy gear. Some of them require notice, although the traction engine must be omitted as having done very little towards the modern motor vehicle, even of the heavier kinds. Some, on the other hand, were much nearer in approach to the now acceptable vehicle, and are worthy of record and careful consideration.

Rickett's road steamers, 1858

The locomotive shown by Fig. 19 was designed and made by Thomas Rickett, Castle Foundry, Buckingham, in 1858, for the Marquis of Stafford.

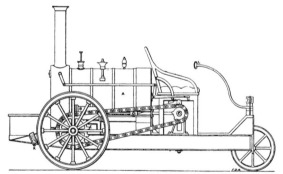

FIG. 19.—RICKETT'S ROAD STEAMER (MARQUIS OF STAFFORD'S), 1858.

It was called a steam carriage, but was really only a steam engine on three wheels, two of which it drove by means of a pitch chain. The boiler A was of the return tube type, with firebox and up-take at the rear, where a

28

platform was made for coal and stoker. In the front was a seat for three, including driver or steerer. The cylinders were placed toward the rear, at B, over castings which formed axle boxes with springs of short range. The crank shaft F ran in and between bearings which were fixed under the seat, to the upper part of two narrow tanks which formed the longitudinals of the under frame. The crank shaft was coupled to the piston rod by connecting rods C, and at one end carried a small sprocket wheel or pinion for a chain which drove the main axle by a wheel 2·5 times the size of the pinion. One driving wheel was fixed to the main axle, and one only was used for ordinary running. On stiff roads or hills, the other driving wheel was driven by a clutch, which had to be thrown out of gear on curves. The steering lever L was coupled direct to the front wheel fork, and the regulating lever G, to the right of the driver's seat. The cylinders were 3-in. diam. and 9-in. stroke, the revolutions about 220 per minute at 10 miles per hour, and the pressure about 100 lb. per square inch. It did not often run at this speed, though it reached 12 miles. The main wheels were of iron, with a wood rim between an inner iron rim and an iron outer tyre. Brake blocks K were arranged to act on both main wheels. The weight was 30 cwt.

In 1860 Rickett made other road steamers of the modified arrangement shown in Fig. 20. The boiler A was similar and the tank frame, steering

Rickett's road steamer, 1860

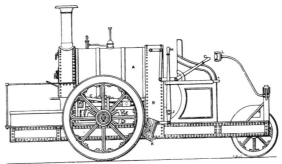

FIG. 20.—RICKETT'S ROAD STEAMER (BOUGHT BY THE EARL OF CAITHNESS), 1860.

lever L and regulator G the same as in Lord Stafford's steamer, but the cylinders were placed at B, inside a downward continuation of the combustion chamber; and a crank F was placed at the rear, connected to the piston rod by the connecting rod C; a spur pinion on the crank shaft, geared with the wheel E on the driving axle.[1] The weight was about 50 cwt.

Rickett made two or three of these engines, and afterwards made another, to act as light omnibus traction engine, for Spain. In this he again departed from his previous method of driving, and having discarded first the chain and then the spur gear, went back to the direct driving, after the manner of 1830. His engine was like a railway engine with steering gear.

[1] *Trans. Society of Engineers*, 1858–62, p. 128: Paper by A. F. Yarrow.

THE INTERMEDIATE PERIOD

It was a road locomotive with horizontal outside cylinders coupled direct to the outside cranks of a pair of rear driving wheels, and on good leaf springs. The cylinders were 8 in. diam., and 22 in. stroke. The driving wheels were 4 ft. in diameter. The engine weighed 6 tons, and would haul about 4 tons at 10 to 15 miles per hour.[1]

These last engines embodied the result of a great deal of experience. They were almost entirely steel and wrought iron, and it is to be regretted that Rickett did not write an account of his experiences, and a well illustrated description of his last engines.

Carrett & Marshall's road steamer, 1862

An engine of somewhat similar type, which attracted a great deal of attention, was made by Messrs. Carrett & Marshall, of Leeds, in 1862. It was a three-wheeled front-steering locomotive engine mounted on the wheels it drove, and with seats in front for seven, most of whose weight was carried on the front steering wheel, which was 6 in. wide. The total weight was nearly 5 tons loaded. The engine had two cylinders of 6 in. diam., and 8 in. stroke, the crank shaft carrying two spur pinions, gearing into wheels five times their size. The driving wheels were 4 ft. in diameter, had steel spokes and rim, with hard wood felloes and iron tyre outside. The engines it will be seen were powerful for the size of the steamer as a carriage, as they ran at 350 revolutions per minute, at a speed of 10 miles per hour. The driving wheels were connected to the axle by clutches, either of which could be withdrawn for turning sharp curves. When put to work the wear of the spur pinions was found to be excessive, no doubt largely due to running on all ordinary curves with both clutches in.

Supported steering forks

The front steering fork was made capable of withstanding the impact on ordinary roads, under its load of over a ton, by an arrangement which may be described with reference to Fig. 25. Beside being held in place by the stem at the top of the fork, support was given to it by extending the wheel axle some distance on either side of the wheel, and towards the ends of this a pair of parallel rods c, with bearings, were connected rearwards to a strong crosshead at F, the crosshead being on a spindle, the lower end of which was firmly carried in the framing, not free as shown in Fig. 25, p. 33. Thus the fork was relieved of the heavy stresses.[2] The engine was built for Mr. Geo. Salt, and afterwards, when he disposed of it, was altered by the insertion of differential gear and other modifications. Its running at high speed caused it to be stopped by the authorities, when it was again altered, and became a small slow speed traction engine.

Yarrow's steam carriage, 1862

The steam carriage illustrated by Figs. 21 and 22 was a great advance on Rickett's, and was made to the designs of Messrs. A. F. Yarrow & Hilditch, and exhibited in the 1862 Exhibition. At that time Mr. Yarrow interested himself in this subject, and read the useful paper before the Society of Engineers, in October, 1862, already referred to. The opposition,

---

[1] *Steam Locomotion on Common Roads :* Fletcher.
[2] *Trans._Loc. Engineers*, 1862, p. 138.

30

however, to anything but horse haulage soon drove him to high speed locomotion on the water instead of on the road. The carriage shown was built to carry thirteen, including the driver and stoker, and is said, when

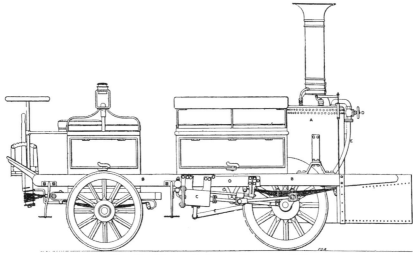

FIG. 21.—YARROW'S STEAM CARRIAGE, 1862.

fully loaded, to have weighed about 2·5 tons, thus leaving about 1·75 tons for carriage, fuel and water. Nearly 2 tons were carried on the drivers.

The driving wheels at the rear were placed within the frame B and

FIG. 22.—YARROW'S STEAM CARRIAGE, 1862.

springs. They were 3 ft. in diameter, and were both fixed to the axle. The gauge, or distance between the wheels, being thus very much smaller than usual, it was contended that differential gear was not necessary for turning corners, as the slipping of one or other wheel was generally negligible, and a cranked axle was thus avoided. The springs rested on the main brass bearings on the axle journals outside the springs, and outside these were the overhanging cranks. The pistons were coupled direct to these as in locomotive practice, a radius rod F, connected to the main bearings, being used to maintain constant distance instead of horn plates. The cylinders c were 5 in. diam. and 9 in. stroke, and were fixed to plates on the outside of the longitudinals of the frame B, the latter being of ash, 4·5 in. deep, lined with iron plate 0·25 in. thick. The spaces under the seats formed water and coke bunkers, the side panels opening outwards. The whole of the working parts, though not so shown, were, it is stated, covered in for protection from dirt and dust.

The boiler A was of the vertical type, with vertical smoke tubes. The shell was of steel, 2 ft. in diameter and 3 ft. 9 in. high. The heating surface was 40 sq. ft., and the grate 21 in. in diameter, giving 2·3 sq. ft. of grate surface.

Yarrow's boiler and engine proportions

These dimensions, according to Mr. Yarrow's rules at the time, would give 5 actual horse power of heating surface, and 6·5 horse power of grate surface. Thus, with a little forcing, the boiler would supply from 6 to 7 actual horse power, or 10 horse power for a short time.

The dimensions for evaporative requirements were adopted on the assumption of an average tractive resistance of one-thirtieth, or 74 lb. per ton of the load, and the maximum tractive power required in the engines was calculated on a maximum resistance of one-fourth the load, or 560 lb. per ton, allowing for gradients of 1 in 10.

Mr. Yarrow adopted for these small boilers 8 sq. ft. of heating surface per actual horse power, and grate surface of one twenty-fifth of the heating surface, or 0·35 sq. ft. per actual horse power required.

The boiler was fed by a pump worked off the main axle, and the exhaust steam passed into the uptake. The steering was effected by a chain wound on and off the lower end of an enlarged part of the vertical steering spindle. The carriage ran well, and there can be no doubt it had points of considerable merit, especially that of simplicity. It will be seen that the carriage had about 0·2 horse power per cwt.

Pattison's steam carriage

Another type of steam carriage of this period is illustrated by Figs. 23 and 24. It was built by Mr. A. Pattison. It differs entirely from that of Mr. Yarrow. The engine and boiler are in the front, and the engine drives through a countershaft with two speed gear, the whole motor and transmission gear being mounted on a turntable race so as to swivel with the steering wheels. The front driving and steering axle runs in bearings c, in hornplates, and it supports boiler and other parts on springs, the fixed radial distance for the gear being maintained by the use of curved horn-plate slots. The turntable ring attached to the boiler, which fits in a corresponding ring in the frame, is toothed around part of its circumference, and a pinion at the lower end of the steering pillar x gears into it. The

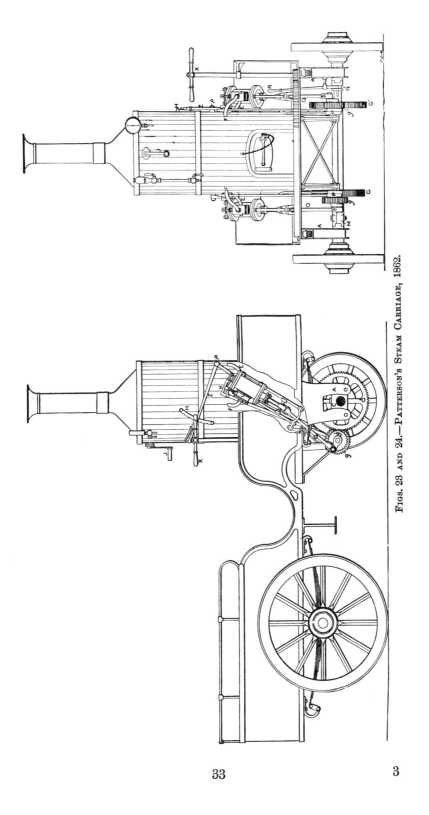

Figs. 23 and 24.—Patterson's Steam Carriage, 1862.

33

3

cylinders and the guide rods were attached to plates fixed to the boiler, steam being supplied to steam chests L by pipes F, which, as well as the exhaust pipes E, ran upwards under the lagging, the former passing inside the boiler to the regulator or steam valve J. Reversing and variation of cut off were effected by the links K, operated by the bell crank lever Z in quadrant H, and pivoted at *p*.

The counter shaft was fitted with two pinions, *g* and *g'*, both fixed, and of different sizes to gear alternatively with the wheels G and G'. The driving wheel on the right hand side could be allowed to run loose by the clutch N and lever R, but the other wheel was frictionally held tight enough to drive ordinarily by the clamp M, but not tight enough to prevent slip in turning corners. The two wheels G and G' were connected, so as to

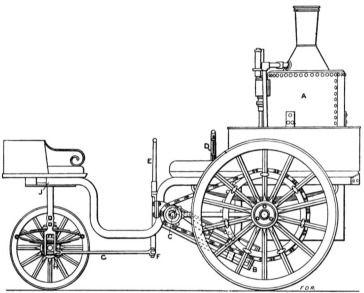

FIG. 25.—HOLT'S ROAD STEAMER, 1866-67.

move together on the axle, for putting one or other of the pairs of gear into operation.

The cylinders were 3 in. diam. and 6 in. stroke, and the working pressure 100 lb. The carriage weighed about one ton, in working order, and carried six to eight people. It appears to have had about 0·3 horse power per cwt. of carriage.

Passing over a steam carriage made by Messrs. Tangye, of Birmingham, one exhibited in the 1862 Exhibition by Mr. Lee, one by Mr. Armstrong, one each by Messrs. Seaward & Co., D. Adamson & Co., Lough & Messenger,[1] we come to another type of small road steamer made by Mr. H. P. Holt in 1866-67, and shown by Fig. 25.

[1] *Steam on Common Roads*, by W. Fletcher.

THE INTERMEDIATE PERIOD

Holt's carriage contained several features of novelty and interest in connection with the construction of a light road vehicle. It was carried on three wheels, but might equally well have been mounted on four, and have been provided with better seating accommodation, without departing from the main features.

At the rear of the vehicle, the frame of which was made of light angle iron, was suspended a field tube boiler, 4 ft. in height and 2 ft. in diameter, with 50 ft. of heating surface and 3 sq. ft. of grate surface. This supplied steam at a pressure of 250 lb. to a pair of double cylinder engines, the engines being fixed side by side on an inclined frame, and having independent crankshafts maintained at a fixed distance from the axle by radius rods. The cylinders were all 3 in. diam. and 6 in. stroke. On the outer end of each crankshaft was a sprocket pinion 8 in. diam., connected by pitch chains to sprocket wheels 24 in. in diam., on the drivers, which were 4 ft. 6 in. diam. The whole of the machinery was boxed in, including two feed pumps worked direct from the crossheads. The exhaust from the engines passed into a cast-iron box which formed a baffle plate at the bottom of the uptake above the fire, where it was superheated and escaped through five jets up the chimney. This arrangement, owing to the large number of beats from four cylinders, made comparatively little noise. Part of the exhaust was used to heat feed-water when required. The feed pumps were of novel and simple design. The plunger uncovered a port near the end of its outstroke, through which water entered from the water tank above it in level. On the return stroke this suction port was covered by the plunger, and the water which had filled the pump barrel was forced through the delivery valve into the boiler. The bunker was at the side of the boiler, and carried sufficient coke for a run of about 40 miles, and the water tank enough for about 20 miles, the coke consumption being about 5 lb. per mile under favourable circumstances. The carriage in running order weighed 30 cwt. and carried six to eight people, including steersman and engine and boiler attendant. It ascended grades of 1 in 14 at about 7 miles per hour, and on ordinary roads ran at from 15 to 20 miles per hour.

The steering wheel was worked by a cross handle on the top of a steering pillar E, the lower end of which was fitted with an arm or cross-head F, transverse to the car. At the ends of this cross-arm a pair of horizontal rods G were jointed, their forward ends engaging with the low ends of the fork at H, which was wide, and at the ends of each limb carried bearings in slots. Upon these bearings leaf-springs rested The stresses resulting from the road resistance and obstacles were thus referred to the strong tube which carried the steering pillar. The upper end of the fork was fixed to a locking ring upon which the fore end of the frame rested.

The reversing levers D of the two pairs of engines could be grasped either separately or together, ordinary link motion being used for this

35

and for variable cut off, the valves being between the two cylinders of each engine. The engines could be used as brakes, and on easy roads one engine was sufficient. A great advantage of the independence of the engines was not only that differential gear was dispensed with, but the engines automatically accommodated themselves with great smoothness to different speeds of the drivers in turning any corners. The engine power was apparently about 0·4 per cwt. of carriage without load.

Thomson's road steamers, 1871

First rubber tyres

Of another type of road steamer it is necessary in this part to give some particulars, namely, the Thomson road steamers, which embodied a great deal of mechanical engineering skill, and showed the advantages of flexibility of driving wheel tread as obtained with rubber tyres.

Figs. 26 and 27 illustrate one of these tractors, the *Ravee*, one of four made by Messrs. Ransomes & Sims, Ipswich, for the Indian Government in 1871. It had two cylinders, 8 in. diameter and 10 in. stroke, geared both 3·75 to 1 and 12 to 1, to 5 ft. 2 in. driving wheels, fitted with solid rubber tyres 10 in. by 5 in. thick, protected by linked steel shoes s. At 150 revolutions per minute the engines gave a speed of 10 miles per hour, or of 3 miles per hour for the slow speed. On the ends of the crankshaft κ were pinions which could be thrown into gear direct with the driving gear wheels P P, or these could be thrown back towards the crankshaft bearings out of gear, and the pinions M M' thrown into gear, driven by the pinion L on the crankshaft, and gearing, with the wheel o on the countershaft carrying the pinions M M'. The boiler A was on Olrick's Field system, with 177 sq. ft. of heating surface and 11 sq. ft. of grate surface. The fire-box was of large size, and the field tubes B descended to about the level of the top of the fire-door. At the bottom of the uptake D was a cast-iron pot E, carried by a central tube, on the top of which was an inlet for the exhaust steam. At H was a regulating screw at the lower ends of the rod *h*, by which the position of the lever c could be set to control the escape of the exhaust delivered into the pot E by a cone in the exhaust nozzle. At the top of the smoke-box c was a perforated plate F'. The single leading wheel was also fitted with a rubber tyre, but a double leaf-spring was also here used. The engines were used to draw two-wheeled omnibuses carrying 65 passengers and mails between two stations in the Punjab, 70 miles apart. Major R. E. Crompton, then Lieutenant Crompton, took these engines over, and was responsible for their running for some time.[1]

Numerous experiments made with engines of this class, fitted with Thomson's wheels, showed that they could haul an unprecedented load in comparison to their own weight, and, correspondingly, could ascend inclines insurmountable with iron tyres. Experiments made by the late Sir William Anderson gave a co-efficient as high as 0·576. A 6-ton engine took a load up an incline of 1 in 9, 20 tons up an incline of 1 in

[1] *Proc. Inst. Mec. Eng.*, Aug., 1879.

20, and took itself up a grass slope of 1 in 4·5. A 7-ton engine, with 5·5 tons on the drivers, was tested by drawing a dead weight over a pulley. On good ground, it gave a pull of 63 cwt., giving the co-efficient

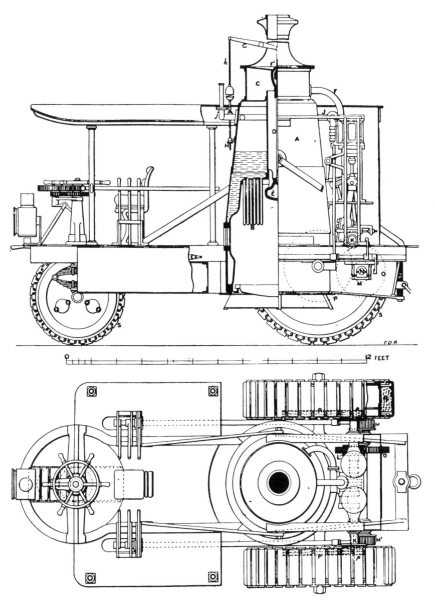

FIGS. 26 AND 27.—THE THOMSON ROAD STEAMER, "RAVEE."

of 0·576, and it gave a pull of 51 cwt. on a less favourable surface, or a co-efficient of 0·466. These figures were afterwards confirmed by the Royal Agricultural Society trials at Wolverhampton in 1871, when the average co-efficients for iron tyres and rubber tyres were found to be respectively 0·5 and 0·3 on dry macadam roads.

A somewhat similar engine to the *Ravee*, but named the *Sutherland*, was tested at the trials mentioned, with results which showed the high tractive efficiency of road steamers fitted with rubber tyres. It was the *Ravee* which made the longest journey ever made on common roads by a self-moving engine, namely, from Ipswich to Edinburgh and back in October, 1871, the engine and carriage it hauled together weighing 19 tons.

Messrs. Robey & Co. made some of Thomson's road steamers with Thomson's vertical pot boiler and link plate shield tyres.

Messrs. Burrell & Sons also made three-wheeled road-engines with Thomson's rubber tyres, locomotive boilers, and with spur gear, as in ordinary load locomotives; and many were sent abroad to haul omnibuses.

Experience with large, loose, solid rubber tyres The cost of renewal of the indiarubber tyres was, however, too great for continued commercial success of the system. A set of three tyres, such as those used on the *Ravee*, cost £241. The tyres were loose upon the iron rim, and as this was flanged, the tyre had to be stretched or sprung over the flange to get it on. In working with the heavy weight of these road steamers, the rubber was pressed out and elongated.[1] Under the load the 5-in. tyre was squeezed to 3·25 in., and as the wheel rolled along not only was the rubber pressed out in advance, but it became closely adherent and stretched tightly on the rear of the wheel, the thickness being reduced by this to 4·5 in., while it resumed its normal thickness in the front. At 8 miles an hour the squeezing and stretching and creeping of the tyre was such that it left the wheel in front by a distance of about 4 in. From this action the rubber tyre was continuously rotating round the wheel in a reverse direction, making a complete revolution rearwards to every 30 to 40 forward revolutions of the wheel. It will thus be seen that the rubber was very heavily worked and rubbed, and hence gradually reached a condition in which it lost its strength and broke. The linked steel shoes prevented this to a considerable extent, but so strong is the tendency to the creep referred to with a loose tyre that even when the rubber ring was placed between the wheel rim and an iron exterior tyre, on Adams's system, the link connecting the outer tyre to the inner rim was frequently broken, and it was difficult to keep them connected. This was also the case when the rubber was used in **Rubber blocks** blocks.[2] These facts give some idea of the difficulties connected with the use of solid rubber tyres under heavy loads, and it is rather curious that Thomson, having twenty-five years before invented the pneumatic tyre, did not make some efforts to develop it into a form suitable for light road steamers, so as to get some of the spring from compressed air instead

[1] *Proc. Inst. Civ. Eng.*, vol. xxxvi.
[2] A large number of different forms of elastic wheels have been made, some of which were described by Head, *Proc. Inst. Civ. Eng.*, vol. xxxvi.

of all from soft rubber. With the pneumatic tyre the rubber would not be "worked" so injuriously as in the solid, and the cost would be much less. The insistent load of the *Ravee* and such engines was too great for pneumatics or for any loose solid rubber tyres, even of very large section, and then the expense of the wheel complete was prohibitive. Messrs. Burrell have, however, successfully used such tyres of large width, by adopting rubber of decreasing hardness from the inner side where they are vulcanised on the wide iron rim.

It may be remarked, in passing, that although almost innumerable attempts have been made during the past twenty-five years to produce a satisfactory metallic spring wheel, none have been permanently successful, and the best traction engine wheel to-day is the Boulton wheel, with wood blocks in pockets of the outer tyre. Some of the spring wheels made by Aveling, Fowler, McLaren and Burrell promised well, but all have in their turn been given up in favour of spring-mounted engines, the best being leaf-springs.

In running the *Fourgon Postes*, between Lyons and Grenoble, doing 500 miles per week, and in high speed traction train work over long distances with bad roads and no roads, and over the Southern Alps, Messrs. J. & H. McLaren had unique experiences with spring wheels, springs of all kinds, and rubber springs. In France the whole of the running was done under time penalties, and in all weathers, and all experience went to prove that only leaf-springs would stand when rigid wheels were used, and that spring wheels were for this heavy work unsatisfactory. In one set of engines rubber buffers were used over the main axles in addition to volute and Timmins's springs, but it was found that the occasional leakage of oil into the rubber buffer-box and the working of the rubber under the load would, after a few days' use, cause the rubber to spue out of the air-hole in the box in the form of rubber ropes.

These engines ran for several months, including winter, at from 8 to 12 miles per hour, including the crossing of several spurs of the French Alps with variable gradients including one of 1 in 10. They were fitted with gas head-lights and side lamps, each engine carrying a receiver sufficient for the double journey, and carrying its own gas-compressing pump. Petroleum lamps were tried, but were frequently put out by the vibration of the engine at the high speed. Much of the experience with these engines was of a nature that would only arise with the heavy form of high speed locomotive, but, on the other hand, many of the lessons learned were not without high value in connection with lighter mechanically-propelled vehicles on common roads.

In 1870 Loftus Perkins made what may be called a steam horse, for attachment to the front of any vehicle. In it the engine, boiler, and driving-gear were all mounted on circular frame carried above a single wheel, with which all turned in steering, as described in the reference to Fig. 27. The wheel was 25 in. by 12 in., fitted with

Pneumatic tyres

Heavy tires vulcanised on rims

Attempts to use spring wheels

McLaren's high-speed road locomotive and spring wheels

Perkins' single-wheel tractor, 1870

39

a Thomson rubber 5 in. by 10 in. The engine was a compound vertical engine fixed on one side of the boiler, and had two cylinders $1\frac{13}{16}$ in. diameter and two $3\frac{1}{8}$ in., all with a stroke of 4 in. He carried a pressure of 450 lb. in a Perkins' tube boiler, 26 in. high, $15\frac{5}{8}$ in. wide, and $20\frac{3}{8}$ in. long, but did not use that pressure on the engines. The engine ran at speeds up to 1,000 revolutions per minute, and drove the rubber-tyred wheel by bevel gearing, as shown in Fig. 28, in which c c are

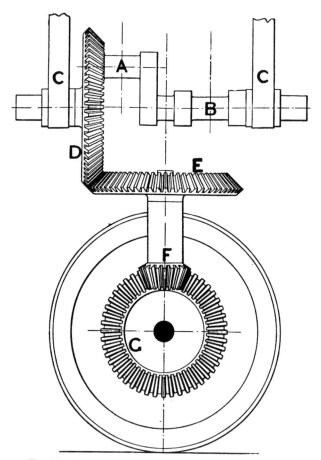

FIG. 28.—GEARING OF PERKINS' SINGLE WHEEL TRACTOR.

parts of the vertical engine frame attached to one side of the boiler. A and B are the two cranks, one web A being let into the bevel wheel D. This drove a similar wheel E, on the lower end of the spindle of which was the pinion F, gearing into the wheel G on the rubber-tyred road wheel. The weight of the steam horse was 30 cwt.

The steam from the engines was passed by a pipe to a condenser connected with a water-tank under the vehicle it hauled, which, in

THE INTERMEDIATE PERIOD

Perkins' experiment, was a light van, and weighed, with a number of passengers, 57 cwt. A trial run to St. Albans and back gave a coal consumption of 2·17 lb. per ton mile. The average steam pressure used in its engines was 250 lb., and the average speed was 3 miles per hour. The grate surface of the boiler was 0·75 sq. ft. The very high pressure, the kind of boiler, and the very small engines are points of interest.

Finally reference may be made to a steam carriage, arranged Irish car fashion, constructed in 1871 by J. G. Inshaw, to a steam brougham by H. A. O. Mackenzie, made in 1875–76, and to the steam three-wheeled dog-cart made by Mr. A. B. Blackburn in 1878, of which the most useful particulars are given in his patent specification No. 3838 of 1877. In this he used a horizontal small tube coil for his boiler, heated by a petrol bunsen burner. It supplied steam to a very small three-cylinder torpedo engine, the exhaust from which passed into a condenser consisting of a long horizontal cage of small tubes cooled by a long winnowing-machine form of fan. Motion was given to the road wheels by gut belt running over small pulleys on a second motion shaft, and thence over guide pulleys to large grooved pulleys on the driving wheels. A petrol vapouriser or air carburettor, heated by exhaust steam, was described. There are several points of interest in this specification. The carburettor will be described when dealing with carburettors hereafter.

Blackburn's steam dog-cart, 1877

41

Chapter III

POWER AND FRICTIONAL LOSSES

Costly waste of power under present road conditions

SO long as our roads are made as they are at present, and as inefficiently maintained, and so long as so many hill-tops are looked upon as the natural and necessary routes over which our roads must be made and every load hauled, so long shall we have to expend many hundreds of thousands of pounds, or even millions, per year in unnecessary power in horses and in engines and in fuel.[1]

At present we must take roads as they are, and provide power accordingly.

Large reserve power necessary

A study of the motor vehicles of various kinds made during the past seventy-five years shows that only those who provided plenty of power succeeded, but the quantity of power used by them is not numerically of much guide to us, although the form they gave to their engines and boilers and methods of arranging them are instructive. There is not much to guide us as to the relation between the actual and the nominal horse power of the engines of the early builders, but we may make some approximate estimates. The engines of Hancock's *Automaton*, for instance, had 9-in.

Power used by Hancock and others 1826-40

cylinders and 12-in. stroke, and at 10 miles per hour made about 70 revolutions, the chain wheel on the axle being taken as same as that on crankshaft, or a piston speed of 140 ft. per minute. The pressure may be taken at 100 lb., although he commonly ran with 70 lb., and judging by the valve gear he seems to have used, and by that which others used, cut off was no doubt very late, the steam supply being regulated by throttling. Thus he may easily have had available for hills a mean pressure of at least 70 lb. pressure throughout the stroke. The actual power of the engine would thus be, for a short time at least, with its two cylinders,

$$\frac{2\,P.L\,2.\,A.N.}{33,000} = \frac{2 \times 70 \times 1 \times 2 \times 63{\cdot}6 \times 70}{33,000} = 37{\cdot}9\,\text{HP}.$$

Allowing a mechanical efficiency 85 per cent., which is probably much too

---

[1] For a discussion on this subject and proof that a hundred millions sterling might be profitably expended on road improvement and construction, see the author's presidential address to the Society of Engineers in February, 1898.

high, the actual power at the crankshaft may have been 32 HP. This was used with a coach weighing, when loaded with passengers, about 4 tons, or 80 cwt, so that Hancock had $\frac{32}{80} = 0.4$ HP. actual per cwt. of coach loaded. When the chain pinion on the crank shaft was smaller than that on the axle the power was still higher.

Gurney used on his coaches, which, when fully loaded, weighed about 3·75 tons to 4 tons, or 75 to 80 cwt., engines with 9-in. cylinders and 18-in. stroke, making, with 5 ft. wheels at 10 miles an hour, 56 revolutions per minute. His power was, at the lower working pressure given by him, about 28 HP., but at the higher pressure, which he had available and sometimes used, and with an assumed mechanical efficiency of 85 per cent., he had **[margin: Power as used by Gurney and Dance]**

$$HP. = \frac{2\,P.\,L2.\,A.N.\,0.85}{33,000} = \frac{2 \times 70 \times 1.5 \times 2 \times 63.6 \times 56 \times .85}{33,000} = 39,$$

so that he had $\frac{39}{75} = 0.52$ HP., or $\frac{39}{80} = 0.49$ HP. per cwt. of coach loaded.

Estimated in the same way, Summers and Ogle had about 0·39 HP. per cwt. of total load, and Maceroni and Squire about 0·43 HP.; Hill's coach of 1840 was exceptionally powerful, and had per cwt. about 0·5 HP. Taking the mean of all these, we have for the coaches of those days 0·45 of a horse power as a maximum. **[margin: Mean power per cwt. of the old coaches]**

Coming to the more recent period, we find considerable variation in this respect, but not more than in the character of the vehicles themselves and in the superiority of some of the work. **[margin: Power used by Yarrow and others, 1860–97]**

The Yarrow and Hedley vehicle (Fig. 21), for instance, had probably not more than 0·19 to 0·2 H.P. per cwt. of total load. Those of Rickett, as made for the Marquis of Stafford and the Earl of Caithness, had such very small seating accommodation in relation to the weight of the vehicle itself that they need not be considered. The carriage made by Patterson (Fig. 23), had about 0·3 HP. per cwt. total load. The *De Dion* omnibus, as tested in the Poids Lourdes trials from Versailles in 1897, had about 0·25 HP. per cwt., and the *Scotte* omnibus about 0·15 HP. per cwt. for about half the speed of the *Dion* omnibus.

FRICTIONAL LOSSES—MECHANICAL EFFICIENCY.

The maximum power provided in any of these vehicles is required only for the heavy gradients met with in almost every district. The work to be done against gravity far exceeds that otherwise to be done by the motors of a road vehicle. This will readily be seen from a consideration of the several resistances, which together show the total power required. **[margin: Power required to meet frictional losses in transmission]**

The first to be taken into account is the friction of the propelling machinery itself. This differs so much that it may vary from 7 or 8 per cent. of the whole of the power represented by the indicator diagrams to as much as or more than 30 per cent. of that power when two or more countershafts, each with gearing, and in some cases with bevel wheels, spur gearing, or belts and chains and used. In this way the power actually exerted at the periphery of the driving wheels may be but a poor representation of that which the pistons are doing.

Unfortunately there is very little really known yet of the mechanical efficiency of the combined transmission gear of any, even the best, of the cars which are running and earning golden praises to-day. The brake horse power of the engine or motor itself is known, and in some cases its indicated horse power is known, and thereby its mechanical efficiency. But however refined the motor may be in this respect, it is only in the best of the cars that corresponding efficiency is found, or may be expected, in the gearing. It is of course more important in the motor, because high mechanical efficiency is to a great extent a measure of practical efficiency or trustworthiness, but economy of power, and thereby of fuel and weight, can only be reached by high efficiency of every working part. In future it may be expected that trials conducted by independent authorities will show not simply that some cars are better hill-climbers than others of nominally the same power, but why it is that such very different average consumption of petrol marks the long journey performance of different cars. A summary of what is known to-day on these points will be given hereafter in dealing with the trials of vehicles in 1899.

Friction of gear and bearings

It would be useless here to dwell upon the friction of a pair of spur or bevel wheels because it can only be estimated from the coefficient of friction of the surfaces of a pair of teeth. Moreover, not only are the proportions of gearing and bearings used in construction abnormal, and dictated in many instances by mechanical exigencies, but the lubrication is often of the most rudimentary kind, the loads per unit of effective surface very variable, and the speeds of wide range. Assuming, however, that the dimensions of bearings are carefully considered, the lubrication improved and the effective surface of a bearing of given dimensions taken into consideration, the following table [1] will be found of some service. It shows the result of experiments with a 2-in. spindle running in bearings of 4 in. in length, with the surface or chord of contact cut away to different widths, with a load of 300 lb., and at a temperature of 104° F., the lubrication being constant and plentiful with good oil from an ordinary syphon lubricator.

From Table I. it will be seen that the co-efficient of friction μ of a gun-metal bearing upon a steel journal 2 in. diameter, with the brass cut away to a chord length of 1·5 in., fell from 0·09 at 10 revolutions per minute to 0·018 at 122 revolutions, and remained constant to 192 revolutions, and then increased to 0·035 at 362 revolutions.

Effect of width of bearing

It will be seen that the friction was lessened by reducing the width of the brass to 1 in. and that it was still further lessened by reduction of width

[1] *Proc. Inst. Civ. Eng.*, lxxxix. p. 454.

POWER AND FRICTIONAL LOSSES

Friction of Bearings. Table I.

Width of brass, 1·25 inch.			1·0 inch.			0·5 inch.		
Revolutions per Minute.	Feet per Minute.	μ	Revolutions per Minute.	Feet per Minute.	μ	Revolutions per Minute.	Feet per Minute.	μ
10	5·24	0·0955	12	6·3	0·1190	9	4·71	0·1190
19	9·95	0·0595	17	8·9	0·0770	17	8·9	0·0910
31	16·23	0·0245	29	15·2	0·0462	27	14·14	0·0735
28	14·66	0·0420	87	45·6	0·0161	58	30·4	0·0385
122	63·90	0·0182	152	79·6	0·0126	163	85·3	0·0140
163	85·30	0·0182	231	121·0	0·0112	198	103·7	0·0140
192	100·50	0·0182	268	140·4	0·0126	262	137·2	0·0161
233	122·00	0·0203	350	183·3	0·0154	309	161·8	0·0175
257	134·60	0·0245	379	198·5	0·0189	367	192·2	0·0210
303	158·70	0·0280						
362	189·50	0·0350						

to 0·5 in., but that the speed at which the coefficient again began to rise occurred earlier.

Table II. shows the effect of increase of load. Here again the narrow surface contact caused less friction, the total resistance R in this table showing this very clearly; but it also shows how steadily the friction remained with a growing load up to 350 lb. with a constant speed of 233

Effect of load on bearing

Friction of Bearings. Table II.

Siphon Lubrication.   Temperature, 40° C.   Revolutions per minute, 233.

Width of brass Angle subtended	1·5 inch. 97°		1·25 inch. 77°		1·0 inch. 60°		0·5 inch. 29°	
Load in lbs.	μ	R.	μ	R.	μ	R.	μ	R.
50	0·1148	5·74	0·0840	4·20	0·0756	3·78	0·0420	2·10
100	0·0651	6·51	0·0504	5·04	0·0420	4·20	0·0252	2·52
150	0·0457	6·85	0·0371	5·57	0·0287	4·31	0·0189	2·83
200	0·0346	6·92	0·0288	5·76	0·0225	4·50	0·0173	3·46
250	0·0277	6·92	0·0231	5·77	0·0180	4·50	0·0168	4·20
300	0·0231	6·93	0·0192	5·76	0·0150	4·50	0·0196	5·88
350	0·0198	6·93	0·0168	5·88	0·0129	4·52	0·0180	6·30
400	0·0175	7·00	0·0154	6·16	0·0112	4·48	0·0199	7·96
450	0·0162	7·29	0·0137	6·17	0·0100	4·50	0·0196	8·82
500	0·0147	7·35	0·0126	6·30	0·0101	5·05	0·0181	9·05
550	0·0139	7·64	0·0116	6·38	0·0092	5·06	0·0183	10·07

revolutions per minute, but how rapidly it increased with the narrow surface above that, until at 550 lb. the resistance and the coefficent were double that of the bearing surface 1·0 in. in width. Under practical conditions the narrow bearing would in all probability have siezed, and

45

generally in practice the bearing and surfaces for the heavy loads would for continuous pressure have to be much larger than those shown in the tables for the higher loads.

Effect of bath lubrication The same bearings running in a bath of oil would run with from one half to two-thirds the friction and some of the highly-finished hardened steel bearings run with less friction than the hard steel and gun metal. Under the practical condition, however, of many of the bearings in motor-cars it may be taken that the frictional resistance of ordinary bearings of steel in gun-metal, or steel on cast-iron, will not be less than from 0·035 to 0·05 of the load upon them.

Friction of ball bearings The value of ball and of roller bearings in reducing frictional losses is particularly noticeable in cycle and motor vehicles. The following table gives the coefficient as found by experiment with a bridge bracket cycle bearing with eleven balls $\frac{5}{16}$ in. in diameter. The load and the speed varied, and it will be seen that within the limits of the test,[1] the coefficient of friction remained constant, the frictional resistance increasing directly as the load.

FRICTION OF BALL BEARINGS.

Ball Bearings Lubricated with White Rock Oil—Temperature 70° F.

Load.	Revolutions per minute, 19.	Revolutions per minute, 157.
lb.	μ	μ
10	0·0075	0·0060
20	0·0075	0·0075
40	0·0071	0·0075
60	0·0070	0·0075
80	0·0071	0·0071
100	0·0073	0·0072
120	0·0071	0·0069
140	0·0072	0·0067
160	0·0072	0·0066
180	0·0072	0·0066
200	0·0073	0·0066

Roller bearings Roller bearings of the well-made forms now obtainable—as, for instance, the Mossberg bearings—must in the future come largely into use, as the frictional resistance of shafts running in them is reduced as compared with ordinary bearings much in the same proportion as it is reduced by ball bearings, namely, from 30 to 40 per cent. Tests made on tramcars have shown the friction of these roller bearings to be 60 per cent. less than that of the same cars with common axle bearings, and the starting effort only 24 per cent.

With end thrust bearings the advantages of the roller thrust bearing are even greater, numerous tests of Mossberg thrust bearings showing

[1] *Proc. Inst. Civ. Eng.,* lxxxix., p. 453.

that whereas the coefficient of friction of ordinary thrust bearings was 0·05 at starting, with roller thrust bearings it fell to less than a tenth of this, and even to as low as 0·0025.

These figures, showing the frictional resistance of different kinds of bearings, are given as an indication of the amount in particular cases of the loss due to bearing friction. It is not considered useful to follow this up by giving any expressions for the friction of bearings generally, because the value of such expressions depend entirely upon their application. A bearing which has the pull upon it of the driving effort and the transmission or driven effort, both in or near the same direction, may have its load double that of the same bearing with the same power transmitted, but with the pull of driver and driven in opposite directions. In the latter case the bearing load may·be almost *nil*. Hence the estimation of the load upon a bearing may be, and often is, of much greater importance than the coefficient of friction relating to it, and may be difficult of estimation. Moreover, the load may be variable, or of the constantly intermittent order, requiring in the one case a maximum of strength and surface, and in the other a minimum may suffice. A bearing which, when carrying a dead load, may be heavily loaded with 450 to 500 lb. per square inch, might carry 1,500 lb. or more if in the position of the brasses at the big end of a connecting rod, and often has to carry even double this when it has the small movement but intermittent load of the small end of a connecting rod. *General considerations concerning bearings and friction*

Thus it requires the training of the mechanical engineer to estimate the sort and quality, as well as magnitude of a bearing's load and work before the amount of the frictional resistances occupy an important part of his calculations.

The frictional losses due to gearing may form, and often do form, a considerable part of the difference between motor power and the power at the road wheel periphery. These losses are almost impossible of estimation, except for each particular case, and under any circumstances the design and workmanship may render the application of any carefully found coefficients quite illusory. The workmanship, the materials, the gear and spindle combination, the speeds and loads, and the lubrication must all be known, and then the judgment of the mechanician must be used in the selection of constants and coefficients for arriving at a sound estimate, a selection upon which all depends. **Estimates of frictional losses in trains of gearing illusory**

For these reasons no space will be occupied in this book on either the form, strength or friction of gearing as text book subjects,[1] although references will be made to gearing of different kinds as actually used.

---

[1] For the results of experiments on journal friction, deduction therefrom, and theoretical expressions, see *Proc. Inst. Civ. Eng.*, lxvi. p. 69; xcvi. p. 427; *Friction and Lubrication*, by R. A. Thurston; *Proc. Inst. Mech. Engineers*, 1883, p. 632; *Manual of Rules, Tables and Data*, D. K. Clark; *Tooth Gearing*, Cromwell; *Machine Design* Unwin.

MOTOR VEHICLES AND MOTORS

Wherever it becomes necessary to allow for the friction of all the running gear of a car for the purposes of approximation to power questions, the mechanical efficiency of the combination from crankshaft to driving-wheel periphery will be taken as 0·70 except where otherwise stated. As will be found hereafter in an analysis of the results of the Automobile Club trials last year, a report upon which was drawn up by the author and Major Holden as judges, 70 per cent. appears to be considerably too high for most of the vehicles. Data concerning power required and provided in modern vehicles will be found in the pages devoted to them hereafter.

Chapter IV

ROAD RESISTANCES

THE road resistances with which motor cars have to contend are generally small as compared with the resistance due to gravity. They vary enormously with different road surface materials and maintenance, and have to be added to the gravity resistance. In any consideration of the power required, the highest ordinary surface or rolling resistance must generally be assumed.

TABLE IV.

Resistance to Traction on Common Roads.

Road Surface Material.	Resistance in lbs. per ton on Iron-tired Wheels.
Asphalte	22 to 28
Wood, hard	22 „ 26
„ soft	30 „ 38
Granite setts	
Macadam, very hard and smooth	40 „ 45
„ good	45 „ 52
„ traffic rolled, wet	52 „ 58
„ steam rolled, new and muddy . . .	58 „ 62
„ new, flat spread	95 „ 105
Gravel	100 „ 140
Granite tramway	12·5 „ 15
Iron plate tramway	10 „ 12

In most cases these resistances increase slowly at higher speeds, and it must also be noted that the resistance on bad, soft and gravel roads will probably be greater with propelling wheels than with most hauled wheels.

All the figures given include the axle friction of a pair of wheels, and in some cases of four wheels. Most of the figures relate to road resistance at walking or slow trotting pace. The draught, as above stated, increases with increase of speed, and experiments made with stage coaches gave an increase in the following proportion on level roads. It may be remembered, however, that Hancock and Gurney found that their coaches ran better above four miles per hour than below that speed.

MOTOR VEHICLES AND MOTORS

TABLE V.

Effect of Speed on Draught.

Gradient.	Draught at 6 miles per hour.	At 8 miles.	At 10 miles.
Level	100	108	116
1 in 60	111	120	128
1 „ 40	140	166	172
1 „ 30	165	196	200
1 „ 26	213	219	225
1 „ 20	268	296	318

These figures must be taken as qualitatively useful rather than quantitatively accurate. They were obtained by Sir H. Parnell many years ago when dynamometer experiments at speed with horse haulage were even more difficult than now. They serve to show, however, the growth of the traction resistance with speed and with moderate hill climbing.

Effect of width of wheels
Some interesting modern experiments with wheels of different widths on the roads which are most met with in new countries have been made by Mr. H. J. Waters, of the Missouri State Agricultural Experiment Station. Some of the results are here tabulated for the information of those who may think of sending motor vehicles to Africa and elsewhere. They are of interest as showing where either narrow or broad tyres may be the better, and why.[1] Where a hard bottom is covered with thick mud or sand the narrow tyre found its way to this bottom with less resistance than the broad tyres. Other generalizations are obvious. See Table VI., p. 51.

Effect of diameter of wheels
The experiments of Morin and of Dupuit, and those of the Royal Agricultural Society and many others, all show that the diameter of the wheels has a very material effect on ordinary roads, and taking them all it would appear that the road resistance decreases nearly as the square root of the increase in diameter of the wheel, although Morin maintained that the resistance decreased directly in proportion to the increase in diameter of the wheel. Thus, according to Morin, the rolling resistance of an iron-tyred wheel 2 ft. in diameter on ordinary roads would be double that of a wheel 4 ft. in diameter, and some of the experiments of the Royal Agricultural Society seemed to support this.[2] It must be supposed that axle friction would partly explain this. Taking, however, the more widely-received idea that the decrease in resistance is in proportion to the square root of the increase of the size of the wheel, then the resistance of a 4-ft. wheel as compared with a 2-ft. wheel would instead of being halved or reduced to 50 per cent. only be decreased to 70 per cent. Thus

$$\frac{\sqrt{4}}{\sqrt{2}} = \frac{2 \cdot 0}{1 \cdot 414} = 1 \cdot 414.$$

[1] *Industries and Iron*, April 7, 1898, p. 273.
[2] *Journal of Royal Agric. Soc.*, 1874, and *Engineering*, July 10, 1874, p. 23.

ROAD RESISTANCES

TABLE VI.

Effect on Draught of Broad and Narrow Tyres on Bad Roads.

		Pull in lb.		Gain per cent. in favour of		Resistance per ton approximate.		
	Character of Road.	1·5-in. Tyre.	6-in. Tyre.	Broad Tyre.	Narrow Tyre.	Narrow Tyre.	Broad Tyre.	Remarks.
1	Hard, smooth, clean, level macadam	99	73	16	—	56·5	41·7	Net load on wagon 2,000 lb. Wagon 1,500 lb. Total, 3,500 lb., or 1·55 tons.
2	Hard, smooth, clean, level macadam	144	123	35	—	82·3	70·3	
3	*Gravel road*, some loose stones	218	164	33	—	124·5	93·7	
4	*Gravel road*, some loose stones	146	100	45	—	81·1	57·1	
5	*Gravel road*, sandy . . .	239	157	46	—	130·6	89·7	Broad wheels weighed 250 lb. more than the narrow.
6	Ditto dry, newly-laid	330	261	27	—	188·5	149·1	
7	*Gravel road*, 1 in. to 2½ in. of slush	262	268	—	2	149·7	153·1	
8	*Gravel road*, 1 in. to 2½ in. of slush	232	241	—	4	132·5	137·7	Speed 2½ miles per hour about.
9	*Dirt road*, hard, dry, level	137	105	31	—	78·3	60	
10	Ditto ditto ditto	178	145	23	—	101·7	82·8	
11	Hard, dry, clay road, with slight ruts	131	76	71	—	74·8	43·4	
12	Dirt road, with 2 in. to 3 in. of dust	91	107	—	18	52	60·1	
13	Hard clay road, sticky surface, ruts 4 in. to 6 in. deep	206	308	—	49	117·7	176	
14	Road with 2½ in. sticky mud	251	325	—	28	143·4	185·7	
15	Clay road, with soft surface 3 in. to 4 in. deep . .	335	436	—	30	191·4	249·1	
16	Clay road, with 3·5 to 4 in. soft, wet mud	287	406	—	42	164	232	
17	Black soil road, stiff mud on top	497	307	62	—	284	175·4	
18	Clay road, dry top, spongy below	473	423	12	—	270·2	241·1	
19	Clay road, dry top, spongy below, with ruts 8 in. deep	596	377	58	—	340·5	215·4	
20	Stiff, muddy, clay road, beginning to dry	325	552	49	—	471·4	315·4	
21	Muddy clay road, slightly frozen on top	549	448	22	—	313·7	256	
22	Soft, sticky, clay road, 18 in. ruts full of water . .	322	·514	—	59	184	294	

To reduce the resistance to 50 per cent. would require a wheel whose increase in diameter is as follows :—

Taking R = resistance of smaller wheel and d its diameter,
r = resistance of larger wheel and D its diameter,

then $\dfrac{R}{r} = \dfrac{\sqrt{D}}{\sqrt{d}} = 2$.   Hence $D = (\sqrt{d} + 2)^2 = (1·414 \times 2)^2 = 8$.

51

MOTOR VEHICLES AND MOTORS

Thus, to halve the resistance the diameter must be increased fourfold according to this deduction. It must, however, be noticed that any of these experimental deductions can only be used with judicious reference to the character of the road surface, for it is perfectly clear that on a very hard, smooth road the reduction in resistance would not be anything like that claimed either on Morin's or on Dupuit's rules, while it is equally obvious that on some soft gravel-covered roads the reduction may be as much as Morin's rule would indicate.[1] Table VI. serves to show how much different widths of wheel may affect results under different conditions.

Draught of a London omnibus or traction resistance on different pavements

Experiments on the resistance to traction of a London omnibus on various roads were made by a committee of the Society of Arts in 1875. The omnibus weighed 2,480 lb., and it was loaded with 3,278 lb. of oats in sacks, the total weight being 5,758 lb., or 2·57 tons. The average speed was 3⅓ miles per hour.

TABLE VII.

Traction Resistance of an Omnibus.

	Resistance in lbs. per ton.
Granite pavement, setts 3 in. to 4 in. wide	17·41
Asphalte roadway	27·14
Wood pavement, not hard	41·60
Good but gravelly macadam road	44·48
Granite macadam, newly laid	101·09

These experiments agree with the figures already given, except that the draught on asphalte seems too high on a London pavement.

Road resistance with rubber tyres

For the purposes of the motor carriage builder, it is not necessary to follow this part of the subject farther, but reference must now be made to rubber tyres, the traction resistance of which is so much less on most common roads than with iron tyres.

ROAD RESISTANCE WITH RUBBER TYRES.

For data on this subject reference may be made to a paper read before the Société des Ingénieurs Civils, Paris, by M. Michelin, in which he gave the results of numerous experiments,[2] the figures giving the total pull on

[1] For further consideration of road resistance questions, see *Proc. Inst. Civ. Eng.*, vol. lxxviii. p. 277; vol. lx. p. 302; *Repair and Maintenance of Roads*, by W. H. Wheeler; "Pavements and Roads," *Engineering and Building Record-Book*, for 1890, pp. 271-330, etc.; *Maintenance of Macadamised Roads*, by T. Codrington, p. 59 (London: E. and F. A. Spon); *Carriage-Ways and Footways*, by H. P. Bulnois, p. 106 (London: Biggs & Co.); "Streets and Highways in Foreign Countries," *U.S. Consular Reports*, 1891; "Construction and Maintenance of Roads," by E. P. North, *Trans. Amer. Soc. Civ. Eng.*, vol. viii.; *Management of Highways*, by E. P. Hooley (London: Biggs & Co.); *Manual of Rules, Tables and Data for Mechanical Engineers*, Clark, p. 962; *Society of Arts Journal*, June 25, 1875; *Construction of Roads and Streets*, by Henry Law and D. K. Clark; *Construction of Roads and Streets*, by William H. Maxwell.

[2] See the *Autocar*, vol. i., 1896, pp. 498-507, 522, 534, 544, and an abstract of M. Michelin's paper appeared in the *Automotor and Horseless Vehicle Journal*, No. 22, p. 391.

the vehicle. To these have been added below the pull in pounds per ton of total load.

The first experiments were made with iron and pneumatic tyres of the following diameters and weights :—

Kind of Tyre.	Diameter of Wheels.		Weight of Wheels.		Weight of Empty Vehicle. lb.
	Front. ft.	Back. ft.	Front. lb.	Back. lb.	
Iron tyres	3·02	3·67	127·6	158·4	1,269·4
Pneumatic tyres	2·95	3·93	85·8	123·2	1,192·4

Condition of Vehicle.	Weight of Vehicle in tons with	
	Iron Tyres. tons.	Rubber Tyres. tons.
Empty	0·565	0·533
With load of 330 lb.	0·714	0·680
With load of 660 lb.	0·860	0·830

The walking speed was 2·65 miles per hour, and the average trotting speed 6·6 miles per hour.

To the weight of the empty vehicle the weight of the driver must be added ; and to make this feature comparable the same driver was used in both cases. These experiments were made over various soils, over different lengths of road, with varying loads, and at different speeds. The results of the trials are summed up in the following tables :—

TABLE VIII.

Experiments in Snow.

Speed and Load.	Iron Tyres.		Pneumatic Tyres.	
	lb.	lb. per ton.	lb.	lb. per ton.
Carriage at a walk, empty.	34·89	61·60	25·23	47·30
Carriage at a walk, load of 330 lb. . .	39·22	55·00	27·96	41·10
Carriage at a trot, load of 330 lb. . .	65·12	91·40	33·59	48·70
Carriage at a trot, load of 660 lb. . .	68·57	79·80	39·51	47·60

These figures show that under all the conditions given, the use of the pneumatic tyre diminishes the tractive effort ; and this economy is greater upon a bad than upon a good road, and it also increases with the speed and the load.

TABLE IX.

Experiments in Mud.

Speed and Load.	Iron Tyres.		Pneumatic Tyres.	
	lb.	lb. per ton.	lb.	lb. per ton.
Carriage at a walk, empty	35·20	62·40	23·10	43·30
Carriage at a walk, load of 330 lb. .	38·06	53·30	27·34	40·20
Carriage at a trot, load of 330 lb. . .	43·01	60·30	28·53	41·80
Carriage at a trot, load of 660 lb. . .	50·73	58·90	31·15	37·50

TABLE X.

Mean of Trials upon Macadam: Dry, New, Dusty, and Well-paved, with Grades varying from 1·2 to 5·8 per cent.

Speed and Load.	Iron Tyres.		Pneumatic Tyres.	
	lb.	lb. per ton.	lb.	lb. per ton.
Carriage at a walk, empty	38·32	68·00	30·91	57·80
Carriage at a trot, empty	44·90	79·50	35·09	65·80
Carriage at a walk, load of 660 lb. .	45·65	53·80	35·51	42·80
Carriage at a trot, load of 660 lb. . .	65·34	75·70	36·08	43·60

Some experiments were made to compare the relative effects of iron tyres and pneumatic tyres with varying air pressure and with plain rubber. The results were as follows:—

TABLE XI.

Speed and Load.	Iron Tyres.	Pneumatic Air Pressure.		Plain Rubber.
		66 lb.	99 lb.	
At a walk, empty	33·64	36·03	34·34	37·62
At a trot, empty	46·64	33·99	41·69	44·15
At a walk, load 660 lb.	44·22	44·75	43·91	56·52
At a trot, load 660 lb.	64·20	45·14	50·64	63·96

From this table it would seem that the plain rubber tyre, while better than iron at a trot, is not so good at a walk. But M. Michelin adds that if the surface is muddy, or covered with snow, so as to remove all elements of elasticity, the full rubber tyre gives better results than an iron tyre, and it must be noted that the size of the pneumatic gave it an advantage over the solid. But the plain rubber is always inferior in this respect to the pneumatic tyre; and for the latter it would appear that the pressure of inflation should not exceed 66 lb. The general mean of the experiments of M. Michelin in traction resistance give the following relative values :—

Pneumatic tyres being represented by . . . . . . 100·0
Full rubber tyres   „   „   .   .   .   .   .   . 129·8
Iron tyres   „   „   .   .   .   .   .   . 132·7

The advantage of rubber and especially of pneumatic tyres are not however, confined to the lowering of traction effort as shown by haulage tests. The absorption at the periphery of the wheel of a large part of the vibration, and the softening of the jar and impact of road inequalities materially modifies the conditions of running of motor vehicle mechanism, increases the life of the whole of the parts below the ordinary springs, increases the adhesion without increase of weight, and of course adds greatly to comfort of riders. ^{Other advantages of rubber tyre}

The lower figures, which have now been given as the resistance with iron tyres and various roads, show the importance of good roads, whether to be used by horses or motor vehicles. Good roads everywhere would save half the present cost of horse haulage.

The higher figures are, unfortunately, those which will have most generally to be used with motor vehicles. Bad roads mean slow speeds. Only those vehicles which are intended for definite districts and town work can take advantage of the low resistance of good roads. ^{Maximum resistance to be provided for}

Chapter V

RESISTANCE DUE TO GRAVITY AND POWER
REQUIRED

Resistance due to gravity on gradients

THIS is the heaviest resistance that has to be provided for in determining the power of a motor for a given vehicle. Its amount is easily ascertainable, and for touring cars is very heavy with our present numerous, unnecessary and wasteful steep hills.

The resistance due to gravity is directly proportional to the steepness or angle of the grade, and may for all practical purposes be taken as equal to that fraction of the total weight of the vehicle which is represented by the fraction expressing the grade, as for instance, 1 in 4 or 1 in 12, namely, the rise is $\frac{1}{4}$th or $\frac{1}{12}$th of the distance traversed, and the gravity resistance is correspondingly equal to $\frac{1}{4}$th or to $\frac{1}{12}$th respectively of the whole load. This errs on the right side, in that it gives figures a very little higher than the true figures.

Table of total power required

Table XII., which has been calculated for motor vehicle purposes, gives the resistance due to gravity at all grades, from 1 in 100 to 1 in 5, and it will be found also to give the power required per ton of vehicle for iron and for pneumatic tyres for level roads, and for all grades and all resistances, the road resistance being taken as 60 lb. for iron tyres, and 45 lb. for pneumatic tyres, and for the purposes of the power columns a speed of ten miles per hour has been taken. Air and wind resistances are not included.

The resistance due to gravity is given correctly in the table as that due to the angle *a* of inclination not expressed as a fraction, but as a sine. The difference is small; for instance, on a gradient of 1 in 10, expressed as a fraction, the gravity resistance would be $\frac{2240}{10} = 224$ lb., but expressed in terms of the sine of the angle the resistance becomes $2240 \times 0.9949 = 222.85$ lb.

It will be seen that this table assumes a road resistance of 60 lb. per ton for iron tyres, and 45 lb. per ton for pneumatic tyres, the advantage at higher speeds and loads being neglected.

The horse power required under any other conditions as to weight,

56

RESISTANCE DUE TO GRAVITY AND POWER REQUIRED

TABLE XII.

Showing Resistance due to Gravity and Total Resistance on average Macadam Roads for a Vehicle weighing 1 ton, at 10 miles per hour, with Iron, and with Pneumatic Tyres, and B. HP. of Engine required to Propel the same assuming efficiency of transmission = 70 per cent.

Gradient.			Resistance due to Gravity. lb. per ton.	Total Resistance with Iron Tyres on average Macadam at 10 miles. lb. per ton.	HP. of Engine required to propel a vehicle with Iron Tyres, weight 1 ton, at 10 miles per hour on average Macadam.	Total Resistance with Pneumatic Tyres on average Macadam at 10 miles. lb. per ton.	HP. of Engine required to propel a vehicle with Pneumatic Tyres, weight 1 ton, at 10 miles per hour, on average Macadam.
1 in "n."	Angle.						
	deg. min. sec.						
Level	0 0 0		0·00	60·00	2·29	45·00	1·72
1—100	0 34 22		22·40	82·40	3·14	67·40	2·57
1—95	0 36 10		23·56	83·56	3·19	68·56	2·62
1—90	0 38 6		24·89	84·89	3·24	69·89	2·67
1—85	0 40 27		26·34	86·34	3·29	71·34	2·72
1—80	0 42 57		28·00	88·00	3·36	73·00	2·79
1—75	0 45 49		29·70	89·70	3·42	74·70	2·85
1—70	0 49 6		31·99	91·99	3·50	76·99	2·93
1—65	0 52 51		34·45	94·45	3·60	79·45	3·03
1—60	0 57 8		37·33	97·33	3·72	82·33	3·15
1—55	1 2 29		40·70	100·70	3·84	85·70	3·27
1—50	1 8 45		44·78	104·78	4·00	89·78	3·43
1—45	1 16 23		49·77	109·77	4·19	94·77	3·62
1—40	1 25 55		55·98	115·98	4·40	100·98	3·83
1—35	1 38 10		63·95	123·95	4·73	107·95	4·16
1—30	1 54 33		74·64	134·64	5·13	119·64	4·56
1—28	2 2 42		79·92	139·92	5·33	124·92	4·76
1—26	2 12 8		85·06	145·06	5·53	130·06	4·96
1—24	2 23 8		93·23	153·23	5·84	138·23	5·27
1—22	2 36 8		101·70	161·70	6·16	146·70	5·59
1—20	2 51 45		111·87	171·87	6·54	156·87	5·97
1—19	3 1 4		117·91	177·91	6·77	162·91	6·20
1—18	3 10 46		124·23	184·23	7·03	169·23	6·46
1—17	3 22 0		131·55	191·55	7·29	176·55	6·72
1—16	3 34 55		139·95	199·95	7·60	184·95	7·03
1—15	3 48 51		149·00	209·00	7·95	194·00	7·38
1—14	4 5 8		159·60	219·60	8·35	204·60	7·78
1—13	4 23 55		171·77	231·77	8·82	216·77	8·25
1—12	4 45 49		185·99	245·99	9·35	230·99	8·78
1—11	5 11 41		202·81	262·81	10·00	247·81	9·43
1—10	5 42 39		222·86	282·86	10·78	267·86	10·21
1—9·5	5 59 49		234·01	294·01	11·20	279·01	10·63
1—9	6 20 4		247·14	307·14	11·70	292·14	11·13
1—8·5	6 43 37		263·29	323·29	12·31	308·29	11·74
1—8	7 7 31		277·85	337·85	12·86	322·85	12·29
1—7·5	7 35 41		296·04	356·04	13·57	341·04	13·00
1—7	8 7 49		316·78	376·78	14·35	361·78	13·78
1—6·5	8 44 45		340·59	400·59	15·51	385·59	14·94
1—6	9 27 45		368·21	428·21	16·34	413·21	15·77
1—5·5	10 18 18		400·81	460·81	17·20	445·81	16·63
1—5	11 18 37		439·31	499·31	19·02	484·31	18·45

MOTOR VEHICLES AND MOTORS

Power required for all resistance speed, road resistance, gravity resistance, and efficiency of transmission between crankshaft and road wheel may be found as follows :—

Let R = the resistance to traction of the vehicle on the road in pounds per ton.

„ G = the resistance due to gravity in pounds per ton.

„ W = total weight on the wheels in tons.

„ V = speed in feet per minute.

„ v = speed in miles per hour.

„ E = mechanical efficiency of transmission from crankshaft to road.

„ P = brake horse power.

$$\text{Then } P = \frac{(R+G)\ WV}{3,3000\ E} \quad (1) \text{ or}$$

$$P = \frac{(R+G)\ Wv}{375\ E} \quad (2) \text{ and}$$

$$(R+G) = \frac{PE\ 375}{Wv} \quad (3) \text{ and}$$

$$E = \frac{(R+G)\ vW}{P\ 375} \quad (4) \text{ and}$$

$$V = \frac{PE\ 375}{(R+G)\ W} \quad (5) \text{ and}$$

$$W = \frac{PE\ 375}{(R+G)\ v} \quad (6)$$

Increase of resistance with speed The figures in Table XII. have been calculated in accordance with the above expressions, and no account has been taken of the increase in road resistance with increase of speed above experimental speeds. This increase is however important, as shown by the figures in tables VIII. to XI., especially with iron tyres and bad roads, and it, with difference in size of wheels, accounts to some extent for the diversity in the results of experiments by different engineers. Dupuis and Morin did not consider the increase was noticeable within the speed limits of their experiments, but there seems no doubt of it. No sufficiently extensive experiments have been made on the point to make it easy to include it in the expression for power required, and it can be best added to the value assumed for resistance by judgment in each particular case, taking into consideration the character of the roads, speed, and size of wheels.

Chapter VI

AIR AND WIND RESISTANCE

IN the foregoing, the wind resistance has not been taken into account. It becomes, however, a very important quantity in connection with high speed or racing cars, and some addition to the power would have to be allowed for it in the case of cars running long journeys on time. The resistance of the air not in movement, but only moved through by the car, must also be considered in the case of cars intended to run at high speeds.

According to the results of experiments by Hutton and others, the resistance increases as the square of the velocity of the body moved through the air. Beaufoy's experiments indicated a lower rate of increase, and Maxim found the same in his flying machine experiments. The resistance calculated from the formulæ for acceleration and velocity of falling bodies appears to be too low, and that derived from some other expressions with certain constants, too high. By taking Hutton's results, however, any error is on the safe side, and according to this, taking r as the air resistance in pounds.

$$r = v^2 A\ 0.0017,$$

v being the velocity in feet per second, and A the area in square feet.

Thus, a car presenting an effective area of 20 sq. ft. and running at a velocity of 10 miles per hour, or of 10×1.466 ft. per second, the air resistance would be $14.66^2 \times 20 \times 0.0017 = 7.30$ lb., or at 20 miles per hour, 29.2 lbs., and at 40 miles per hour, 116.8. Thus, in a car running at 25 miles and meeting a very high wind at the same speed, there would be a wind pressure on a man's face of from 2 to 3 lb. At 20 miles an hour the resistance of the air is equal to two-thirds of the road resistance of a four-passenger rubber-tyre car; and in a high wind of 40 miles per hour, the air and wind resistance together will exceed the road resistance by about five times. That is equal to changing a demand for 8 HP required, into 40 HP, which again is equivalent to changing a gradient from about 1 in 100 to about 1 in 7.5. The resistance of the air and of winds must, therefore, be taken into account in estimating the power required, and they may be taken as a quantity added to r. It is, however, obvious that no car could be fitted with power sufficient to maintain a high speed against a head wind of equal speed, nor does any one wish to ride under a combined velocity air and resistance of 80 miles an hour, or even of 60 miles. (*See* chapter on "Car Performance and Mechanical Efficiency."

59

Chapter VII

THE MODERN PERIOD AND MODERN MOTOR

Originated in France and Germany

THAT progress waits on freedom has rarely been more conspicuously shown than by the wonderful progress, and the great industry which has sprung up in France in five years, as a result of the freedom of the Frenchman to use his own roads. Great Britain had for half a century been as near to a practical self-moved road carriage as France was when Serpollet, Bollee, Scotte, and De Dion and Bouton began a few years ago, and before the imperishable work of Gottlieb Daimler placed in the hands of Levassor the high speed internal combustion motor, and Benz had shown its practical use. England had also the use of the Daimler motor, and it knew something of what Carl Benz, of Mannheim, was doing in the work that has placed him on an equality with, or only second to, Daimler, and Panhard and Levassor; but it would have been as little use to an Englishman to make a motor carriage with it, as it would have been for him to build an ironclad in the middle of Salisbury Plain. The Frenchman could run his vehicle and improve it into a practicable form, while to the English engineer, the roads being consecrated to the enjoyment of those who wanted to use only horses, they were proscribed, and the endeavour to bring into use a mechanical vehicle was looked upon by the privileged and ignorant, in the light that their predecessors saw blasphemy in the introduction of a winnowing machine that dressed corn, not by the " uncertain winds of heaven," but by artificial blast.

Freedom on continental roads not productive of success before Daimler and Benz motors were made

It must, however, be noted that free as were the roads of France for the trial of mechanically propelled vehicles, and ingenious as were the steam vehicles of Bollee, Serpollet, Le Blant, and others, it was not until the advent of the Daimler motor and the Benz motor cars, that any real, rapid and continuous progress in the construction of a motor carriage was made. It remained for the day of power by very high speed of rotation, and the facility of construction and use of the mineral spirit engine, to commence the great movement of to-day. The appearance, successful trial, and relegation to obscurity, more or less complete, of the steam vehicle in various forms, marked its history in the 1860-90 period almost as much as at a much earlier period, it did in England; but in the latter country, the agricultural road locomotive made great strides. The inference would appear

60

to be obvious, namely, that advance became possible with the birth of the practical high speed spirit motor—firstly, from the hands of Daimler, as the original producer of the most used very high speed motor; secondly, with the motor and carriage of Benz; and next with the Daimler motor in the hands of Panhard and Levassor and, subsequently, of many others.

Not until a trade was growing up in France did the Englishman pluck up courage and demand and obtain, in November, 1896, freer use of the roads which are his. Even now he who uses a motor carriage, a vehicle which is the result of so much ingenuity, skill, and indomitable perseverance, is more highly taxed and hemmed in by restrictive rules, regulations and penalties, than the most reckless, unprincipled, and ignorant drivers of half-taught and wayward horses. *British roads freed, under restriction, in 1896*

But although the history of the endeavours of the engineer and mechanician previous to the Act of 1896 offered little encouragement to any to enter upon road locomotion problems, the new era opened up and made possible in the beginning by the work of Daimler, Levassor, and Benz, will ere long, be characterised by so many demonstrations of the advantages of motor vehicle employment, that the present exclusion of mechanical road vehicle users from just and impartial treatment in Great Britain will be removed.

The Daimler and other Mineral Spirit Motors.

Gottlieb Daimler, who had for some years been occupied on gas engine construction, turned his attention to the production of small light petrol motors, made highly powerful by their capability of running continuously at very high speeds of rotation. (Daimler died 6th March, 1900, born at Schorndorf, 1834.) *The Daimler high speed motor*

As a result of much labour and experiment, he produced and patented (No. 9112, 1884) a small high-speed gas engine, with cylinder small in proportion to the stoke of its piston, running at a sufficiently high speed to ignite the charge by the heat of compression, aided by an ignition tube for maintaining the regularity of ignition, when the walls of the combustion space became too cool either by lowering of speed or by the infrequency of the charges when on light work. The ignition tube, or "priming cap" as he called it, was of course required for starting, and until the cylinder was hot.

Early in the following year he patented (No. 4315, 1885) his well-known single cylinder, enclosed crank and fly-wheel engine, which has been the parent of all the Daimler engines of the many forms now made under so many different names. In this engine the inlet and exhaust valves were brought together, one immediately over the other, the inlet being automatic, as in the first engine, and the exhaust actuated by a rod worked by a double cam groove in the outer face of one of the enclosed fly wheels. This cam groove is the equivalent of the peripheral double cam groove used in the Peugeot Daimler of to-day. Daimler, however, fitted it *Daimler motor, 1885 type* with a switch, actuated by a simple governor, also in the side of this fly wheel, by means of which the feather running in the groove was shunted so as to run in a nearly circular path, and thus give no motion to the exhaust valve. Thus, when the speed exceeded the normal, the exhaust valve remained closed, no new charge entered, and the speed again falling, the governor shifted the switch and allowed the exhaust valve to be opened. For starting, the exhaust valve was held open by a support easily moved out of position when the engine had started, so as to allow the valve to be worked. The cylinder was cooled by an enclosed fan wheel, which sent air round the cylinder within a jacket. The crank-chamber was practically air-tight, and acted as a pump chamber, with the piston as plunger. A small inlet valve was provided in the case, and a valve was fitted in the centre of the piston, as shown in Figs. 30 and 31, which in this respect is similar. When the piston descended, the piece t, sliding on the stem of the valve g, and surrounded by the spring which kept the valve ordinarily on its seat, came into contact with the upper part of a fork p, formed by an upward projection from the centre of the crank-chamber casting. If the exhaust valve were open, the superior pressure of the air in the crank-chamber could open the valve g against the resistance of the spring q, and the consequent inrush of air helped to give a complete discharge of the products of combustion of the previous stroke. This latter arrangement is, however, not found necessary, as will be seen from a description of the latest engines.

Daimler motor and bicycle, 1885 Six months after patenting this single cylinder engine, Daimler patented an application of it to a bicycle (No. 10786, 1885) as shown in side view in Fig. 29, and thus gave the first suggestion of its applicability for

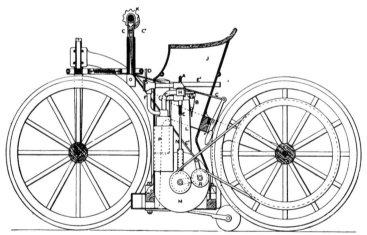

FIG. 29.—DAIMLER'S MOTOR BICYCLE, 1885.

motor-cycle purposes. The bicycle was not beautiful in appearance, but it contained several points of much interest. In it a little high speed motor similar to that one just described was used, and for it he devised the first of the carburettors, of which there are now so many for carburetting air with mineral and other spirit for motor purposes. The cylinder L was air-cooled in the manner already described. In the engraving M is the crank-chamber of the motor, N the rod for giving motion to the exhaust valve in the box H, which also contained the air admission valve A, E being the exhaust valve spring. At P was the carburettor, a pipe F from which allowed the passage of the oil vapour to inlet valve A, a regulating valve at D admitting more or less air and controlling admission of the mixture to the valve A. The pipe C, not connected to H, as appears, admitted air to the carburettor. At B was a petrol lamp burner for heating the ignition tube. The exhaust passed from below the exhaust valve into a silencer, and thence away to the rear by the pipe E'. The motor, driven by a round leather band running over a small pulley on the crankshaft, tightened by the jockey pulley R, which was dropped by the cord c', when the roller K on the steering handle was turned for the purpose of putting the brake on by pulling on cord c. Thus as the brake was put on the motor ceased to drive. There were two rollers, one on either side, as seen on the ground near the driving wheel. Normally they were off the ground, but could be one or both depressed by the foot for standing or for aid in steering. The pulley on the crank-shaft was driven by a frictional drive which was adjustable as to frictional hold, and would slip when the resistance became sufficient to be likely to stop the engine.

The carburettor will be described hereafter, with reference to Fig. 35, after the engines which led up to those now so largely used, have been described.

Daimler Carburettor 1885

After numerous other designs and patents, Daimler patented (No. 10007, 1889) the double inclined cylinder engine, illustrated by Figs. 30, 31 and 32, which was made in large numbers, and was at first applied to motor cars, a few of which are at work even yet. It was known as the V type.

Daimler V type engine

This engine had, as patented, but not as afterwards used, the air inlet valve in the piston, as already described with reference to an earlier engine. In other respects the engine differed from those preceding it. The cylinders A B were inclined at about 15°. The exhaust valve h was still worked by the sliding piece $k\,k$ in the cam grooves i and i' as before the groove i', being nearly concentric, and of somewhat greater radius where the two grooves intersect, than elsewhere. The groove i is also only excentric at and near where the two grooves intersect, where its radius is smaller. The consequence of the arrangement of these grooves is that the sliding pieces alternately follow the two paths. The upper piece k will, when the intersection next reaches it, pass into the inner groove, and that in the inner circle pass into the outer groove. Hence the exhaust valve gets a lift at alternate revolution only.

The governing, by shunting the sliding pieces into a concentric continuation of the inner groove, was replaced by a centrifugal governor $o\,o$, which acted as a lever n, at the top of which was a T piece s. When the normal speed of the engine was exceeded, the piece s moved inwards towards the cylinder, and the arm ᴍ of the bell crank lever, formed with the exhaust-valve pusher *m*, see Fig. 32, came into contact with it in its upward move-

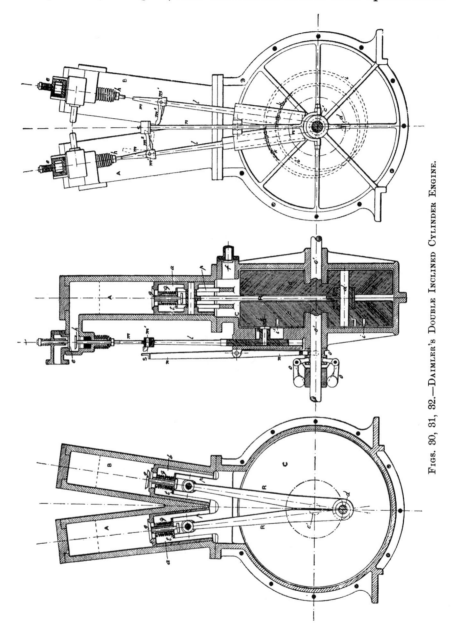

FIGS. 30, 31, 32.—DAIMLER'S DOUBLE INCLINED CYLINDER ENGINE.

ment, as shown by dotted lines in Fig. 32. The pusher *m* was thus deflected, and missed the end of the valve stem *n*. The two connecting rods grasped the same crank pin *d*, one rod working on the outside of the other, as seen

64

THE MODERN PERIOD AND MODERN MOTOR

in Figs. 30 and 31. The two parts of the crankshaft $c'c'$, with their crank disc fly wheels c, were connected by the pin d, and steadiness was secured by long bearings for the crankshaft. The disc fly wheels had but little clearance in the crank-chamber, which was made in two equal parts, centrally bolted together. At f was the inlet valve for the air drawn in by the rise of the piston, and afterwards expelled through the piston valve and exhaust. The air-admission valve is seen at e.

In the patent which describes these engines, Daimler described a double cylinder side-by-side engine, with one double parallel legged connecting rod, the big end of which was of the width of the whole distance across the rods where they entered the two cylinders.

It also describes the arrangement of double opposite cylinders and cranks, shown in Fig. 33. In this the crank-chamber c acted as receiver of the air drawn in by the pistons as in the previous engines, air entering at f. The arrangement secured a partial mechanical balance, and has often since been proposed.

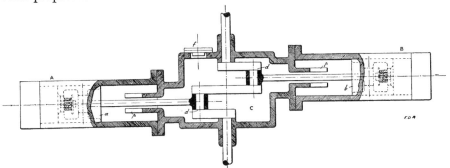

FIG. 33.—DAIMLER'S DOUBLE OPPOSITE CYLINDER ENGINE.

These engines, many of which were made for launches or fixed engine purposes, finally led to that which has become known the world over as the Daimler motor. It was introduced into England about 1892, and was sold here for the above purposes, but was not broadly known until the successes of the Panhard and Levassor, and the Peugeot carriages, 1894, 1895, and 1896, drew wide attention to them as Daimler carriages, in distinction from the steam carriages. Other motors appeared, such as those of the Dion motor tricycles, based on Daimler lines, and these and other motors, including the Benz, will be described hereafter when dealing with the carriages they propel. A description of the principles and mode of operation of the Daimler motor as now made will prepare the reader for a ready comprehension of the whole of the several kinds.

All these modern gas and oil engines are really hot-air engines, in which the expansion in volume of air when heated is employed to give rise to pressure on a moving piston, that expansion being effected in the cylinder containing that piston by the explosive or rapid combustion of a small charge of combustible, such as ordinary coal gas, or the vapour from

Principle, construction and working of mineral spirit motors

naphtha or petroleum spirit, or from petroleum when vapourised under higher temperature in presence of air. The heating, expansion, and cooling of the air are all thus done in the working cylinder. This mode of operation is distinguished from the earlier hot-air engines of Stirling, Ericsson, and others, in which two pistons were used, one acting merely as a displacer piston for passing cooled air, which had done work on a working piston, back into a hot chamber where it was heated, and, being again heated, escaped to the working piston which was in a position to be pushed out, while the displacer piston was almost still.

These engines were necessarily slow-speed engines, were very large for their power, and were very wasteful as heat engines in spite of certain theories of the harmful hypothetic order.

Functions of the Daimler motor parts The modern light oil, spirit, or gas engine, will be readily understood from the following description, aided by reference to the diagram, Fig. 34, which, while being very nearly an actual sectional illustration of the present Daimler motor, shows the parts arranged as to relative position, so that all their functions may be explained without reference to numerous drawings. A description of its parts, and explanation of their function, will serve for motors generally.

In this Fig. 34 A is the motor cylinder, in which works the piston B. This piston is shown at rather more than half its outward downward stroke. It is connected by the connecting rod C to the crank D, which turns in the direction shown by the arrow d. During its descent the piston B draws air into the upper part of the cylinder through the passage E. To reach this passage the air must pass through the valve F, which it reaches by travelling through a somewhat tortuous passage. It enters from without into an adjustable inlet G, whence it enters the chamber H, which is in part an air dust filter. From this it passes down the pipe h into the small chamber surrounding the tube J. From the mouth of the tube h (shown black) and the chamber J, it passes through the annular space formed by the ring k round the tube J. As shown by the arrows, it then passes through the pipe K to the valve F. The valve is shown as closed by the spring f, but when the piston descends as described, the valve is sucked downward, and the air passes into and fills the cylinder A.

The float-feed spray-making carburettor In passing through the annular space at k, the rushing current of air creates a parted vacuum in the tube J, the lower end of which dips into the petrol in the passage communicating with the small oil tank L. By this means a small quantity of oil is drawn up scent-spray fashion, and rushes with the air into the cylinder. Being broken up into a fine spray, it is instantaneously converted into vapour or mist in its passage to and in the cylinder.

The cylinder now, then, is full of air, with which is mixed the petrol vapour, or in other words carburetted air.

The return or rising stroke of the piston now takes place, the carburetted air is compressed into the small volume represented by the space

66

a, the port space E, and the space between the inlet valve F and the exhaust valve. When compression is complete the pressure is about 45 lbs. per square inch. At or near the moment that the piston reaches its topmost or innermost position, the temperature of the air and vapour mixture being raised by its compression, it is easily ignited by the incandescent walls of the ignition tube U, into which some of it enters under compression. The ignition is thus effected just as the piston is ready to begin its down or out stroke, and the temperature of the air is increased to about 1,800° F. This is equivalent to increasing the volume of the air many times, say twenty times; and hence, as the increase of volume is prevented, except by the movement of the piston, the latter is forced to move to the other end of the cylinder by the great increase in pressure. When the piston reaches the end of its outward stroke, or a little before it, the exhaust valve w is lifted by the rod x against the resistance of the spring *w*, and the products of combustion of the vapour and air forming the last working charge escape into the passage w′ beneath it. Thence they pass to the other side of the cylinder and escape by the exhaust pipe w.

The rod x has at its lower end a narrow chisel-shaped end, which rests in a notch at *x*′ on the lever Y, which is pivoted at *y*, and at its lower part carries a roller z. This roller rests upon the cam *m*, which is upon a spindle, driven by a wheel *g*, indicated by dotted lines, this wheel being driven by one half its size indicated by the dotted circle *g*′. The cam *m* thus makes one revolution for two of the crank D.

The piston now returns to its topmost position, carried thereto by the momentum of the fly wheel. During nearly the whole of this up stroke the exhaust valve remains open, as it is important to expel the whole, or as much of the products of combustion as possible.

The piston is now, then, at the point from which we started, and on its descent again draws in air and vapour, and the process is repeated.

In the feed tank L is a float M, which moves freely on the small valve spindle N, the lower end of which sits in a valve seat and controls the admission of oil into the tank from the source of supply. This valve is operated by the float as follows. Resting on the top of the float at the two opposite edges are the weighted ends of the two small levers *n n*, which are pivoted at *o o*. Their inner ends engage with a grooved collar on the valve spindle. When the float falls, in consequence of the consumption of the petrol and its fall in level, the weighted ends of the levers *n n* fall too, and on *o o* as fulcra they lift the collar on the valve spindle. The petrol is thus free to flow into the tank until the float M is again restored to the position shown, in which the valve sits on its seat and the passage is closed. Thus, a constant level in the float tank is maintained. This float feed is the invention of Maybach, and will be described hereafter with reference to its earlier forms.

The oil supply is brought by the pipe o from the reservoir situated in some convenient part of the carriage. It passes through a filter P, consist-

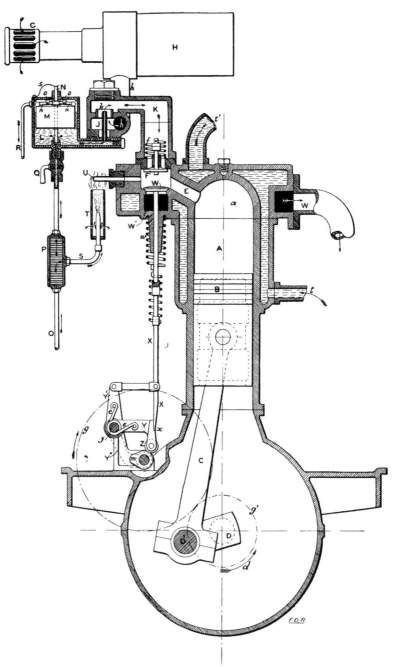

Fig. 34.—The Daimler Motor, 1899.

ing generally of a large number of small discs of fine wire gauze, the filter usually lying horizontally, instead of vertical as shown. From the filter a supply of petrol is taken at s for the burner T, by which the ignition tube U is heated. On the top of the float feed vessel is a light spring s, which is used only when the engine is standing to prevent ingress of petrol.

The speed of the engine is controlled by a governor which acts upon a downward extension of the lever Y'. The governor acts by sliding another cam, not shown, along the spindle which carries the cam m. This second cam varies in excentricity from about the amount of excentricity shown in cam m to no excentricity, or to a cylindrical form. When, then, the downward extension of Y' rests upon the cylindrical part of this cam, it receives no motion, and Y' also stands still in the position shown. But when the engine runs above normal speed the large part of the cam is put by the governor under the downward extension of Y'; the upper part is pushed towards the cylinder, and the rod x with it. The lower end at x, therefore, misses the notch, and Y moves up and down without lifting x and without opening the exhaust valve. The cylinder thus remains full of products of combustion, and when the piston next descends there will be no inrush of air through the valve F, and consequently no admission of petrol vapour. The lever Y, with the roller z and the downward extension of Y', are pressed toward their respective cams by the one spring e. There being thus no new charge in the cylinder, the engine must make another revolution before it can get one, namely, the upward stroke to expel the old products, and the downward stroke to draw in the new charge. The next half revolution must be made to compress the charge, and then, the piston being at the top of its stroke, a new working stroke takes place. If, however, the speed had not fallen to the normal, the governor would not have moved the cam, and at least one more revolution would take place before a new charge could be admitted.

It will thus be seen that the engine makes four strokes, or two revolutions, at least, for one working stroke, the cycle or system of operations being that originally described by Beau de Rochas and first used in the Otto gas engines.

The very high temperature, due to the combustion of the charge in the cylinder, would heat the latter also to a very high temperature were it not for the artificial means adopted for keeping it sufficiently cool. This cooling is effected by a slow current of water entering at t, passing round the cylinder in the water jacket shown, and out at t', the casting containing the valves being also water-jacketed.

From 30 to 45 per cent. of all the heat generated by the combustion of the petrol vapour is carried away by this cooling water, so that it will be seen that a considerable quantity of water will be required or the water must be cooled by some artificial means. Several kinds of coolers are used for this purpose, which will be described hereafter. In some cases, however, the users of motor carriages allow the water to reach the boiling point,

The governor

The water-jacket cylinder cooling

69

beyond which it cannot rise. Having reached this point, the heat of the cylinder is used in vapourising the water. This goes off as " steam," and is sometimes the cause of complaint, and is therefore not a desirable method ; but it saves trouble as to water, because it takes about fifteen times more heat to vapourise 1 lb. of water from the boiling point under atmospheric pressure than it does to raise it from, say, 150° to 212°. That is to say, that suppose no water-cooling apparatus to be employed, and suppose that the jacket water is not allowed to rise above 150° F, then water tank capacity fifteen times greater will be required than if the jacket water be allowed always to reach and continue at 212°. Some form of cooler is, however, always employed, and this extends the time within which the water carried will reach the boiling point or even a rapidly vapourising temperature. Many forms of coolers are now in use, some of which will be described in connection with the vehicles on which they are used. See also Chapter XVIII., p. 338, on " Cylinder Cooling."

It will be seen that, with the exception of the means and apparatus for converting the petrol into vapour, the engine is precisely a gas engine. The similarity is more obvious with those engines, such as the Benz and the De Dion, which, instead of the spray-making carburettor described, employs a petrol surface evaporator or carburettor. These carburettors provide a supply of strongly carburetted air, which the engine receives, just as a gas engine receives gas and mixes it with air enough for combustion more or less rapid according to requirements.

Daimler's surface evaporator carburettor, 1885 They may be understood in their action from Fig. 35, which shows the carburettor designed by Daimler, and patented in 1885, for use with motors of the motor bicycle which the patent covered, and for the engines which soon began to be largely used as fixed engines and launch engines.

The lighter oils, such as petroleum spirit, now called petrol, or gasolene, or benzoline, will all evaporate readily in presence of air, and the more readily in presence of air moving more or less rapidly. In presence of dry air it will evaporate into and be taken up by that air until the air is saturated, which at 50° F. is when the air contains 17·5 per cent. of the hydro-carbon vapour, that is, a mixture of one volume of oil vapour to about 6·7 per cent. of air. This mixture will burn and give a very good white light. It is a rich mixture of oil vapour and air, which, diluted with a considerable quantity of air, will burn with explosive rapidity under the circumstances of its combustion in a gas or oil cylinder.

For the production, then, of this carburetted air, Daimler invented the apparatus shown in Fig. 35, and described it in his patent, No. 10786 of 1885, for a motor bicycle, and in No. 7286 of 1886. It consists of a vessel A, which is about two-thirds filled with petrol. In this is a float D, cylindrical in exterior, and a conical central opening forming a basin, in which the petrol rises through a small hole B, to the level of the petrol outside the float. In the centre of the float is fixed a tube E, which fits freely in the tube F above, fixed to the cover of the petrol vessel. The float is thus free to rise or fall with the level of the petrol. The lower end of the tube E is perforated where it is immersed in the petrol in the conical basin c. Above the oil vessel A is a small vessel i, which forms a reservoir for carburetted air. From this, by attachment to the pipe G, this carburetted air is drawn by the piston of the motor cylinder. When this is drawn out there is a corresponding inrush of air from without down the pipe F. This new air bubbles through the petrol in the basin c, any petrol in the form of heavy spray being caught by the baffle plate K and lip, and caused to fall back into the basin. The air, now heavily charged with petrol, is drawn from the space H into the reservoir i through the holes in the top of the petrol vessel, and through the wire gauze at L, nothing but carburetted air passing. On the top of the reservoir i is a safety valve M, held down by a light spiral spring N. The object of this was, in the event of a back ignition from the cylinder from any cause striking back through the pipe G, to prevent a flame extending beyond the gauze partition L by providing an easy outlet for any rise in pressure. At P is a tubular guide for o, a small rod which rests in the baffle plate K of the float, and extends upwards to form an indicator of the level of the oil in the vessel. The upper part of the rod is surrounded by a ∪-shaped guard and guide Q, of a sufficient length to register the whole range of the rise and fall of the float, and thus show the quantity of oil at any time present in the vessel.

By this arrangement of carburettor, it will be seen that the main body of the petrol was not aërated, and that which was acted upon by the air was maintained at a uniform level in its basin, and kept sufficiently separate from the other. It will be seen also that this apparatus provided for an engine a supply of carburetted air always ready, and prevented the entrance of petrol not vapourised into the cylinder.

Carburettors of this class have received many modifications since this one was made by Daimler, some of which will be referred to hereafter in connection with the carriages they are fitted to, but they remain in principle the same. See Chapter XVI., p. 320, on "Carburettors."

The float-feed carburettor commonly used with all Daimler motors was invented by William Maybach (No. 18072, of 1893). After the description given of the float-feed carburettor of Fig. 34, these will be readily understood. With the descent of a piston in a cylinder F, the air-admission valve c moves inwards, and air is drawn through the passage B by entrance at B', as indicated by the arrow. After its entrance at B', it passes through the restricted annular opening formed by the end c' of the pipe c in the hole surrounding it. The top c' is a little above the bottom of the passage B, and it has but a very small hole.

At D is a float in a petrol vessel A, and having in its upper part a small vertical valve stem. This projects upward into a supply pipe E. When the consumption of oil at c' reduces its level in A, and therefore its rate of flow, under the inductive persuasion of the inward blast of air

Maybach's spray-making float-feed carburettor, 1893

71

entering at B', the float falls, the valve on its top leaves the seat it ordinarily presses against, and petrol can again enter through the pipe E.

Fig. 37 shows a slight modification of the arrangement shown in Fig. 36, one in which the air and oil spray or carburetted air passage B is formed with a similar passage for the hot exhaust products, the small sections showing two alternative forms of these two passages. This form was proposed for vapourising the heavier oil or kerosene; but, as far as I am aware, nothing or little has been done in this direction, and the form

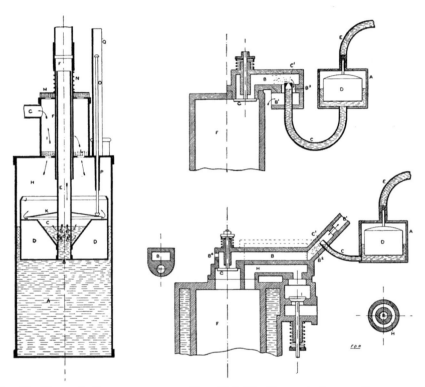

FIG. 35.—DAIMLER'S SURFACE CARBURETTOR, 1885.

FIGS. 36 & 37.—MAYBACH'S SURFACE CARBURETTORS, 1893.

shown is not likely to be useful for that purpose. On variable work, the heat from the exhaust would, and has always proved to be, too irregular to provide oil vapour or oil gas of uniform density, and the variable strength of this would cause very variable combustion and products thereof, making the motor quite unsuitable for most motor vehicle purposes.

In another modified form for petrol, Maybach admitted air into the passage between valve G and jet c' by a controlled valve, which by admitting more or less air at this position varied the quantity of petrol drawn in from c', and thus the strength of the mixture.

Fig. 38.—Benz Motor Tricycle, 1885.

Fig. 39.—Benz Motor Tricycle, 1885: View of Machinery at Back.

73

MOTOR VEHICLES AND MOTORS

It will be noticed that in the float-feed carburettor, illustrated in Fig. 34, the float does not directly carry the valve, and that the valve seat is below the float instead of above it. The method shown in Fig. 34, of actuating the valve by the fall of the float, is not only more convenient for arrangement, but is more effective, as the loaded levers actuating the valve do not dance and dither as the float alone did.

The Benz Motors, and Motor Cars.

Benz motor tricycles, 1885

Sequence of description of the typical motor and its parts has led me from sequence in chronology of invention of the few motors or vehicles which, though not being made, are, or may be, of much interest.

Carl Benz, of Mannheim, took out a patent (No. 5789 of 1886) for an oil spirit motor tricycle carriage which he had made in 1885. In this a single horizontal cylinder water jacketed engine with a vertical crankshaft was used. (See Figs. 38 and 39.) It must be looked upon as the embryo of the Benz cars now known so well, and made in very large numbers. The specification shows two principal forms of it. Figs. 38 and 39 are from photographs recently taken of this interesting car of 1885, and described in the main particulars in the above-mentioned specification. Figs. 40 and 41 are from the latter, but with modifications.

In these the piston in the cylinder A is connected to the vertical crankshaft B, having a fly wheel C running in a horizontal plane. On the top end of the crankshaft is a bevel pinion gearing with a bevel wheel of double its size on a short horizontal shaft, which thus runs at half the speed of the crank, and, by a crank pin at the end, communicates motion to the exhaust valve. At the other end of this valve-motion shaft is a pulley D, which by belt communicates motion to the second motion shaft E E, which is in two parts, by a fast pulley F, the valve-motion shaft being used for conveying the power of the engine. F′ is a loose pulley, and within F and F′ is the balance or differential gear. The shaft E E is divided between the fast and loose pulleys, as shown by Fig. 42, which is a transverse section near the bar G of Fig. 41. Two bracket bearings H H carry these inner ends of the shaft E E. The bar G is for shifting the belt by the fork J. It is actuated by the lever K on the spindle L, on which is a bevel wheel M, which moves a bevel pinion N on a short spindle. At one end of this short spindle is a crank pin O, which engages with the fork P formed on the bar G. By pushing the lever forward the pin O is pushed against the longer line of the fork, and the strap put on the tight pulley. When the lever is pulled backward the pin O pushes against the shorter line of the fork, and then misses or gets past it, the bar having by this time been moved far enough to have pushed the belt on the loose pulley. No further motion of the bar G takes place upon a further backward movement of the handle K ; but its movement to k′ puts a brake block at the end of the spindle L into contact with the brake pulley Q on the side of the fast pulley. The steering was effected by a link and rack and pinion.

74

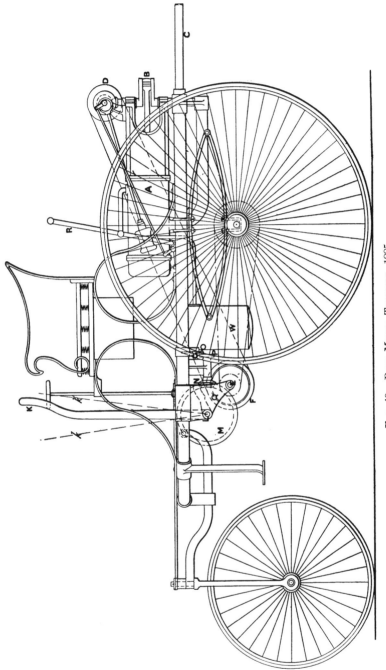

Fig. 40.—Benz Motor Tricycle, 1885.

75

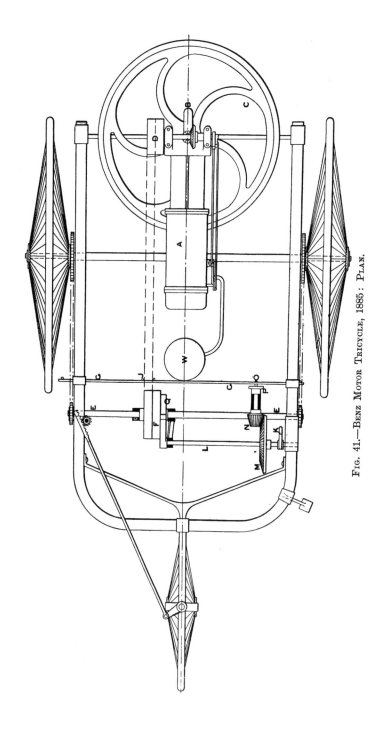

Fig. 41.—Benz Motor Tricycle, 1885 : Plan.

76

THE MODERN PERIOD AND MODERN MOTOR

For cooling the jacket water, Benz described in 1886 the natural circulation arrangement shown by Figs. 40 and 43, in which R is a grid of pipes arranged above the cylinder with internal baffles, so that the hot water which passed from the top of the cylinder A by the pipe T traversed

Benz water cooler, 1885

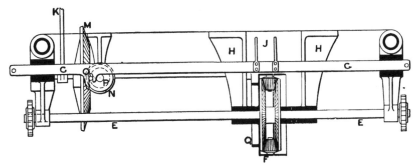

FIG. 42.—BENZ MOTOR TRICYCLE: TRANSVERSE SECTION AT COUNTER SHAFT.

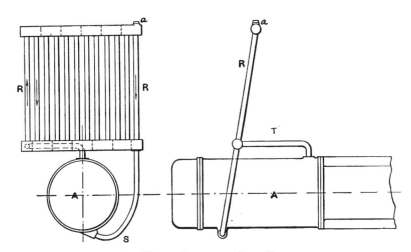

FIGS. 43.—WATER COOLER FOR BENZ TRICYCLE.

the consecutive tubes in series, and finally descended cooled, or cooler, down the pipe s to the bottom of the cylinder. Steam formed escaped at *a*. In Figs. 38 and 39, *i.e.* in 1885, this was not used, a small tank being then, as now, used by Benz.

A carburettor or vapouriser was placed at w, spare petrol being carried in a separate tank above. The vapouriser is of considerable interest even now, and is shown by Figs. 44–49, as described in the English patent of 1886. It consists of a vessel F, containing the light oil, with means for heating it slightly to encourage evaporation. For this purpose some of the exhaust is admitted by the pipe *p* into the chamber U. Air is admitted through an adjustable valve *r* into the annular chamber Q, which is of very small width at the bottom part, where the inner wall of the

Benz surface carburettor 1885

77

chamber Q dips into the petrol. The air is thus forced or drawn through the petrol at considerable speed, and converts it into a spray. To prevent spray from passing upwards with the vapour, layers of thin metallic plates snipped all round, as in Fig. 49, were used. At the top of the vapour vessel was a safety valve F', held down by a light spring. Over the vapouriser was a petrol supply tank o, from which a pipe t, with a cock u, led petrol to the bottom of the vapouriser through a glass gauge g. As well as the

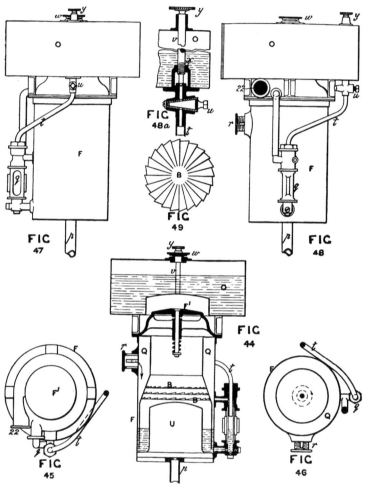

FIGS. 44–49.—BENZ CARBURETTOR: PATENT OF 1886.

cock u there was a regulating valve v, with a screw at x and milled head at y. The valve r was usually set so that a non-explosive mixture of air and vapour or carburetted air was formed. The carburetted air passed away by the pipe 22, taken off tangentially from the top of the generator.

Benz "mixer," 1886

Between the inlet valve on the engine and the vapouriser there was a "mixer," shown by Fig. 50. In this A is the pipe leading to the inlet

78

valve on the cylinder, and B the pipe from the vapouriser. C is a socket carrying internally a thin pipe D, finely perforated. It also carries the larger pipe, which extends upward into the socket E, and is perforated at the lower part F. The air thus enters at F in numerous fine streams, and meets the finer streams of carburetted air coming out of the pipe D. G is a separate movable shutter tube, which may be placed where shown, allowing the most air to enter the holes in F, and thus give the weakest mixture; or it may be anywhere between that position and C, so as to vary the air let in, and thereby the strength of the mixture.

It was thought that by having the stronger mixture or incombustible carburetted air brought up to within a little of the cylinder, an accidental ignition during the admission or compression strokes would not result in

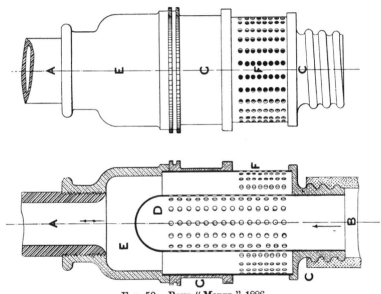

FIG. 50.—BENZ "MIXER," 1886.

explosion or back ignition beyond the mixer. Inasmuch, however, as the carburetted air, as it came from the vapouriser, was not always incombustible, back ignitions did occur, and it became necessary to place a pad of several layers of wire gauze in the socket of the mixer on the vapouriser side. This pad is held between two flanges in the mixer, as will be seen hereafter with reference to the Benz vehicles.

The product of the carburettor may be lowered in richness by the following causes, which may vary the working of the motor or necessity for air adjustment:— **Variation of product of carburettor**

(1) Failure to keep supply of petrol in the vessel, and thus reducing the quantity through which the air passes, and also its volatility—heavy oil remaining.

(2) Fall in temperature of outer air, and insufficient heating from exhaust.

(3) Change in the hygrometric state of the air, as in fog.

(4) Unintentional adjustment of air admission, or accidental change by vibration and other origin.

Benz two-speed and free motor gear, 1885 In another part of the specification referred to, Benz illustrated and described a very ingenious arrangement of two-speed and free running gear for the motor, all with consecutive movements of the same handle, part of the gear being similar to that used in the later Benz cars for the slowest motion for hill climbing. Altogether this is a most interesting specification, taken either as a precursor of a great business in its subject, or as containing numerous things of mechanical interest.

Benz second motor tricycle carriage Fig. 51 is a reproduction from an engraving, lent me by Mr. Benz, of

FIG. 51.—BENZ SECOND MOTOR TRICYCLE CARRIAGE.
Made in 1886, and ran at 10 miles an hour maximum.

the second type of car he made. It differs from the first mainly in the seat or body. This car ran at about 10 miles per hour.

Benz third motor carriage Fig. 52 is from another engraving of the third type of car he made. In it will be seen more nearly the embryo of the cars now so well known as the Benz, the greatest difference in appearance being in the single front wheel. This car ran at from twelve to fifteen miles per hour. Benz seems to have given his cars more liberal size of engine for such small cars than many succeeding makers in their first efforts.

Butler's motor tricycle, 1884-5 The next instructive attempt to produce a motor vehicle propelled by an oil—or rather a mineral spirit—motor was that of Edward Butler, one view of which, partly in section, is given in Fig. 53. Butler's patent is No. 15598 of 1887; but he had made a similar tricycle at an earlier date. He exhibited one in the Inventions Exhibition of 1885, and drawings of it are said to have been exhibited in the Stanley Cycle Show towards the end of 1884.

As at first made, the motor had double-acting cylinders, coupled direct to the single driving wheel; but this arrangement required a slow speed of rotation, which for several reasons would not work well; and the two pairs of valves and the magneto ignition arrangement were also unsatisfactory. Subsequently the pistons and valves were altered for single acting, and instead of direct driving, a form of epicyclic gear in the hub of the single driver, with a ratio of 1 in 6, was used. If this arrangement had been used at first, straight connecting rods instead of those shown

Fig. 52.—Benz Third Motor Carriage.
Made in 1888, for running at 12-15 miles an hour.

could have been used. A bichromate battery and sparking coil were sub-stituted for the magneto machine. The cylinders were water cooled, the guard over the driving wheel being hollow and forming a water reservoir. The cylinders were at first 2·5 in. diameter and 8-in. stroke, the stroke being reduced to 5 in. when the higher speed of about 500 revolutions per minute was adopted with the gearing of 1 to 6.

The vaporiser was of the spray-making type, as afterwards used by Butler in his little domestic gasoline motor.[1] Solid rubber tyres were used, 1·5 in. diameter on the driver, and 1·25 in. diameter on the steering wheels, which were on Ackerman axles, connected and actuated by a strong handle on both sides.

In the cylinder A, was a piston B, with tubular piston rod connected to a cross-head C, which worked in a tubular guide with side slots D, through

[1] *The Engineer*, vol. lxx. p. 65.

which passed the crosshead pin, embraced by the forked end of the hollow connecting rod E. The crank E was fixed on a tube on the driving-wheel axle, and drove through speed reducing gear in the wheel hub. The pistons were double acting, the instroke drawing in air at the front and afterwards compressing it, the mixture then passing to the back end of the cylinder, in which alone combustion took place. The ignition took place at about one-fifth on the outstroke. Rotary valves were employed, as seen in section in Fig. 53, below the cylinder. The air, before being drawn into a form of spray-making carburettor, entered an air purifier at H and heater at J. The heater was simply a chamber into which exhaust pro-

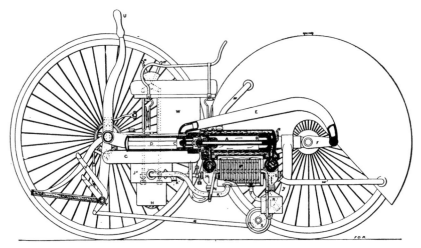

FIG. 53.—BUTLER'S PETROL MOTOR TRICYCLE, 1884-87.

ducts were delivered by the pipe K, and through which air pipes passed on their way to the vapouriser. Thence the vapour passed into a storage chamber P, and from it to the air and vapour adjusting chamber S, by way of pipe Q and the vessel R, in which was a series of sheets of gauze to prevent back ignition. The heated air is drawn in by the front side of the piston through the pipe L, and the air-valve chamber M, and thence through the pipe N into the receiver P.

The cylinders were jacketed, and cooling water was carried in the tank w and the hollow wheel case w′, between which and the tank there was a connecting pipe w.

As a means of easily starting the motor, or manœuvring in streets by preventing its driving, the roller x on an arm pivoted below the letter *y*, could be depressed by the lever v, thus lifting the wheel from the ground. The brake was applied by a foot lever v′, which pulled down a vertical rod seen behind it.

There is a great deal in this specification which is of mechanical interest, and showing much knowledge of the behaviour of air and spirit vapour, and of light oils, under some conditions. There were, as might generally be expected, far too many parts, and the proposed rotary valves would in all probability have proved a failure for the motor ends of the cylinder. The necessity, assumed or real, for avoiding the Beau de Rochas or Otto cycle, also introduced complications.

Figs. 54 and 55 illustrate a simple arrangement of petroleum motor tricycle made and described by Mr. Jas. D. Roots, in specification No.

Roots motor tricycle, 1892

FIG. 54.—ROOTS' PETROLEUM MOTOR TRICYCLE, 1892.

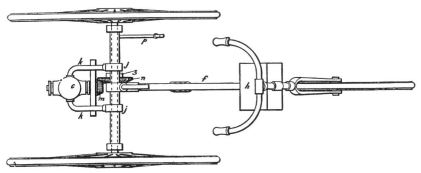

FIG. 55.—ROOTS' PETROLEUM MOTOR TRICYCLE, 1892: PLAN.

23786 of 1892, which also particularly describes a motor. This, however, it is not necessary to deal with here, as it was without doubt what would now be considered a crude form of petroleum engine, in which the front of the piston acted as air pump and compressor, the Otto cycle not being used. The motor, as shown in the specification, is an inverted cylinder engine; but that shown in the tricycle has the cylinder at the bottom, with the crank chamber at the top. Modifications to suit this would of course be required. The merit consists chiefly in the simplicity of the general

MOTOR VEHICLES AND MOTORS

arrangement; but the simplicity which is apparent in the drawings would appear less real if all the necessary details for the working and manipulation of the motor were shown. The vertical cylinder motor was pivoted by arms *k* to a cross tube of the frame at *j j'*, and by virtue of the struts, *g* of the frame was held in its proper position. On the crankshaft of the motor was shown a bevel pinion *m* gearing with a bevel wheel *n* on the tubular part of the driving axle, the boss *z* of the wheel *n* being intended to form part of a differential motion. The lever *p* is described as controlling it. In the front, *f* represents a water tank, the water for cooling circulating to and from this tank and the jacket through the pipes *g g* and *f f*.

It will be noticed that the motor is not pivoted, or its shaft radically connected to the driving axle. This would result in cross stresses and grinding and breaking of the bevel teeth every time the tricycle ran across any common obstructions, and brought the inertia of the motor mass into play on the parts that carry it.

This short account of some of the first of the mineral spirit and petroleum motor cycles and cars must be concluded with a mention only of the little experimental car made by Mr. J. H. Knight, well known in connection with his inventions and work in connection with the development of oil engines from 1887 onwards. Early in 1895 this little car was made and ran with petrol, then he fitted one with vapourisers to use petroleum, after the manner of his engines for stationary purposes, but found that the very variable conditions of working of a motor vehicle gave petrol the advantage of complete combustion unattainable under the same conditions with the heavier oils.

Leaving now those motors and vehicles which were made in the earlier days of the present great advance, we must turn to those which have become the most successful to-day, after only about six years of what may be called commercial development and production. Although a few steam carriages were made in the period referred to, notably those of M. Bollée, M. Leon Serpollet, and the Count de Dion, and M. Bouton and M. Le Blant, they have been few in number, and will be dealt with hereafter when treating modern steam carriages generally.

The Panhard & Levassor Daimler Motor Carriage,
and its performances.

Panhard and Levassor motor carriage

This carriage must be admitted to contain the design, and to have quickly reached the practical success which set the world on the tip-toe of expectancy for the greater successes that were to follow. MM. Panhard & Levassor had for many years been accustomed to a good class of mechanical engineering, namely, wood-working machinery, of which they were among the leading constructors in France. This experience, and the possession of the requisite facilities in works, machine tools, materials, and workmen, gave MM. Panhard & Levassor great advantages in their efforts to produce a satisfactory car from the first, and this was supple-

84

THE MODERN PERIOD AND MODERN MOTOR

mented by their having acquired the Daimler motor rights and the assistance of Daimler in perfecting the motor. Peugeot, Freres, and others, who were not like M. Roger using the Benz engine, had to get their motors from Panhard & Levassor.

The car shown by Fig. 56 was in 1894 the result of some years of work by MM. Panhard & Levassor, and in that year it took part in the Paris-Rouen race, 79·4 miles, organized for the *Petit Journal* by M. Pierre Giffard. This race was won by a De Dion and Bouton steam tractor, to which was attached an ordinary landau. It made an average speed of about 12 miles per hour, and was exhibited in the first exhibition of motor cars in England, namely, that organised at Tunbridge Wells by Sir David Salomons in October, 1894. It was, however, only 5 minutes, 10 minutes, and 23 minutes ahead of the two first Peugeot cars and a Panhard car respectively; and these two makers, both using Daimler motors, divided the first prize of £200. This was in July, 1894. The Panhard & Levassor 1894 car Paris-Rouen race

At that time MM. Panhard & Levassor had made nearly 350 Daimler motors and about 90 cars, most of which were in use in and about Paris, and about a dozen were ready for delivery at the end of July. MM. Peugeot alone had been supplied with 80 Daimler motors. They had thus at the end of 1894 accumulated unapproached experience not only in the construction of the motors for all kinds of carriages, but in the construction and running of these with all the gear and apparatus in the carriages. This may be taken as an explanation of the vigorous survival of the type of *ensemble* they had at that date reached.

M. Levassor drove one of these vehicles in the Paris-Bordeaux race in June, 1895, 735 miles, in 48 hours 48 minutes, at mean speed of 14·9 or, practically, 15 miles per hour on the whole run, and a maximum of hourly mean of 18 miles. The carriage without supplies weighed 604 kilogs., or a little under 12 cwt.; it carried Levassor and one mechanic, and won the race. Peugeot cars made similar but not quite equally good runs. From that race in 1895, until the present time (November, 1899), the triumph of the mineral spirit motor vehicle has been complete, steam cars having been completely displaced as long-distance high-speed light vehicles. Levassor in the Paris-Bordeaux race, 1895

In 1896 again, in the race from Paris to Marseilles, a distance of 1,711 kiloms., or 1,060 miles, when thirty-two vehicles started, the Panhard and the Peugeot vehicles took the first and second places, the former covering the distance in 67 hours 43 minutes, or an average speed of 25·2 kiloms. or 15·62 miles, over the whole of that long run. The Delahaye also made good running on this race. There were three steam cars in the thirty-two entered. All failed from one cause or another, one however only because of the failure of its pneumatic tyres for which it was too heavy. Another, a De Dion brake run by the Marquis de Chasseloup-Laubat, was, after complete overhaul and some improvements, winner of the Marseilles-Nice Turbic race in January, 1897, 144½ miles (233 kiloms.) in 7 hours 45 minutes, or a mean of 30 kiloms., or 18 miles per hour. The Paris-Marseilles race, 1896

85

When fully loaded this brake with passengers weighed nearly 3 tons, but on this run it reached higher speeds than had previously been made, as much as 36 miles per hour being reached for short distances. This was the last run of a steam car in any of the French races, M. Serpollet not having entered. The second prize was won by a Peugeot car. Higher

Paris-
Dieppe and
Trouville
races, 1897
Paris-
Amsterdam
1898
Versailles-
Bordeaux
1899

and higher speeds have been since made on the Paris-Dieppe, 1897, about 40 kiloms. or 25 miles mean speed; Paris-Trouville, 1897, 170 kiloms. at about the same speed; Paris-Amsterdam, 1898, mean of 44·7 kiloms., or 27·7 miles; 1899, Versailles-Bordeaux race, 555 kiloms. or 344 miles without a stop, at a mean of 48·76 kiloms., or 30·2 miles for the whole, while for day-stretch the mean was 60 kiloms. or 37 miles, and the speed reached 80 kiloms. or nearly 50 miles. This was done by Charron on a Panhard & Levassor Daimler motor car, carrying two persons, and weighing about 1 ton, and fitted with a 12 to 15 horse motor.

I will not give here further particulars of these and other races, except the round France tour of 2,300 kiloms. last; the foregoing being sufficient to show that nearly the whole of the great tests of endurance and speed, tests of greater severity perhaps than to which any other machinery in the world was ever put, have been successfully withstood by carriages driven by gearing operated by high-speed spirit motors, mostly of the Daimler type, as used by Panhard & Levassor, and similar motors as used by Peugeot and Freres. More recently the Mors, the Delahaye, the Bollée, the de Dietrich, and others, have been making wonderful records, but always the light oil or spirit motor.

The lines therefore laid down by Panhard & Levassor were well chosen, and may in the first instance be described with reference to the side view, Fig. 56, then by reference to the English Daimler cars, built on the same lines, and afterwards by reference to the Panhard racing car of 1898–99. In Fig. 56, A is the crank box of the inclined cylinder, V-type

The Pan-
hard &
Levassor
Daimler
motor car
of 1894

Daimler motor, as described at p. 62. The motor cylinders are seen behind the carburettor B, which was of the type shown by Fig. 35, and behind the exhaust silencer c. It will be found that all these details have been modified, the one by the use of the float-feed spray-making carburettor of Maybach (Figs. 34 and 36), another by the use of the parallel cylinders in line with the centre line of the carriage, and the other by putting the exhaust silencer under the carriage towards the rear, and in fact using a double silencer. The petrol cylinder was placed at D, and a feed for the ignition tube lamps was placed at E. At F was a mixture regulator. In line with the crankshaft of the motor was a shaft carrying parts of the main clutch which engaged with conical surfaces on the rear face of the fly wheel K. This shaft carries the three shifting wheels L, gearing with the spur wheels above, by which either of three speeds could be used. The three wheels or pinions L were shifted by the lever N, pivoted at n, and on the lower end of which was a horizontal link, which connected it to the collar at p. In the position shown, the middle

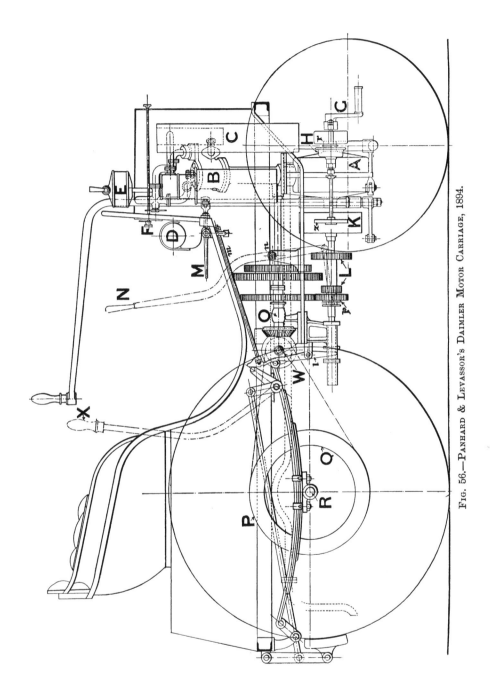

FIG. 56.—PANHARD & LEVASSOR'S DAIMLER MOTOR CARRIAGE, 1894.

87

speed is in gear. At M is a pedal, communicating with the lever l by means of the rod m. By depressing this pedal the lower forked part of the lever l pushes the collar against which it bears, rearwards. By this means the main clutch at K is pulled out of contact and the shaft carrying it and the gear L is freed, so that when the speed is being changed by shifting the gear L, only the inertia of that shaft has, or need have, to be overcome by the gear teeth in making the change from one speed to another.

Changing speed   This is one of the most troublesome things an automobilist has to learn. He arrives at a hill and his motor begins to slow down. He knows he must lower the ratio of his gear before the motor's speed has been so reduced that it may stop. He runs as long as he can on the speed in gear, then almost simultaneously, but yet in succession, depresses his pedal and disengages his clutch, throws the lever N over to the speed he wants, and then again lets the clutch go into gear. This may be learned, and the teeth may happen very often to be opposite tooth spaces when they come together sideways. Sometimes they are not, then an instant is lost, and when they do engage, after a lot of side grubbing of the teeth, the difference between the speed of the car and that at which the motor wants to drive it with the gear that has been shoved in, is illustrated by nearly or quite broken teeth, or perhaps only by a jerk of the car, half enough to dispossess it of its passengers. Changing gear is more easily done on an increasing speed than on a decreasing speed, and much depends on the rapidity of succession of the three actions, and particularly of the gear-lever change action.

The gearing L referred to, then, drives a second shaft, which runs in bearings, one of which is seen at O. Outside this bearing is a mitre pinion, which gears with one of two corresponding mitre wheels on a differential gear box (as will be explained with reference to Figs. 65 and 66). By this means the transverse shaft W is driven in either direction for forward or backward movement. From sprocket-chain pinions, on the ends of shaft W, the wheels P on the driving-wheels are driven, these wheels running loose on the axle R like ordinary carriage wheels. The wheels P have cast with them rings Q for the receipt of brake straps, which are operated by a pedal not shown. A hand brake acting on each driving-wheel tyre is actuated by the lever X. At Z is a friction wheel running on the outside of the fly wheel K and driving a small circulating pump r. G is the starting handle on the end of the crankshaft.

The arrangement of engine and gear is, as will be seen hereafter, in principle the same as that of Amedée Bollée, as used in his design of the steam omnibus made in 1885 at Mans (Fig. 321), who in this followed James of 1824; but it is impossible to give too much praise to the late M. Emile Levassor for his work in the application and development of the Daimler motor and his design and construction of the Panhard & Levassor arrangement of gear and apparatus.

88

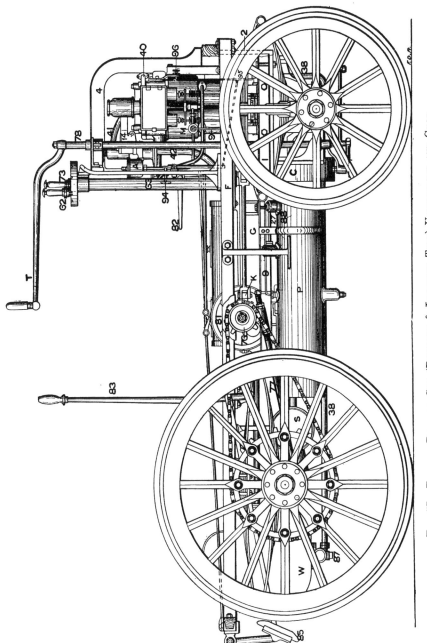

Fig. 57.—English Daimler Car (Panhard & Levassor Type) Underframe and Gear.

89

Built upon precisely the same general lines as the Panhard & Levassor last described, is the English Daimler. The latter, however, comprises parallel cylinder motor, the carburettor of the Maybach Daimler carriage, described hereafter, and other improvements made during the past three years. The English Daimler cars are illustrated by the following twenty Figs.

The English Daimler Motor, 1898-99

Figs. 57, 58 and 59 show a side elevation, plan, and front elevation respectively of the standard underframe and gear manufactured by the Daimler Co., of Coventry. To it may be attached any form of body carrying from two to eight people, or loads up to 1 ton.

Underframe

The underframe consists of a rectangular structure F of wood encased in channel iron and two longitudinal channel bars 1, bolted to hangers 2 in front and to the flat iron crossbar 3 at the back. This underframe will be more fully referred to hereafter.

Motor

The motor M, which is placed in the front of the vehicle and covered by the sheet iron box 4, is of the Phœnix Daimler, as it is called, type; that is like that shown in Fig. 33, but with two parallel vertical cylinders each 3·962 in. = 90 mm. bore by 4·75 in. = 120 mm. stroke, and develops as now made about 6 B.HP. at 700 revolutions per minute. It is constructed, as shown in Figs. 60 and 61, of three principal castings, namely, the cylinder heads 5, the cylinder bodies 6, and the crank case 7, which is in two halves (see Fig. 60), the arms 8, Fig. 61, by means of which the motor is carried from the channel bars 1 of the frame, being cast in one with the upper half of the crank case. As will be seen from the section of the motor, Fig. 61, the cylinder heads and the bodies up to the end of the stroke are water jacketed. The joint x between the two requires to be made with great care in order to prevent water getting into the cylinders or " blowing " from one to the other, either of which faults cause great loss of power through the imperfect combustion which results. It is made by grinding the surfaces together with fine emery powder and water after they have been machined as true as possible and scraped.

Petrol supply

The supply of petrol is carried in the tank P, Figs. 57, 58 and 59, suspended below the frame. The oil is forced, by means of pressure derived from part of the exhaust, as described later on, up the pipe 9 through the cleaning vessel 10, Figs. 60 and 61, which is filled with discs of wire gauze, and thence by pipe 11 to the float vessel A (see Figs. 34, 57, 60 and 62).

Carburettor

From here it is fed into the chamber 12, Figs. 60 and 62, where it meets with air drawn in through the adjustable air-admission head 13, Figs. 34, 63, 64 and 60, on the cylindrical vessel L on each suction stroke of the motor. The vessel is seen in most of the Figs. last referred to (see Figs. 57 to 61). The mixed oil vapour and air from the carburettor, Fig. 62, pass along the pipe 14 into the admission chamber 15, Figs. 60 and 61 of the motor.

Valves

The distribution of the valves, etc., for one of the cylinders is shown by Fig. 61, that for the other cylinder being exactly similar in all respects.

90

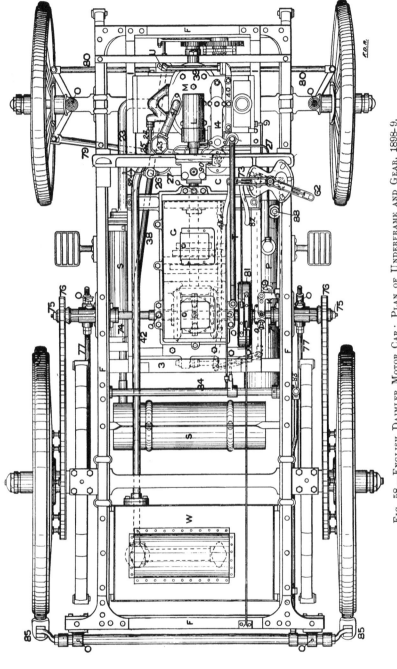

FIG. 58.—ENGLISH DAIMLER MOTOR CAR: PLAN OF UNDERFRAME AND GEAR, 1898-9.

91

On the suction stroke of the motor the admission valve 16 opens against the resistance of the spring above it, and the charge is drawn into the cylinder from the chamber 15 through the port 18. On the return stroke the valve 16 closes and the charge is compressed into the combustion chamber 19 and port space 18 (which are now made smaller to give a compression pressure of 60 lbs.) ignited by the platinum ignition tube E heated by the burner 19, Figs. 60 and 61. After the piston has completed its working stroke, the exhaust valve 20 is raised by a cam on the shaft 21, Figs. 60, 61 and 64, which is geared one to two with the engine shaft, and

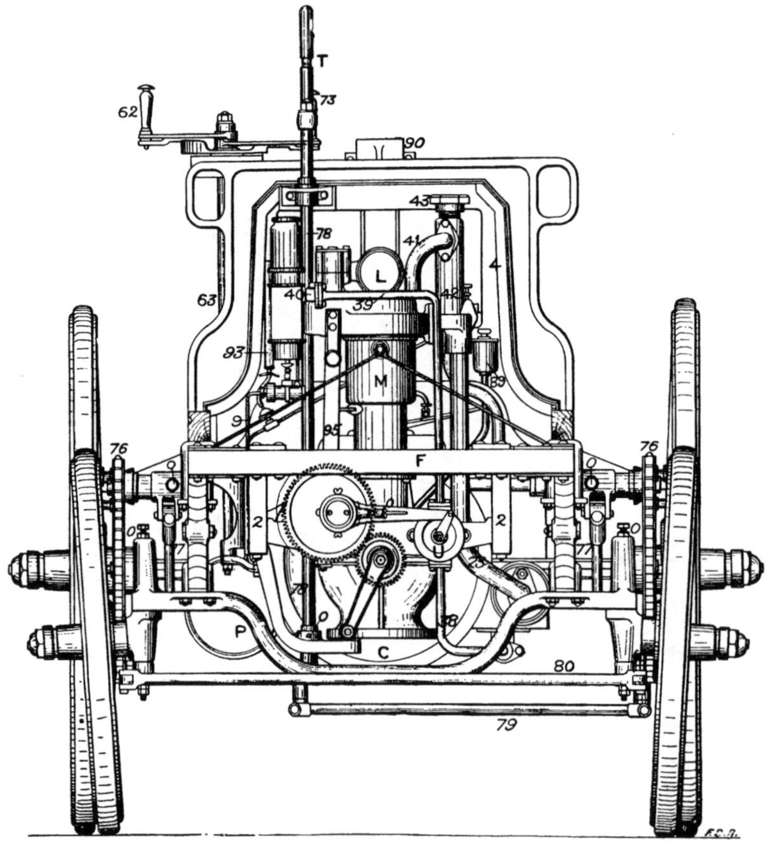

FIG. 59.—ENGLISH DAIMLER CAR: FRONT VIEW OF UNDERFRAME, HOTOR, AND GEAR.

the burnt gases escape into the chamber 22 under the valve, and thence round the cylinder head and out by the pipe 23 into the silencers S S, Figs. 57 and 58, and thence into the atmosphere at the back of the vehicle.

Maintaining pressure in petrol tank by exhaust gases

A portion of the exhaust is diverted through the pipe 24, Figs. 58 and 63, near the cylinder head, and passes through the cleaning vessel 25, Fig. 58, into the chamber 26, containing a valve which allows a small part of the exhaust to pass along the pipe 27, Figs. 57 and 58, into the petrol tank P,

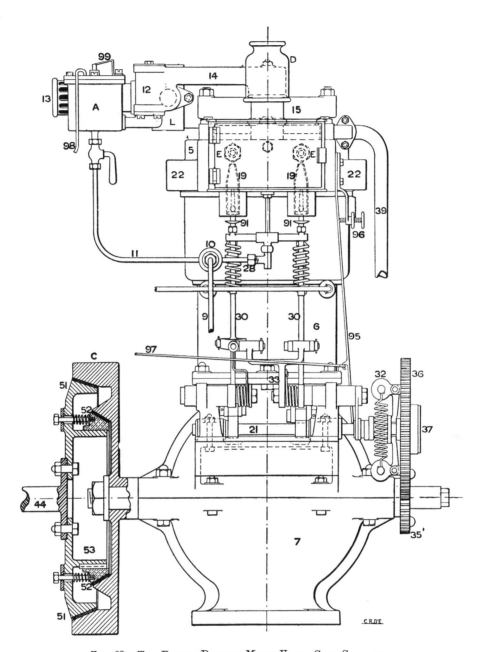

FIG. 60.—THE PHŒNIX DAIMLER MOTOR-VALVE GEAR SIDE, AND
PANHARD & LEVASSOR CLUTCH, 1899.

so maintaining the pressure required to force the petrol up to the float vessel A.

Burners

The burners 19, which heat the ignition tubes, take their supply of oil from the same tank P by means of the branch pipe 28 from the cleaning vessel 10, through which the main supply passes to the float vessel A. The cranks in this engine are placed at 180° apart, so that the weight of the pistons, etc., is balanced.

Cycle of operations

The cycles in each cylinder follow one another thus :—

Cylinder 1.—Suction.	Cylinder 2.—Compression.
Compression.	Explosion.
Explosion.	Exhaust.
Exhaust.	Suction.'
Suction.	Compression.

So that there are two explosions followed by two exhausts, and so on. The cams on the second motion shaft 21, Figs. 60 and 61, are set at right angles, so that the one exhaust valve begins to lift as soon as the other closes. The cams work on rollers at the ends of the bent arms 29, Fig. 61, which, being lifted, raise the jointed rods 30, Figs. 60 and 61, and so lift the exhaust valves against the pressure of the springs 31, Figs. 60 and 61.

The governor

By means of the cam, controlled as to position by the governor 32, which operates the cam sleeve sliding on the shaft 21 when the speed of the engine exceeds the normal, the vertical arm 33, Figs. 60 and 61, is moved angularly inwards, pushing the rods 30, first one and later on the other towards the cylinder, so that the notches in the hit-and-miss arms 29, as they rise, miss the ends *p* of the rods 30, and the valves remain closed until the speed of the motor is reduced to normal, when the governor pulls back its cam, and the lever on the outer end of the spindle which works the arm 33 runs on a circular collar and rods 30 fall back into place again.

The cam shaft 21 is driven from the crank shaft by means of the machine-cut bronze spur wheels 35, 36. On the face of the wheel 36 is a boss on which works an excentric 37, Figs. 60–63, which drives the semi-rotary pump U, Figs. 58, 59 and 63, which maintains the circulation of water round the cylinder jackets.

Cooling water circulation

Returning now to the general drawings, Figs. 57, 58 and 59, the cooling water is carried in the tank w at the back of the vehicle ; it flows along pipe 38 to the pump U, by which it is forced up the pipe 39, entering the jacket at 40, Figs. 58 and 59, near the admission valves. After circulating round the cylinders, it emerges by pipe 41, and flows back to the tank by way of pipe 42. It will be seen that this pipe is enlarged where it is joined by pipe 41, and that the upper end forms a small vessel (see Fig. 59), which is closed by the screw cap 43. This can be easily removed by hand by the driver, so that he may see whether the water is circulating from pipe 41 and down pipe 42 to the tank, and thus ascertain whether the pump is working properly. The aperture at the top of pipe 42 is also used for filling the tank.

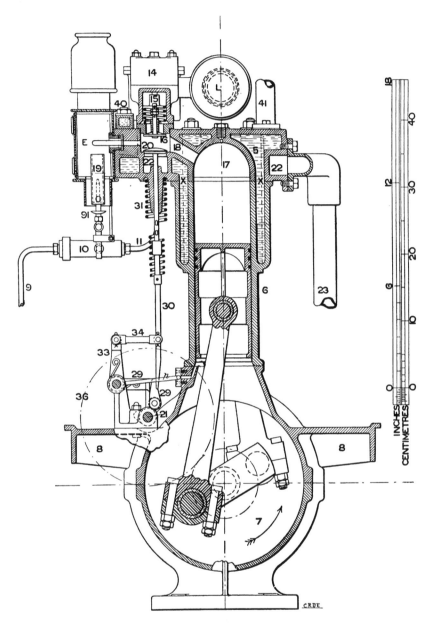

FIG. 61.—THE PHŒNIX DAIMLER MOTOR : TRANSVERSE SECTION, 1899.

**Trans-
mission
gear**

The power is transmitted from the motor M to the clutch shaft 44, Figs. 60, 65 and 66, through the friction clutch c, Figs. 57, 58, 59, 60 and 63, which is controlled by the pedal 45, Figs. 58 and 67. By depressing the pedal 45, the arm 46, Figs. 65 and 66, on the short cross shaft at the back of the gear case (see also Fig. 58) is moved towards the motor. This causes the forked arm 47, Fig. 65, to move in the opposite direction,

**Main
clutch**

pulling the bar 48, and with it the brass 49 and clutch shaft 44, against the pressure of the spring 50. This movement causes the cones 51, 52, Fig. 60, to disengage from the friction surface on the motor half of the clutch c, Figs. 57, 60–63, so allowing it to run free of the gear. On releasing the pedal, the spring 50, Fig. 65, pushes the clutch shaft towards

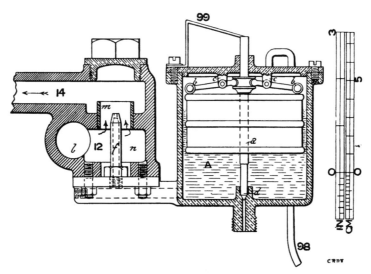

FIG. 62.—FLOAT-FEED DAIMLER-MAYBACH CARBURETTOR.

the motor, first causing the inner cone 52, which slides on feathers on the drum 53, to come into contact, the friction being sufficient to absorb only part of the power, and then the outer cone 51, the angle of which is smaller, comes into contact, so that the whole power is put on gradually and the motor not brought up so suddenly, as it would be very often with the single outer clutch cone.

The portion of the clutch shaft 44 within the gear case c, Fig. 65, is square, and on it slides the sleeve 54, which carries four spur pinions, 55, 56, 57, 58; between the pinions 56, 57 is a groove, in which works a fork on the end of the arm 59, Fig. 66, which is welded into the sliding bar 60. By moving the bar 60 the pinions are caused alternately to engage with spur wheels on the parallel countershaft 61.

**Speed-
changing
gear**

This speed-changing gear is controlled by the hand lever 62, Figs. 57, 58 and 59, working on a sector on the standard 63, conveniently placed in

96

front of the driver. There are four notches on the sector, into any one of which a spring-held key under 62 may be dropped, so holding the lever 62 in place. The spring for this key is in the handle 62, and is released by depressing the knob at the top, Fig. 59. These four notches, counting from the outermost inwards towards the driver, correspond to positions of the sliding sleeve 54, Fig. 65, which puts the pinions into gear with the wheels on shaft 61 in the following order : slowest speed, 55–64 ; second speed, 56–65 ; third speed, 57–66 ; fourth and highest speed, 58–67. Motion from the handle 62 is conveyed to the rod 60 by a horizontal rod on an arm at the bottom of a vertical spindle in the standard 63, Fig. 59, and part of this rod marked R is seen at Fig. 66, where it is connected to the lever L and thence to the bar 60.

The power is transmitted from shaft 61, Figs. 58, 65 and 66, to the cross countershaft 69 by means of the bevel wheel 70, carried at the end of 61, and with which either of the bevel wheels 71 or 72 may be put into gear according as the vehicle is required to move in the forward or backward direction, the whole of the parts between the letters H H, Fig. 66, with the differential gear D being moved. The bevel wheels 71 and 72 are built up of renewable-toothed wheel rings, bolted to flanges on the ends of the tubular sleeves H, which are free to slide on the solid shaft 69. To the inner faces of the flanges are also bolted the two halves of the differential gear box D, the spindle carrying the bevel pinions of the differential gear being rigidly held between the two halves of the box when these are bolted together, while the bevel wheels are free to slide on feathers on the two halves of the shaft 69. **Reversing gear**

The reversing gear is controlled by the hand lever 73, Figs. 57, 58, and 59, moving over part of the sector already mentioned on the top of the tubular standard 63, and operating a tubular spindle therein, at the lower end of which is an arm connected by a link to the bell crank K, Figs. 57 and 58, the movement of which is communicated through the collar O, Fig. 58, to the sliding sleeves H and bevel wheels 71, 72, the whole moving transversely, as before stated.

In the sector over which the reversing lever 73 moves are three notches, into any one of which it may be held by a spring key, like that of lever 62. On the middle notch both the lever wheels 71, 72, are out of gear with 70. In the front notch the wheel 71 is moved into gear with 70, as seen in Fig. 66, giving the forward direction of motion to the vehicle. On the back notch wheel 72 is moved into gear with 70, giving the backward direction of motion.

At the ends 74 of the solid shaft 69, Figs. 58 and 66, are bosses, on the face of each of which is a projection or tongue fitting into a groove or slot in corresponding bosses on the short shafts 75, Fig. 58, which carry the sprocket pinions 76 at their outer ends. These tongue projections and sockets and the short shafts are used to permit of the shaft 69, with its gear, being lifted out of the frame for inspection. **Transmission to chain pinions**

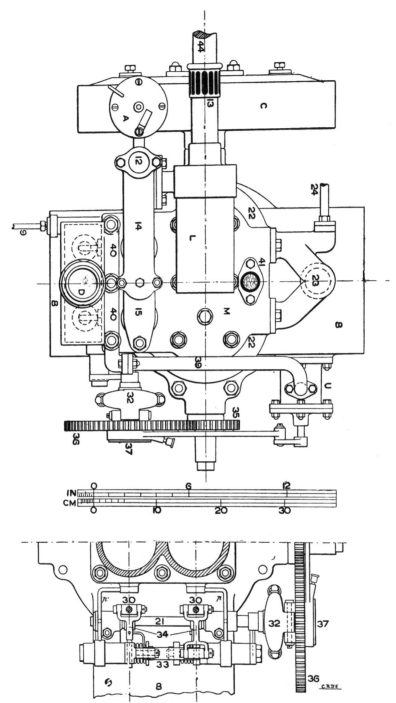

FIGS. 63 & 64.—PLAN OF PHŒNIX-DAIMLER MOTOR AND GOVERNOR GEAR.

98

The sprocket pinions 76 drive by means of pitch chains on to the sprocket-wheel rings, carried by studs from the spokes of the driving wheels.

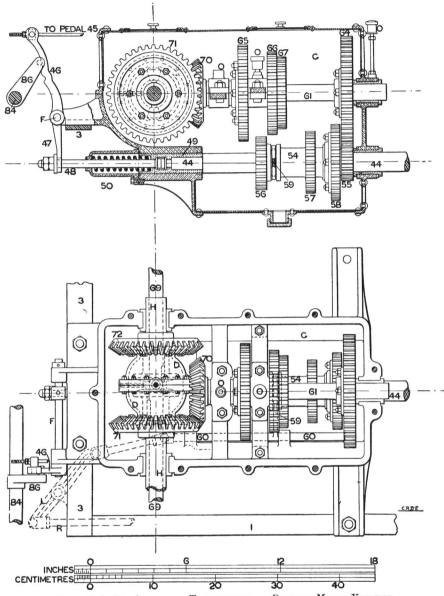

FIGS. 65 & 66.—SPEED AND TRANSMISSION OF DAIMLER MOTOR VEHICLES.

The slack of the chains may be taken up as required by the chain tighteners 77. (See description hereafter given of the Panhard & Levassor Racer.)

The road wheels run as do ordinary carriage wheels, but have parallel boxes, on the ends of the fixed axles. They are constructed with wooden spokes and rims, with solid rubber tyres gripped in iron bands.

Steering gear

The steering wheels work on long vertical cone pivots (see chapter on axles, etc.) at the ends of the fixed front axle, which is cranked to pass below the motor, as seen in Fig. 59. The vehicle is steered by the tiller T, Figs. 57, 58 and 59, which works the vertical spindle 78, at the lower end of which is a radius arm, as seen in Fig. 57 and by dotted lines in Fig. 58, the movement being transmitted to an arm on one of the wheel pivots by the link 79, Figs. 58 and 59, and across to the other by means of arms in front of the pivoted axles connected by the bar 80, which is in tension when the car is running.

Brakes

There are three brakes. One is a band brake working on a drum 81,

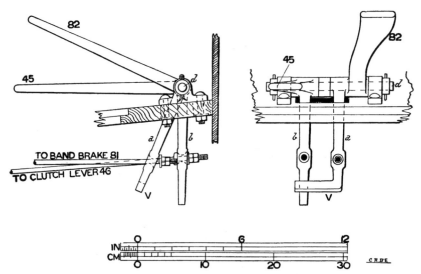

FIG. 67.—CLUTCH AND BAND-BRAKE CONTROL PEDAL GEAR, DAIMLER MOTOR VEHICLES.

Control pedals

Figs. 57 and 58, keyed to the hollow spindle of the bevel wheel 71, Fig. 66, on the counter-shaft 69. This brake is worked by the pedal 82, which, as seen by Figs. 57, 58 and 67, by means of the projecting horn v, also actuates the pedal 45 and releases the clutch c. Thus, although the clutch may be released or one part released by the pedal 45 for varying speed or slowing down without actuating the brake, the brake cannot be worked without putting the clutch out of gear. Thus from Fig. 67 it will be seen that by depressing pedal 45, as shown, the arm b pulls upon the clutch lever rod, leaving the brake pedal 82, its arm a, and band-brake rod unaffected. But when pedal 82 is depressed, the part v pushes the arm b of pedal 45, and pulls the rod of the clutch simultaneously. The other brakes are spoon brakes, controlled by the hand lever 83, Figs. 57 and 58, on the end of the cross-bar 84, Fig. 58. By moving this forward, the rod 84, pulling on a pendant arm on the spindle in bearings o at the back of the car, partially rotates it, and so the spoons 85 are pressed against the tyres of the back wheels. At the same time that these spoon brakes are applied, the arm 86 on the bar 84, Figs. 65, 66, and 58, strikes the arm 46, Figs. 65 and 66, which controls the clutch, and the motor is put out of gear, and remains so as long as the spoon brakes are on, a pin on the end of the arm 86 engaging in the notch shown in the lever 46.

Before starting the motor the water tank w is filled up through the cap-covered aperture 43 until it overflows at 87, Fig. 57, from pipe which reaches nearly to the top of the inside of the water tank w. **Preparing for running**

The petrol tank P is filled through the screw-capped aperture 88, space being left above the petrol for the air under pressure maintained by the exhaust, as already described.

FIG. 68.--ENGLISH DAIMLER MOTOR WAGONETTE, 6 B.HP., 1898-9.

The lubricator 89, Figs. 58 and 59, provides oil for the crank case, and the double lubricator 90, Figs. 58 and 59, for the cylinders. Several other lubricators are marked o, most of which are of the Stauffer kind. They require to be attended to frequently, but not necessarily before every short run.

To start the ignition tube burners 19, Figs. 60 and 61, a small quantity of methylated spirit is poured into the cups 91 below the burners and ignited. This heats the burners sufficiently for vaporizing. The relief pressure cock 92, Fig. 58, on the exhaust pressure valve chamber 26 is closed, and then the pressure in the petrol tank raised by means of the hand pump 93, Fig. 59. The petrol then flows by the pipe 9, Fig. 57, 58 and 61, to the burners, and being vaporized by the heat from the spirit burnt in the cups 91 may be ignited by a match, and burns with a clear blue flame, heating the ignition tubes to a bright red. Everything being

101

Fig. 69.—English Daimler Motor Siamese Phaeton.

Fig. 70.—English Daimler Motor Cranford Wagonette.

ready, one or two turns of the engine, free as it is of all gear, by a handle on the front end of the crank shaft, is sufficient to set it going, and this is easy, even with a fairly high compression of about 40 lb.

FIG. 71.—ENGLISH DAIMLER MOTOR OMNIBUS, 12 HP., 1898-9.

The screw 94, Fig. 57, on the splash-board controls the speed at which the governor acts. This is effected by pulling more or less upon the rod 97, Fig. 60, which holds the flat spring 95 away from the sliding sleeve actuated by the governor. By removing the existence of this spring the

Regulating speed of motor

103

governor is enabled to push the sleeve inwards sufficiently to cause both valves to be cut off at a very slow speed of rotation, and the engine stops. When the screw 34 is released, the spring 95 presses against the sleeve with a force which is adjusted by the screw 96, Fig. 60, near its upper end, so that the power of the governor is only able to overcome the spring pressure when the speed is greater than the normal or, say, 700 revolutions per minute. Some modifications of this will be described hereafter with reference to other cars.

FIG. 72.—DAIMLER MOTOR, 12 B.HP., GOVERNOR GEAR-SIDE : 1899 TYPE.

Various forms of English Daimler motor vehicles   A few illustrations may now be given of the different forms of carriages made by the Daimler Motor Company at Coventry. The underframe motor and gear is the same in Figs. 68, 69, and 70, the bodies only being different. All are fitted with the nominal 4½-HP. motors, which now give about 6 HP. on the brake at about 700 revolutions per minute. That is to say, the "works" in all are as shown by Figs. 57–66. The bodies are nearly all, it may be mentioned, by Messrs. Mulliner.

The omnibus shown by Fig. 71 is fitted with precisely similar under frame and gear arrangements, but stronger, and the motor is of the four cylinder 12 B.HP. instead of two cylinders.

THE MODERN PERIOD AND MODERN MOTOR

These are only a few of the types of vehicles made with the same motors and general arrangement of gearing. They include every form of luxuriously fitted carriage, private omnibus, sporting car, light delivery van, or lorry, and it was one of these nominal 4½-HP. light vans, carrying a ton, which was awarded the first prize after the trials by the Royal Agricultural Society in 1897, when the cost of fuel was found to be 0·46*d*. per cargo ton mile, or less than a halfpenny per ton of goods carried a mile. Cost of fuel

Fig. 73.—Daimler Motor 12 B.HP., End Elevation, 1899 type, showing Electric Ignition Gear.

The 12-HP. frame and motor, as used in the omnibus, Fig. 71, is the same as that used in the 12-HP. wagonettes like those of the Hon. J. Scott Montague, J. R. Hargreaves, J.P., and others.

Figs. 72 and 73 show the side and end views of the newer form of the 12-HP. Daimler motors (16 B.HP.), as now made by the Daimler Company in this country, as well as by the Daimler Company of Cannstatt. They show the bed plate or frame casting, as made at Cannstatt, with plummer-block caps to fix to longitudinal frame tubes fixed at the ends 12-HP. Daimler motor, 1899

105

only of the under frame. By this arrangement a slight amount of elastic freedom is obtained, to the advantage of the under frame and other parts, and at the same time the whole motor can be lowered from its carriers by undoing the eight bolts of the four plummer-block caps and the pipe connections. These views will be perfectly understood by reference to the description of the motor, Figs. 34, and 60–64. The magneto-electric ignitor shown in Fig. 73 will be described in full hereafter.

Complication of the gearing
From this description of the Panhard & Levassor and the English Daimler motor vehicles, it will have been seen amongst other things that there is what is at present a necessary complication of transmission gear as a means of getting different speeds. The same quantity of gearing, and perhaps several clutches, in a stationary machine for bringing about the same changes could not be considered a complication, and there would be no objection to it. When, however, it becomes necessary, for the sake of simplicity and lightness, to change speeds, by pushing the teeth of spur wheels sideways into gear while running, the mechanical engineer naturally feels that the arrangement, if not barbarous, is of the makeshift order, although it is an arrangement similar to that which works perfectly well in a traction engine which is stopped when change of speed is required.

Art in use of speed change gear
Well, it is a makeshift; but the successful use of a makeshift may in many cases give birth to an art or a pleasant emulative occupation, as in rowing, or driving horses. So the skilful motor car driver learns to use his main clutch and his speed change lever so perfectly, that the teeth slip into gear without quarrelling, and the car hardly knows that its acceleration in feet per second per second has suffered a change. For all that, the art is like some of the older and out-of-date occupations—such as the snuffer makers—it is only admirable so long as necessary, and if a more perfect speed gear could be obtained it would be hailed as a boon.

Effect on gear of changing speed
Some reference may here be made to that which most learners of motor car driving do not sufficiently understand with regard to changing speed.

In some of the ordinary English Daimler cars, with 3 ft. 3 in. drivers, the speeds are 3·11, 5·99, 8·64 and 12·2 miles per hour, with a motor running at 700 revolutions per minute. The speeds, and speed of rotation of motor and gearing at these speeds, will then be roughly as follows:—

Miles per hour.	Ratio of speed of counter-shaft and speed of motor.	Revolutions per minute of counter-shaft.
3·11	1 to 3·2	220
5·99	1 „ 1·67	420
8·64	1 „ 1·15	610
12·20	1 „ 0·75	870

Thus, to change from 3 miles per hour to 6 miles, a pair of spur wheels has to be thrown out of gear, and a pair thrown in by a stroke of the hand, which will raise the speed of the counter shaft from 220 to 420 revolutions, or by 200 revolutions per minute, or reduce the difference between its speed and that of the engine by that amount. To avoid the shock resulting from so great a sudden change of speed, the main clutch is pulled out of gear, and it, with all upon the clutch shaft 44, Figs. 65 and 66, immediately falls sufficiently in speed to bring the peripheral speed of the wheel 56 nearly into unison with that of wheel 65 that it may be pushed into gear. Now the car is running at 3 miles per hour (or as much less as road or hill resistance may have reduced it to, since the clutch was taken out), but the gear is in for 6 miles per hour. Hence, if the clutch-shaft fell to the speed that would permit the 6-mile gear to slip in easily, it will now be running at $220 \times 1.67 = 367$ revolutions per minute, and the main clutch has to be called upon to grind or rub this up to 700, or rub out a difference of $700 - 367 = 333$ revolutions. Upon the care with which the clutch is kept in order and is used will depend the smoothness or jerk with which the change is made. If the engine could be lowered for the moment to the 367 revolutions per minute by throttling or otherwise, and then raised to its normal speed after the gear is shifted, the change could of course be much more smoothly effected.

In changing speed, especially from the first or lowest to the second (a change which involves an increase of 100 per cent. in car speed), the shocks and stresses thrown upon gear and car by inexperienced drivers may do more damage in one minute than good running would do in a year. For this reason the learner should at first be careful to avoid unsuccessful attempts to increase the speed on a hill before the reduction of steepness is considerable. In changing from the 8·6-mile to the 12·2-mile speed, the car speed increase is only 42 per cent., but the acceleration is over 80 per cent. of that from 3 to 6 miles. The difference, however, between the speed of the motor and of the clutch shaft, after the gear is changed, is less than that belonging to the change from 3 to 6 miles per hour, namely, a difference of 330 revolutions. At the top or highest speed the difference is $700 - (610 \times 0.75) = 242.5$, and hence the imperfect use or bad action of the clutch, by too sudden action or by seizing, cannot cause so great an increase in acceleration of the car in feet per second. The more nearly the velocity of the driver and driven coincide, the less, of course, will be the impact on sudden throwing into gear or sudden seizing of the clutch whatever those velocities may be.

Changing from a higher to a lower speed for hill ascent requires still greater care and skill to prevent damaging effect on the machinery.

From a consideration of the foregoing it is obvious that the whole of the fly-wheel power should be in the part c of the clutch; while the parts 51, 53 should be as light as possible consistent with strength, in order that the inertia and momentum stresses should be a minimum.

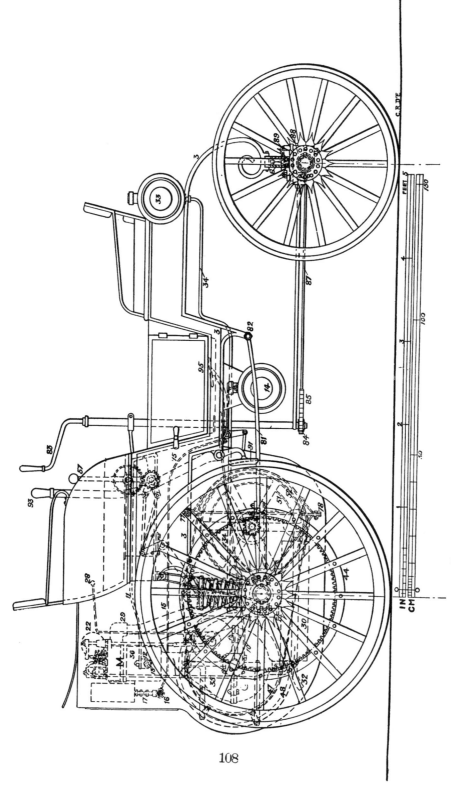

Fig. 74.—Belt-driven Chainless Daimler Motor Maybach Carriage.

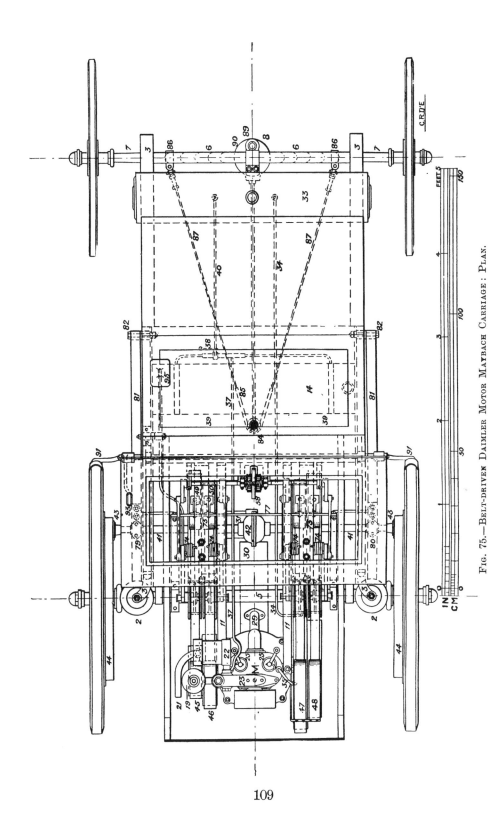

Fig. 75.—Belt-driven Daimler Motor Maybach Carriage : Plan.

109

MOTOR VEHICLES AND MOTORS

Belt transmission gear

As one method of avoiding the source of the objections to change gear, its cost in tooth rings when in the hands of learners, and its noise especially when not exceedingly well made or when worn, belt gear has been tried by some makers and with considerable success.

The Daimler-Maybach belt-driven car

Amongst the first vehicles so made was the Cannstatt or Maybach-Daimler carriages, exhibited in the Crystal Palace Motor-Carriage Exhibition in 1896, of which the author was the reporting judge. This car is illustrated by Figs. 74–79, and of it a good many have been made with different forms of bodies.

Frame

The framework and body are constructed almost entirely of wood, and are carried from the rear axle on two sets of double helical springs 1, which are held at their ends between the shallow cups 2, bolted to extensions of the iron skeleton 3 of the frame, and those marked 4, clipped as shown in Fig. 76, to the back axle 5. In front the body is carried by the single plate spring 6, attached to the scroll-shaped prolongations of the skeleton frame 3, and to the front axle 7 at the swivel joint 8.

Motor

The vertical double-cylinder Daimler motor M is carried from the woodwork of the frame by the flat iron bars 10 and 11 (one of each of which is seen in side elevation in Fig. 74). The lower bars 10 are bolted to the crank case below one of the bearing bolts on each side, and the upper bars 11 are fixed to the pivots 12, which work on the studs 13 in the sides of the cylinder casting. The motor is thus carried in a very flexible manner. When the carriage is in motion the vibration is almost imperceptible, and not objectionable even when standing still.

Petrol supply

Oil is carried in the tank 14, Figs. 74 and 75, and by means of pressure obtained from part of the exhaust, as previously described with reference to Figs. 58 and 63, is forced up the pipe 15, Fig. 74, into the cleaner 16, Figs. 74 and 76. From here part passes up the vertical tube 17 to the tube ignition burners, and part through the pipe 18, into the float vessel 19, Figs. 75 and 76, from which it flows to the spray-making jet in the mixing chamber 20, where it meets with the air sucked into the pipe 21, and through the silencing vessel 22, Figs. 74 and 75. The mixed air and oil vapour pass through the pipe 23 into the chamber 24, beneath which are the admission valves for both cylinders.

Exhaust valves

The exhaust valves 25, Figs. 75–77, are worked by the short arms 26 on the top of vertical rods 27. These depress the valves against the resistance of spiral springs, Figs. 74 and 75. They are actuated through a system of links by cams on a shaft geared 1 to 2 with the engine shaft. By pressing the knob 28, Fig. 74, the exhaust valve hit-and-miss gear may be thrown out of action, and the motor prevented from working. In other particulars the motor is the same as the Daimler motors described already. The exhaust gases pass down the pipe 29 into the silencers 30 and 31, Figs. 75 and 76, and escape at the back of the vehicle through the pipe 32.

Cooling water and cooler

The cooling water for the cylinders is carried in the tank 33 in the front of the carriage. It flows by gravity down the pipe 34, Figs. 75–77,

110

and is discharged into a channel or trough-shaped interior of the fly wheel 47. Here the rotational velocity imparted to the water causes it to sweep round with the inside of the wheel as a semi-solid. Part of this water is caught up or " turned " off by the mouth of the pipe 35, Figs. 76 and 77, acting like a hollow turning tool, that which it turns off being forced upwards into the jacket 36, entering at the top as seen in Figs. 75, 76 and 77. After circulating round the cylinders it passes out from the top of the cylinder

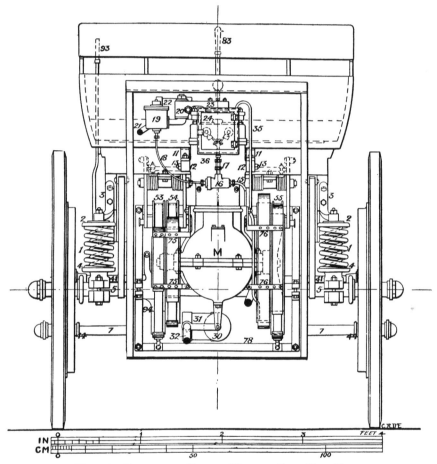

Fig. 76.—Belt-driven Daimler-Maybach Car: Back View.

by the pipe 37, Fig. 75. By means of the cock 38, Fig. 75, it may either be diverted into the foot-warmer 39, forming the footboard, and then forced **Foot** up into the tank 33, or it may be allowed to flow straight to the tank **warmer** through the pipe 40. About 4 galls. of water are carried, which suffices for a run of six hours at normal speed.

The engine drives by means of belt gear on to the countershaft 41, **Trans-** which carries the differential gear 42, and at each end the spur pinions 43, **mission**

111

which gear into the internally toothed rings 44, bolted to the wooden spokes of the rear wheels. There are four belt pulleys 45, 46, 47 and 48, Figs. 75 and 77, mounted in pairs on the engine shaft, one pair at each side of the crank case. These drive the pulleys 49, 50, 51 and 52, which are mounted in pairs on the countershaft 41. The belts pass over jockey-pulleys 53, 54, 55 and 56, by raising or lowering any one of which the belt passing over it is made tight or slack. All the jockey-pulleys are down as shown in Figs. 74, 76, so that all the belts are slack and the motor

Speed
change gear is out of gear. The mechanism for raising any one of these jockey-pulleys and thus tightening the belt passing over it, the others at the same time remaining slack, is controlled by the handle 57, at the driver's right hand, Figs. 74, 75, 77 and 79. This gear acts in the following manner: the handle 57 is formed in one with the quadrant wheel 58, the lower half of which is toothed; this wheel is keyed to the short shaft 59, which slides in bearings 60, so that the wheel may be moved either to the right or left against the pressure of the springs 61. Suppose the carriage is required to travel at its slowest or second speed, the handle 57 must be pulled to the left; this will put wheel 58 into gear with the pinion 62, which is keyed on the end of the shaft 63, carried on bearings 60 and 65, Figs. 78 and 79; to this shaft are also attached the crank arms 66 and 67, Fig. 78. By means of rods 68 and 69 the motion of these arms is imparted to bell-crank levers 70 and 71, which carry the jockey-pulleys 53 and 54 independently at the ends of their long arms. The bell cranks are attached to the ends of a rod working in bearings bolted to the iron plate 73, Figs. 75 and 76, springs 74 being coiled on the rod in such a manner as to always tend to lift the jockey-pulleys. The wheel and pinion 58 and 62 being now in gear, the shaft 63 may be rotated in either direction; if the slowest speed is required, the handle 57 is pulled backwards; this causes the shaft 63 to rotate in the forward direction, and with it the crank arm 67, Fig. 78. Through the rod 69 the motion is imparted to the bell crank 70, the jockey-pulley 53 being raised, and the belt on pulleys 45 and 49 tightened. It will be noticed that, while performing the above operation, the crank arm 66, which controls the jockey-pulley 54, has rotated through the same angle as the arm 67, but the "off" position of this arm is at such an angle below the horizontal that the only result of raising it through an angle equal to that moved through by 67 is to cause a very small downward movement of the bell crank 71 and jockey-pulley 54, followed by a slight upward movement, neither of which are sufficient to affect the belt 46, 50. When the second speed is required, the handle 57 is first moved to the vertical position, which lowers the jockey 53, slacking and putting the belt 45, 49 out of gear, and then the handle pushed forward, thus rotating the shaft 63 in the backward direction, and causing the crank arm 66 to lift the jockey 54, Figs. 77 and 78, and so tighten the belt 46, 50. This rotation of the shaft does not move the bell-crank 71 sufficiently to affect the belt 45, 49. The third and fourth or highest speed-gear pulleys 48, 52 and 47, 51 are controlled in

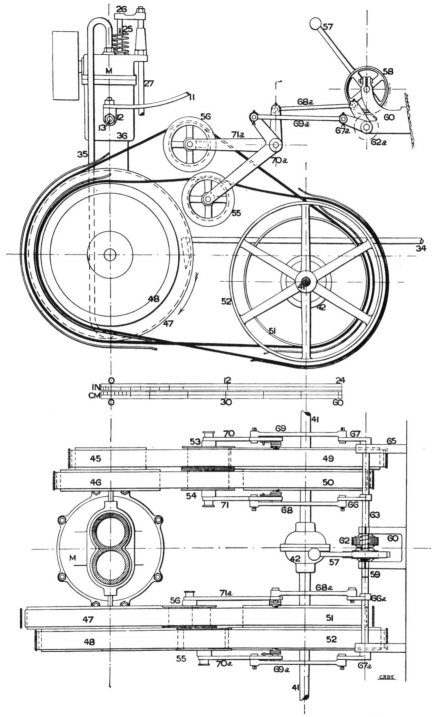

FIGS. 77 & 78.—BELT DRIVEN DAIMLER-MAYBACH CAR: ARRANGEMENT OF BELT-DRIVING, AND SPEED-GEAR, AND CYLINDER-COOLING SYSTEM.

an exactly similar manner, except that the handle 57 is pushed to the right causing the wheel 58, Fig. 79, to gear with the pinion 62a, keyed on the shaft 63a, through which the jockeys 55 and 56 are controlled. Referring particularly to Figs. 77 and 78, the handle 57 is now in the backward position, crank 67a is lifted, but not high enough to pull rod 69a and pulley 55 hard up against the belt; but the backward movement of the handle has lifted pulley 56 high enough to tighten the belt. Now, by pushing handle 57 to the vertical, both crank arms 66a and 67a will be thrown backwards and lowered, and crank arm 66a will be just about in the position shown of 67a. Under these circumstances pulley 56 will be lowered to the level now occupied by 55, and 55 will be just about in its present position, because the crank arm end 67a has moved only from a position a little above to one

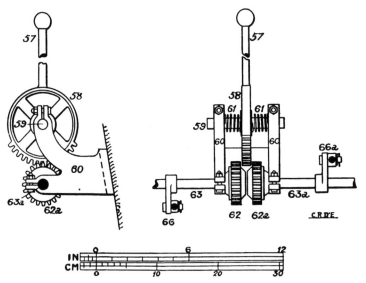

Fig. 79.—Daimler-Maybach Car Speed-Changing Gear.

a little below a line connecting the crank arm pivot and the bell crank. Hence the crank arm 79a remains practically where it was. Now, continuing the movement of handle 57, crank arm 67a descends, pulls rod 69a, and lifts jockey-pulley 55, pulley 56 being still further dropped.

Protection of belts   These four sets of wheels correspond to speeds of 4, 7, 11 and 16 miles per hour of the carriage. There is no reversing gear. The outer portion of the circumference of each of the belt pulleys is surrounded by an iron sheath, held about ½ in. from the surface of the pulley, so that, when the jockey-pulleys are down, the belts are only able to fall as far as the sheaths, and they consequently endeavour to expand longitudinally, so pressing against the sheaths all round instead of rubbing on the pulleys. Those which surround the pulleys on the engine shaft are carried from the crank case by the flat iron bars 75, 76, Fig. 76, the others from the crossbars

114

77, 78, which are carried from the axle 5, as seen in the side elevation. The jockey-pulleys are made of wood and are flanged at both sides to prevent the belts slipping off. The countershaft 41 is carried in bearings 79, 80, which are on the same forging as the lower cups of the springs 1, the front ends of these forgings being bolted to the flat iron bars 81, Figs. 74, and 75, which are free to turn on the pins 82, fixed to extensions of the skeleton frame 3; this provides at once the radius arm for maintaining the distance between the pinions and gear wheels and suitable carriage for the countershaft 41. Jockey-pulleys

The steering gear is operated by the tiller and vertical bar 83, at the lower end of which is fixed the sprocket pinion 84, over which runs the chain 85, the ends of which are attached to clips 86 on the front axle by the rods 87, and universal joints. The axle is clipped at the centre by a locking plate 88, which is free to turn on the pin 89, carried by the plate 90, on which the spring 6 bears. Steering gear

There are three brakes: The shoe brakes 91, on the extremities of a cross-bar, are pressed on the tyres of the rear wheels by a forward movement of the hand lever 93. Another brake 94, operated by the pedal 95, bears upon a portion of the surface of the pulley 49 not surrounded by the belt above mentioned. Brakes

The wheels of the carriage have wooden spokes and felloes shod with steel rims or tyres, to which are vulcanised solid rubber tyres. Road wheels and tyres

Several of these carriages have been in use in England since 1896, and they are liked because of the smoothness of the running of the belts, which are kept well out of mud and wet. The four speeds secure easy change to suit varying road resistances, and this gear is economical in running and in maintenance. Belt slipping and squeaking occurs when the speed changing is too suddenly made, but causes no trouble, and with ordinary care and inspection of the belts, unexpected breaking need not occur. It will be noted that tightening of the belts pulls them partly off the pulleys and reduces the arc embraced. As, however, the difference of diameter of any pair of pulleys is small, this reduction can be afforded, and it has the advantage that the jockey pulleys act on the inside of the belt, which is therefore not subject to contrary flexure, and the jockey pulleys do not affect the belt jointing or fastenings. Lumpy fasteners should not, however, be used, and the choice of fasteners as well as of belt needs, as it does for all belt driven cars, to be previously made. In the use, too, of any of them, the automobilist cannot too often be reminded that " a stitch in time saves nine," and perhaps a long walk as well. The carriage in use

THE PEUGEOT MOTOR CARRIAGE OF 1896.

At the time Messrs. Panhard & Levassor were accomplishing feats with Levassor's Daimler motor car, Fig. 80, which has become the standard of all their essentials of the best design, Messrs. Peugeot Frères, of Valentigney, were making cars of the design shown by Figs. 80, 81, 82, The Peugeot-Daimler motor car. 1896 and 83. In this car, as will be seen, there was a strong likeness to the Panhard & Levassor in the transmission gear, in the use of a surface vapouriser with a Daimler motor, and Stud or Ackerman steering axles. In some respects the carriage was of superior arrangements for appearance, but the position of the motor at the rear, where most of the passenger weight came, put an unnecessary load on the hind wheels, and for good steering too little on the front wheels. The Peugeot position for the motor is moreover inferior with respect to convenience of the driver. Nevertheless these carriages performed some remarkable feats, and in several ways are of much mechanical interest. As compared with the recent cars by the same firm, it differs mainly in the position and type of motor in different arrangement of gearing, and in dispensing with the surface carburettor.

Frame   The under-frame of the Peugeot car is tubular, and not of the simple form of the Panhard-Levassor, but is higher at the back and front than at the centre, and it has some important members at a lower level. The rear part is carried on long leaf springs, with range sufficient to allow them to straighten. The front is carried on one long cross spring, as shown in Fig. 83. The upper and lower members of the under-frame form a truss the parts of which are maintained or braced by the pieces carrying the bearings of the clutch and countershafts. The fore-carriage gear differs in arrangement from those of other makers, though its essential features have been since embodied in some American electrical cars. A transverse member of the front of the frame is pivoted to the centre bridle of the spring, the ends of which are connected to the main fore axle by cranked hangers. At the extreme ends of the fore frame are two quadrants, against which the ends of the main axle rest, and are guided, leaving the axle free to move up and down angularly, by a lift from either wheel, without affecting the frame.

Motor   The motor of the date of this carriage was the **V**-type Daimler, as already described, but without the valve in the pistons for the scavenger charge.

Carburettor   The carburettor used on some of the cars was of the Daimler surface type, described with reference to Fig. 35, with slight modifications. At the date, however, of the carriage illustrated, the Messrs. Peugeot were also using, as seen in Fig. 81, the spray-making float feed carburettor of Maybach already described, Fig. 62, but with an adjustable air admission above **Objection to** the spray instead of the solid cap, and this has survived. The objections to **surface** the surface evaporation were those which are found to exist with them even **carburettor** as now used in other forms in motor cycles. The lighter constituents of the mineral spirit pass off first, and leave a heavier residue, which may reach a density of 0·72, and with this, especially in cold weather, evaporation and carburetting takes place slowly, and a sufficiently rich mixture cannot be obtained until this residue is drawn off and new spirit substituted. The level of the spirit was also difficult to maintain. These difficulties were avoided by using the float feed spray-making carburettor, the only disad-

116

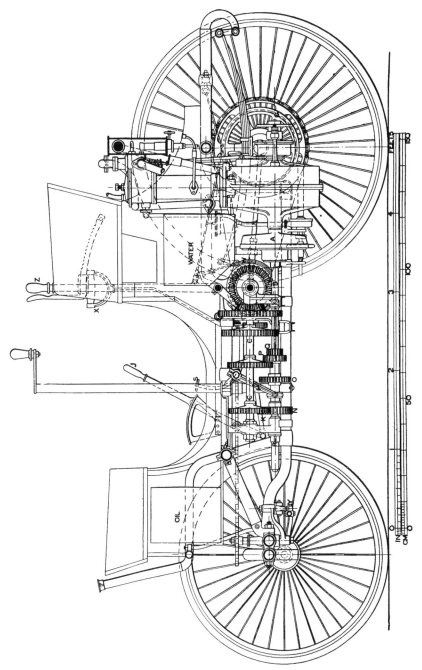

Fig. 80.—The Peugeot-Daimler Motor Carriage, 1896.

117

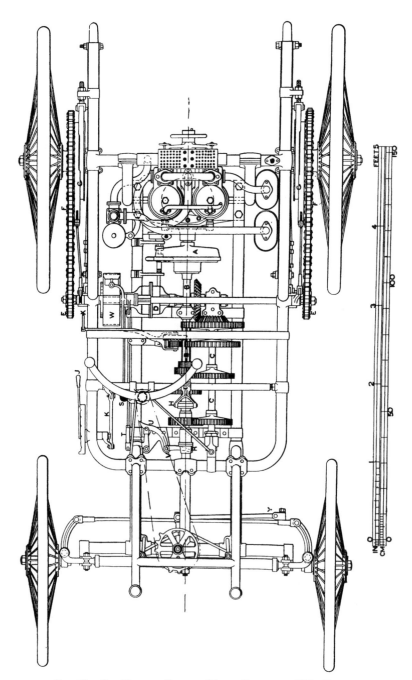

FIG. 81.—THE PEUGEOT-DAIMLER MOTOR CARRIAGE, 1896: PLAN.

118

vantage of which is that the motor cannot be run through so large a range of speed by throttling, as when it works off a surface carburettor, which has some vapour storage capacity.

The ignition was by Daimler tubes heated by lamps, as in Figs. 34 and 61, and similar to those which will be fully described with reference to recent Panhard & Levassor carriages. Ignition

The power of the motor was transmitted by means of a clutch A, Figs. 80, 81 and 83, to a square clutch shaft B, carrying four Peugeot transmission gear

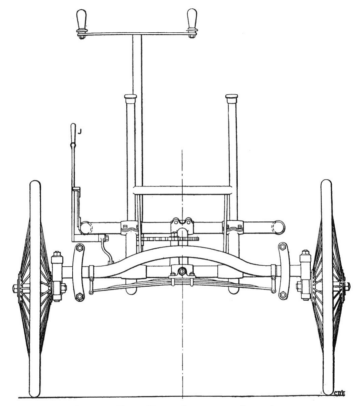

FIG. 82.—PEUGEOT CAR: FRONT SPRING AND AXLE ARRANGEMENT.

pinions of different diameter, which geared selectively with four wheels on the second motion shaft C, for four speeds, and with one separate wheel for reversing. At the rear end of the second motion shaft was a bevel pinion gearing, with a bevel wheel on the transverse shaft, which carried the differential gear D, and at its ends, the pitch chain pinions E E, which communicated motion to the driving wheels by chains F on sprocket wheels on the driver limbs, these sprocket wheels also carrying brake pulleys, on which the brake straps G G acted, and were actuated by the lever Z. The second motion shaft is not

119

directly over the clutch shaft, and this gives the gearing the appearance in
Figs. 80, 81, of being of wrong proportions. The gear wheels being a
considerable distance apart, the forked lever H, Fig. 81, which is worked
by the hand lever J, the link K, and an arm on a transverse shaft L, must
be of long range as seen. The spur wheels shown in gear are those of
nearly equal size, for the fourth or highest speed. By moving the lever J
forward either of the other three pinions, O P Q, were put into gear alternatively.

At the front end of the clutch spindle was a sleeve R, in which was a
spring acting against a collar for pressing the clutch into gear. For
throwing this clutch out of gear the pedal S was used. By depressing this
pedal the rod T was pulled, and the end of this was connected to a bell
crank lever. On the end of one arm of the latter was carried a smaller
roller acting against a bent lever U, pivoted at V, the outer end of the bent
lever being connected to the sleeve on the end of the clutch shaft. By this
circuitous path the clutch was put out of gear or allowed to go into action.
By further depression of the pedal S, a brake W on the transverse shaft is
applied.

Reversing For reversing, the speed change lever was set between the first and
second speed notches, and then the handle X put forward. This shifted an
independent spur wheel, movably mounted on a bracket (as indicated in
the centre of the plan, and in the elevation, Figs. 80 and 81), and put it
into gear with a pinion on the clutch shaft, and a wheel on the second
motion shaft. The reverse motion was thus given to the latter. Of course
the clutch pedal was depressed when any of these changes were made.

Steering gear The steering was effected by a double handle on the top of a spindle in
a hollow vertical pillar, motion from the bottom end of it being communicated to a sprocket wheel on the foregear. This wheel gave motion
to an arm and link Y, and thereby to the rod I, connecting the short steering axle arms in the usual way.

Peugeot clutch The cone clutch A is shown in detail by Fig. 83, p. 122. The hollow
cone A is keyed upon the motor shaft. The inner cone C is secured to the
clutch shaft B by four large feather-keys D, which permit it to move on
the shaft. At a larger radius are four bolts E, which pass through the
flange F at the end of the shaft, and through the inner face of the inner
cone, in each case through holes nearly a fit with the bolts. Within the
casting forming the inner cone the holes for these bolts are enlarged so
as to accommodate strong spiral springs of short range, these springs
pushing against the bottom of the enlarged holes, and against the flange
F. When the two conical faces of the cone, which are leather-covered,
first come into contact, they do so under the pressure of these four springs,
so that only a part of the total pressure is brought into use at first, the
object being to secure a gentle preliminary action, as and for the purposes
already described with reference to the English Daimler motor cars. The
additional pressure is put on by the spring at the end of the clutch shaft

120

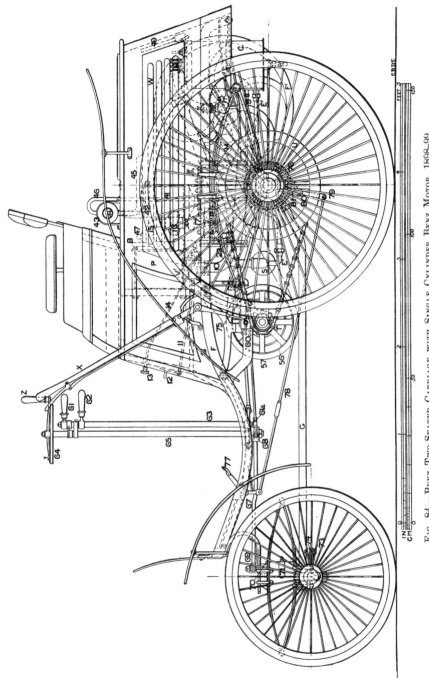

Fig. 84.—Benz Two-Seated Carriage with Single Cylinder Benz Motor, 1898-99.

121

when the pedal is completely released. It will be seen that the end of the motor shaft is enlarged, and receives a bush, shown white, in which the end of the clutch shaft runs.

Such then were the main features, and it may also be said the many features and parts of the Peugeot car of 1896. Only to a small extent are they recognisable in the Peugeot of to-day. The more recent form, with horizontal motor and simpler parts, will be dealt with with other later carriages; but the old one is of great interest as showing the great advance which has been made, and how much easier motor carriage construction should be now that the difficult scheming of simplifications has been done.

THE BENZ MOTOR CARRIAGES OF 1896–99.

The Benz carriages, 1896–99 Turning now to the Benz carriages of 1896 or 1897 to 1898–99, as illustrated by Figs. 84–98.

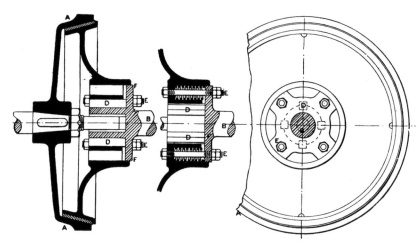

FIG. 83.—PEUGEOT CAR: MAIN CLUTCH.

The elementary features of the design of the single-cylinder Benz cars may be said to have been fully foreshadowed by his designs of ten years ago. The horizontal engine, the belt driving, electrical ignition, surface carburettor, cycle wheels, chain drive—all have survived in the cars of which he has made such an enormous number. Although he has constantly been making improvements, it is only recently that he has made an organic departure; namely, in the construction of a powerful racing car, unless we except the use of a double cylinder engine, with the necessary modification that change involves.

Figs. 84, 85 and 86 show a side elevation, plan, and end elevation of a two-seated Benz carriage of the 1898 pattern.

Frame There is no underframe proper independent of the body, the two being

122

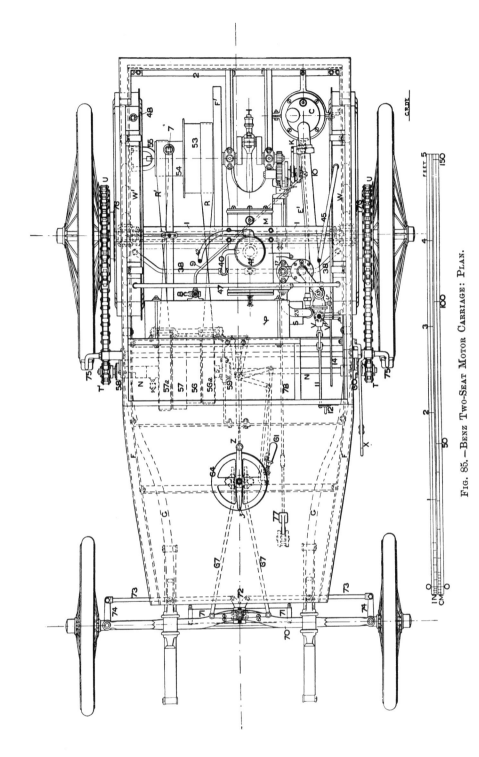

FIG. 85.—BENZ TWO-SEAT MOTOR CARRIAGE: PLAN.

123

built together. The skeleton on which the body is built consists of an outer structure F of wood, faced inside and outside with $\frac{3}{16}$ in. flat iron, on carriage builder's methods, and connected across by stays, by which the various mechanical parts are carried.

Motor   The motor M is enclosed in the boot of the body at the back of the seats. It has a single horizontal cylinder 4·376 in. = 110 cm. bore × 4·375 in. stroke, and develops about 3 B.HP. at 600 revolutions per minute. It is shown in detail by Figs. 87–93. It is constructed of three principal castings, the cylinder body 3, which is water-jacketed throughout its whole length, Figs. 89 and 90; the valve chamber 4 also water-jacketed, Fig. 89; and the front extension 5, which carries the bearings of the crankshaft Figs. 87 and 88, the side arms of which are of channel suction, and are prolonged to carry the motor from the angle bar 2, Fig. 85, at the back of the frame. It is also carried by means of the lugs 6 on the cylinder body, which are bolted to the cross **T** bar 1 of the skeleton frame, Fig. 85.

Petrol supply   The petrol tank P is carried from the back of the box under the seats of the body, Figs. 84, 85 and 86. It is filled through the aperture at the end of pipe 7, Fig. 86, which is closed by a screw cap, the level of the oil in the tank being shown by the gauge glass 8, at the base of which is the cock which controls the flow of petrol by way of pipe 9, Figs. 85 and 86, **Carburettor and connections** to the carburettor c, seen in all the general drawings, and in detail in Fig. 91. The level of the petrol in the carburettor is regulated by a float A, Fig. 91, to which is attached the light tubular rod B, which is free to slide in a guide D in the cover of the vessel; when the level of the petrol falls below a certain limit, the float A, also falling lifts the plate E, which covers the nozzle F, and the petrol flows in until the correct level is again restored, when the nozzle is closed by the plate. The petrol in the carburettor is warmed and partially vapourised by means of part of the exhaust from the motor, which is diverted along the pipe E′, Figs. 85, 91 and 96, and enters the carburettor below the false bottom G, Fig. 91, and escapes through holes in the bottom H. On each suction stroke of the motor, air is drawn in through the gauze-covered openings in the head J, Fig. 91. It passes down pipe K, and sweeps over the surface of the petrol, producing a vapour which is richer in hydrocarbons than is necessary for obtaining the required combustion in the motor cylinder. The vapour is sucked up the pipe L, which is surrounded by the cylinder M, and passes along the pipe 10, Figs. 48, 85–91, to the conical valve chamber v, Figs. 84 and 85, in the top of which is an air aperture v′. The valve in this chamber has two passages—one for the carburetted mixture, and one for pure air sucked in at v′; it is operated by the jointed rod 11, from the pointer 12, Figs. 84 and 85, which moves over a sector on the front of the seat-box. This sector **Vapour supply control** has points marked on it indicating the relative quantities of gas and air, or, in other words, the quality of the mixture passing into the motor cylinder. In the back of the valve chamber v at v is a throttle valve, controlled from the knob 13 at the end of the rod 14, Fig. 84. A rack is cut

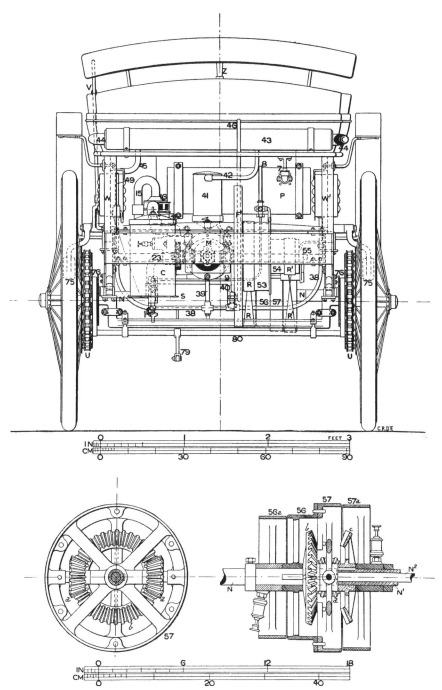

FIG. 86.—Benz Two-Seat Motor Carriage: End View.
FIG. 96.—Pulleys and Differential Gear.

125

in the end of rod 14, near the knob 13, and this may be engaged with a catch in the guide piece let into the woodwork, and the throttle held in any required position, so that the quantity of the charge passing from the

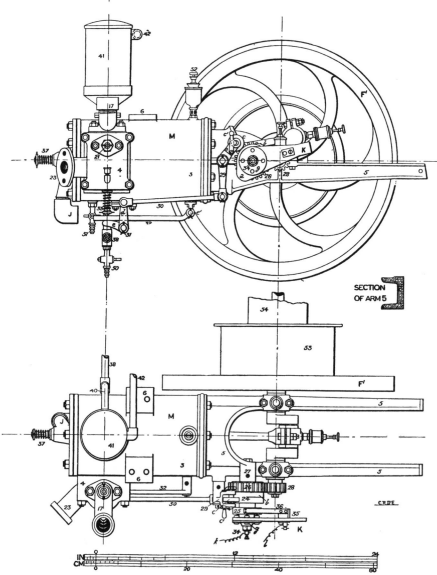

FIGS. 87 & 88.—BENZ 4·375 SINGLE-CYLINDER MOTOR: ELEVATION AND PLAN.

valve chamber to the cylinder may be regulated. This arrangement for controlling the throttle valve has, however, been improved in more recently made Benz vehicles, and a lever guided by a quadrant substituted for the

126

THE MODERN PERIOD AND MODERN MOTOR

Throttle valve

knob 13. After passing the throttle valve *v*, the charge is sucked up the pipe 15, Figs. 84, 85 and 86, through the cleaning chamber 16, which is filled with discs of wire gauze, and thence into the admission valve chamber 17, Figs. 84–86, and also Figs. 87 and 88, which show in detail the valve arrangements of the motor.

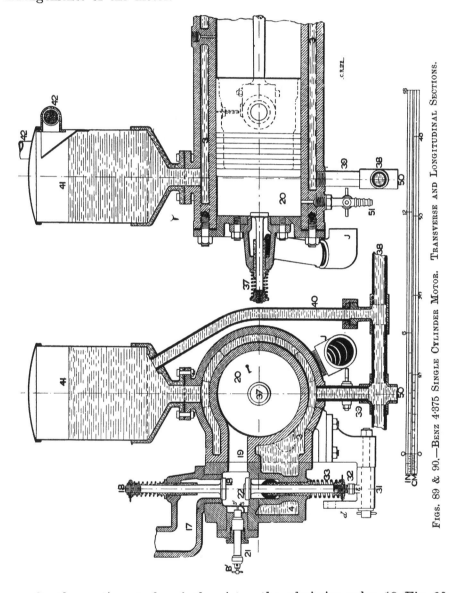

FIGS. 89 & 90.—BENZ 4·375 SINGLE CYLINDER MOTOR. TRANSVERSE AND LONGITUDINAL SECTIONS.

On the suction stroke of the piston the admission valve 18, Fig. 89, opens and allows the charge to pass, by way of port 19, into the cylinder. On the next stroke it is compressed into the combustion chamber 20, and port space 19, and fired by an electric spark between the points of the

Compression and ignition

127

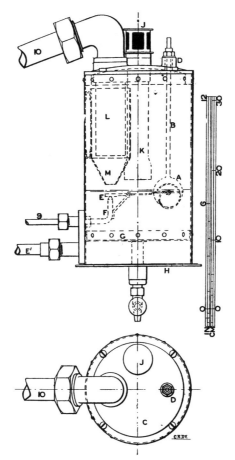

Fig. 91.—Benz Carburettor.

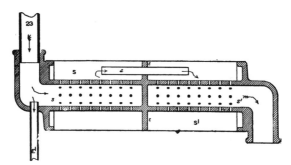

Fig. 96.—Benz Exhaust Silencer and Branch to Petrol Tank.

128

ignition plug 21, Figs. 89, 92 and 93, the point at which the charge is fired being determined by the setting of the contact maker κ, Figs. 85, 87, 88 and 92. On the completion of the working stroke, the exhaust valve 22, Fig. 89, is lifted, and the products of combustion escape by way of pipe 23 to the exhaust silencer s, Figs. 84, 86, and thence into the atmosphere. The silencer is shown in section in Fig. 96. The exhaust enters by pipe 23 into pipe s, and thence by numerous holes into chamber s. From this it passes by pipe a into chamber s', and then to the exit by passing through a large number of small holes.

The exhaust valve is operated by means of a double cam 24, Fig. 88, which, with the electric contact disc 25, is carried on a stud shaft screwed into the boss 27, on the motor casting 5, Figs. 85 and 88, the cam and disc being geared 1 to 2 with the crankshaft, by means of the spur wheel and pinion 26 and 28. **Exhaust valve of cam**

On the cam 24 is a large cam a, Figs. 87 and 88, which operates the exhaust valve under ordinary conditions of running; and a smaller cam b, for the purpose of making it easier to start the motor by lifting the exhaust valve and letting a small portion of the compressed charge escape just before the ignition point is reached. This is effected by moving the small hand lever c', Figs. 87 and 88, in the direction towards the crankshaft, and so pushing the sliding spindle which carries the roller c inwards, and causing it to occupy a position at which it may be operated by both the cams a and b. As it is shown in Fig 88, the exhaust valve will be lifted as in ordinary working once every other revolution, the large cam a actuating the bent lever 29, through the roller c, at the end of its short arm. The motion is communicated by the link 30 to the arm d, Figs. 87 and 88, on the sleeve 31, which is carried on a stud fixed to a bracket on the cylinder, as shown by Fig. 89. On the sleeve 31 is a tongue e, which bears against a hardened steel piece let into the free vibrating lever 32, Figs. 87 and 88, pivoted at e'; and on the upper face of this is another hardened piece, against which the end of the exhaust valve rod is pressed by the spring 33. The motion imparted by the cam to the sleeve 31 causes the tongue e to lift the lever 32, and with it the exhaust valve 22. The exhaust valve is closed by the spring 33, which, acting through the link work, also keeps the roller c pressed against the cam piece 24. **Compression release**

The electric ignition contact device is shown in Figs. 85, 87, 88 and 92; the disc 25, which revolves with the cam 24, is made of vulcanite, and has fixed on a portion of its circumference the metal piece f Fig. 92, which is electrically connected to the stud on which the cam revolves. The lever κ, by means of which the point at which ignition takes place is varied, is also made of vulcanite, and is carried by the metal boss g, which works on the end of the stud and is frictionally held by the nut 34, Figs. 87 and 88. Near the end of lever κ is a projection 35, which carries the flexible metal contact arm 36, the free end of which bears on the circumference of the disc 25. The electrical connections are shown by Fig. 92. **Electric ignition apparatus**

The accumulators A A and induction coil B are carried in the box below the seats of the carriage. The free positive pole of the cells is connected by wire *j* to one end of the primary circuit of the induction coil, the other end being connected by wire K to the binding screw at the end of the stud on which the cam 24 and disc 25 revolve, and consequently is electrically connected to the metal piece *f*, Fig. 92, on the circumference of disc 25. Once every two revolutions of the crank shaft of the motor the metal

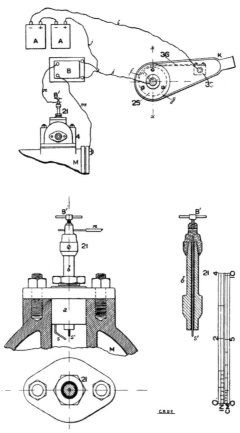

FIG. 92.—DIAGRAM OF CONNECTIONS OF ELECTRIC IGNITION APPARATUS, BENZ MOTOR.
FIG. 93.—IGNITION IN POSITION AND IN SECTION.

Advance and delay of ignition piece *f* comes into contact with the spring arm 36, which is connected to the free positive pole of the accumulators by the wire *l*, so that as long as the plate *f* and arm 36 are in contact the primary circuit of the induction coil is completed and the current flows round it inducing a high potential current in the secondary circuit in the coil, one pole of which is connected by wire *m* to the motor casting, which is electrically connected to the sparking point *s*, Figs. 89 and 93, by contact of the holder piece *a*, Fig. 93;

130

and the other pole is connected by wire n to the binding screw B', Figs. 92 and 93, at the end of the insulating china part b of the plug through which passes the contact wire s'. The high potential spark passing between the points s and s', fires the charge in the motor cylinder. It will be seen that the contact piece f on the disc 25 moves in a fixed relation with the position of the piston in the motor cylinder, and it is so placed as to be centrally on the vertical line xx, Fig. 92, when the piston is at the inner end of its stroke, and if the arm K is so placed that the end of the flexible metal piece 36 comes into contact with f, when this is in the vertical position, the charge will be fired exactly on the inner dead centre; but if the arm K is rotated either up or down, the charge will be fired before or after the dead centre, more or less according to the amount by which the arm is displaced, se that the best point for ignition may be fixed according to the work which the motor is required to perform. That is to say, either advance or retarded ignition may be obtained at will.

As will be seen in Figs. 87, 88 and 90, there is a valve 37 of similar construction to the admission valve 18 in the cover at the back of the cylinder. This valve works automatically in conjunction with the throttle valve v previously described, by means of which the quantity of the charge passing into the cylinder is varied. In the earlier Benz motors the valve 37 was not used, with the result that as the passage past the throttle valve was reduced the duration of the suction stroke was insufficient to enable a full charge to be drawn in past the contracted aperture, and consequently the compression stroke began with a partial vacuum in the cylinder. To obviate this the valve 37 was added. It opens against the resistance of a spring, which is a little stronger than that controlling the admission valve 18, so that it only opens when the quantity of the charge drawn in past the admission valve is insufficient to fill the cylinder at the pressure which that valve determines. It then opens and admits pure air sucked in through pipe J, until the pressure in the cylinder is that determined by spring 37, or not much below that of the atmosphere.

Supplementary air valve

There is no governor on the Benz motor, the speed being controlled by the point or period of ignition, and for ordinary running by means of the throttle valve, which enables the driver to vary the speed of the motor by varying the quantity of the mixture, from about 250 revolutions to 900 revolutions per minute, without leaving his seat.

Governing

The water-jacketing of the cylinder and valve chamber is seen in section in Figs. 89 and 90. The cooling water is carried in the tanks w, Figs. 84, 85 and 86, one at each side of the body. They are connected below by the cross pipe 38, in which is a branch pipe 39, Figs. 86 and 89, connecting it to the bottom of the cylinder jacket, and also a branch pipe 40 connecting it to the lower part of the vessel 41 on the cylinder over the combustion chamber. From the upper part of the vessel 41 above the maximum water level, is a pipe 42 connecting it to the condenser 43, Figs. 84, 86 and 90, which consists of a closed cylinder through which passes the pipe 44,

Jacket-cooling water

131

which is open to the air at both ends. The condenser thus consists of an annular chamber, the inner cooling surface of which is formed of the tube 44, through which air is forced to flow by the diagonal form given to the ends in contrary directions. From the bottom of this condenser a pipe 45 is connected to the top of tank w. The pressure in the condenser is prevented from rising above that of the atmosphere by means of pipe 46, which passes down below the motor and is open at its lower end. The water tanks w are also connected by pipe 47, Fig. 85. It appears to be very doubtful as to whether any true circulation of the water is set up in this complicated system of pipes and vessels, what happens when the motor is working being probably as follows. On starting the motor the tanks w will be nearly full, the water being cold and standing at about the level shown in vessel 41, Figs. 89 and 90. After the motor has been working for a short time, the water in the jackets and vessel 41 will begin to boil, and the steam rising up the pipe 42 into the condenser 43 is partly condensed, and runs back by way of pipe 45 into tank w, and the rest escapes into the atmosphere by pipe 46. This action will continue until a considerable part of the water is boiled away, when the tanks w must again be filled. Although the precise method of the circulation is doubtful, it is obvious that there is one place where the water is much more rapidly heated than it is at any other place, and Mr. Benz has taken care to provide so many passages that there must be freedom of movement for the water in whatever direction it happens to be acting. The tanks are filled through the aperture in tank w' closed by the screw cap 48, the level of water in both tanks being indicated by the gauge glass 49 at the back end of tank w. When required, the water may be emptied from the whole circulating system by opening the cock 50, Figs. 87, 89 and 90. The cock 51, Figs. 87 and 90, below the combustion chamber, is for the purpose of relieving the pressure in the cylinder when it is required to move the piston freely by hand.

The lubricating oil for the cylinder is supplied by the sight feed lubricator 52 near the front end of the piston travel. Large lubricators are also fitted on the crankshaft bearings and connecting rod head.

The crankshaft of the motor is extended to one side, and on it are keyed the flywheel F' and the belt pulleys 53, 54, Figs. 85, 86 and 88, the overhanging weight being supported by the ball bearing 55, which is carried from the skeleton frame.

The belt pulleys 53, 54, drive by means of crossed belts R and R' on to the fast and loose pulleys 56, 56a, 57, 57a, on the cross countershaft N, which runs in the three ball bearings 58, 59, 60, and which carries the differential gear and, at its outer ends, the sprocket pinions T, which drive by means of pitch chains on to the sprocket wheels U.

By means of the belt driving gear two speeds may be given to the countershaft N, the high speed by shifting belt R from the loose pulley 56a on to the fast pulley 56, and the slow speed by moving belt R' from

Water connections

Circulation

Belt transmission gear

Speed change gear

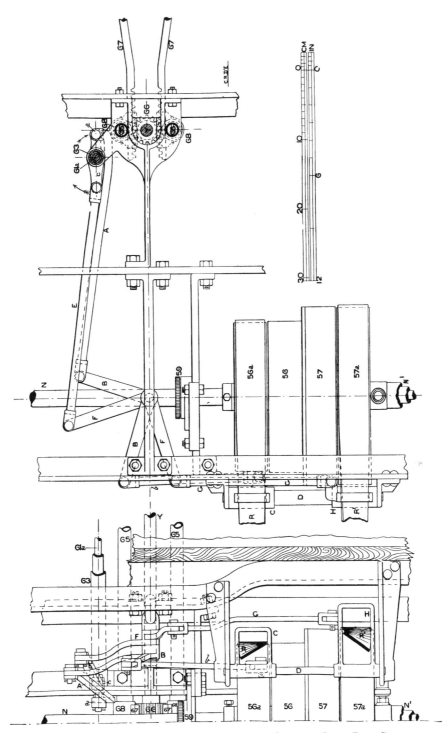

FIGS. 94 & 95.—BELT SPEED CHANGE GEAR AND STEERING GEAR, BENZ CAR.

133

the loose pulley 57a to the fast pulley 57. These movements are controlled by the handles 61, 62, Figs. 84 and 85, placed near the steering hand lever z. The high-speed belt R is controlled by the handle 61, which operates a vertical spindle 61a, Figs. 84 and 94, at the lower end of which is an arm a, connected by a rod A to one arm of the bell crank B, the other arm of which is connected by a link b, to the belt shifter c, within which runs the upper half of the high-speed belt R. When the handle 61 is moved to the left away from the driver, the arm a, moving in the direction shown by the arrow, Fig. 94, causes the belt shifter c to slide along the guide bar D and the belt is moved from the loose pulley 56a on to the tight pulley 56. When the slow speed is required the belt R must first be brought back on to the loose pulley 56a by moving the handle 61 to its original position, Fig. 94, then the handle 62 must be moved outwards causing the arm c, at the lower end of the tubular spindle 63, to move in the direction of the arrow, Fig. 94, the motion being imparted by the rod E to the bell crank F, and from it by rod G to the belt shifter H, which controls the slow speed belt R', causing the belt to be moved from the loose pulley 57a on to the tight pulley 57.

Differential gear The differential gear on the countershaft N is arranged inside the belt pulleys as shown by Fig. 96. The pinions a are free to revolve on spindles fixed in bosses in the belt pulley 57. They gear with bevel wheels b and c, of which b is keyed to the solid part of the countershaft N, and c is keyed to the tubular part N^1, which carries at its outer end the sprocket pinion T^1, and is free to revolve independently of N on the projection N^2, which is of reduced diameter and tapers slightly towards its outer end. The bevel wheels of the differential gear are of bronze. The loose pulleys 56a, 57a, are of slightly less diameter than the tight pulleys 56, 57, in order to reduce the tension on the belts when not working.

Steering gear The steering is controlled by the hand lever z, which, with the pointer z, is moved over the ring 64, Figs. 84 and 85, mounted at the top of the fixed tubular standards 65; see also Figs. 94 and 95, between which works the vertical shaft Y, Fig. 94 of the steering gear. At the bottom of the shaft Y is keyed a pinion 66 gearing, with racks at the ends of the bars 67, which move in guides 68 at the bottom of the standards 65. The other ends of the bars 67 are connected to the crosshead 69, Figs. 84 and 85, mounted on a short stud spindle working in a bearing in the crossbar 70, which ends near the spring bridles, and is fixed to the ends of the forward scrolls. This spindle is connected to a similar one below, by means of the double plate spring 71. The lower spindle works in a bearing attached to the front axle, and carries a **V**-shaped arm 72, connected by links 73, which are in compression to the arms 74 on the vertical part of the short steering wheels. By moving the steering lever z over the ring 64, the pinion 66 gearing with the racks at the end of bars 67, pulls the one and pushes the other, the motion being transmitted, through the spring 71 and links above described, to the wheel pivots. It will be noticed that the pointer z moves in the direction which the carriage will take. The road wheels are constructed with direct steel spokes and metal rims with solid rubber tyres.

There are four brakes. The two spoon brakes 75, which may be pressed on the tyres of the drawing wheels by means of the hand lever x, and the band brakes 76 acting on drums bolted to the arms of the sprocket wheel o. These are operated by the pedal 77, by depressing which the rod 78 is pulled, causing the lever 79, Figs. 84 and 86, to move forward. This rotates the crossbar 80, Fig. 80, and causes the short arms 81, one at each

Brakes

FIG. 97.—BENZ "IDEAL" TWO-SEAT MOTOR CARRIAGE, 1898–99.

end, to move in the opposite direction, tightening the bands, one end of each of which is held by the arm 81, while the other is free to move between collars on the end of the crossbar 80.

The whole of the foregoing description of the Benz single-cylinder motor vehicle applies to the double-cylinder motor vehicle, except of course as far as readily understood modifications are required, such as in the duplicated carburettor and silencer, and water attachments and connections are necessary.

Benz double-cylinder motors

The general appearance, arrangement, and type of the Benz carriages, the construction and working of which have been described, are shown by Figs. 97 and 98.

Types of Benz cars

Fig. 97 is from a photograph of the "Ideal" single-cylinder motor carriage, and Fig. 98 shows the carriage of the four-seated dog-cart pattern with double-cylinder motor.

A very large number of the single-cylinder motor carriages have been made, and have proved on the whole very popular, more especially since various small but important improvements have been made in the strength of connections and steering gear in the electric ignition mechanism, in the method of fixing the sprocket wheels to the driving wheels, in bearings and fixtures of the transverse chain pinion shaft. They are not intended to be high-speed heavy-work cars, and although at slow speeds and at starting the single-cylinder motor gives a pulsatory movement to the car,

FIG. 98.—BENZ DOUBLE-CYLINDER MOTOR FOUR-SEAT CARRIAGE, 1899.

the running is quite comfortable at higher speeds. The Benz cars are moderate in price, and with the most recent improvements provide, when properly cared for and maintained, satisfactorily for the requirements of ordinary users.

The recent Benz racing car of 14 HP. is illustrated in perspective by the engraving Fig. 98A from a photograph. It is propelled by a double cylinder motor of the Benz pattern already illustrated, and combined belt and spur gear and chain transmission, giving four speeds, the maximum of which is stated to be from 34 to 37 miles per hour. It took the first prize in the Berlin-Leipsig race of 115 miles in September, 1899.

Fig. 98a.—Benz 14 HP. Racing Car, 1899. (*See p.* 186.)

Chapter VIII

RECENT HIGH SPEED LONG DISTANCE PETROL MOTOR VEHICLES

THE vehicles illustrated and described in the last chapters belong to a period covering at least eight years. The master mind having conceived the parent ideas, and invented the parts and combination of parts for putting those ideas into practice, improvements follow as a natural result of use. If the originator becomes a user and is in a position to follow up his first creation, he may make the first and most important improvements, and so secure to himself some of the fruits of his labours of origination, which cannot be secured under the patent laws.

In this way, Messrs. Panhard & Levassor and Messrs. Peugeot have to a very great extent retained the lead they took with the vehicles they ran in the races of 1894–95. The precise arrangement of their machinery had not until recently been copied by others than the licencees under the Daimler motor patents, but the general construction and experience gained with these vehicles of the originating firms, has led to the birth of many motor vehicle manufacturing firms.

The two firms mentioned have embodied their great experience in their most recent carriages, and in particular they have paid most minute attention to the cause of every defect and to their removal by higher or more suitable class of materials and workmanship, and by devices for securing greater security and continuity of action.

It will not be convenient to adhere further to any chronological order of presentation of different makes of vehicles, as the dates of vehicles now to be dealt with overlap.

It will be noticed that with all the more recent vehicles the small leading wheels are abandoned, in favour of wheels more nearly the size of the drivers. This is a most desirable modification, and is a return to the practice of Messrs. Peugeot of 1895. There is no reason when the Ackerman steering axles are used, for adopting the small front wheels, which were almost a necessity in horse-drawn vehicles. On the other hand, there are the strongest reasons for using even larger wheels in front

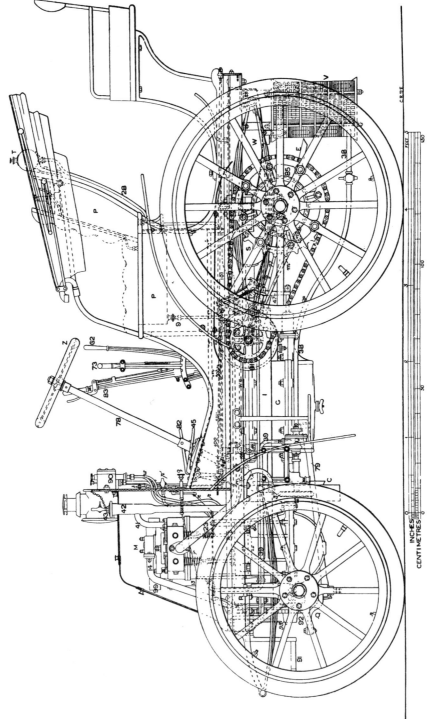

Fig. 99.—Panhard & Levassor Daimler Motor Carriage, 6-H.P. Racing Type.

138

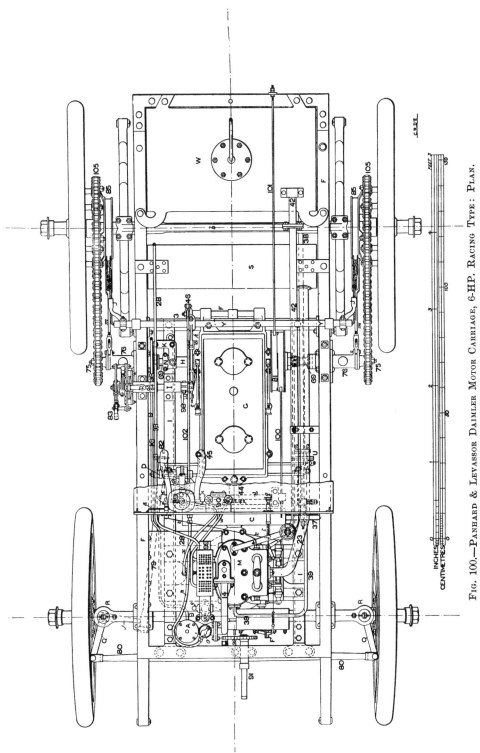

Fig. 100.—Panhard & Levassor Daimler Motor Carriage, 6-HP. Racing Type: Plan.

139

than can generally be used as drivers. The smaller the wheel the more violent the concussion in passing over any obstruction, or over what is more common on our badly-maintained roads, the succession of saucer- and dish-shaped holes that result from traffic over roads full of unnoticed slight defects which soon develop. As then it is essential that steering wheels should run as steadily as possible, they should be large, and not after the manner of the old pony chaise or other horse vehicles, made small so as to lock under the body.

Panhard & Levassor Racing Carriage.

A Panhard & Levassor 6-HP. racing carriage of the 1898 type is illustrated in general arrangement and in detail by Figs. 99 to 115. It is almost identical with that of the more powerful 1899 racers, except that in these four-cylinder motors are used. The mechanical parts of this car are almost the same as those of the standard underframe and gear of the English Daimler Co., which is itself a copy, with slight modifications, of an earlier vehicle, manufactured by Messrs. Panhard & Levassor, similar to that shown by Fig. 56. The engravings, Figs. 99 and 100, show a side elevation and plan of the car. Having been designed primarily for the purpose of speed, it has all its parts constructed in the lightest possible manner consistent with sufficient strength to withstand the wear and tear of long runs at high speeds. It is intended to carry two persons, but a light seat for two more can be attached at the back, as shown in Fig. 99. For four-cylinder 12-HP. motor, see pp. 104–5 and frontispiece.

Underframe The underframe consists of a rectangular structure of wood similar in form to that of the English Daimler; but the channel iron encasing the latter has been dispensed with, careful design having made it possible to carry the whole weight of the car on a plain frame of wood only 5 cm. = 2 in. nearly, square, seen in section at F in Figs. 101 and 103, thus saving considerably in weight and securing an elastic flexibility, which can only be obtained in metal frames by special mechanical contrivances. The motor M and gear-case G are rigidly attached to the two longitudinal angle-iron bars 1, 1, which are suspended in front by the hangers 2 (see Figs. 99 and 102), and bolted to the cross angle-iron bar 3 at the back. The whole is carried from the axles by two single-plate springs in front, with long cantilever hangers in front and two double-plate springs at the back.

The Motor Phœnix-Daimler The motor M (see Figs. 101–111) is an improved form of that described in connection with the English Daimler; it has two parallel vertical cylinders, each $3\frac{19}{32}$ in. or 3·59 in. = 91 mm. bore by 5 in. = 127 mm. stroke, and develops 6 HP. at 700 revolutions per minute. The accompanying drawings have been marked as far as possible with the same figures and letters as used for the English Daimler, illustrated by and described with reference to Figs. 34, 60, etc. The cylinder heads 5 and bodies 6 are of cast iron, and the crank case 7 and carrying arms 8, of cast aluminium. The petrol tanks P, Fig. 99, are carried in the body of the vehicle below,

140

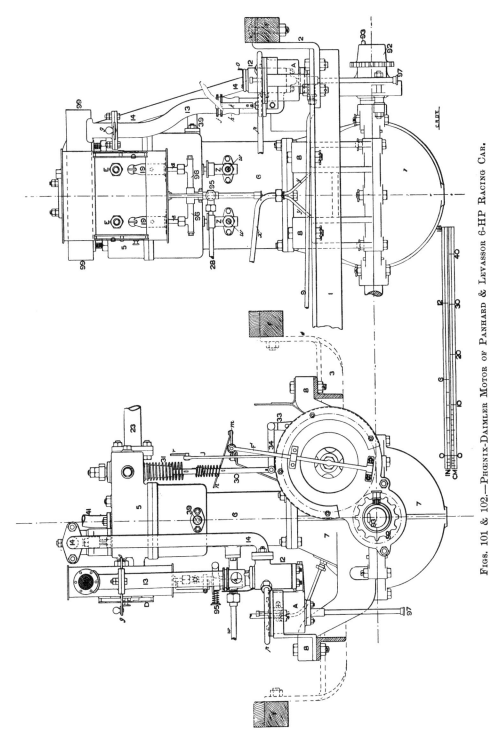

Figs. 101 & 102.—Phenix-Daimler Motor of Panhard & Levassor 6-HP Racing Car.

141

Carburettor and at the back of the front seats; the petrol flows by gravity through the pipe 9, Figs. 99 and 104, to the float vessel A of the carburettor, which is shown in detail in Figs. 104 and 105, its principal parts being of cast aluminium. The float vessel A contains a light float, which moves freely up and down on the steel needle valve spindle a. Upon the float rest the weighted ends of the levers b, pivoted between projections c, their inner ends resting in the grooved collar on the needle a. As shown in Fig. 104 the float-vessel is sufficiently full of petrol. The float, in endeavouring to rise, by further petrol admission, lifts the levers b, causing their inner ends to press down the needle, and in this way closing the inlet d. As the level of the petrol in the vessel A falls, the float sinks also, and with it the weighted ends of levers b, so that the needle valve is lifted, and more petrol flows into the chamber e from the tank P, until the float vessel is again nearly full, as shown in Fig. 104. At each suction stroke of the motor a small quantity of petrol is sucked into the carburettor n, m, 12 through the nozzle f.

Air supply The air supply is drawn in part from pipe 99 above the ignition tube burners, Figs. 99, 100, 101, 102 and 106, where it is dried and heated, the quantity taken from this pipe being regulated by the slide g, Figs. 101 and 102, near the top of the air admission pipe 13; at the bottom of this pipe, near the carburettor, is also another air-regulating aperture h, Figs. 101, 104 and 105, which can be opened more or less by means of the revolving regulator j, held by the spring catch k. The air passing into the pipe 13 **Car-** enters the carburettor at l outside the downwardly projecting bush m, **buretting** which forms the annular chamber n, and sweeping up round the nozzle f as shown by arrows, Fig. 103, vaporizes the petrol, the mixed air and oil vapour passing out by way of pipe 14 into the admission chamber 15 of the motor, Figs. 106 and 108. Besides the slide g, Figs. 101 and 102, and revolving regulator j with spring k, Figs. 104 and 105, there are two apertures for admitting air to weaken the mixture. These are the hole in the screw cover of the carburettor, Fig. 104, covered more or less by the slide o, and the pipe p, Figs. 100 and 102, the opening of which is regulated by the cock p' on the dashboard in front of the driver, Fig. 103.

Admission The arrangement of the valves, etc., for one cylinder, is shown by **and exhaust** Figs. 106 to 109, that of the other being exactly similar. On the suction **valves** stroke the admission valve 16, Figs. 106 and 108, opens against the resistance of the spring above it, and the charge passes into the cylinder by way of the port below. On the next stroke the charge is compressed into the combustion chamber 17, and port spaces opening into it, and fired by **Ignition** the ignition tube E. After the working stroke is completed, the products of combustion are expelled by way of port 18, past the exhaust valve 20, and through the passage 22, Figs. 106 and 109, to the exhaust pipe 23, which is connected to the silencer s, Figs. 99 and 100. The platinum ignition tubes E, Figs. 102 and 106, are heated by the burners 19, enclosed in a sheet-iron box. These are fed from a separate small tank T, at the top

142

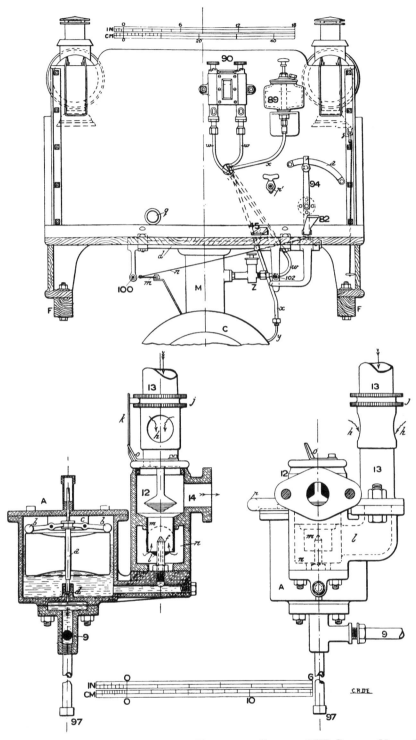

Fig. 103.—Inside View of Dashboard and Fittings of Panhard 6-HP. Daimler Motor Car.
Figs. 104 & 105.—Float-Feed Spray Carburettor, Panhard 6-HP. Carriage.

143

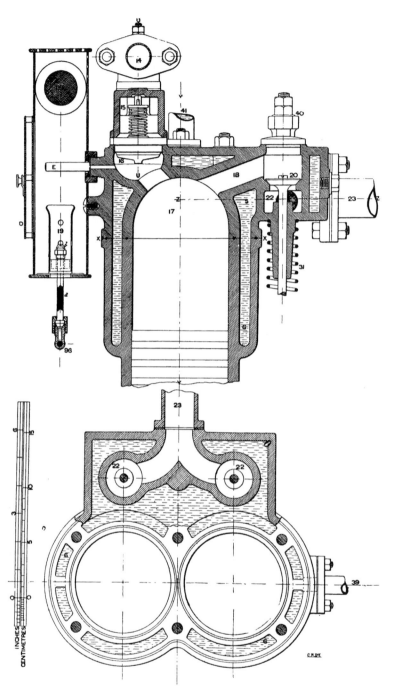

Fig. 106.—Vertical Section through Cylinder and Valves of Phœnix Daimler Motor of
Panhard 6-HP Carriage, showing Water Jacket, Valves, and Lamp.

Fig. 107.—Transverse Horizontal Section on Lines, X X and Z Z, Fig. 102, of Phœnix
Daimler Motor, showing Water Jacket and Exhaust Passage under Valves.

144

of the back of the driver's seat, Fig. 99. The petrol flows along the pipe
28, Fig. 100, past the regulating valve 95, Fig. 102, and divides along the
pipe 96, one end of which is seen in section at centre of burner in Fig. 106,
from here it flows up the pipe *a* (see also Fig. 102), containing a wick of
cotton wool or asbestos wrapped in wire gauze, and when the burner has
been heated escapes in the form of gas from the nozzle *b*, air being drawn
in through holes in the outer tube 19, at the mouth of which the mixed
gas and air burn with a very hot and almost invisible blue flame.

Contrary to the arrangement of the English Daimler motor, the
pistons of both cylinders work on one crank, the cycles following one
another thus :—

Cylinder 1.—Suction.	Cylinder 2.—Explosion.
Compression.	Exhaust.
Explosion.	Suction.
Exhaust.	Compression.
Suction.	Explosion.

so that there is one explosion every revolution, which gives a more equal
turning moment. The more recently constructed English Daimler motors
follow the French practice in the above respect. The exhaust valves 20,
one of which is seen in Figs. 106 and 109, which are made of steel, are
worked by cams *a a* on the shaft 21. Fig. 111, at 180° (the letters *a a* in
Fig. 111 are somewhat like a letter *d*). The shaft 21 is geared 1 to 2 with
the crank shaft by the machine-cut bronze spur wheels 35 and 36, enclosed
in the brass case B, Fig. 111. The cams run against rollers *s s'* in the ends
of levers 29 and 29A, Figs. 110 and 111, which are pivoted at *p p*, and held
down on the cams by the springs B, B'. The stepped ends c, Fig. 110, of
these levers engage with the flattened ends of the rods 30, 30A by means
of which the exhaust valves are lifted against the pressure of springs 31,
Figs. 106 and 109. The speed of the engine is governed by causing the
rods 30 and 30A to miss the steps *c*, Fig. 110, at the ends of levers 29 and
29A, so leaving the valves closed and thus confining the burnt gases in the
cylinders. This is effected by means of the governor, which consists of the
egg-shaped weights 32, Figs. 110 and 111, pivoted at *d*, on the inner face
of the gear wheel 36, and held together by the springs E. Each ball has
an arm *e*, the end of which is rounded and free to move in a parallel slide
at *e'*, Fig. 111; one on each side of the sleeve *f*, which is free to slide
along the shaft 21. At the opposite (inner) end of the sleeve is a
cylindrical part *g*, against which is pressed the end of the lever *h*, which is
mounted at one end of the short spindle, carried by the bearing c. At its
other end the spindle c carries the vertical arm 33, upon which the spring
b acts. So long as the speed of the engine does not exceed 700 revolutions
per minute, the lever *h* bears against the cylindrical part *g* of the sleeve *f*;
but when this speed is exceeded, the governor balls, flying further apart,
cause the arms *e* to move the sleeve *f* inwards, and the end of lever *h* runs

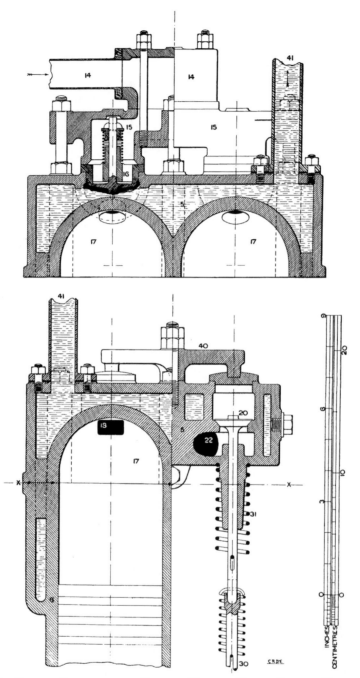

FIG. 108.—VERTICAL SECTION THROUGH CYLINDER HEADS OF PHŒNIX-DAIMLER MOTOR, TAKEN AT RIGHT ANGLES TO THAT OF FIG. 102, ON LINES U U AND Y Y.

FIG. 109.—VERTICAL SECTION THROUGH CYLINDER AND EXHAUST VALVE CHAMBER AT RIGHT ANGLES TO FIG. 102, AND WITH ONE CYLINDER REMOVED TO SHOW SECTION OF EXHAUST VALVE CHAMBER AND PASSAGE IN ANOTHER PLANE.

on the cam part *j*, which pushes it outwards once every revolution. This motion is imparted to the spindle working in the bearing c, and causes the vertical arm 33 to move inwards towards the cylinder, with the result that the valve rod 30 connected to it by the link 34, Fig. 110, also moves inwards, and the step c on the end of the lever 29 is missed by the end of the rod 30, and the valve is not fitted. If the speed of the motor still continues to increase, the balls fly further apart and the sliding sleeve *f* is pushed inwards until the end of the lever arm *h* runs on to the larger cam *k*, Fig. 111, giving the lever *h* a greater movement, so that the arm 33, moves sufficiently far inwards to cause the end of the slot *l* at its upper end to press against the joint pin of the link 34*a*, so that the valve rod 30*a*, is also moved inwards and misses the step c on the lifting lever 29*a*, thus leaving the second exhaust valve closed. (In the parts marked 29*a*, 30*a*, and 34*a*, the letter *a* has been made to appear like a letter *d* with a short vertical number. The difference between the *a* and *d*, by the same hand will be seen by reference to the *d* at the pivots of the governor arms, Fig. 107.) When the motor slows down again, rod 30*a* first falls back into position and then rod 30.

There are two arrangements for controlling the action of the governor from the driver's seat. For the purpose of "slowing down" the motor, without causing it to stop, the lever 94, mounted at the rear of the dash- board, Fig. 103, which in normal working conditions is clipped on the sector *a*, near its outer (right-hand, facing dashboard) end, is moved to a more or less nearly vertical position, as shown, according to the speed at which the motor is required to run. This lever has an extension to the front of the dashboard, and to it is connected the wire *n*, by which the bell crank *m*, seen in plan in Fig. 100, also in Figs. 110 and 111, is pulled, the motion being transmitted by a wire and spring connector *o*, Fig. 111, from the other arm of the bell crank to the lever F', Figs. 100, 101 and 111, which is pulled inwards against the pressure of a flat spring. A boss G on the lever F', Fig. 111, bears against the end of a steel pin, which passes freely through a hole in the cam spindle 21. Through the cam sleeve *f*, and a slot in the cam spindle, there is a key or pin; the slot being long enough to allow the key and the sleeve to slide longitudinally on the shaft. Against this key the steel pin pressed by the lever F', at G, presses, so that the inward movement of the lever F' pushes the sleeve *f* in the same direction, causing the end of lever H to run on the cam *j* or *k*, and so preventing the exhaust valves from being lifted as described in connection with the governor. When the pull on the wire *n* is released, by moving the control lever 94 on the dashboard to the outer end of the sector, as shown in Fig. 103, the springs on the governor balls cause the arms *e*, Fig. 111, to again move the sleeve *f* outwards, and the end of the lever *h* runs on the cylindrical end of the sleeve, as shown in Fig. 111.

The other governor control is known as the accelerator, it is worked from the knob *q* of Figs. 99, 100 and 103, on the dashboard. This operates the rod r, by pulling which against the pressure of spring t, the upper end of the vertical lever J, Fig. 110, pivoted at K, is pulled inwards, causing the lower forked end, which is faced with hard steel, to bear against the inner end of the sliding sleeve f, so preventing the governor from acting and allowing the motor to run uncontrolled up to its terminal velocity under the conditions, and to reach as much as 1200 revolutions per minute. The motor may thus be kept in gear for running downhill at high speeds.

Water jacket and cooling water

The cooling water for the cylinder jackets is carried in the tank w, Figs. 99 and 100, at the back of the vehicle. It is filled through the pipe 42, the top of which is closed by the screw cap 43. The water flows from the tank by pipe v, Fig. 99, round the cooler v, which consists of a considerable length of pipe, about 55 ft. of 0·75 in. diameter, covered with thin aluminium ribs, between which the air rushes when the car is in motion, thus carrying away a considerable part of the heat absorbed from the cylinders. The water emerges at the bottom of the cooler and flows by gravity along the

Circulating pump

pipe 38, to the small rotary pump u, Fig. 99, which is driven off the clutch fly wheel c, by the rubber-tyred friction wheel 37, which is held against the fly wheel by spring pressure. From the pump the water is forced up the pipe 39, Figs. 99 and 100, and enters the lower part of the jacket on the front cylinder. The water surrounds the cylinders and head, as shown by Figs. 106–109, and emerges from the head over the back cylinder by pipe 41, which joins the filling pipe 42, through which it flows to the tank w. The stream of water up pipe 41 and down pipe 42 may be seen by removing cap 43, when the pump is working.

Lubrication

The lubricating oil for the cylinders is contained in the lubricator 90 on the dashboard, Figs. 99, 100 and 103, and flows to the cylinders by pipes w seen in the same engravings, entering the cylinders' bodies at the connections z, Figs. 102 and 103, which also carry the kerosene cups, from which a small quantity of kerosene or ordinary lamp oil can be forced into the cylinders from time to time for cleaning and lubricating purposes. The crank-case lubricator 89, seen in plan in Fig. 100 and in Fig. 103, is placed by the side of that for the cylinders, and feeds through pipe x, the oil being directed into the crossheads by means of the branch pipe y, seen in Fig. 102.

Transmission gear

The engine drives the clutch shaft 44, Figs. 100 and 112, through a friction clutch c, similar to that described in connection with the English Daimler, except that the inner sliding cone 52 has been dispensed with owing to the lightness of the car. The speed gears contained in the gear case G, which is of aluminium, are arranged exactly as shown in Figs. 65, 66, of the English Daimler, the cross countershaft 69 being also similarly constructed. The controlling levers are however arranged in an entirely

Control gear

different manner at the driver's right hand, as shown in Figs. 99, 100 and 112. The innermost lever 73 controls the reversing gear; it may be held in any one of three notches seen in Fig. 99, corresponding to the forward direction of motion, out of gear and reverse. By means of a short tubular shaft in

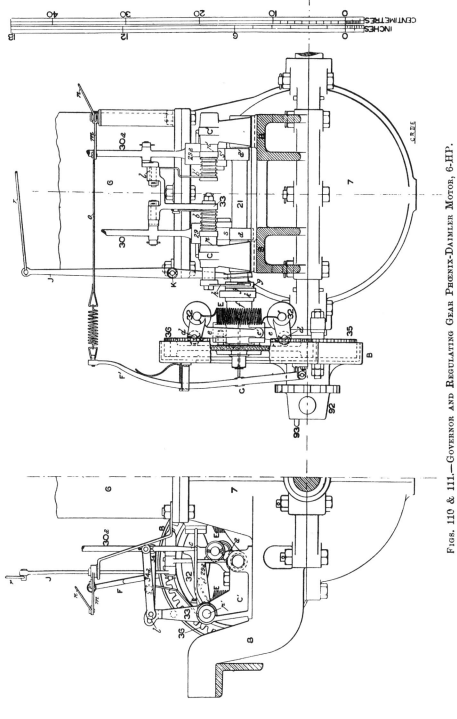

FIGS. 110 & 111.—GOVERNOR AND REGULATING GEAR PHŒNIX-DAIMLER MOTOR, 6-HP.

149

the bearing 63, Fig. 100, it operates the downwardly projecting lever 64, Fig. 112, connected by a bridle to one arm of the bell crank K, Figs. 100 and 112.

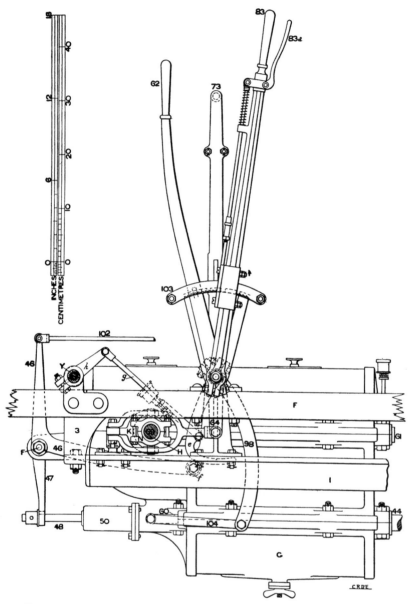

FIG. 112.—SPEED CHANGING, REVERSING AND CLUTCH LEVERS AND CONNECTIONS. PANHARD 6-HP. DAIMLER MOTOR CARRIAGE.

The other arm of the bell crank operates a crank, the ends of which work on pins in the split collar N, Fig. 100, which grips a boss on the sliding sleeve H, carrying the bevel wheels and differential gear on the cross-shaft 69, as described for the English Daimler, so that either of the bevels may be geared with that on the end of the countershaft 61, and held in place by means of the collar N, and links and levers connecting it to the hand lever 73. The second lever 62, Figs. 99, 100 and 112, is for shifting the speed change gears ; it is held in one of four notches on the outer side of the sector 103, seen in the plan, Fig. 100, and in Fig. 112, by a small projection on its inner side. It is connected to a hollow shaft, passing through the reversing lever shaft and the bearing I, Fig. 100, and carrying on its inner end the curved lever 98, Figs. 100 and 112, connected to the sliding shaft at 60, Fig. 112 (see also the English Daimler car), which carries the fork for moving the spur pinions on the clutch shaft 44, by the short cross link 104, as already described with reference to the English Daimler carriages.

Speed changing

It will be seen that the various parts of the reversing and speed-changing controlling gear are much lighter than those on the English-made Daimler car, but in general the reduction in weight is judiciously made so that necessary strength is not sacrificed.

The power is transmitted from the countershaft 69, Fig. 100, to the road wheels by means of sprocket pinions and pitch chains, to sprocket-wheel rings on the driving or road wheels, Figs. 112 A, shown to a scale of one-fourth full size, the tongue and grove clutch by which the counter-shaft 69 transmits motion to the short sprocket pinion spindles 75. Reference has already been made to the method of construction when describing the English Daimler cars which are fitted with the same device.

Trans-mission to road wheels

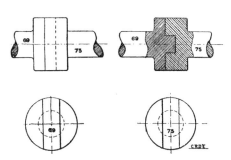

FIG. 112A.—COUNTERSHAFT AND CHAIN PINION, SHAFT TONGUE AND GROOVE CONNECTION OF PANHARD & LEVASSOR CARRIAGES.

It makes it easy to get the central part of the countershaft out for the inspection or renewal of the gear upon it, without taking out the short spindles 75 or removing their bearings. I have commented else-where upon the use of these short chain pinion spindles depending entirely upon the lateral stiffness of the frame sides for the maintenance of their position. They would seem to run satisfactorily in spite of adverse expectations, but the fact that chains leave the wheels sometimes without explanation may be due to springing of all the parts which this broken shaft arrangement facili-tates. The sprocket-wheel rings are held by only four studs to four spokes of the drivers, the spokes being swelled where the studs pass through them and are fixed by one nut each. The rings stand some $2\frac{1}{2}$ in. from the spokes, so that there must sometimes be a considerable twisting stress as well as transverse stress upon the spokes. Such a means of attachment offends

Chain adjustment

an engineer's notions of the requirements, but these things do not give way, probably due to the accommodating elastic flexibility of the spokes.

The slack in the chains is taken up by the tighteners 77 (see Figs. 99 and 115). It will seen that these connect the back axle to the fixed bearings 76, Fig. 115, which is a part vertical section to a larger scale, looking from the near to the off side of Fig. 99. The bearings 76 carry the short sprocket, pinion spindles 75. Radial distance between these and the axle is maintained, with freedom for the latter to move radially in a vertical plane, by the bars 77.

To the ends of the axle, outside the point of attachment of these adjustable radius-bar chain tighteners, are bolted the lower halves of the double-plate springs carrying the rear part of the vehicle. The upper halves of the springs are bolted to the broadened flat ends of the cranked cross bar b, Fig. 115, which is free to move longitudinally in the guides c, attached to the woodwork of the under-frame, thus allowing the back axle to be moved and the chains tightened as required by means of the right and left hand adjusting nuts in the radial tightening bars 77.

Road wheels The road wheels are of very light construction with wooden spokes and rims and $2\frac{5}{8}$ in. Michelin pneumatic tyres.

Brakes The vehicle is controlled by three brakes, the hand brake 81 on the countershaft 69, operated by depressing the pedal 82, Figs. 99, 100 and 103, keyed to the cross-shaft d, Figs. 100 and 103, on the far end of which is a lever connected to one end of the band of the brake by the wire 100, the outer end being held by the light rod 101, and by a lug attached to the back of the frame. This pedal also actuates the clutch pedal 45, so that **Brake gear** the engine is taken out of gear before the brake is applied. The clutch pedal 45, which may be worked independently of the hand brake, is connected by the rod 102 to the lever 46, Figs. 100 and 112, on the end of the short cross shaft F, at the back of the gear case G. This shaft carries an arm 47, Figs. 99 and 112, the forked end of which operates the clutch shaft in the manner described in connection with the English Daimler. There are also the two band brakes 85, Figs. 100 and 115, acting on drums bolted to the sprocket-wheel rings on the road wheels. These brakes are operated by the hand lever 83, which is the outermost of the three at the driver's right hand. The lever 83 may be held in any position on the sector 107 by means of a ratchet and catch worked by the handle 83 a, Fig. 112, which is gripped with the brake lever handle, when this is to be moved. The brake lever actuates a bar running through the tubular shaft of the speed changing lever, and carrying on its inner end the bent lever e, Fig. 112, which serves two purposes; first, on moving lever 83 forward, the quadrant part of lever e running on the roller f, depresses the lever 46, which operates the shaft F, so putting the clutch out of gear by lever 47 and rod 48, see also English Daimler. At the same time the link g, Figs. 112 and 115, and arm h rotate the hollow shaft Y, which carries at each end a lever J (see Figs. 99 and 115) · over rollers and between checks on the end of

152

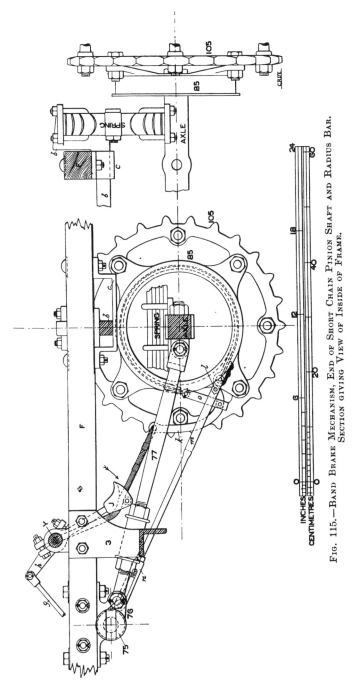

FIG. 115.—BAND BRAKE MECHANISM, END OF SHORT CHAIN PINION SHAFT AND RADIUS BAR. SECTION GIVING VIEW OF INSIDE OF FRAME.

these levers J, and through the hollow shaft Y, passes a wire rope, the ends of which are attached to the bent levers k, one of which is seen in Fig. 115, the arrangement for the other brake being identical. The bent lever k, is pivoted at l, within the long forked end of the bar m, adjustable by the

153

shakle n; this bar m, also carries one end of the band of the brake, the other end of the band being connected to the lever k by the link o. The movement imparted to the hollow shaft causes the levers J to rotate in the direction of the arrow, Fig. 115, so tightening the rope, and pulling the bent levers k forward and tightening the bands. It will be seen that this brake action is quite satisfactory for powerful action in the forward direction of movement of the car, the brake band being held by a tangential pull on the rod m. In the backward direction however the pull of the upper part of the brake band is on the light lever k, so that it may be easily possible that stoppage on a very stiff hill might be accompanied by a crippling of this, and recourse must be had to the pedal-acted brake 81, the action of which upon the driving wheels is through the chains, the failure of either of which might be disastrous. To guard against any such disaster however, a sprag E, Fig. 99, is used, and it can be dropped to the ground ready for action by means of the hand ring and cord z on the dashboard, Fig 103. The brake drums are of bronze and the bands of steel faced with a specially woven fabric which is not readily destroyed by the heat generated by friction on the drums.

Steering gear The steering wheels are carried at the ends of the fixed front axle in vertical cone pivots R. The steering is controlled by the hand wheel z, Fig. 99, on the sloping bar 78, at the lower end of which is a worm wheel, Figs. 113 and 114, engaging with a sector, both being enclosed in a cast-iron casing bolted to the angle bar I of the frame. By revolving the hand wheel z, the worm moves the sector up or down, rotating the horizontal shaft, which carries the lever D at its outward end. By means of the adjustable link 79, the motion is transmitted to the right-hand wheel pivot through the arm j, Fig. 113, and across to the other wheel by the link 80 (see also Fig. 100) connecting the arms Q. This steering gear has the advantage that the shock, when running over even large obstacles on the road, cannot cause the driver to lose control of the hand wheel, thus securing greater safety and precision at high speeds, with sufficient quickness of action. It is not equal to the bar or tiller steering for obstacle races, or even for constant driving in crowded traffic.

Starting the motor The motor is started by the handle 91, Figs. 99 and 113, which always hangs in the position shown. It drives the boss 92, Figs. 101–113, through a chain gear. By pressing in the tongue 93, the boss is held as a free wheel or automatic clutch to the engine shaft, by which it drives in one direction only, so that, so long as it is rotated faster than the engine shaft, it drives it, but when the engine starts it overruns the boss, so pushing out the tongue, or pawl, and leaves the starting gear free.

Before the motor is started, the lamps 19 are heated by a hand spirit lamp, the spirit cup, as employed on the English Daimler, being dispensed with.

Preparation for running The oil-plug valve 95 is then opened slightly and closed again, allowing a small quantity of petrol to pass. This vapourises sufficiently to start

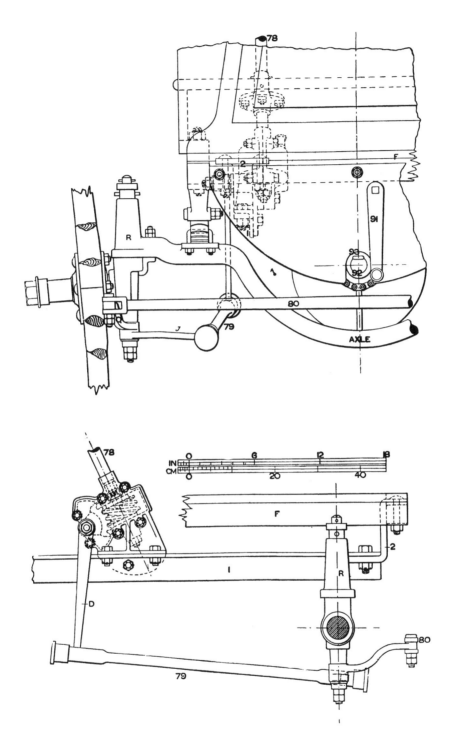

FIGS. 113 & 114.—STEERING GEAR, FRONT AXLE AND CONNECTIONS.
PANHARD-DAIMLER MOTOR, 6-HP CARRIAGE.

155

the lamps, then the plug valve 95 can be opened full, and the full power of the lamps obtained for heating the ignition tubes E. These things being in order, it must next be seen that the reversing lever 73 is in the middle notch, at which position everything is out of gear. The speed lever 62 on the notch nearest the driver's seat, *i.e.*, the slowest speed, and the brake lever 83 at the front end of the sector, all as shown by Figs. 99 and 113, and the governor control lever 94, Fig. 103, must be placed at the right-hand end of the sector. When this has been done, the air admission apertures h, and the cock p', must be closed and the slide g, Figs. 101 and 102, at the top of the air suction pipe pushed in, and the tap 9 on the pipe from the bottom of the petrol tank P, Fig. 99, opened. The tongue 93 is then pressed in, and the hand starting lever 91 turned until compression is felt in one of the cylinders. One sharp turn is then given, and the charge will be ignited by the ignition tube E, and the motor started. The handle 91 must be held steady for a moment, to ensure that the tongue is forced out, and then dropped, and as quickly as possible the revolving air admission adjusting head h, Figs. 102 and 104, turned so that the apertures are opened, and the slide g pulled out for the same purpose. The hole o for supplementary air admission in the top of the carburettor, Fig. 105, must also be opened a little. The lubricators 89, 90 are then opened.

Variation of mixture   Thus, for starting these motors, arrangements are made so that as rich a mixture as can be fired is alone admitted to the cylinders, and it will have been seen from the description just given that the whole of the air admission was cut off, with the exception of that which passes through the gauze-covered openings 99, Figs. 101, 102 and 106, which admit the air into the heated chamber, and thence through the smaller of two holes in the slide g, down the air admission pipe 13 to the vapouriser. As soon as the motor is started a much weaker mixture must be used, and for diluting with air from without the first inlet employed, after the slide g is opened, is the adjustable inlet head by which air enters through the more or less covered holes h in the pipe 13. Further dilution of the charge, by opening the holes h more fully, may be made when necessary, as, for instance, when the air is clear and dry, and on very light running. The adjustments so far referred to are as a rule quickly made and settled before the carriage starts running. The practised ear soon learns to detect any insufficiency **Varying mixture while running** of air admission at one or other of the air openings. For varying the richness of the charge during running, and also for adjusting it to suit the rate of admission of the petrol through the jet piece f, the cock p', Figs. 99 and 103, connected to a pipe p, Figs. 101–104, which admits air into the upper part of the vapouriser, is employed. It will be understood that the admission of air, into the part marked 12, Fig. 103, will not only dilute the charge otherwise adjusted, but will reduce by its own volume the quantity of air passing the jet piece f; a very small admission of air at this position will thus greatly modify the richness of the charge. It may

156

be here mentioned, although it is not a part of the preparation for running, that any petrol vapour which liquefies falls back to the bottom of the carburettor in which the jet *f* stands, and ultimately becomes vapourised.

To stop the motor, all that is necessary is to move the small controller lever 94 on the dashboard to the vertical position, as shown in Fig. 103, which causes the governor to cut out both cylinders by leaving the exhaust valves closed. The lubricators 89, 90 are at once closed, and the petrol tap 9, under the seat, also closed. Stopping motor

The burners are, by those experienced with the running of these cars, usually blown out, and the plug valve 95 left open until a slight gurgle or coughing noise is heard at the burners, accompanied by petrol, a small quantity of which squirts out of the nozzles, and then the valve is screwed up tight. The object of this is to keep the gauze and asbestos core of the burner and the burner itself free from hard carbon deposit, which is more likely to occur if the burners are simply put out and nothing done to cool them quickly, and at the same time leave the core moist. **Extinguishing burners**

Messrs. Panhard & Levassor fit the cars with electric ignition when required, but hitherto have preferred the tube ignition.

No one will deny that there is in this carriage many hundreds of proofs of ingenious and painstaking scheming of details and arrangements to produce the composite and successful whole. The parts now appear to be the obvious parts in their obvious places, and it is difficult for any but those who have tried to produce a similar result, to realise a tithe of the perseverance, combined with ingenuity, mechanical skill, and material facilities, which have been necessary to achieve this result. There are a few parts which are the subjects of patents, but for the labour, thought, and patient deduction from experiment and trial, which in the end have led to the combination of parts for the many functions successfully performed, there is but little protection by patent and no reward, except so far as the honours of success may be counted as reward. **Great ingenuity of the combination.**

The exceptional position as owners of engineering works which Messrs. Panhard & Levassor enjoyed, when the latter commenced and followed up the problem he solved, has given them advantages which now enable them to command an unrivalled and lucrative business; but the man who is inventor only has but a poor reason for devoting attention to such a problem. There should be better means of giving protection to the producer of ingenious combinations of mechanism, such as that of the vehicle last described, a combination which has made that possible which was unachieved before, and which the world wants. If it were a mere cycle lamp, with every feature and principle old, the combination would, in some courts at least, be held to be subject matter for an invention protected by the Patent law, but not so with a combination which in comparison makes the cycle lamp the merest trifle from every point of view.

Chapter IX

RECENT HIGH SPEED LONG-DISTANCE VEHICLES
(*Continued*)

The Peu-geot cars A T the time when the Panhard & Levassor cars were making for their constructors a great reputation during 1894 to 1896, when the feats performed in France with these petrol-motor vehicles were hardly believed in this country, and were with prejudice believed to lead to nothing, MM. Peugeot and Frères were the most successful rivals. They used a similar motor of Panhard & Levassor's make; but, as has been shown by Figs. 80–83, they placed it at the rear instead of in the front. They used a similar arrangement of gearing, after having tried frictional gearing, the mechanism generally being very much that of Levassor. The departures from the designs of the earlier firm will have been seen from the description to be mainly in the kind of frame, the method of suspension, the steering gear, and the size of the leading wheels.

The Peugeot Racing Carriages

Peugeot racing carriage

General Figs. 116–123 show a side elevation, plan, and back view of a Peugeot carriage, with 8-HP motor, to seat three persons. The motor, which is of the horizontal double-cylinder type, is enclosed by the body at the back of the principal seats. It drives through a friction clutch and spur gear on to a countershaft which extends across the carriage below the underframe. This countershaft in turn drives a similar one, by means of one of five sets of spur wheels and pinions, which give four forward speeds and one reverse speed to the vehicle. The power is finally transmitted from the second countershaft to the rear road wheels by means of pitch chain gears.

The steering of the vehicle is on the Ackerman principle, and is controlled by means of a conveniently-shaped hand bar in front of the driver's seat, a double pitch chain and sprocket wheels are employed to transmit the movement from the standard operated by the hand-steering bar to a second standard which carries the lever arm operating the steering wheels, so that the shocks and vibration are not transmitted to the steering bar.

There are three hand brakes controlling the forward movement of the

158

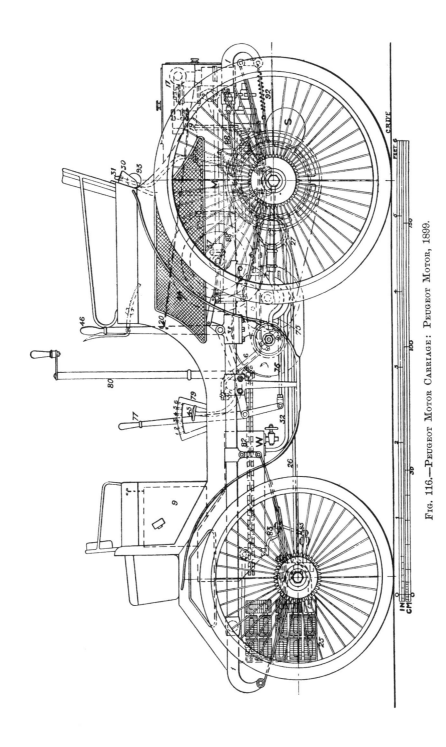

Fig. 116.—Peugeot Motor Carriage: Peugeot Motor, 1899.

MOTOR VEHICLES AND MOTORS

vehicle. Two of these act direct on the road wheels, while the other, which is employed for all ordinary occasions, is placed on the countershaft, from which the rear wheels are driven. It is so arranged that the friction clutch is always disengaged before the brakes can be applied.

The oil and water tanks are carried below the small front seat, so that their weight is carried principally by the steering wheels. They are each of about 12 gallons capacity, so that very long continuous runs may be undertaken. A radiator for cooling the circulating water is suspended in front below the frame.

Underframe
This is constructed chiefly of steel tubes, and consists of the side tubes 1, 1, Figs. 116, 117 and 118, to which are brazed the cross tubes 2, 3, 4. There are also two longitudinal channel bars 5, 6, which carry the motor and gear case. They are bolted to projecting lugs on the cross tube 3 in front, and to the brackets 7, 8, which are brazed on the bent tube 4 at the back. The underframe and body are carried from the axles by means of long single-leaf springs.

Motor
The motor M, which is shown in detail in Figs. 119–122, is of the Peugeot horizontal double-cylinder type, and develops 6-HP. at about 680 revolutions per minute. The cylinders are each 96 mm. = 3·75 in. bore by 132 mm. = 5·168 in. stroke, both pistons acting on one crank, so that there is one working stroke for every revolution.

The petrol supply is carried in the tank 9, Figs. 116 and 117, below the front seat. This tank feeds the float vessel 10, Fig. 117, of the carburettor, and the ignition tube burners 11, 11, Figs. 119 and 120, through the pipe 12. The float vessel and carburettor, Fig. 122, which are of the Daimler-Maybach type, are carried from the air supply pipe 13 by means of the connection 14, Fig. 114. The supply of air to pipe 13 is regulated by the slides 15, 16, by means of which the ends of the pipe may be opened or closed to any desired extent. The air sucked in at 15 is heated in the portion of the tube 13, where it is surrounded by the cage 17, placed directly above the burners 11, 11; and passes down the pipe 14 into the mixing chamber 18 of the carburettor, the quantity admitted being regulated by the slide 19, Figs. 116 and 122, which is controlled from the driver's seat by means of the milled hand wheel 20, Fig. 116. On each suction stroke of the motor the charge is drawn from the chamber 18, and through the pipe 21 into the admission chamber 22 of the motor.

Carburettor
The float-feed carburettor is seen in detail in Fig. 122. The action of the float A is the same as in the Daimler-Maybach float-feed spray-making carburettors already described, but the pivoted arms E, which engage with the collars on the needle valve B, are below, instead of on the top, of the float, the weight of the needle valve being sufficient to tend always to make it sit upon its seat against the lifting tendency of the levers. When the supply of petrol falls, the float A sits upon the outer ends of the arms E, and these lift the valve B, and permit the inflow of petrol through the pipe 12. On the outstroke of the motor piston, air is drawn in through pipe 14,

160

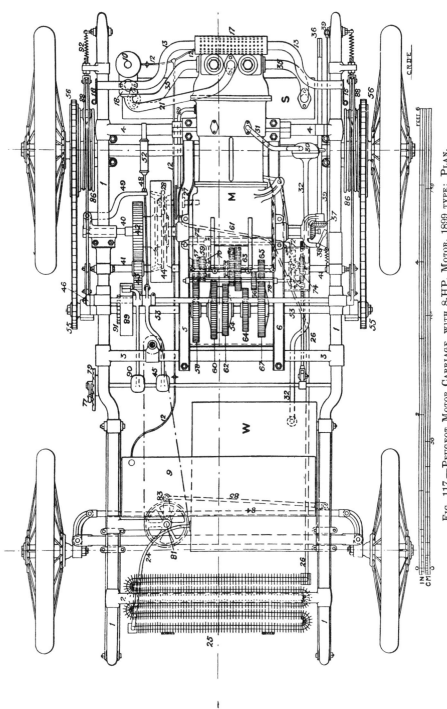

Fig. 117.—Peugeot Motor Carriage, with 8-H.P. Motor, 1899 type: Plan.

past the slide 19, on its way to the cylinder through pipe 21. At c it passes the petrol spraying jet which is fed through the passage D, and by induction draws therefrom a spray which is broken upon the end of the plug F, and carried with the air which is carburetted by its vaporization. At G is a pipe for emptying the carburettor float vessel, and at H is a cock for emptying the carburettor.

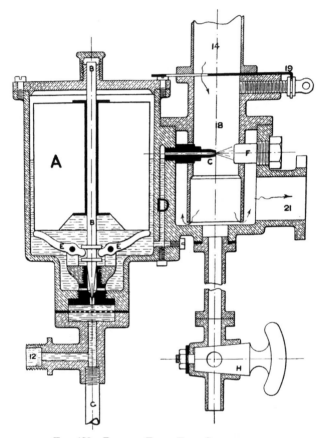

Fig. 122.—Peugeot Float-Feed Carburettor.

Valves
Admission valves
The action of the valves is identical with that of the Daimler motor. On the suction stroke the admission valve A, Fig. 119, opens against the resistance of the spring above it, and allows the charge to pass from the chamber 22 by way of port B into the cylinder. On the return stroke it is compressed into the combustion chamber c and port space B, and fired by the ignition tube D. After the working stroke is completed, the exhaust valve E is lifted, and the products of combustion escape by way of port F and pipe 35, Figs. 117 and 118, into the silencer s.

162

The exhaust valves are operated by means of the cam G, Figs. 119 and 121, which is mounted on the centre of the crank shaft between the connecting rod heads. The cam G consists of a drum, on the surface of which is a groove, as seen in Fig. 121, in which runs the tongue H, pivoted at the end of the arm J, which is mounted on the longitudinal solid shaft K, Fig. 119. At the back end of this shaft is a bell crank lever L, at the ends of the arms of which are pivoted the pushers *l l*, held in position by springs *s s*. **Exhaust valves and valve gear**

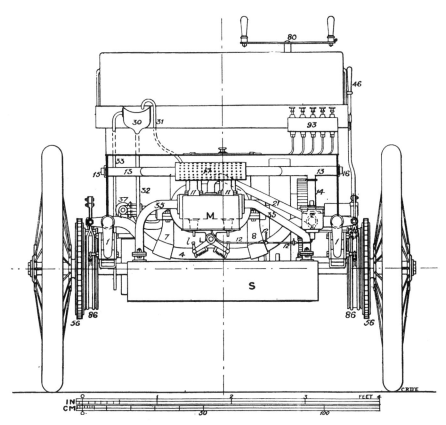

FIG. 118.—PEUGEOT MOTOR CARRIAGE: BACK.

As the crank shaft revolves, the pivoted tongue H, sliding in the groove of the cam G, causes the arm J, which stands vertically when H is at the crossing point of the groove, to rock to a maximum distance to one side and back to the vertical for one revolution of the cam, and then to repeat the same motion on the other side of the vertical position for the next revolution, and so on. It will be seen that the two grooves in the periphery of the big cam G cross each other just above the point occupied by the tongue, as shown in Fig. 121, and that the tongue being of some length the grooves act as rails and facing points, so that the tongue is directed at

163

the crossing so as to follow the continuation of the groove it occupied before reaching the crossing. The change over from one to the other is determined by the change in angular position, which it gets in running in the part of the groove remote from the crossing. The movement of the arm J is imparted to the longitudinal shaft K, and causes the arms of the bell crank lever L to rise and fall alternately, so that the pushers *l l* strike the ends of the spindles of the exhaust valves E E, and each valve is held open alternately for half a revolution once every revolution of the engine shaft.

Double groove cam

The governor N is mounted on the crank shaft at one side of the crank case, Fig. 121. When the normal speed is exceeded, the balls *n n* fly apart, and pull the sliding brush o inwards. This operates the bell crank P, causing the sleeve Q, Fig. 119, which slides on the shaft K, to be pushed in the direction of the arrow, Fig. 119, against the resistance of the spring R. At the back end of the sleeve Q is an arm Q′ sliding on a guide *s*. The arm Q′ has projections *q q* at its end, one on each side, so that looked at in plan it is shaped like the letter T. The movement of the sleeve Q, above described, causes the pieces *q q* to be projected into the angular path of the projections *t t*, at the ends of the spindles which carry the pushers *l l*, so that these are caused to tilt inwards and to miss the ends of the exhaust valve spindles, so that the valves are not lifted until the speed of the engine is again reduced to the normal. The action of the governor on the exhaust valve spindle-pushers *l l* will be more clearly seen from the detail of governing gear shown below Fig. 116, in which the governor has pushed the sleeve Q of Fig. 115 away from the crank shaft by means of the bell crank lever P, so that one end *q* of the piece Q′ has come into contact with the tongue *t* of the pusher *l*, which but for the interposition of *q* would have been position for pushing the exhaust valve stem.

Governor

The cooling water for the cylinder jackets is carried in the tank w, Figs. 112 and 113, below the petrol tank *q*. The water flows from the tank by way of pipe 24, round the cooler 25, and thence by pipe 26 to the rotary pump 27, which is driven from the engine fly wheel by the friction wheel 28. The pump forces the water up into the cylinder jackets through the pipe 29, entering at 30, Fig. 121. After circulating round the jacket it is delivered through pipe 31 into the vessel 30. From this it flows down the pipe 32, back into the tank w. Should the flow of water into the vessel 30 be too rapid for pipe 32 to carry it off, it overflows through pipe 33, Fig. 118.

Water jacket and cooling water

The exhaust gases from the engine pass through the pipes 35, into the silencer s, and thence escape into the atmosphere at the back of the vehicle from a large number of small holes at the bottom.

Silencer

The motor is started by turning a key handle on the shaft 36, Fig. 113, at the inner end of which is a bevel wheel 37, gearing with a similar wheel which is permissively keyed to the engine shaft by pulling the wire 39, by means of which the bell crank 38 presses the spring-resisted key inwards on the end of the crank shaft, and thus fixes the bevel wheel, which

Starting handle

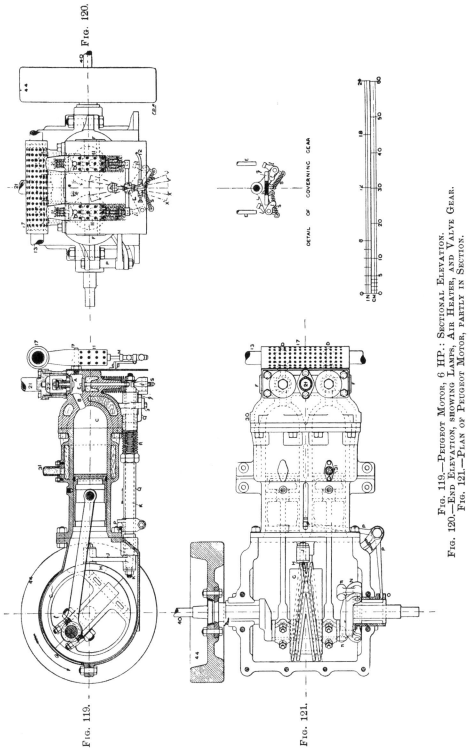

FIG. 119.—PEUGEOT MOTOR, 6 HP.: SECTIONAL ELEVATION.
FIG. 120.—END ELEVATION, SHOWING LAMPS, AIR HEATER, AND VALVE GEAR.
FIG. 121.—PLAN OF PEUGEOT MOTOR, PARTLY IN SECTION.

DETAIL OF GOVERNING GEAR

FIG. 120.

FIG. 119.

FIG. 121.

165

ordinarily runs loose on the shaft; the action of this loose wheel is similar to that already described with reference to Figs. 101 and 113.

Transmission gear

The extension 40 of the crank shaft, Figs. 117 and 121, drives the countershaft 41 by means of the spur wheels 42, 43, and is put into or out of gear with the engine by means of the cone friction clutch 44, which is the same as that already described with reference to Figs. 80 and 83.

Main clutch

The clutch is controlled both by the pedal 45 and by the brake lever 46. When the engine is put out of gear for the purpose of operating the speed changing device, the pedal 45 is depressed, pulling the wire 48 and lever 49. This causes the finger at the outer end of lever 49, which works in a groove turned in shaft 40, to move outwards, and thus pulls the clutch cone out of gear. On removing the foot from pedal 45, the spring 52 pulls the lever 49 back, and the finger pushes the clutch cone again into gear, in the position shown in plan in Fig. 117.

By means of one or other of five sets of spur wheels and pinions, the countershaft 41 drives a parallel shaft 53, which carries the differential gear 54, and at either end the sprocket pinions 55 of the chain driving gear. The chain wheels 56 are attached to, and drive through the hubs of the rear road wheels.

Speed gear

The change gear wheels 57, 58, 59, 60, 61, 62, 63, 64, give four changes of speed for the forward direction of motion of the carriage; and the wheels 65, 66, 67 give one reverse speed. The wheels 57, 59, 61, 63 are mounted in pairs, each sliding independently on a feather on the countershaft 41. At one side of each of the two pairs is a grooved box, in which engage the forked ends of striking rods 70, 71. Rod 71 is not clearly shown in Fig. 117, as it is directly below the shaft 41. Rod 70 is fixed on the bar 72, which is free to slide in bearings at either end. One end of the bar is bent at right angles, the end being flattened, and engages in the groove of a cam disc 73, the groove of which is zig-zag in one fourth of the circumference, the rest being normal to shaft 41, on which the cam is free to turn.

Speed-changing gear

On the box of the cam disc is a pinion 74, with which gears a rack cut on the end of rod 76, Fig. 116, which can be moved backwards or forwards by means of the hand lever 77.

The rod 71, which controls the pair of wheels 61, 63, is fixed on bar 72A, similar to 72, and the reversing pinion 66 is carried on a movable bar 78, the end of which engages in the groove of the cam disc similarly to that of 72 and 72A.

When the hand lever 77 is at the position marked 2 on the sector 79, Figs. 116 and 117, the ends of all three bars 72, 72A, 73 are in the normal part of the cam groove, and all the wheels are out of gear. At portion 3, as shown, the cam has been rotated, and the end of bar 72 carried along the groove to the innermost point of the zig-zag, thus pushing the bar 72 a sufficient distance to put the wheels 57, 58 into gear as shown in Fig. 117, which gives the slowest speed to the carriage. During this operation the ends of the bars 72A, 78 have remained in the normal

166

part of the groove, and consequently the wheels which they control have not been moved.

A further movement of the hand lever 77 to the position 4 causes the end of the bar 72 to be carried first into the normal position, thus putting 57, 58 out of gear, and then into the outermost part of the zig-zag, which puts 59 into gear with 60, corresponding to the second speed of the carriage. During these movements the ends of bars 72A, 78 have still remained in the normal part of the groove. On moving lever 77 to position 5 the end of bar 72 runs into the normal part of the groove, putting 59, 60 out of gear, and continues there during any further movement of the lever 77 towards the driver. Just as the end of bar 72 is carried into the normal portion of the groove, that of 72A is moved into the zig-zag, and 61 gears

Fig 123.—Peugeot Motor Carriage, 6 HP.

with 62, giving the third speed. The final position, 6, of lever 77 puts these out of gear and causes 63 to gear with 64, giving the 4th and highest speed to the vehicle.

To reverse the direction of motion the hand lever 77 is first moved forward to position 2, when all the wheels will be out of gear, and then to portion 1, which continues the forward rotation of the cam disc and causes the end of bar 78 to travel outwards along the zig-zag, putting 66 into gear with 65, 67, the intervention of wheel 66 causing the reversal of direction of rotation. **Reversing**

The steering of the carriage is controlled by the curved hand bar at the top of the steering pillar 80, the motion being imparted to the short vertical spindle 81, Fig. 117, by means of the double set of sprocket wheels and pitch chains 82. The bent lever 83 on the lower side of this spindle is connected by rod 85 to the bar 84, and the motion imparted to the front wheels which run on stud axles pivoted at the ends of the fixed axle. **Steering gear**

There are three brakes; two are band brakes 86, which act on drums **Brakes**

167

bolted to the arms of the chain driving wheels. They are controlled by the lever 46, the movement of which also puts the friction clutch out of gear. The bands of the brakes 86 on the driving wheels are of leather, faced with wooden blocks, and are attached to the wire ropes 88, Figs. 116 and 117. When these brakes are not in use the bands are held off the drums by means of the springs 92. It will be seen that these brakes are of the Lemoine or Spanish windlass kind. There is also the band brake 89, Fig. 117, on the countershaft 53, which is operated by means of the pedal 90. The backward movement of the vehicle is prevented by means of a pawl which may be put into gear with the ratchet 91 on the countershaft 53, this taking the place of the sprag used by other makers. The lubricating oil for most of the principal working parts is fed from the combination lubricators 93 at the back of the driver's seat.

The wheels of the carriage are constructed with direct steel spokes and steel rims, with Michelin pneumatic tyres.

Hooded Victoria Fig. 123 shows one of these carriages, but fitted with touring and hooded body for Sir David Salomons. The wheels are fitted with solid tyres.

Chapter X

MORS MOTOR CARRIAGE AND MOTOR

FIGS. 125 127 show a side elevation, plan, and back end elevation of one of the larger types of Mors carriage. It is provided with seating capacity for four or five persons, the principal seats being back to back on the lines of a dogcart. In its mechanical design it bears some resemblance to the Benz vehicle, but at the same time it may be said to embody very many of the improvements which are felt to be wanting in the design of the simpler and less costly carriages. Fig. 124 is from a photograph of a similar vehicle fitted with tangent spoke wheels.

The principal working parts are enclosed in the body of the vehicle below the seats, the body being so arranged that by removing four nuts and one pipe connection the whole of the superstructure may be lifted off, exposing the machinery, as seen in Figs. 126 and 128, for the purpose of inspection or repairs.

The motor, which is placed on one side at the back end of the boot, drives fast and loose pulleys by means of open belts mounted on a countershaft, to which the power is transmitted through the intermediary of a cone friction clutch. This countershaft carries the differential gear, and the power is transmitted from it to the rear road wheels by means of sprocket wheels and pitch chains. The driver sits on the left-hand forward seat, and has the steering bar directly in front of him, and on either side of the steering pillar are the hand levers for controlling the belt driving gear, which gives two forward speeds to the carriage, and one reverse speed, which is controlled by a separate lever at the driver's left hand, as is also the lever controlling the arrangement for taking up the stretch in the belts. By means of thumb-screws at his right hand the driver has control over the petrol supply and air supply to the carburettor, and by means of a hand lever the power of the motor may be varied to any desired extent between maximum and minimum by throttling the charge entering the motor cylinders. There are two pedals, one on each side of the steering pillar, by means of which the motor may be put out of gear with the countershaft which drives the road wheels, and a brake applied to check the forward motion of the carriage. There are two other powerful band

169

brakes applied direct to the road wheels by means of hand levers on either side of the vehicle. A sprag is also provided, which obviates the danger of

FIG. 124.—THE MORS FOUR-CYLINDER 7·5-H.P. TOURING CARRIAGE, 1899.

the carriage moving backwards if from any reason it should stop on an incline. A large store of both petrol and water are carried, sufficing for

170

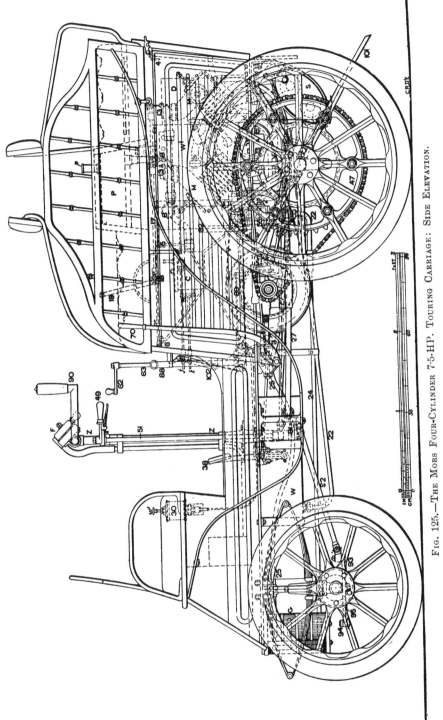

FIG. 125.—THE MORS FOUR-CYLINDER 7·5-H.P. TOURING CARRIAGE: SIDE ELEVATION.

171

long-continuous runs. The firing of the charge in the motor cylinders is effected electrically, a dynamo and accumulators, which may be charged *en route*, being carried. The weight of the carriage in running order is rather less than one ton.

Underframe This underframe takes the form of a skeleton frame, to which the woodwork of the lower part of the body is attached. It consists of a rectangular structure of flat iron bars strengthened at the back end by the transverse angle iron 1, Figs. 126 and 128, and also by the angle stays 2, 3, which are made use of for supporting the various mechanical parts. The sides of the boot of the body are constructed of cork, faced inside and out with sheet iron and edged with angle iron, the sides being connected across above by the angle-iron tie bars 4, 5, 6. The construction of the upper detachable seat portion of the body is best seen in the back end elevation, Fig. 127, which shows the method of bolting it to the sides of the boot. The back is closed by two wooden flaps, the outer of which, when let down, forms the footboard for the back seats. The frame and body are carried from the axles by means of double-plate springs in front and single-plate springs behind.

Motor The motor M, Figs. 125–128, and in detail in Fig. 129, is of the Mors four-cylinder type, developing about 7·5 HP., at 800 revolutions per minute. The cylinders are arranged in pairs placed opposite one another on either side of the crank case, and inclined at an angle of 45° to the crank shaft A, Fig. 129, which passes symmetrically through the length of the crank case, and in which are formed two cranks at an angle of 180°, the pistons of each pair of cylinders on opposite sides of the crank case being coupled to one crank. This arrangement is effected by the use of a connecting rod with ordinary big end on one side, working inside a forked connecting rod for the piston of the opposite cylinder. The four cylinders, which are all of identical construction, are each of 70 mm. = 2·75 in. bore by 89 mm. = 3·5 in. stroke. Their construction is best shown by the side elevation of the engine, Fig. 129, in which one cylinder is shown in section and the other in outside elevation. It will be seen that the cylinder body and the head or combustion and jacketed end are formed in one casting, which secures the important advantage of doing away with the troublesome head joint, which has already been referred to, and which, if not very perfectly made, allows leakage from the water jacket into the cylinder, and causes great loss of power. In the Mors cylinder this cannot possibly take place, as the water joint J is formed in the side of the jacket, and the cylinder end is closed by an ordinary cover B, which can be easily and effectively ground in place and bolted down. The combustion chamber and port space 7 alone are water jacketed, the body of the cylinder being cooled by means of radiating ribs, as shown in Fig. 129.

Petrol supply The petrol supply is carried in the tank P, Fig. 125, in the upper detachable part of the body between the backs of the seats. A spare tank, from which tank P may be replenished, is sometimes carried in the box at

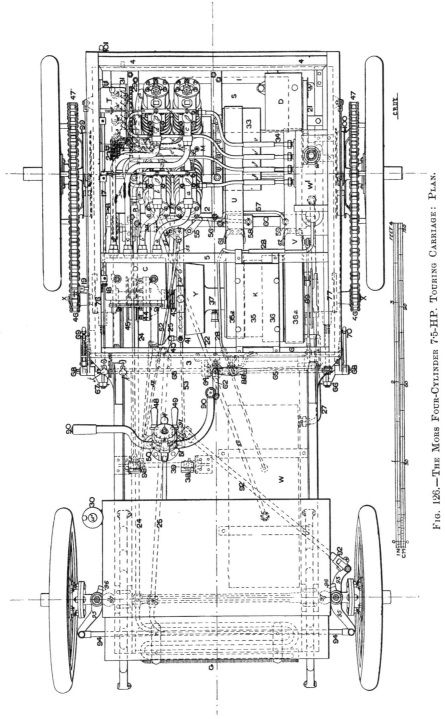

Fig. 126.—The Mors Four-Cylinder 7·5-H.P. Touring Carriage: Plan.

173

the front of the body. The tank P is filled through the aperture closed by the screw cap *p*, Figs. 125 and 127. The petrol flows by way of the pipe 8 into a float vessel inside the carburettor box, c. In the supply pipe 8 at the side of the carburettor is a small cock, operated by means of the thumb-screw *j* at the driver's right hand, so that the supply may be adjusted as required.

Carburettor

The carburettor c, which in principle is identical with the Maybach float-feed carburettor previously described, consists of an outer rectangular metal box c, inside which is a cylindrical float vessel, the level of the petrol in which is regulated by a float and needle valve. From the lower part of the float vessel there radiate four small pipes, conveying petrol to four jets just within the mouths of the four pipes *a*, *b*, *c*, *d*, Figs. 126 and 128, which lead to the admission valve chambers of the motor cylinders (see Fig. 125). On the suction stroke of each piston air is drawn into the box c through a rectangular gauze-covered aperture 9 at its front end, the quantity of air passing being regulated by means of the thumbscrew *k*, placed near *j*, which regulates the petrol supply. The air drawn into the box c rushes round the float vessel and along one or other of the admission pipes *a*, *b*, *c*, *d*, sweeping over the petrol jet in its mouth, and being thus carburetted passes along the pipe into the admission chamber 10 of the motor cylinder, Fig. 129. The admission valve 11 opening on each suction

Valves

stroke of the piston allows the charge to pass, by way of the passage 7, into the cylinder. On the return stroke it is compressed into the combustion chamber and port place 7 and fired electrically at *h'* at a predetermined point before the dead centre by the sparking plug E hereafter described. At the completion of the working stroke the exhaust valve 12 is lifted by means of a cam on the countershaft 13, which is geared one to two with the engine shaft, and the exhaust gases pass by way of the pipes *e*, *e¹*, *e²*, or *e³*, as the case may be, into the silencer s, which is suspended below the frame at the back of the vehicle and out of the bottom of which the exhaust gases pass quietly into the atmosphere.

Variable speed of motor

The speed and power of the motor may be varied through a wide range by throttling the charge entering the cylinders. This is effected by means of a throttle valve at the end of each of the admission pipes where they enter the admission chambers. One of these valves is seen in section

Throttle valves

in Fig. 129. It is formed by the cylinder 13 closed at the top and fitting accurately into the casting 14, which forms the admission chamber and serves the purpose of holding the admission valve in place. In the side of the cylinder 13 is a heart-shaped opening which, when the cylinder is rotated, is moved backwards or forwards across the mouth of the admission pipe c, so reducing the area of the passage as required or closing the passage altogether, so that the cylinder is put out of action. The motion of the cylindrical valve 13 is guided and regulated by means of the small set screw 15, the inner end of which has the thread turned off and passes through a slot in the side of the cylinder, so keeping it in place. It will be seen from Figs. 128 and 129, that each of these throttle valves has a lug and pin 16, projecting on one side at the top. These are connected in pairs by connecting rods d', and each pair is connected by a link to an arm on the longitudinal spindle 17, Figs. 126 and 128, which is carried in bearings from the angle-iron edging at the top of the side of the body. By means

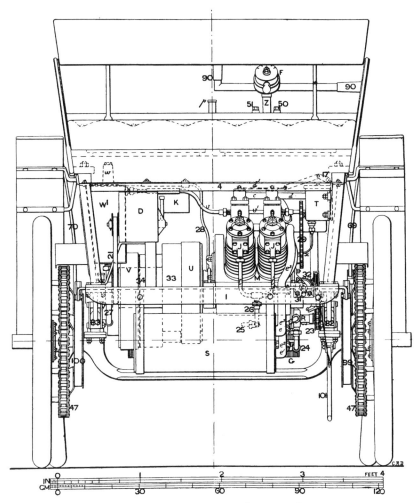

FIG. 127.—MORS 7·5-HP. TOURING CARRIAGE: BACK VIEW.

of bevel gear 18, at the front end of spindle 17, it is geared to a short cross spindle, at the outer end of which is the hand lever 19. By moving this lever backwards or forwards the driver is able to adjust the throttle valves for both pairs of cylinders simultaneously to any desired degree, but the hole in the throttle valve of one pair being set later than the other, the driver may cut out one pair only, so halving the power of the motor.

175

Cleaning valves and cylinders

It will be seen that there is a small hole closed by a sliding cover at the top of each of the admission chambers, Fig. 129, the object of these being to enable a small quantity of petrol to be injected from time to time to keep the admission valves clean. A small aperture 20, closed by a set screw in the cylinder cover, serves the same purpose for the body of the cylinder.

Electric ignition

The electric ignition apparatus comprises both dynamo and accumulators, a coil, switch, and a sparking plug and make and brake circuit apparatus for each cylinder. The dynamo D, Figs. 126 and 127, is enclosed in a wooden box placed at the back corner of the boot opposite the motor, and carried from the underframe. It is driven from a pulley at the end of the motor shaft by means of the leather belt 21, the speed at which the dynamo runs being about twice that of the motor. The coil, which is enclosed in a wood box K, is carried from the cross angle-iron stays 5, 6 joining the sides of the body. It consists of a simple coil of insulated wire wound on an iron core, thus forming an electro-magnet when connected up

Electric fittings

to either the dynamo or accumulators. When the circuit is broken within the combustion chamber of any one of the cylinders, the self induction of the coil produces a spark of ample power to ignite the charge. The accumulators, of which there are two sets, each of four elements, are carried in the box under the small front seat. They are divided into two sets because the dynamo is only sufficiently powerful to charge four cells and at the same time supply the necessary current for the ignition apparatus. The circuit switch F is mounted at the top of the steering pillar. When the motor is to be started, the switch is set so that the two sets of accumulators are connected up to the coil K, and supply the current necessary for firing the charges before the dynamo is run up to full speed. When this takes place the switch is moved to the position which breaks the accumulator circuit and connects up the dynamo to the coil, and there are two other positions of the switch which connect up one or other set of cells to the dynamo so that they may be charged ready for starting the motor, and at the same time the supply continues from the dynamo to the sparking apparatus. There is also a stop position of the switch, which cuts out both

Sparking plug

dynamo and accumulators. The sparking plug E, and make and brake circuit apparatus, are best seen by reference to Fig. 129. The plug E consists of an inner rod h, terminating inside the cylinder in a boss h'. It is held firmly in and insulated from the outer part m of the plug which is screwed into the cylinder cover. The rod h in each of the four plugs is connected by an insulated wire l to one end of the coil K. Inside the cylinder, in contact with the boss h', is an arm n at the end of a short spindle n' which passes through a gland in the cylinder wall and carries at its outer end the finger p, the end of which is held by a spring q attached at its lower end to a plate r fixed to the cylinder casting. It will be seen that the tension of the spring q always tends to keep the arm n pressed against the boss h' of the sparking plug E. The arm n is caused at predetermined intervals to break contact with the boss h' by means of a cam

176

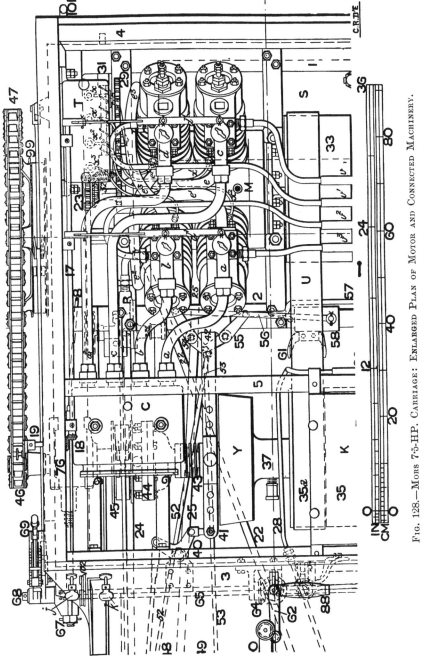

FIG. 128.—MORS 75-HP. CARRIAGE: ENLARGED PLAN OF MOTOR AND CONNECTED MACHINERY.

177

12

on the second motion shaft 13 of the motor. This cam bears against a roller at the end of the spring-held rod s, the end of which, when pushed up by the cam, strikes the end of the finger p and, overcoming the tension of the spring q, causes the spindle n' to make a small part of a revolution, and the end of the arm n breaks contact with the boss h'. The igniting spark is thus caused.

Electric circuit   The electrical circuit is as follows: The negative terminal of the dynamo, and also that of the cells, is connected to the motor casting, and is therefore in electrical connection with the arm n of the contact breaker in each cylinder. The + terminal of the dynamo or that of the cells is con-

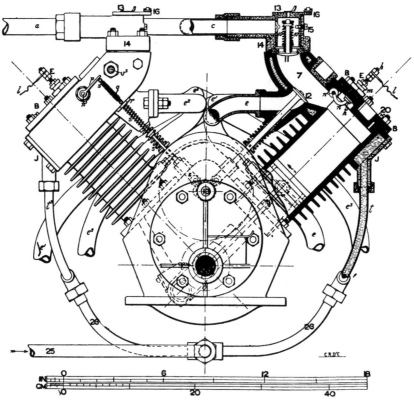

FIG. 129.—MORS FOUR-CYLINDER 7·5-HP. MOTOR.

nected through the switch F on the steering pillar to one terminal of the coil K, the other terminal of which is connected by a separate wire to the inner insulated rod $h\,h'$ of and in each sparking plug. The cams on the shaft 13 of the motor which operate the contact breakers are so arranged that previous to the instant when the charge in any one of the cylinders is to be fired, the arms n of the contact breakers in the other three cylinders have been caused to part contact with the bosses h' of their respective sparking plugs.

178

These breakings of contact do not, however, produce any sparking because the circuit is still completed through n and h' of the fourth cylinder, so that when the instant arrives for the charge in this cylinder to be fired, the break of contact between n and h' causes a powerful self-induction spark to pass between these contact points, as the current has now no other circuit round which to flow.

The cooling water for the cylinder head jackets is carried in the tank w suspended below the frame at one side in the front of the vehicle. The water flows from the tank by way of the pipe 22 to a small rotary pump R, which is driven off one end by the crankshaft by means of sprocket wheels and pitch chain 23, Figs. 123 and 124. The water is forced from the pump along the pipe 24 round the radiator G, Figs. 125 and 126, which is placed in the extreme front of the vehicle below the frame. After circulating round the radiator it flows along pipe 25, which terminates in a T piece below the motor, Fig. 129, and the water here divides along the pipes 26 and enters the jackets at J by way of four branch pipes t, t^1, t^2, t^3. After circulating round the jackets it emerges near the upper end of each port and flows by way of the pipes v, v^1, v^2, and v^3, Figs. 126–128, into the vessel w′ attached to the side of the body, and thence by way of the pipe 27 back into the tank w. If by any chance the water is pumped into the tank w′ more rapidly than it can flow out by way of the pipe 27, the overflow pipe 28 carries off the surplus back to the main tank. The large tank w is filled through the tank w′, at the top of which is an aperture closed by the screw cover w. **Water jacket**

The lubrication of the cylinders is effected automatically by means of four small force pumps enclosed in the oil tank T, which is attached to the side of the body. The pumps are driven off the cam shaft 13 of the motor by means of sprocket wheels and pitch chain 29, Figs. 127 and 128. They draw oil from the tank T and force it along the four small pipes x, x^1, x^2, x^3, Fig. 128, into the cylinder bodies near the lower end of the piston travel. An efficient lubrication of the pistons, which is very necessary, as the heat from the cylinders is only dissipated by means of radiating ribs, is thus secured. The positive feed-pump lubricator is shown in detail on a large scale in Fig. 130. In this, however, the sprocket wheel is shown as a grooved pulley A. A sprocket wheel at A on the spindle B, which carries a worm C, drives the worm wheel D, which turns the spindle having journals E in bearings F, and which is excentric, along the part G between the bearings. On the part G are four small double excentric straps H, carrying pivoted levers J, in the ends of which are adjustable screws K, the positions of which determine the stroke given to the plungers L by pressure of K on the plunger heads. The lower end of the plungers L, which are lifted by the springs O pressing against the pin M in the plunger heads, rises above the packing leather N on the top of the valve seat P, and oil flows in to the ball valve which is supported by the spring Q. On the descent of the plunger the oil above the valve is forced past it to the **Lubrication**

179

distributing pipe R, which is the same as the pipes marked x in Figs. 126, 127 and 128.

Oil is supplied to the crank case by means of a separate lubricator 30, Fig. 121, placed within reach of the driver, so that by depressing the plunger w' from time to time, he may force a jet of oil to the cranks.

Starting the motor The motor is started by means of a key handle, which is slipped into the outer end of the spindle 31 which projects at the back of the vehicle; the spindle is then pushed inwards against the resistance of a spring seen in Fig. 124, which causes a bevel wheel 32 at its inner end to gear with a similar bevel on the end of the crankshaft. By pressing upwards a finger a', Fig. 123, a catch is dropped over the spindle 31, which is now held so that the bevels are in gear; two or three turns are now given to the key handle, which will start the motor, causing it to overrun the handle and releasing the catch a', so that the bevels are put out of gear by the pressure of the spring on the spindle 31.

Transmission gear As previously mentioned, p. 169, the power is transmitted from the motor shaft to a countershaft by means of open belts. On the motor shaft is keyed the two-speed cone 33, 34, the larger pulley of which also forms the fly wheel of the engine. The pulleys 33, 34 carry belts U, V, which drive fast and loose pulleys 35, 35a, 36, 36a, which are mounted on a sleeve 37, on the countershaft x x, Fig. 126. At one end of the sleeve 37 is carried the outer fixed half of a cone friction clutch Y, through which the power is transmitted to the chain pinion shaft x x. The friction clutch Y is operated by depressing a pedal 38, Figs. 125 and 126, keyed to a spindle 39, which is carried in bearings below the underframe. At the further end of this spindle is a lever arm connected by a link to one arm of a bell crank 40 (see plan, Figs. 126 and 128), the other arm of which operates the bar 41, which is pivoted at 42, and forked to engage with a groove turned in the boss of the inner sliding cone of the clutch Y, near 43. By depressing the pedal 38 the bar 41 is caused to move the inner cone of the clutch away from the fixed member against the resistance of the spring 43, so putting the belt transmission out of gear with the countershaft x x; when the pedal 38 is again released the spring 43 puts the clutch into gear.

Speed changing On the countershaft x x is carried the differential gear 44, one side of the casing of which forms the drum of a band brake 45. On the ends of the countershaft, outside its bearings, are keyed the sprocket pinions 46, which drive by means of pitch chains the sprocket-wheel rings 47, which are carried by six studs from the spokes of the rear road wheels.

The belt-shifting gear, by means of which the belts U, V are moved from the loose to the fast pulleys, or *vice versa*, is controlled by the hand levers 48, 49, Figs. 125 and 126, which are mounted at the upper ends of the vertical spindles 50, 51, one on either side of the steering pillar z. These hand levers move over sectors which, with the upper bearings of the spindles 50, 51, are carried from the fixed standard z. At the lower

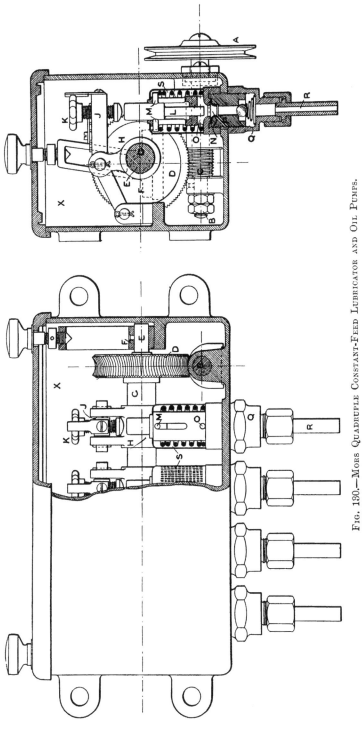

FIG. 130.—MORS QUADRUPLE CONSTANT-FEED LUBRICATOR AND OIL PUMPS.

181

ends of the spindles 50, 51 are short arms, connected by rods 52, 53 to similar arms at the lower ends of the short vertical spindles 54, 55, which work in bearings carried from the cross angle bar 2 of the under-frame. At the upper ends of the spindles 54, 55 are also mounted short arms, connected by rods 56, 57 to the blocks 58, 59, which are free to slide on a rectanglar guide bar 60, which is mounted on the upper surface of the angle bar 2 of the frame. To the sliding blocks 58, 59 are attached the forks 61, 62, within which pass the belts u, v. In order to bring the high-speed belt u into action, the hand lever 48 is moved outwards over its sector, which causes the block 58 and fork 61 to slide along the guide bar in the direction towards the belt v, so causing the belt u to move from the loose pulley 35a on to the fast pulley 35; it will again be brought back on to the loose pulley by moving the hand lever 48 to the off position, as shown in Fig. 126. Corresponding movements of the hand lever 49 will cause the low-speed belt v to be shifted from the loose pulley 36a to the fast pulley 36, or *vice versa*. It will be observed that before either of the hand levers 48, 49 is moved outwards over its sector, the other handle is in the off position, which indicates that the belt which it controls is running on the loose pulley. As it is sometimes desirable to have the driving belt in a position partly on the loose pulley and partly on the tight pulley, so that the belt slips considerably, and the full power is not transmitted, it is arranged that the hand levers 48, 49 may be held fixed at any position on their sectors. This is effected by having a ratchet cut on the under face of each sector (see below z, Fig. 125), in which is engaged, by spring pressure, a toothed block at the end of a small hand plate pivoted below each handle 48, 49. This plate being gripped with the handle disengages the catch when the lever is to be moved, and on being released holds it fixed in any position on the sector, and consequently the belt-shifting fork may be held in any position between full on and off.

Belt tension adjustment Owing to the liability of the driving belts to stretch from various causes, a belt-tightening arrangement is provided, which is always under the control of the driver, and it is so arranged that it is not necessary for him to leave his seat. The tightening of the belts is effected by moving the complete countershaft x x and its bearings, parallel to the engine shaft A, so as to increase the distance between their centres. The mechanism by means of which this operation is performed is controlled by the hand lever 62, Figs. 125, 126, 131 and 132, which is mounted at the upper end of the vertical bar 63, in the centre of the carriage, between the front seats. At the lower end of this spindle is a bevel gear 64, which operates a tubular shaft 65, extending across the whole breadth of the carriage between the bearings 66, 67. These bearings carry the solid spindle 68, which passes through the length of the tubular shaft 65 and carries the brake levers 69, 70, Figs. 125 and 126, at its ends, as hereafter described.

At the ends of the tubular shaft 65, which will be understood to
182

butt up against the inner faces of the bearings 66, 67, are worms 71, 72, which operate the worm wheels on the spindles 73, 74, which are

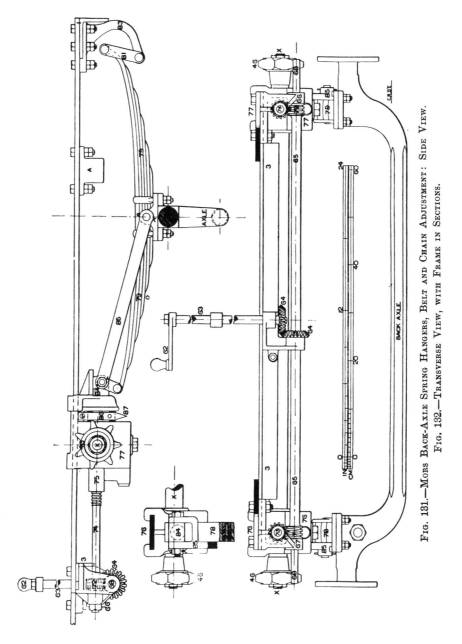

FIG. 181.—MORS BACK-AXLE SPRING HANGERS, BELT AND CHAIN ADJUSTMENT: SIDE VIEW.
FIG. 132.—TRANSVERSE VIEW, WITH FRAME IN SECTIONS.

parallel with the length of the vehicle. One of these spindles is clearly seen in the side elevation, Fig. 131. A screw thread is cut on its back end, which enters a boss 75, cast on the front end of the countershaft bearing 77.

The arrangement of the corresponding spindle 73 and bearing 76 being also as above described. The bearings 76, 77 of the countershaft x x slides upon the longitudinal bars of the underframe, upon which they are held after the manner of a slipper crosshead, as shown in detail by the small engraving between Figs. 131 and 132.

The sliding movement is given to both bearings 76, 79 equally by rotating the hand lever 62 at the top of the spindle 63. The rotation is transmitted to the tubular shaft 65 by means of the bevel gear 64, and from this shaft, through the worm gears 71, 72, to the spindles 73, 74. As these spindles are restrained from movement in a longitudinal direction, their screwed end working into the bosses at the ends of the bearing 76, 77, cause these, with the countershaft x x and all its gear, to slide forwards and backwards along the side bars of the frame according to the direction in which the hand lever 62 is turned, thus making it possible to tighten or slacken the driving belts u, v independently of the driving chains, the fixed distance between the countershaft x x and the back axle being maintained by the radius rods 85.

Chain adjustment The chains are adjusted independently of the belt-driving gear in the following manner: The back axle, which is best seen in Figs. 131 and 132, is sprung by means of the leaf-spring 78, 79 directly below the side bars of the underframe. The back ends of these springs are slung by links 80, 81 from the extremities of the hangers 82, 83, which are bolted to the underframe. The front end of each spring is slung from the end of a short bolt or spindle 84 (the outer upward end of which is flat and rests against the projecting bracket above the figure 84) connected by the radius bar 85 to the back axle, and its round part is free to slide in bosses at the back end of the bearing box of the countershaft x x. A screw thread is cut on the forward end of it, and on this part, and in a fork or jaw, is a round nut 86. In the periphery of this nut there are holes into any one of which a tommy may be inserted for turning it, and into which a pin on the spring 87 may engage to prevent its turning by vibration. It will be seen that by turning the nut 86 on each side of the carriage the sliding spindles 84 will be caused to move backwards or forwards, and with them the whole back axle with its wheels and carrying springs. Thus the driving chains may be adjusted by varying the distance between the countershaft and axle independently of any movement of the countershaft x x, which carries the sprocket pinions.

Reversing gear In connection with the belt-driving gear there is also an arrangement for giving a reverse motion to the carriage by means of the slow-speed belt v. This is operated by means of the handle 88, Figs. 125 and 126, near the belt-adjusting lever 62 before described. By drawing up the handle 88, and then turning it through a right angle and pushing it forward, a band brake 89 is applied, which holds fixed the carriers of the spindles of the pinions of epicyclic gear inside the pulley 36a (see also Fig. 133). This pulley, which carries the external toothed ring of the gear, and is in ordinary working the loose pulley of the slow-speed belt v, now drives through the pinions of the gear. These pinions revolve on their now fixed spindles and on the spur wheel on the sleeve 37, and consequently a reversed direction of rotation to that of the pulley 36a is transmitted through the friction clutch y to the countershaft x x, and thence to the road wheels. On restoring the handle 88 to its original position, the epicyclic gear is again put out of action, and the pulley 36a runs loose on the sleeve 37.

The arrangement of the reversion gear, shown in Fig. 133, is from the patent drawings, and is almost the same as in the car illustrated.

When the brake band on the pulley B is applied, the pulley F, and with it the sleeve c and the arms D carrying the stud spindles of the pinions E E, are all fixed. When then the pulley L is turned, the spur wheel A on its boss turns the pinions E E, and they give motion to the internal toothed ring of the pulley G, which is fixed to the clutch sleeve H. The direction of rotation of the pulley G and clutch sleeve is thus opposite to that of the pulley L which is driving them.

In the arrangement shown in Figs. 125 and 128 the relative positions of the fast and loose pulleys are changed, but the principle of the reversing device is the same.

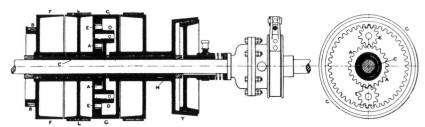

FIG. 133.—REVERSING GEAR OF MORS CAR.

Steering gear

The steering of the carriage is controlled by the hand bar 90 at the upper end of a vertical spindle, which passes down the fixed standard z, Figs. 125 and 126. At the lower end of the spindle is an arm 91, which is connected by the bar 92 to an arm 93 on the pivot of one of the front wheels, and this pivot is connected to that of the other wheel by means of the bar 94, joining the bent arms 95, which project in front of the pivots, so that the bar under all ordinary conditions of running is in tension. The front axle, which is fixed, is straight, and is connected to the under-frame by means of double-plate springs. The extent to which the front wheels may be turned is determined by contact of the projecting stops 96, Fig. 126, on the wheel pivots with the front axle at 97. The wheels of this carriage are of wood, but the Mors vehicles of this type are more generally fitted with the tangent spoke wheels, as seen in Fig. 124. The pneumatic tyres are of the Michelin construction.

185

Brakes   There are three brakes for controlling the forward motion of the vehicle. That which is used for ordinary occasions is the hand brake 45 on the countershaft x x, Fig. 126; it is applied by depressing the pedal 98,

FIG. 134.—THE MORS 116-HP. RACING CAR, 1899.

which at the same time operates the pedal 38, and puts the friction clutch y out of gear before the brake is applied. If this brake is not sufficiently powerful to check the motion of the carriage, the band brakes 99, 100 are applied by means of the hand lever 69 at the driver's right hand.

MORS MOTOR CARRIAGE AND MOTOR

These brakes act on drums, which are bolted to the sprocket-wheel rings on the rear road wheels. It will be seen that there is a hand lever 70, similar to 69 at the opposite side of the carriage. This may be used by the person sitting on that side, and if applied at the same time as 69, secures equal action of the brake 100, and relieves the spindle 68 on which they are mounted of the twisting strain which is put on the spindle when the brake 100 is applied by the lever 69 at the opposite end of the spindle. The sprag 101, Figs. 125, 126 and 127, is connected by a cord passing over rollers to the hand lever 102 at the driver's right hand. By moving the lever forward, the end of the sprag is lowered to the road surface, and prevents the carriage moving backwards down an incline.

The Mors carriage illustrated is one of several kinds made by the Mors **Other Mors** **cars** Company in Paris-Grenelle, the smaller vehicles being fitted with horizontal motors with two cylinders on opposite sides of the crank. The four-cylinder motor runs very steadily, and most of the vehicles fitted with them are arranged for high speeds.

Figs. 134 and 135 are from photographs of one of the recent Mors racing **Mors racing** **cars** cars with which very high speed records have been made. They differ in arrangement of machinery from that of the general purpose cars, and approach very closely to the lines of Levassor's main principles of design. The motor is a vertical four-cylinder engine of 16-HP. placed in the front of the car, and the transmission is by means of gearing. The car weighs only about a ton, and has done some remarkable running this spring.

FIG. 135.—MORS 16-HP. RACING CAR.

187

Chapter XI

THE GOBRON-BRILLIE MOTOR CARRIAGE

The Gobron-Brillié touring car

THE motors of the various carriages hitherto described have not been designed with any special reference to neutralization of vibratory efforts, except in so far as the use in some cases of cranks at 180° may be said to secure this. The comparative freedom from objectionable vibration of some of the more recent car motors has been due more to lightness of parts and high speed, with consequently small dis- **Vibration** turbing effort per working stroke, than to any special care in other respects. Experience does not show that there is very much room for improvement in this respect, although it may be admitted that the irregular effort of the piston during a working stroke, when the motor is running light and the car standing, is objectionable in many of the cars of older make, in which the perfection of workmanship of the main parts and the governor adjustment were not equal to those of the recently constructed cars. See chapter on motor balance and vibration.

Among the more serious efforts to produce a very steady, running motor are those of the Société des Moteurs Gobron et Brillié, one of whose cars is illustrated by Fig. 136, which is from a photograph of a touring car, and in detail by Figs. 137–144.

Underframe The underframe consists of a main side-trussed tubular structure, transversely stayed by members, which carry the motor and transmission gear; the body sits upon small face plates, with intervening rubber cushions, attached to the upper tubular member of the side frames. The motor is placed over the hind axle, and transmission is effected by spur gearing on the chain pinion shaft, and on an intermediate shaft. On the latter is a friction clutch by which the communication of motion to the former is controlled. The chain pinion shaft is driven through a differential gear, which it carries, and upon its ends are the chain pinions by which motion is communicated to the driving wheels by chains running on sprocket-wheel rings fixed to the spokes of the drivers by seven stud bolts.

The Brillié motor The motor itself has two cylinders, each containing two oppositely reciprocating pistons connected in pairs to cranks at 180°, so that both

188

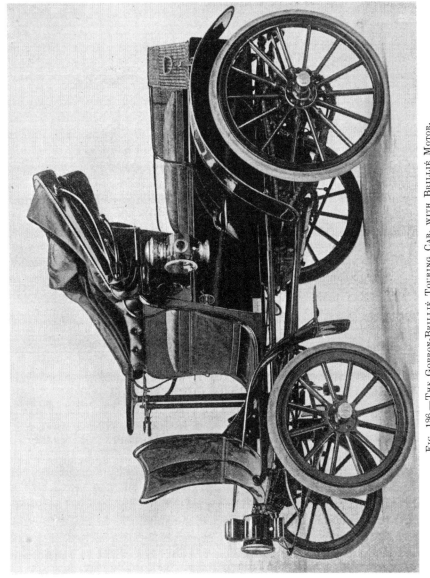

FIG. 186.—THE GOBRON-BRILLIÉ TOURING CAR, WITH BRILLIÉ MOTOR.

189

the upper pistons and both lower pistons have simultaneous balanced movements. The motor cylinders are 80 mm. = 3·15 in. diameter and the same in stroke, and it gives 6 HP. at a speed of 700 revolutions per minute.

The steering gear is also novel in some respects, and comprises an ingenious if somewhat complicated arrangement, by means of which the usual variable relative angularity is given to the two steering wheels by apparatus which is practically an automatic self-holding steerage. The carburettor, petrol feed, governor and throttling arrangements are combined, and form another departure from the ordinary practice.

The main features of the motor are shown by Figs. 137 and 138, which are respectively longitudinal and transverse vertical sections. With the exception of the necessity for a crosshead connecting the two upper pistons and their connecting rods, it is a very simple motor, although the whole of the connections are not shown in the engravings, nor is the case which continues the enclosure of the engine from the part where the crank case stops, near the figures 6, 7, Fig. 137.

As in the Linford and other engines of years ago, the two pistons, working in opposite directions in one long cylinder, secure very high velocity of expansion of the heated gases, and one set of valves serves two pistons. The total expansion is, of course, the same as with the ordinary arrangement, only the rate at which expansion is effected is doubled, and this should give higher duty. The two cylinders give an impulse for each revolution with a four-stroke cycle, as in Panhard & Levassor cars.

It will be seen that the two upper pistons 1, 2 are connected to the crosshead 3, which has short rods formed upon it, and pivoted in the piston in the usual gas-engine way. This crosshead is a rigid casting, which maintains the parallelism of movement of the parts connected to it, including the two connecting rods 6, 7, which are firmly secured into its ends in the same plane as the two short connecting rods. So long therefore as the two rods 6 and 7 are of precisely the same length, there will be no cross stresses on the pistons 1, 2, even when the explosive combustion occurs in the space c of one cylinder only.

Valve gear   By means of the gearing, indicated by dotted lines in Fig. 138, the smaller wheel being shown in Fig. 137, a spindle carrying cams D is rotated at half the speed of the crankshaft. The cams give a fixed range of lift to the exhaust valves E through the rods 11, the lower ends of which are pivoted to a spindle carrying rollers which run on the cams. The lower ends of the rods are guided by radius arms pivoted at 12, and they thus have a slight angular motion, which is allowed for in the holes in the case through which they pass.

Valves   The air and vapour admission valves A have external springs, according to the Patent Specification, No. 29074 of 1897, as shown in Fig. 138, but the preferred method is that shown in Fig. 139, where the spring is enclosed and removable valve seats used. The casting forming the cover of this valve, and the passage for the air and vapour from the

190

admission throttle valve 14, and vapour admission 7, is held in place by a bridge and clamping screw not shown.

The carburettor, as it is usually called, is, in fact, a rotative petrol **Carburettor**

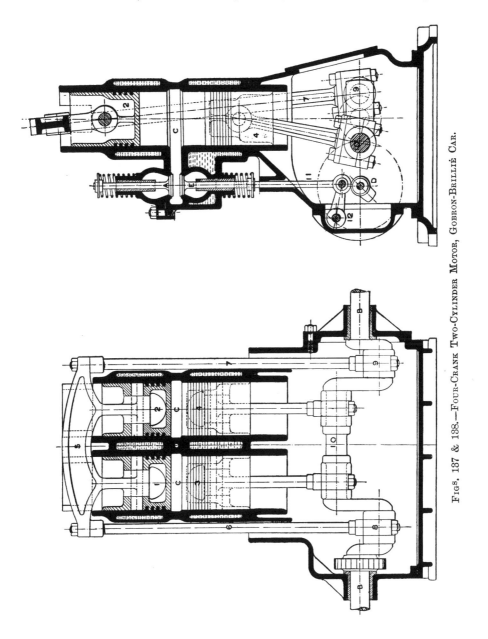

FIGS. 187 & 188.—FOUR-CRANK TWO-CYLINDER MOTOR, GOBRON-BRILLIÉ CAR.

measured feed apparatus and atomiser. It is described in Patent No. 17094 of 1898, and will be readily understood from Figs. 139 and 140, which show it as constructed. In the former is a side view showing the operating mechanism, and in the latter the petrol measuring and air admission arrangements are shown by vertical and horizontal sections.

Petrol food measurer

The measurer consists of a conical-seated multiple bucket 5, held upon its seating, formed in the casting 2 by a light spring 3, which sits upon the little disc 4 upon a small central spindle. Oil is fed in at the entrance 16, and thence finds its way all round the larger part of the bucket, and enters by the annular passage 5, exactly opposite which is the outlet passage 7, near which the bucket fits in the same way that a plug fits in an ordinary shell cock.

The horizontal section is taken on the line of x x the vertical section, and in it will be seen a passage with an air admission pipe 6. When the air and vapour admission valve A admits a charge, the greater part of the air is drawn through the admission valve 14, Fig. 139, but at the same time a small quantity of air enters the pipe and passage 6, and carries with it the small quantity of petrol which is in that bucket which is opposite the passages 6 and 7, into the pipe 15, passing on its way a small grating at the inner end of the passage 7, not shown in the drawings. An air escape is provided by the capped upper extension 17, which also forms a funnel for the admission of cleaning petrol, which may be run out by means of the screw plug valve 18, and outlet 19.

The governor

The governor contains several new features. On the crank shaft B is placed an eccentric, with strap and arm jointed to a swinging lever 22, pivoted at 25, as seen in Fig. 139. On the pivot spindle 25 is a loose disc 24, acting as a weight whose inertia is employed as now to be described. Motion has to be communicated to the ratchet wheel on the end of the bucket spindle by the pawl 8 through the rod 21, actuated by the pivoted hit-and-miss lever 35. On the lever 22 is a bell-crank pusher lever 32, on a pivot 33; this lever is caused to press with a given force upon a pin 26, by means of a coiled spring at 33, the pin being fixed in the disc 24, and projecting through a large hole in the lever 22. On the end of the pin 26, at the back of the disc 24, is a spring 27 (see also Fig. 139A), fastened to the lever 28; this prevents the rotation of the disc, and gives it a certain persistence of position.

The operation of the governor is as follows: So long as the speed of the engine is normal, the spring at 33 is sufficient to enable the pusher lever 32 to push the disc by means of the pin 26, and to maintain the position on the lever 22 shown in the engraving, the inertia of the disc at this speed of movement not being sufficient to deflect the spring at 33. When, however, the speed of the engine materially exceeds the normal, the inertia of the disc 24 is sufficient to push the upper part of the bell crank further backwards before the disc gets into motion, the pin 26 moving across towards the opposite side of the large hole seen in the lever 22, Fig. 139. The other end 34 of the bell crank pusher is thus lifted, and it misses the finger on the lever 35, and hence the stroke of the lever 22 is made without giving motion to the rod 21, or the petrol-measuring bucket, the engine then gets no petrol until the speed has fallen to the normal. At 36 is a projection from some part of the motor which prevents descent of the lever 35 below the position shown. It has been mentioned that at 14 there is an air throttle valve, and by this the speed of the engine may be varied between about 300 revs. per minute, and the maximum

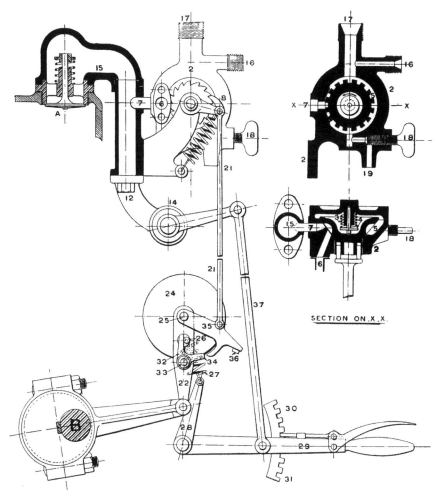

SECTION ON X.X.

FIG. 139.—PETROL-FEED CARBURETTING APPARATUS AND GOVERNOR OF GOBRON-BRILLIÉ CAR.
FIG. 140.—VERTICAL AND HORIZONTAL SECTIONS OF INTERMITTENT PETROL-FEED APPARATUS.

speed of, say, 1000 revs. per minute. The normal speed of the engine being 800. When the air supply is lessened or increased by means of lever 29 and rod 37, the tension on the spring 27 is at the same time increased or decreased, thus when the lever is moved to the position 31 on the quadrant and the persistence of the inertia disc 24 is increased, the governor comes into play at a lower speed, and the quantity of air entering at 14 is diminished. The throttle valve 14 is similar in form and operation to that shown at Fig. 34 of the Daimler motor, and Figs. 60 and 62 of the Phœnix-Daimler motor of Panhard & Levassor.

The Brillié steering gear
The steering gear, illustrated by Figs. 141 and 142, is also the subject of a patent, namely, No. 1524 of 1899, its object is to give a variable multiplication of the ratio of movement between the steering hand wheel and the front wheels as they are moved for turning shorter and shorter curves. From the plan, Fig. 142, it will be seen that two arms project from the stud axles for steering by the rod H H, as in the steering gears previously

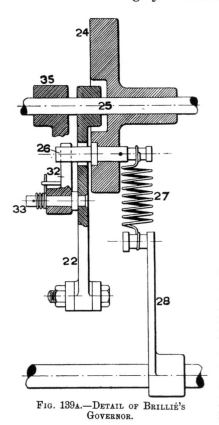

FIG. 139A.—DETAIL OF BRILLIÉ'S GOVERNOR.

described, one of the arms being extended to G, for the rod Q, which is actuated by the arm M, which receives motion differentiating from that of an ordinary arm, or from the steering wheel V, in the manner now to be described. Attached to the car is a fixed steering pillar T, within which works freely the steering spindle A carrying the steering wheel V; at the bottom of the fixed steering pillar is a fixed pinion K, and at the bottom end of the steering spindle itself is an arm L, carrying at its outer extremity a vertical bushed bearing carrying a spindle D, on the upper end of which is a spur quadrant B gearing with and having an epicycloidal motion upon the fixed pinion K, as the arm L is given angular motion by the steering wheel. At the lower end of the spindle D, is a short arm M, and pin C fixed in line with the centre of the quadrant, the pin C gives motion to the rod Q double pivoted at G and it has, as the arm L is moved from the position shown in the engraving towards the letter Q, movement which follows a cycloidal curve imparted to it. The actual movement is plotted on Fig. 142, where dotted lines on the steering wheel show angular movements of about 10°, each of course appearing as equal. Now, for either of these movements of 10° of the hand wheel there will be a different amount of movement of the pivot at the end G of the steering rod Q, or of the stud axles as shown by the dotted lines radiating from the vertical pivot F of the left-hand axle. It will be seen that the 10° of movement of the hand wheel is accompanied by an almost imperceptible movement at G, while on the other hand when 30° of movement of

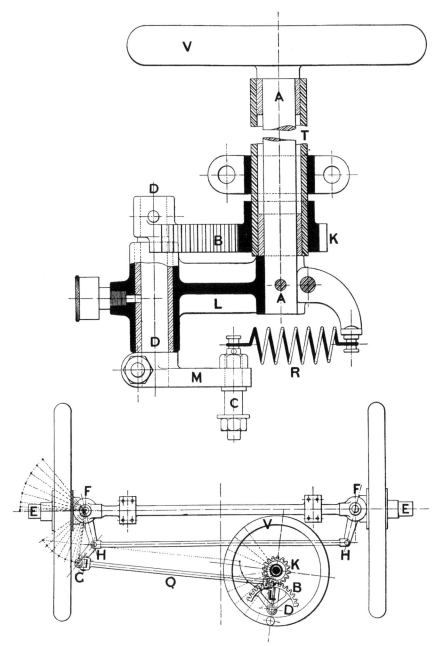

FIG. 141.—ENLARGED DETAIL OF GOBRON-BRILLIÉ STEERING DEVICE.
FIG. 142.—PLAN OF GOBRON-BRILLIÉ STEERING GEAR.

195

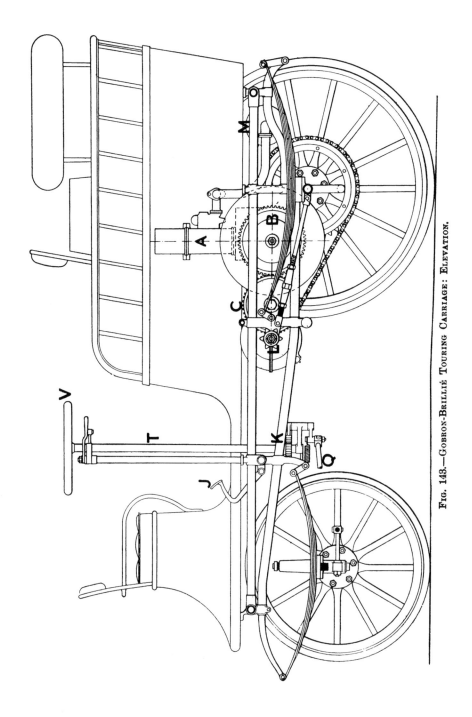

Fig. 143.—Gobron-Brillié Touring Carriage: Elevation.

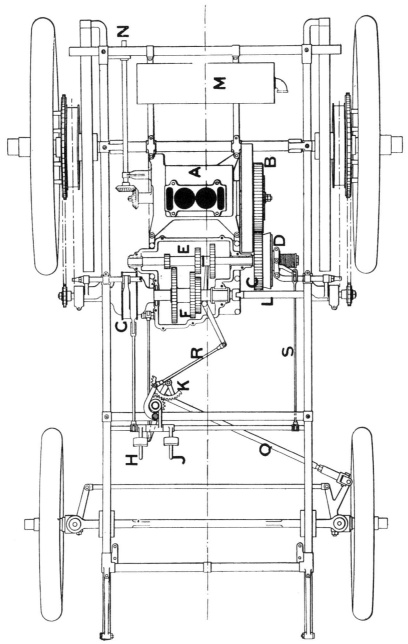

FIG. 144.—GENERAL ARRANGEMENT OF GOBRON-BRILLIÉ CAR: PLAN.

197

the hand wheel have been made, the motion of the steering axles is multiplied several times.

On the wheel is plotted the cycloidal curve described by the pin c at the wheel end of the steering rod Q, and an inspection of this will show that at the beginning and end of this cycloidal path the motion at Q in the direction of the length of the rod Q must be small. The spring R tends to keep the spindles A, D and the pin c all in line, and also to take up any slack that might otherwise cause a rattle. It will be seen that at most of the positions of the hand wheel v the steering arm M is practically locked in the position given to it by the wheel, because a push from the road wheels acts at a disadvantage, which is represented by the length of the pitch radius of the quadrant into the radius of the arm M.

Transmission gearing The general arrangement of the motor, machinery, and frame will be gathered from Figs. 143 and 144, which give a plan and elevation of the main features. In this case, however, I have not been able to make these drawings complete as to air, oil, water, and mixture connections, as in the case of the Daimler, Panhard, Pengeot, Benz, and Decarville; but the knowledge of these latter will enable any reader, with the aid of the separate Figs. 137–144, to understand what these arrangements must be.

In Figs. 143 and 144, the structure of the frame is clearly seen. The motor A drives by spur wheel B the spur wheel C and clutch D, the intermediate shaft on which are the spur wheels E gearing selectively with the spur wheels F on the sprocket pinion shaft, the two parts of which receive their motion from the differential gear A. The motor is started by putting a handle on the end of the spindle N, which is fitted at its other end with a bevel pinion and automatic release of an ordinary kind. The silencer is placed at M.

At H, J are pedals which operate the hand brake on the outside of the differential at A and the main clutch C respectively by means of connecting rod s. The speed gear is operated by the handle of a vertical spindle by the side of the steering pillar T, the connecting link for the purpose being seen at R. The part L of the chain pinion shaft is connected to the part in the gear box by means of a joint coupling.

But few of these vehicles have yet appeared in this country; that which was exhibited at the Richmond Show of the Automobile Club in 1899 did not take part in the trials, but was awarded a silver medal for novelty and ingenuity of the parts now described, and for appearance.

Chapter XII

THE CANNSTADT DAIMLER MOTOR VEHICLES.

IN the works of the Daimler Moteren Gesellschaft, at Cannstadt, a number of Daimler motor vehicles have been made, including the Daimler-Maybach car of 1896, with suspended motor and belt drive, as already fully described, and of which several are in use in this country ; the later type of gear-driven cars with motors, as described with reference to Figs. 144A, 144J ; the heavy lorries, two of which were tested during the Automobile Club trials of 1899 ; and, lastly, racing cars up to 24 HP.

The general appearance of the 24-HP. racing car is shown by Fig. 144A, which is from a recent photograph of the car ordered by Count Zborowski for use in this country and in the French races.

Racing car 24 HP.

This is a fine-looking car, geared to 46 miles per hour, a speed which may be considerably exceeded on good level road with governor cut out. On its lower gear it will ascend gradients of 1 in 15 at about 20 miles an hour. The gear is of the kind next to be described with reference to Figs. 144B–144J. It carries petrol and water for any distance that is likely to be covered in a day, the water consumption being very small indeed. Three brakes are provided, and on the wheels are large pneumatic tyres.

The general arrangement of the under frame, motor and gearing, and water cooler will be gathered from that of the Cannstadt Daimler Victoria, illustrated by Figs. 144B–144E, details of which, and of the Cannstadt wagonettes, are shown by Figs. 144F–144J.

Geared Cannstadt cars

In several important points these cars and their gearing differ from the Daimler motor cars of Panhard & Levassor, and of the English Daimler Company.

From the illustrations it will be seen that the motor M, which is of $5\frac{1}{2}$ brake HP., is attached by clamping lugs to longitudinal frame tubes F, which are themselves carried at the front by cap plates resting on the springs and at the rear by hangers from the body at the back of the well, the body being carried about 18 inches further back on long double-leaf springs, so that the motor-carrying tubes are to some extent independent of the carriage body. Thus some of the vibration due to the working of the

Mounting of motor

199

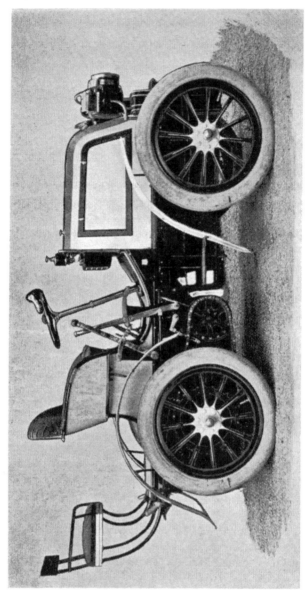

Fig. 144a.—The Cannstadt Daimler 24-HP. Racing Car, 1899–1900.

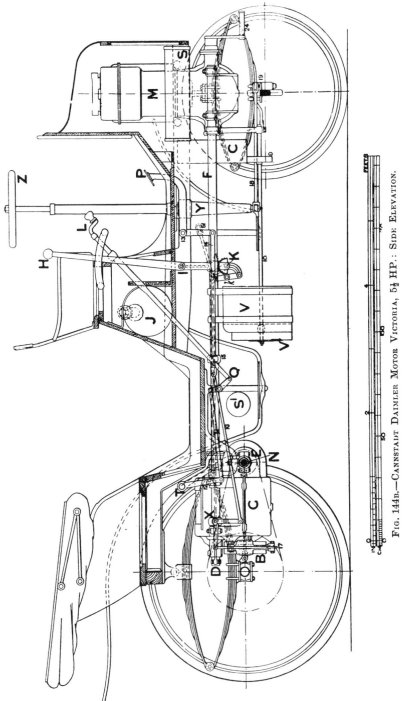

Fig. 144B.—Cannstatt Daimler Motor Victoria, 5½ HP.: Side Elevation.

motor is absorbed or intercepted by this tubular support, which has very slight elastic flexibility.

Transmission gear

The crank extension or clutch shaft D is connected to the speed gearing, which will be hereafter described in the box G. Below the extension of the shaft D in this box is a short parallel spindle D^1, Figs. 144E to 144J, on which are the fixed speed change spur wheels, and at the forward end a bevel pinion E^1, gearing into a corresponding wheel E^2 on the differential gear in the box E on the transverse chain pinion shaft N N. Motion is communicated from the shaft N to the driving wheels by chains running on sprocket rings carried by stud bolts through the spokes.

Speed change gear

The method of operating the speed-changing gear is shown by Figs. 144F–144H, but the arrangement of the parts in these views differs a little from that of the Victoria. From these figures it will be seen that on the extension of the clutch shaft D within the gear case G, there are mounted two sets of spur pinions, 36, 37 and 38, 39, which may be caused to slide along the shaft on the feather d; and corresponding to the four spur pinions are spur wheels 40, 41, 42, 43, keyed to the spindle D^1. The position occupied by either set of spur pinions on shaft D is controlled by the hand lever H, at the driver's right hand (see also Figs. 144B–144D). The lever H is mounted on a cross spindle I, and is guided at its upper end by a grid sector 33 carried from the body of the carriage. At its lower extremity is a toothed sector H', which, by causing the spindle I to slide longitudinally in its bearings i and i', may be engaged with the teeth of either of the racks at the extremities of the rods 2, 3, which are free to slide longitudinally in their bearings in the gear case G. On the rack rods 2, 3 are downwardly projecting arms, which are forked at their lower extremities, these forks engaging with collars e and e', formed on extensions of the boxes of each pair of spur pinions. Projecting from the controlling lever H, where it joins the spindle I, is a second arm 34, having a rounded head, which, whenever the lever H approaches the vertical position, comes into contact with the roller 35 on the extremity of the bar 4, Figs. 144C and 144F–144H, which is pivoted at its front end from the gear case G, and at the back on the upper extremity of the lever 32. This lever is itself pivoted upon a thrust bearing cap 50 on the end of the shaft D, and by which the latter is pulled or pushed for operating the clutch C. The lower end of lever 32 is held in position by link 5. When the projection 34, by the operation of the lever H, retires from the roller 35, the spring at F, acting on the rod 6, pulls the lever 32, which, through the thrust pivot 50, puts the cone C into gear. When the lever H is in the position shown, everything is out of gear; by forcing the lever to the right and then pushing it to the end of a of the outer slot of the quadrant, the toothed quadrant H' enters into gear with the rack 3, and causes it to move in the opposite direction to that given to the lever H, and the pinion 36 is thereby put into gear with the wheel 40, which gives the lowest speed. To put the wheels in gear for the next speed, the lever H is removed from a across the quadrant to the position a', having,

Four speeds

202

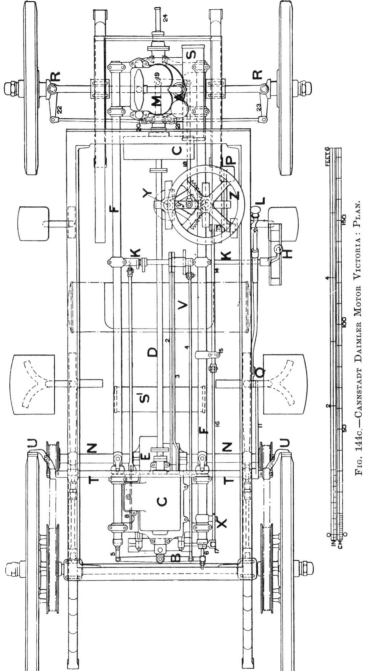

FIG. 144c.—CANNSTADT DAIMLER MOTOR VICTORIA: PLAN.

203

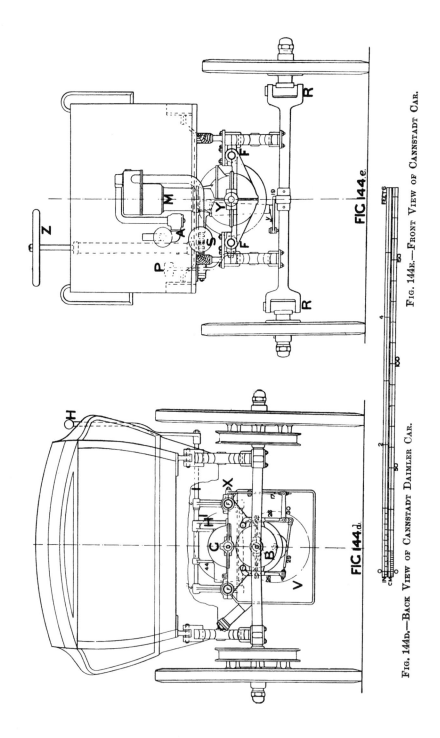

FIC 144.e.

FIG. 144E.—FRONT VIEW OF CANNSTADT CAR.

FIC 144.d.

FIG. 144D.—BACK VIEW OF CANNSTADT DAIMLER CAR.

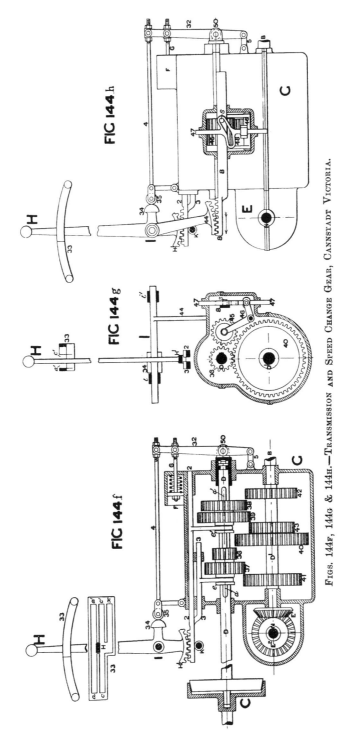

FIG 144 h

FIG 144 g

FIG 144 f

Figs. 144f, 144g & 144h.—Transmission and Speed Change Gear, Cannstadt Victoria.

205

in passing from one position to the other, momentarily thrown the clutch out of gear, and then put the pinion 37 into gear with the wheel 41. By returning the lever H to its original central position, and then pulling it to the left into the middle slot of the quadrant, the toothed sector H' at bottom of the lever enters into gear with the rack bar 2, and then by moving the lever to the end c of the slot, or to the end c', the pinions 38 or 39 are put into gear with the wheel 42 or 43 respectively, the last mentioned giving the highest speed. Thus four speeds are available. For **Reverse motion** changing the direction of motion, the lever H is pulled over into and along the third slot x in the quadrant 33, which puts the toothed quadrant H' out of gear with the racks 2, 3, and puts the toothed quadrant 44 into gear with the rack bar 8, and pulls the bar in the direction of the arrow. When this occurs, the pin s slides along the bent slot 48 carried by the spindle 47. As the bar 8 cannot move except in the direction of its length, its movement in the direction of the arrow lifts the slot bar 47, Fig. 144H, and in doing so raises the horizontal part of the bell crank lever 46, Fig. 144G, and thus puts the long spur pinion 45 into gear with the slow-speed pinion 36, and with the slow-speed wheel 40, its intervention reversing the direction of rotation which wheel 40 would receive if pinion 36 were put directly into gear with it. The toothed ends of the rack bars 2, 3, 8 are all supported by friction rollers on the spindle K.

Spoon brakes There are three brakes; two of them are brakes U U on the shaft T, actuated by the handle L on the end of a screw spindle and tubular rod ending at the pivot Q on one end of a double-armed lever. The latter is connected with the rod 11 to the arm 12 on one end of the way shaft T. The other is a well-designed friction brake seen at B, and in detail in Figs. **Rigid friction brake** 144I and 144J. At P Fig. 144B, will be seen a pedal pivoted at 13, and by a downwardly projecting arm and rod 14, guided at 15, connected to the rod 16, by which the vertical arm of a bell crank lever pivoted at x is operated. Pendant from the horizontal arm of the bell crank lever x is a rod 17, seen also in Fig. 144I. From this latter figure it will be seen that when the rod 17 is by these means lifted, the bell crank lever 30 pulls the rod 29, and equally and oppositely pulls the levers 25, 26, pivoted at 27, 28, and thereby forces the brake blocks w upon the brake pully B. When the pedal P is used for operating this brake, a slotted rod 51 connected to the top of the arm K operates the rod 4, and thereby the lever 32, and thus puts the clutch c out of gear whenever the brake is used.

Steering gear The steerage connection from the steering wheel z, on the lower end of the spindle of which is a pinion gearing with a toothed quadrant on the top of a spindle, passing through the tubular support Y, and having upon its lower end an arm y. This is connected to the rod 18, and thereby to the arms 22, 23 on the pivoted axles by means of the bell crank pivoted at 19, and the rods and short links 20, 21.

The supply of petrol is carried in a cylindrical tank J, and from it the float-feed carburettor at A is fed.

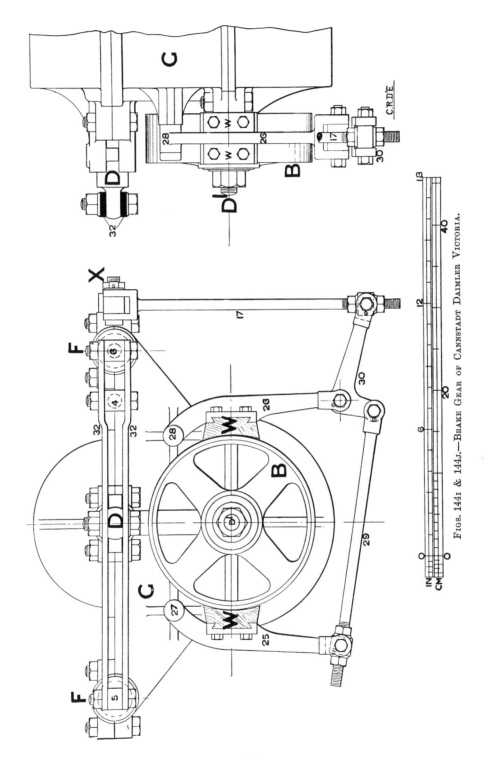

Figs. 144i & 144j.—Brake Gear of Cannstadt Daimler Victoria.

207

Water
cooler

Cooling water is carried in the tubulous water cooler at v, and water circulation is kept up through pipes not shown by a small semi-rotary pump. The water in the tubulous condenser is cooled by the rush of air maintained through its tubes by means of a fan v'.

Two silencers are employed, one seen at s and the other at s'.

The Cannstadt wagonettes have all their machinery mounted upon the frame formed by the longitudinal tubes and cross connections to springs, and the body is thus free from important fixtures; but the arrangement of the slotted quadrant for the speed change gear, and of part of the steering spindle column and other handles, is such that a good deal more is attached to the body than is ideally desirable. The friction brake is one of the best employed by any of the makers, but as it is only applied through the chains it is not as valuable as it would be if applied to the main wheels.

Generally tube ignition is used on these cars, but the Bosch igniter is sometimes fitted.

Chapter XIII

THE VALLÉE, DELAHAYE, CROWDEN, AND OTHER CARS

THE cars which have now been fully described are those which have been most before the public, or have attained high distinction. There are many others, the mechanical features of which are similar in essentials to those already illustrated, or of which the chief points may be understood from a brief description, aided by fewer drawings than have been necessary for the others.

Among these latter is the recent form of M. Vallée's car, in which a somewhat noteworthy departure has been made towards simplification, by adopting only one speed reduction and no speed-change gear. All variations of speed are obtained by varying the speed of the motor by running with a throttle valve, and the power of the motor, said to be 16 HP. in the racing vehicle illustrated by Fig. 143, is transmitted to the driving wheels by one wide belt, as seen in Figs. 146 and 147. The motor has four cylinders c c, 110 mm. or 4·32 in. in diameter, and 200 mm. or 7·87 in. stroke, and is placed on the front of the frame, which is tubular. The motor runs normally at 600 revs. per minute, and at this speed the driving wheels make 260 revs. per minute, the pulleys A, B, being in diameters as 5 to 11·5. Electric ignition is used, and the magneto-electric apparatus is placed in the box D. Mechanical tremblers are used, operated by four cams adjoining the exhaust-valve cams, which run at one-half the speed of the crankshaft, and lift the tremblers successively. The cams are mounted on the separate spindles, and insulated contact piece carriers are mounted on levers at one of the ends of each of these. These levers are connected, and by a rod are moved simultaneously from the seat, so that the period of contact and of ignition can be varied at will.

For starting purposes, the exhaust valves at L are held open by separate cams, which can be put into position temporarily for this purpose, so that for starting the compression pressure is lowered.

The leather belt E is about 250 mm., or nearly 10 in. wide, and its tension maintained, increased or decreased by moving the rear axle backwards or forwards with its springs in the slotted carriers G G' by means of the lever H and quadrant J, after the manner of the Bollée voiturette.

The Vallée racer

209
14

Fig. 145—The Vallée 16-HP. Racing Car.

210

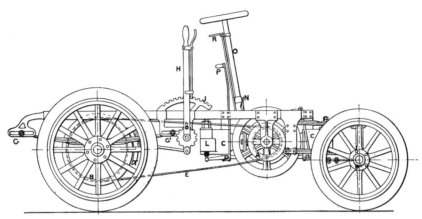

FIG. 146.—VALLÉE 16-HP. RACING CAR: ELEVATION.

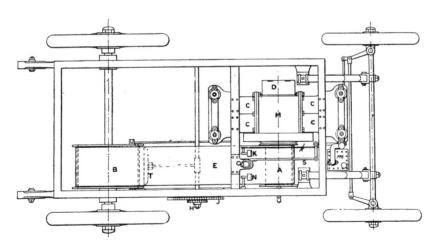

FIG. 147.—VALLÉE 16-HP. RACING CAR: GENERAL ARRANGEMENT AND PLAN.

The pulley B encloses the differential motion, a reversing gear, and an internally-acting hand brake, actuated by the pedal N.

Having four cylinders and electric ignition, the range of speed of the engine is very large, with of course any degree of variation within that range. Outside it, for very slow speeds of motor, the belt is allowed to slip. The throttle valve is actuated by a pedal K, which by means of the connecting rod k controls or arrests the supply of mixture from the carburettor M. On the pillar o of the steering wheel is the handle R for actuating the reversing by means of the rod r. By means of the handle P and its connections, the ignition period is advanced or retarded.

When the rear axle is pulled forward by the handle H, the driving pulley is pulled into the brake block T.

The car was tested at the Paris-St. Malo and Paris-Ostende races when it behaved remarkably well, ran very silently, and covered the distances of 370 and 322 kiloms. in 8 hr. 41 min. and 7 hr. 47 min., or at an average speed of 42 kiloms. or 26 miles per hour, the maximum speed not being so much above the mean as usual, but the hill-climbing speed unusually high, and well it might be with about 16 HP. and a load of only two persons.

A good deal may be urged in favour of the simplicity secured by the arrangement of a big powerful single belt, and easy means of varying the tension upon it, and prevention of excessive tension due to atmospheric changes; but it must be remembered that the system means the employment of a very powerful motor. For high-speed purposes this has much to recommend it, as the extra cylinders and valve gear will in most cases be preferable to multiplicity of gear. For moderate speeds, touring and light car -use, however, it does not commend itself, or will not, until more simple multiple cylinder motors than are yet made are available.

The Delahaye carriages, in their different forms, are also well known. They are belt-driven cars, propelled by a double-cylinder horizontal petrol motor, easily accessible and well made. Ignition is electrical with retarding arrangement used at starting. A mixture controlling device is used, by which the speed is controlled through a considerable range. It is actuated by a pedal, and is used for all speed variation on ordinary running. The design of the carriage generally is good, but the mechanical arrangements will not be here illustrated.

The following are also among those whose names, as manufacturers of cars of the type dealt with, are favourably known in France :—

MM. Gauthier-Wehrlé, with horizontal opposite cylinder motors placed transversely, and with open-gear driving.

The Bolide carriages, by M. Léon Lefebvre, with horizontal cylinder motor and part belt and part spur-gear transmission.

M. Amédée Bollée, with horizontal motor and part belt driving.

The Ducroiset Company, with horizontal cylinder motors, placed in front, and belts. The Hirtu car made in England by Messrs. Marshall & Co., Clayton, Manchester.

Georges Richard, maker of several forms of carriages.

MM. de Diétrich & Co., carriages of several forms driven by Amédée-Bollée motors.

In America there are several makers of petrol motor vehicles, with seats for two, four, or more. Among these are :—

The Duryea, with motors of two more Duryea forms, some of which have been patented in England, including No. 1140 of 1895, No. 7036 of 1896, and No. 14301 of 1896.

The Winton Company; the Haynes-Apperson Company.

Fig. 148 gives a perspective view of the underframe gearing and

FIG. 148.—CROWDEN'S 10-HP. MOTOR AND CARRIAGE: UNDERFRAME.

motor of a six-seated car made by Mr. C. T. Crowden, Leamington. The motor has twin cylinders, with a central fly wheel on a shaft with overhanging cranks. Motion is transmitted from the crankshaft to a second motion shaft by a chain on a sprocket wheel next the fly wheel, the chain driving a sprocket wheel loose on the shaft. On the sleeve carrying this sprocket wheel is another such wheel driving a lay shaft below the engine, pulleys on which give motion to belts driving pulleys on the second motion shaft, for three different speeds and reversing. By means of jockey pulleys either of the belts are selectively used. A band brake is applied to the outside of the differential gear case.

213

The cylinders are placed in a sheet-copper tank of water instead of being jacketed in the ordinary way.

The valves are all in separate valve boxes, which can be easily removed for renewal or repairs.

A surface vapouriser is used, and by means of a throttle valve, controlled near the driver's seat, the speed of the motor may be varied through a considerable range, electric ignition being used.

By means of a three-way valve, actuated from the seat, a further

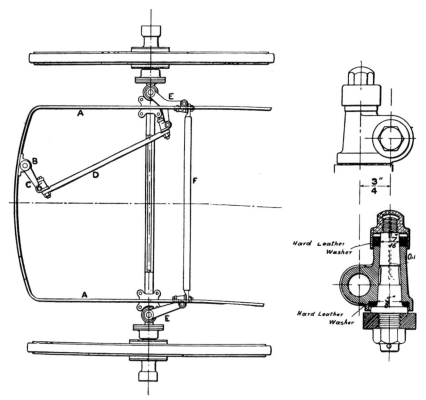

FIG. 149.—CROWDEN'S STEERING GEAR.
FIG. 150.—CROWDEN'S DOUBLE-CONE DUST-PROOF STEERAGE COUPLING JOINT.

variation of power is available. Either of the cylinders may be used alternately, one cylinder being sufficient to drive the car on level roads at a good speed, or both cylinders may be used, the change being effected as quickly as desired.

The steering wheels are of the same size as the rear wheels, and the pivots of the steering connections are all of a double conical dust-proof enclosed kind, illustrated by Fig. 150, the steering connections by means of this pivot being shown by Fig. 149. In this A is part of the light flat bar frame used by Mr. Crowden, to a part of which is attached a bearing B for the foot of the steering spindle. On this spindle is an arm C, connected to rod D through one of the double-cone connectors. The other end of this rod is similarly connected to the double arm E on the steering axle. The

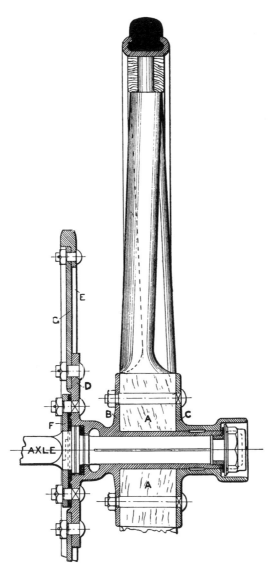

FIG. 151.—CROWDEN'S WOOD-SPOKE WHEEL FOR MOTOR VEHICLES

arms E and E′ being rearwards, the bar F connecting them is normally in compression, but as it is made stiff by the use of a short tube, the advantage of the rearward arms is secured without detriment. The connectors at the end of the bar F are also of the kind shown by Fig. 150.

MOTOR VEHICLES AND MOTORS

A very simple arrangement of hangers for the hind axle springs has been devised by Mr. Crowden, so that the distance between pinions and sprocket wheels of the driving chains may be adjusted to the chains without affecting the belts.

Attention should be particularly drawn to the construction of the road wheels, which are a very ingenious modification of the Hancock wheels. The roots A of the spokes are of the full width between the flanges c and B of the naves, and are narrowed from this root to the ordinary width of spokes, by cutting away from one side only. Two spokes cut out of one piece of stuff with practically no waste. They should be split if made of oak, which is usual. The excess of width of root thus stands on one side of the spoke, and by placing the spoke alternately one way and the other in the naves, an excellent staggered spoked wheel of great strength is secured. The construction of the wheel is shown by Fig. 151.

Chapter XIV

MODERN LIGHT PETROL MOTOR VEHICLES

THE whole of the modern petrol vehicles which have now been described have been of the larger, more powerful, and more expensive types, and some of them have been of the high power and small seating capacity for racing purposes. Several of those which have been illustrated are fitted either with the bodies shown in the engravings, or

FIG. 152.—DECAUVILLE LIGHT MOTOR CAR.

with the different forms of bodies which constitute them motor vans, motor lurries, and light motor omnibuses.

In the earlier stages of the modern developments the tendency was to pay most attention to these larger and more powerful vehicles, chiefly as a result of the inducements of the French races. During the past two years, however, more and more attention has been given to the production of lighter and less powerful vehicles, capable of a moderate speed, to carry two, or at the most three, persons, and to be satisfactory in appearance as a motor carriage, and not to be merely a carriage without horses. Among the vehicles of this class is that of M. Decauville.

THE DECAUVILLE LIGHT MOTOR CAR.

Figs. 152 and 153 give perspective views of one of these cars, and Figs. 154, 158 show a side elevation, plan, and end elevation of the motor and gear of the two-seated car.

De Dion and Bouton motor

The motor M has two vertical cylinders, each of which is very simi-

FIG. 153.—DECAUVILLE LIGHT MOTOR CAR: BACK VIEW.

lar in construction to that of the 1¼-HP. type, de Dion-Bouton tricycle motor. They are mounted on a common crank case K, to which each cylinder is bolted by two long vertical bolts, which pass between lugs cast on either side of the cylinder head, so that they keep the head joint tight, as well as bolting the cylinder to the crank case. The motor is carried from the longitudinal tubes F of the frame by means of the four arms A, which are cast with the crank case K, which is made in two parts, joined **Valves** by longitudinal bolts, as seen in Figs. 155 and 156. The cylinders and valve chambers are cooled by numerous radiating ribs, no water being carried on the car. The petrol supply is carried in the tank P, on the left hand

218

side of the frame viewed from the back. This tank is connected by a pipe 2 with the vessel c of the carburettor, which in itself forms a considerable reservoir of petrol. The glass gauge 3, at the back end of the vessel c, indicates the level of the petrol, both in it and the tank p. Above the vessel c of the carburettor, which is of the surface type, is a chamber c′, at the top of which are valves 3, 4, the former for regulating the quality of the carburetted mixture, and the latter for regulating the quantity of the charge passing into the motor cylinders. There are also two adjustable apertures 5, which admit air into the carburettor. These are set before the carriage is started. The air valve 3 is controlled by the hand lever d, and may be adjusted by the driver from his seat, as are also the throttle valves 4, which are both operated from the longitudinal spindle E, by means of the levers and links, 7, 8, 9. A balance weight 10 is mounted on the spindle E, in order to counterbalance the weight of the levers 7 and links 8.

The charge is drawn from the carburettor chamber c′, past the throttle valves 4, and along the pipes 11, into the admission chambers 12 of the motor cylinders. The admission valves are held on their seats by the light springs v, Fig. 154, and are opened in the usual way by the atmosphere pressure, which comes into play when the partial vacuum is formed within the cylinders on the suction strokes of the pistons. It will be seen that these valves are at the outer ends of the valve boxes Q, which are cast in one piece with the cylinder heads, so that when a fresh charge enters, it sweeps over the exhaust valves, thus aiding in keeping these cool, which is one of the chief difficulties met with in air-cooled motors. The admission valves are conveniently accessible by removing the screw-covers 14, and the exhaust valves are directly below the similar covers 15. The exhaust valves are lifted by means of cams, which are driven by spur gears at half the speed of the crankshaft, and which actuate short vertical rods which press against the spring-held stems T of the exhaust valves. One of these is seen in the end elevation, Fig. 156. The arrangement of the exhaust valve gear is closely similar to that of the de Dion tricycle motor, see pp. 258, 268. The motor works on the ordinary Beau de Rochas cycle—suction, compression, explosion, exhaust. The pistons are coupled to cranks at 180°.

The exhaust products are expelled by way of the pipes 13 into the cylindrical silencing vessel s, from which they escape by holes in the back end, as seen in Fig. 156. A small portion of the exhaust is led off by the pipe J, Figs. 154 and 155, from the junction piece x of the pipes 13, at the front end of the silencer s; the exhaust vapour passing up the pipe J, enters a chamber k at the front end of the vessel c of the carburettor. By means of a valve adjusted by the milled head k′, a certain quantity of the hot vapour is allowed to pass through a pipe near the bottom of the vessel c, and escapes at l, at the back end of the vessel. This warms the petrol, so that it is more readily vapourized. The charge in the motor cylinders

Carburettor

Regulation

Throttle valves

Exhaust valve

Silencer

Electric ignition

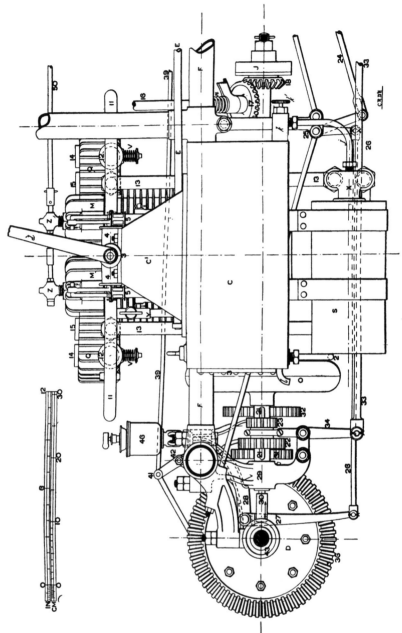

Fig. 154.—Decauville Light Car: Side Elevation of Motor and Mechanism.

220

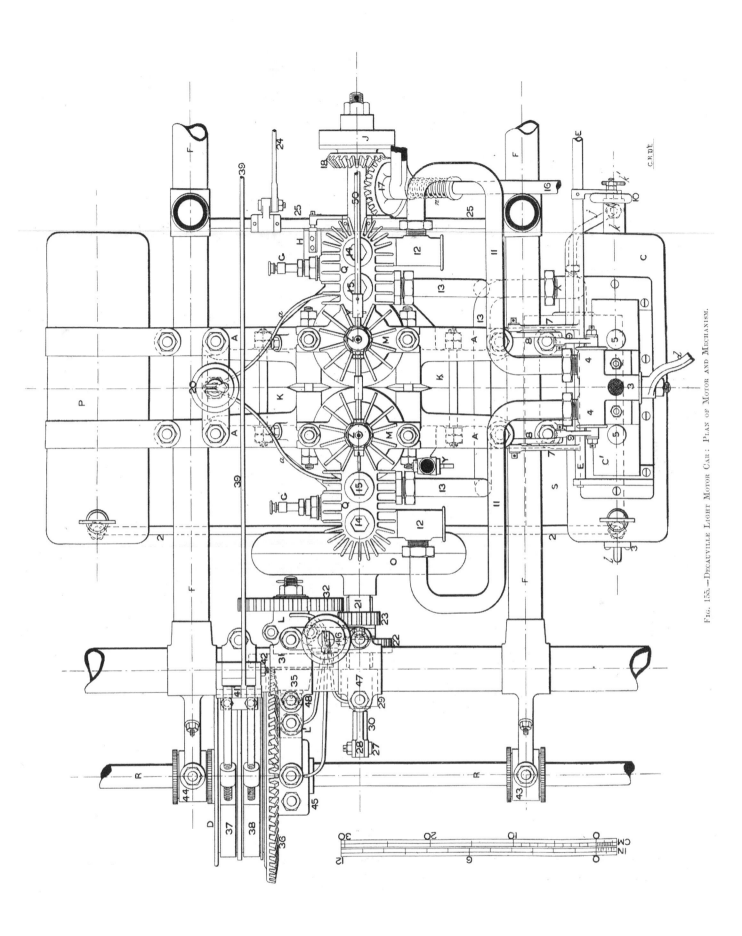

FIG. 155.—DECAUVILLE LIGHT MOTOR CAR: PLAN OF MOTOR AND MECHANISM.

The material originally positioned here is too large for reproduction in this reissue. A PDF can be downloaded from the web address given on page iv of this book, by clicking on 'Resources Available'.

is fired electrically by means of the sparking plugs G, accumulators and
induction coil being carried in the body of the car. The point at which
the charge is fired may be varied by the driver from his seat by adjusting

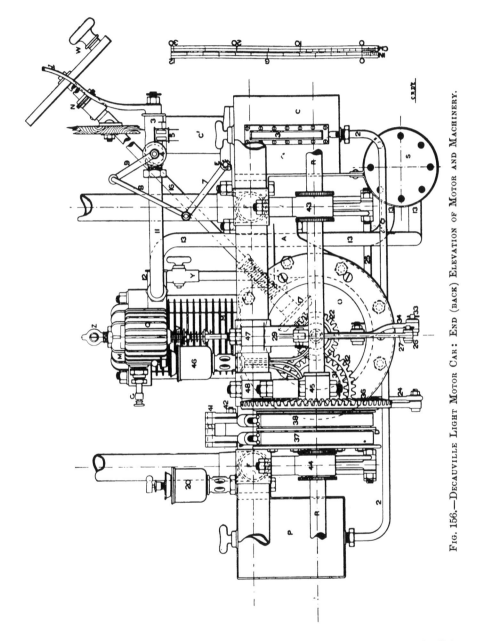

FIG. 156.—DECAUVILLE LIGHT MOTOR CAR: END (BACK) ELEVATION OF MOTOR AND MACHINERY.

the contact device H, Fig. 146, which is similar to that used by de Dion,
see pp. 262, 268.

The motor is started by the driver from his seat by means of the   **Starting
from seat**

hand-wheel w, Fig. 156, which is mounted at the upper end of the spindle 16, carried in bearings attached to the wood structure of the carriage body. At the lower end of this spindle is mounted the bevel wheel 17, which, by pressing down the spindle 16 against the resistance of the spring *n*, is put into gear with a similar bevel wheel 18, mounted on the box J of a silent pawl clutch at the front end of the crankshaft of the motor. When the bevel wheels 17, 18 have been put into gear, a turn of the milled head N, below the wheel w, will hold the spindle 16 in place, and by revolving the wheel w, the motion is communicated through the pawl J to the crankshaft, and the motor started. As soon as this takes place the speed of the motor overruns the case of the clutch J, and putting it out of gear,

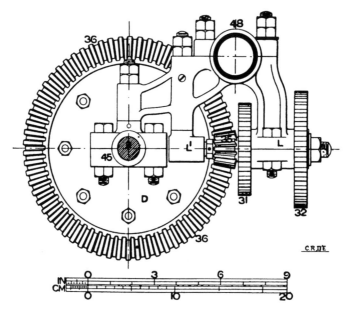

Fig. 157.—Bevel Driving Gear of Decauville Light Car.

the bevel wheels 17, 18 are no longer drivers. The bevel wheels 17, 18 may now be disengaged by releasing the catch N, which allows the spring *n* to push the spindle 16 up into its original position, as shown in Fig. 156. Previous to starting the motor the relief pressure cocks z at the tops of the cylinders are opened. Both are controlled by the rod 50, which is actuated from the driver's seat.

Lubrication The cylinders of the motor are both lubricated by a sight-feed lubricator 20, which feeds through the small copper pipes *a*. Oil is supplied to the crank case *k*, through the pipe and cock *y*.

Transmission gear The power is transmitted from the crankshaft of the motor through a friction clutch o, which also forms the fly-wheel of the engine, to a hollow

222

spindle 21, upon which are carried the spur pinions 22, 23. The friction clutch o is controlled by a pedal in front of the driver, the movement of which is transmitted through the rod 24 to a downwardly projecting arm on the cross spindle 25 A second lever arm on this spindle is connected by the rod 26 to a lever 27, which is pivoted at the extremity of the projection 28, Fig. 154. This forms part of the bracket 29, which is clamped to the back cross tube of the frame and forms the bearing for the back end of the spindle 21, Fig. 154. The lever 27, near its pivoted end, enters a slot cut in the end of a spindle 30, Figs. 154 and 155, which passes through the tubular spindle 21, and which, when moved inwards in the direction of the arrow, Fig. 154, disengages the clutch o, and puts the motor out of gear with the transmission mechanism.

The spur pinions 22, 23 are cut upon the common boss, which is **Driving gear** free to slide on a feather on the spindle 21, and by suitable movement they may be caused to gear with the spur wheels 31, 32, which are keyed on a countershaft, carried in bearings L and L', Figs. 155 and 157. The movement of the pinions 22, 23 is controlled by a hand lever near the driver. This lever operates the longitudinal bar 33, Figs. 154 and 156, which is connected to the lower end of the pivoted lever 34, the upper end of which operates a collar working in a groove cut in the boss between the pinions 22, 23. By moving the pinion 22 into gear with the wheel 31, the **Speeds** high speed is given to the vehicle, and by gearing 23 with 32 the slow speed. Intermediate speeds are obtained by varying the adjustment of the throttle valves controlling the charge entering the motor cylinders, or by varying the point of ignition. Some of the cars of the kind now being described are fitted with three sets of speed gears, instead of two, as here shown, but inasmuch as the high-speed wheels have to be put into gear in changing from the low to the intermediate, or *vice versa*, the arrangement in this respect is not a good one.

The power is transmitted from the countershaft carrying the spur wheels 31, 32, to the rear axle R, by means of a bevel pinion 35, Fig. 157, which gears with the bevel wheel 36. This transmits the power through the differential gear, enclosed in the box D, to the rear axle R. On **Differential** the outside of the differential gear box D, are two band brakes 37, 38, **gear** which are operated by a pedal, connected by the tension rod 39 to a pin 41, carried on the frame tube at 42, and which is connected by adjustable rods to the bands of the brakes 37, 38, the other ends of the bands **Brakes** being fixed to the frame tube.

The back axle R, runs in ball bearings 43, 44, which are carried **Ball bear-** by the rear frame tube, and there are also two similar bearings near **ings and** the hubs of the road wheels, which are seen in Fig. 153. Near the **steel gear** bevel wheel 36, the axle is also supported by a plain bearing 45. The gear is all of steel, and machine cut; the bushes of the bearings 29, L L', 45 are also of steel. They are all lubricated by the sight-feed lubricator 46, which is carried on a metal plate between the inner bolts of the

223

clamp caps 47, 48, which hold the brackets of the bearings 29, L L' to the back tube of the frame.

Steering gear and fore carriage   The car is very neatly carried in front by a single spring, as seen in Figs. 152 and 158. The extremities of this spring D, rest on the top of the pillar A of the stud axle E. This pillar passes through a long vertical boss B on the front tube F of the car frame. From the centre of the front tube F, the frame is suspended from the spring by a bolt and bridle, so that under the resistance of the spring, the tube F, with the boss B, is free to slide up and down on the pillar A, and within the cup cover C, which is fastened to the top of the pillar A. The upward extension of the boss B, within the cover cup C, ends at b, the space above b acting as a dashpot sufficiently to prevent upward shock during the free play of the springs on rough roads. At the bottom of the pillar A, where it enters the boss G, is a rubber or leather washer, which softens the impact downwards, when

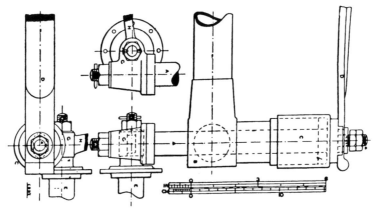

Fig. 158.—Front Spring and Steering Axles of Decauville Light Car.

the spring is depressed through its full range. At the lower end of A, is a cone-shaped extension which fits tightly into the conical boss G, on one side of which is a lug, through which the fixed end of the axle E passes, and is rigidly fixed by a nut. Extending from each of the lugs G, is an arm H. These arms are connected by short rods to a **T**-piece and bell crank, from one end of which is a connection to the steering pillar.

A large number of these cars has been made, and their handiness and the ease with which they are manipulated has made them favourites with those who do not object to the rattle of the gearing, nearly all of which is exposed, or to the method of changing speed. They have been very successful in races with vehicles of their type, and they were awarded a silver medal at the Richmond Show of the Automobile Club, in 1899. The Société Decauville Aîné is now also making a four-seated car, propelled by a vertical petrol motor of 5 HP., placed in the front after the manner of the Daimler cars. The transmission is by spur gearing, giving four speeds.

MODERN LIGHT PETROL MOTOR VEHICLES

Mors Light Two-Seat Car.

Fig. 159 shows the exterior of a Mors light two-seat car, fitted with a **Motor** horizontal engine over the front wheels, having two cylinders on opposite

Fig. 159.—Mors Light Two-Seat Petrol Motor Car, 4 HP.

sides of a single crank, the crankshaft being continued by a clutch shaft down the centre of the car. From this the chain pinion shaft with differential motion is driven by bevel gear and spur change gear, motion being conveyed to the road driving wheels by chains in the usual way. The motor is of 4 HP., and is supplied with petrol vapour by a float-feed spray carburettor after the manner of that used in a multiplied form in the large car already illustrated. Ignition is by electric means, and separate adjustments of the main air supply and of the carburetted air are provided, so that the motor may not only be run at various speeds by throttling, but the mixture may be independently adjusted.

The cylinder is partly cooled by a water jacket about the combustion and valve chamber, and partly by radiating ribs, water circulation being maintained by a small pump, the heated water passing through a cooler on its way to the tank.

Speed gear   Three speeds and reverse movement are provided, and the handles for the control of these and the ignition switch are all brought to the steering pillar. Two pedals provide for the working of the main clutch and the band brake on the differential gear box respectively. Two band brakes on the sprocket wheel rims are worked by a lever at the side of the off seat. The weight of the car is given as under 7 cwt.

It was brought out early in 1899, with the intention of supplying the smaller-sized vehicle required by many who did not like two-seated motor cycles, and at a price lower than that of the larger carriages. The behaviour of the first car run in this country was not encouraging, the gearing being noisy, and the motor from one cause or another requiring frequent attention from the driver. Quite possibly these latter defects were only those of imperfect adjustment and imperfect knowledge of the peculiarities on the part of the driver, but the reduction of the noise of the gearing would not be so easy.

The Daimler Motor Company's Critchley Light Car.

Fig. 160 illustrates a light car made by the Daimler Motor Company from the designs of their works engineer, Mr. J. S. Critchley. It is a two-seated car, fitted with a Daimler motor of either 3 or 6 effective HP., similar to those described in Chapter VII.

Here, however, all similarity ends, as the power is transmitted by means of two belts, which are of considerable length. These belts drive a short transverse shaft parallel with the main axle at two different speeds. On this shaft is a pair of strong spur pinions, alternately used to drive spur wheels on the main axle, giving two speeds. Thus there are four speeds, and a simple arrangement of idle spur wheels gives a reverse movement.

Underframe   The arrangement of the motor and of the gear referred to is shown by Figs. 161–163.

The underframe, to which the motor and body, and the light bogie to which all the mechanism is attached, is of light channel iron, and the motor A is suspended from two three-armed brackets B B, so that it can

Bogie and speed gear   be pushed forward at will, to put more or less tension upon the belts c c', which, running on the pulleys D D', transmit motion to the fast and loose pulleys E E', fixed and loose respectively upon the shaft F, which is supported by bearings G G' and the central bearings of the bracket H, all of which are carried on the two-wheeled bogie frame J beneath the main frame. As the belts are shown, both are on the loose pulleys, and D' acts as fly wheel as well as driving pulley. Within the arms of the bracket H are the two spur pinions I and K, for low and high speeds respectively, which gear with the wheels L and M, both of which are fast to the hollow spindles N N, which give motion to the differential gear at O, which in its turn gives motion to the divided main axle P, which runs in long bushed bearing brackets Q Q. The bushes being about two-thirds the length of the

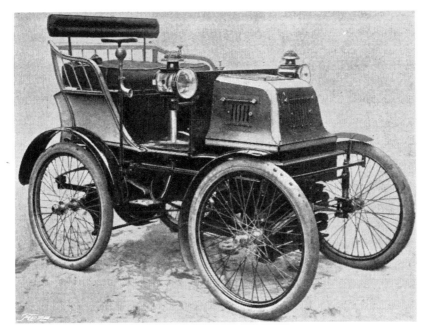

Fig. 160.—English Daimler Motor Co.'s Critchley Light Car.

tubular bosses of the brackets, the smaller ends of the hollow spindles N N occupy the remaining one-third of these bearings.

The action of the gearing is as follows : Starting with the slowest speed, the belt C is put on to the fast pulley E by means of the belt-shifting handle R, spindle R^1, arm R^2, and transverse bar S, upon which is the belt striker at T ; the spur pinion I having previously been put into gear with the spur wheel L, by means of the lever U attached to the hollow spindle U', on the lower end of which is the arm V. This gives angular motion to the lever upon which is the sliding piece W. This lever is pivoted at X, and its forked end embraces the collar Y, and thus throws the two pinions I and K alternatively into gear with L or M. The

Changing speeds

227

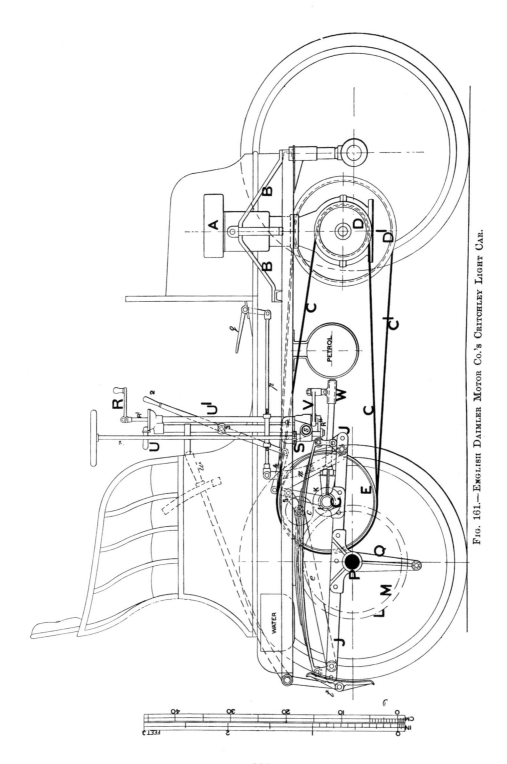

FIG. 161.—ENGLISH DAIMLER MOTOR CO.'S CRITCHLEY LIGHT CAR.

228

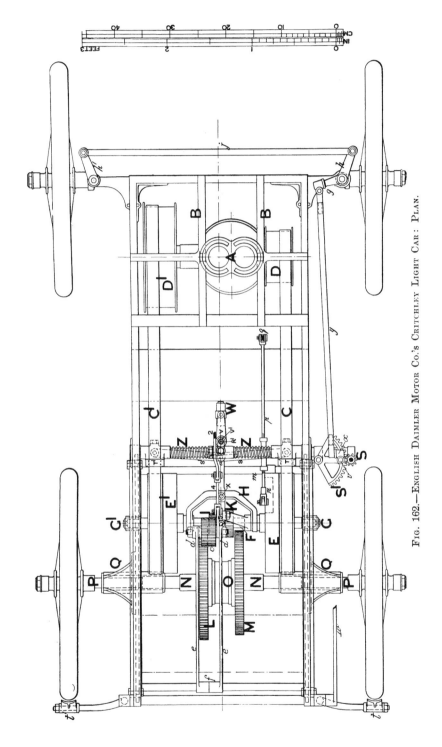

Fig. 162.—English Daimler Motor Co.'s Critchley Light Car : Plan.

229

pinion I being now in gear with L, the slowest speed is given to the main axle.

With the belt c remaining in the position shown, but with I taken out of gear and K put into gear with M, the second speed is obtained. A third speed is obtained when I is in gear with L, and the belt c' put upon the fast pulley E'; and the highest speed is obtained with that belt acting as driver and K put into gear with M. At z z are springs which tend always to keep the belts on the loose pulleys, and against the resistance of these the strikers T T' have to be moved by means of the bars s s' and the handle U; the bars s s' are slotted where they engage with the central pin A, so that they are only pulled and not pushed by that pin. The result of this arrangement is that the belt is always off, or nearly off, one fixed pulley before the other belt begins to cover the alternative fixed pulley.

Reversing motion
At c is the idle spur pinion previously referred to for reversing. It is carried on a small spindle fixed at d d, in a double lever e e, which is pivoted at f, at the back of the bogie frame. This lever e is brought into position for reversing by a hand lever 2, pivoted at 3, and engaging at its lower end with a link 4, which operates the bell crank 5, and through it pinion c, when the speed pinions I, K are in the position shown in Fig. 162. This gives a reverse motion at the slowest speed.

Brakes
At M is a transverse shaft on which is an arm n, provided with a brake block inside the rim of the pulleys E. This brake is actuated by a rod p and a pedal q, with bell crank lever connection. There are also two spoon brakes t t', actuated by means of a lever w.

Belt tension adjustment
The motor is pushed forward into the belts with sufficient force to give the desired tension on the belts by means of an adjustable screw rod not shown.

By reference to Figs. 160 and 162 it will be seen that the brackets Q Q extend downwards, and have at their lower ends bosses of considerable length connected by a stiff stay bar, so that great rigidity is given to the whole of that part of the structure which carries the bearings of the main axle, in the centre of which the driving effort is exerted.

Steering gear
The steering of the car is done by a small wheel on the vertical spindle r, on the lower end of which is a pinion s, gearing into a quadrant s', on which is a pin v, embraced by a coupling piece x (like that described with reference to the Crowden car) at one end of the bar y. A similar coupling piece is used at the other end of the rod, on the arm g of the bell crank h, on the vertical pivot of the stud axle. The other stud axle is simultaneously moved by the connecting rod j, on the arms h h' (For these parts and the front spring and hanger arrangements see also Fig. 163.) The steering hand wheel is small, and the ratio of its movement to that of the steering axles is small too. Hence a very small movement of the wheel is accompanied by too much movement of the car across the road. A steering handle-bar more like that of the bicycle or of the Peugeot car form would be preferable.

Of the motor and its feed and exhaust connections and method of working, it is not necessary to say more than they are the same as those described in Chapter VII., with the exception of the suspension from the cylinder, which is somewhat like that of the Daimler-Maybach carriage.

From the end of the crankshaft everything is different, and provided that the arrangement for pushing the belt pulleys into the belts is such as to secure rigid parallelism of the crankshaft with the second motion shaft, and some elasticity in the push, it ought to work well. Even if the belts are all that could be wished, or exactly the right length, it might be well to have a pair of jockey pulleys, not for regular use, but to take up the excess of slackness of one belt over the other, when it occurs, and thus prevent excessive tension on one of them.

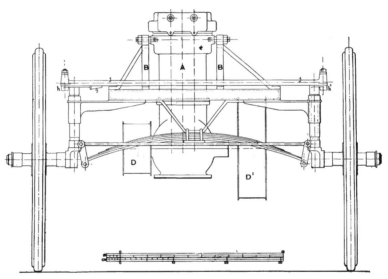

FIG. 163.—ENGLISH DAIMLER MOTOR CO.'s CRITCHLEY LIGHT CAR:
FRONT VIEW OF FRAME.

If this be done it might be said that there would no longer be any necessity for the suspension mounting of the engine, and this would appear to be true except in the case of driving by a single belt, when the suspension arrangement might be used without any jockey pulley. The tension in this case would be put upon the belt by the screw adjusting rod, with the intervention of a short range stiff spring in compression, so as to give the belt a little less rigid centres that would allow for belt inequalities.

Generally it may be said that belts for this class of car should be wider than those commonly used, and always rather thin, supple belts.

The belts will not run well or the shafts remain rigidly parallel,

231

except with very rigid engine-adjusting arrangements, if there is much difference in their tension, but this difficulty might be easily overcome by very light, adjustable jockey rollers. The only other way out of this difficulty is the use of one belt only, and more gearing for speed range; but for a car of this kind the spur-wheel speeds already arranged would probably be quite sufficient for most users if it had a slightly smaller difference in pulley diameters than that of the lowest speed shown, and a little larger diameter of cylinder, and a good electric ignition, so as to run on a throttle valve when the grades allowed it.

The arrangement as it is, secures a good mechanical job in the construction of simple, plain, strong gearing, well carried in an independent little bogie. It secures stiffness and rigidity of relative positions of parts carrying bearings and gear, and the whole of the work is cheap machining. The whole car, with slight modifications and lightening of several parts, could be very cheaply made, and with the selection and proper usage of really suitable belting properly joined, and cared for, it should be a satisfactory car to run.

THE DARRACQ-BOLLÉE CARRIAGE.

The cars illustrated by Figs. 164–167 are made by MM. A. Darracq & Co. from the designs of M. Leon Bollée, whose name has been so well and long known as that of an original designer of motor vehicles. Figs. 164 and 167 show carriages with two different forms of bodies, which are known as the Duc-Tonneau and the Duc-Darracq respectively. Both are carried by precisely the same underframe, gear, and motor, the arrangement of which is shown by Figs. 165 and 166, which are respectively views of the whole of the working parts from above and from below.

The frame is built up of tubes, and is carried on double-leaf springs on the front axle, which carries stud axles at its ends with forwardly projecting steering arms. The frame is supported on single-leaf springs at the rear.

Motor
The motor, which is started by a handle on the fly wheel on the near side of the frame, is a modification of the well-known Bollée form, having a single air-cooled ribbed cylinder, on the single-way crankshaft of which, is a five-cone pulley c, with epicyclic reversing gear at the fly wheel at one end. A crank arm in line with the crankshaft proper is driven by the crank pin for giving motion to a governor. A float-feed carburettor is used,

Transmission gear
and tube ignition at B. From the cone c on the crankshaft, motion is conveyed to a second motion shaft behind the main axle, there being on this shaft a corresponding five-cone pulley E, a clutch s, and a pinion gearing with a differential gear, within the case G. The belt is shifted from pulley to pulley by a belt striker mounted on an inclined axis which is seen in Fig. 166, one end of the striker being seen in that same figure, and the other end in Fig. 165. This striker is operated from the hand wheel A, under which will be seen a disc D, having five small projections. This

232

disc and the hand steering wheel are clearly shown in the perspective view,
Fig. 164. By lifting this disc and turning it to either of five positions,

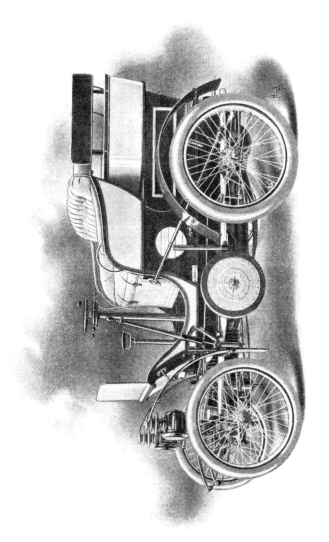

FIG. 164.—DARRACQ-BOLLÉE DUC-TONNEAU MOTOR CARRIAGE.

motion is conveyed by a pitch chain on a sprocket wheel at the end of the
steering pillar, and this conveys motion to the wheel D′, Fig. 166. This in
turn, by the rack shown, gives motion to the belt-striker arm, the bearing

or pivot of which being inclined causes the strikers to follow the taper of the cone pulleys in the opposite directions, hence the belt is easily shifted for either of the five speeds.

Steering gear   The steering arrangements will be easily seen from Figs. 164, 165 and 166, a rod with a rack, actuated by a pinion at the bottom of the spindle on which is a steering wheel A, giving motion to a bell crank arm and a

FIG. 165.—DARRACQ-BOLLÉE CARRIAGE: VIEW OF UNDERFRAME, MOTOR AND GEAR FROM ABOVE.

connecting rod *d*. This rod connects the two arms on the short steering axles. There are two pedals P P', one of which actuates a pair of hand

Brakes   brakes, one near either driving axle and the other for actuating the main clutch for varying speed, or slowing, stopping, or changing speed. At Q is

Reversing   seen the lever by means of which the epicyclic gear near the fly wheel is operated. The lever L actuates the brake upon the outside of the clutch body. Diagonally placed on the same side as the fly wheel is a rod with a screwed end actuating a sliding carriage by means of which may be varied within a small range the distance of second motion and driving axle from the crankshaft for tightening the belt. The silencer is seen at H.

The motor is 5 HP. and runs at 800 revolutions per minute. It is arranged where it is easily accessible, but requires more covering in

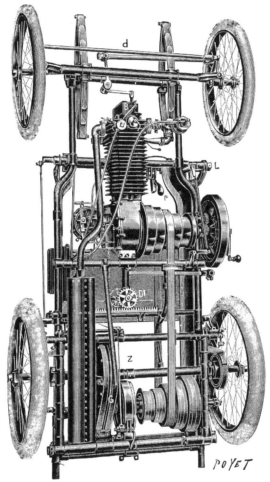

FIG. 166.—DARRACQ-BOLLÉE MOTOR CARRIAGE: VIEW OF FRAME AND MACHINERY FROM BELOW.

than it, as an air-cooled motor, is likely to get. The arrangement of the machinery, the adoption of five-cone pulleys and means of striking a belt thereon, present several points of much interest, and very comfortable carriages of various designs are adapted to the one form of underframe and gearing shown. Sufficient time has not elapsed, however, to permit reference to experience with them.

235

The lighter forms of the carriage weigh only about 1,100 lb., or, say, half a ton.

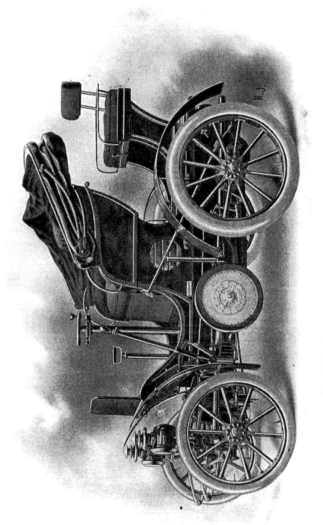

FIG. 167.—DARRACQ-BOLLÉE MOTOR CARRIAGE.

THE RENAULT LIGHT MOTOR CAR.

The Renault car   Figs. 168–170 illustrate the main features of what is known as the Renault Motor Voiturette. On a racing occasion this car ran from Paris to

Rambouillet, a distance of 64·62 miles, in 2 hours 49 minutes, or at the mean speed of 23·75 miles per hour. It is a neat little car driven by a

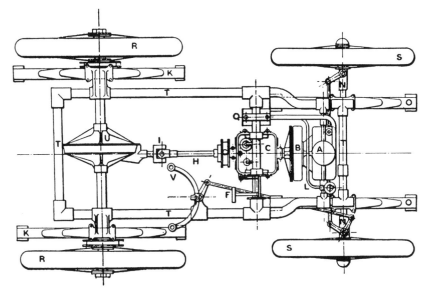

Fig. 168.—Renault Light Motor Carriage, with Dion Motor: General Arrangement.

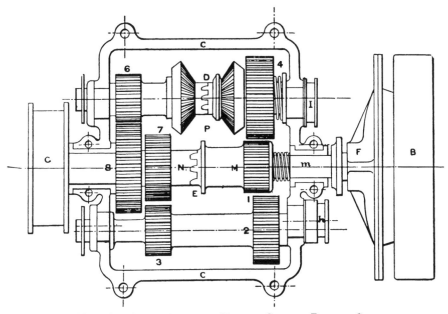

Fig. 169.—Change-Speed and Reverse Gear of Renault Car.

2¼ HP. de Dion et Bouton motor placed at A, in front of the car, and driving by means of a clutch B the gear in the gear-box C, by which either

Renault transmission gear

237

of several speeds may be adopted and transmitted to the universal joint within the drum G, which, as will be seen by Fig. 170, provides a hand-brake pulley. The power is transmitted from the gear by the rod H and the universal joint I to the bevel pinion and bevel wheel and differential gear in the box U on the main axle. The frame T is of tubular construction, narrowed at the front axle, and it carries, by means of strong brackets, the bridles for the springs K. The steering arrangements are very simple, consisting simply of a cycle handle V, and connecting rod at the lower end of the steering pillar, coupled up in the way seen in Fig. 168 by two rearward arms on the vertical pins in the stud axles at the forked ends N of the straight front axle. The chief point, however, of this notice of the Renault car must be a reference to the speed-change and reversing gear, which, while being very much like that commonly used on lathes of some

Renault speed gear

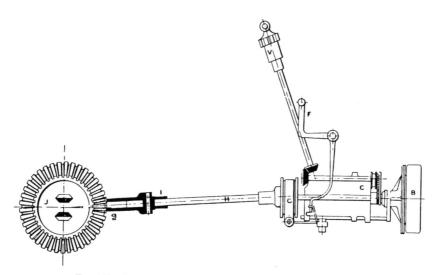

FIG. 170.—DIAGRAM OF TRANSMISSION GEAR OF RENAULT CAR.

kinds, has modifications for the purpose of this application which give it interest. The diagram, Fig. 169, is not complete in all details, but it permits of an explanation of the gear; B being the clutch as before, it will be seen that a band on the pulley G is operated by a hand lever F, Fig. 170, with certain connecting links. At the lower end of a spindle V is a bevel pinion acting on a corresponding pinion on a horizontal spindle. When driving at high speed the drum at G runs at the speed of the motor, the motion from it being conveyed through the sliding clutch M, and its corresponding part N, to the spindle 8, Fig. 169. When necessary to change the speeds, the clutch sleeve M is separated from N at E, compressing the spring near m, and at the same time the eccentric spindle, on which the pinion sleeve carrying the pinions 2 and 3 works, is turned sufficiently to put the pinion 1 into gear with the pinion M, and through it the pinion

238

3 into gear with the pinion 7, thus conveying to the spindle 8 a slower speed. For a still slower speed the eccentric H is turned back again, that on the spindle on the opposite side of the box turned through half a revolution by means of chain gear indicated in Fig. 170, and the pinion M is thus brought into gear with the wheel 4, and the pinion 6 into gear with the wheel 8, and the still slower speed obtained. For reversing the motion of the car a bevel wheel, not shown, is forced into position between the bevel pinions at P, which are separated at the claw clutch D, and reverse motion at the slowest speed thus obtained. The clutch B is controlled by a pedal F, Fig. 170, which at the same time actuates the brake band G already referred to. The whole of the details of the gear-changing mechanism, by turning the two eccentric spindles, cannot be shown without further drawings, but those given are sufficient to show its character, and to show how a completely boxed-in set of gear, suitable to a light car operated by a small motor, may be constructed.

The English patent for this gear is No. 3981 of 1899. The application for this patent affords an instance of the objections to our " provisional protection" system, a system which opens the door to much very undesirable procedure. Renault's application was made on February 22nd, 1899, and indexed as of February 22nd, under the title of "Improvements in pivotal supporting devices for window sashes, doors, and the like," and was indexed as such in the first quarterly index of 1899. Acceptance was not advertised until it appeared in the quarterly journal for August, when, under the same number, and of the same date as above, the title appeared as " Improvements in driving and speed-changing mechanism for motor cars." Thus no less than six months of delay of publication of real subject of the application and the misdirection of a misleading title resulted.

The Accles-Turrell Light Car.

Among the attractive English light cars now offered to the public is that known as the Accles-Turrell car. It is illustrated in perspective by Fig. 171, and Fig. 172 is a plan diagram of the arrangement of its motor and machinery.

The frame is of light shallow channel iron, connected transversely by pieces of the same material, the width of the frame being lessened in front to meet the requirements of the general design and the use of large steering wheels.

The motor has a single horizontal cylinder, 3·75 in. diameter, and the same in stroke, cast in one piece, with its valve seat casings and jacket which surrounds the valves as shown by Fig. 175. It is placed in the front of the car, with its cylinder A, Fig. 172, in rear of the crank case B, supported at b, the cylinder being firmly carried by the transverse support a, which is made up of two pieces, the channel bar mentioned thus forming a box section.

The Accles-Turrell car

239

FIG. 171.—THE ACCLES-TURRELL LIGHT PETROL MOTOR CAR, 3½ HP.

240

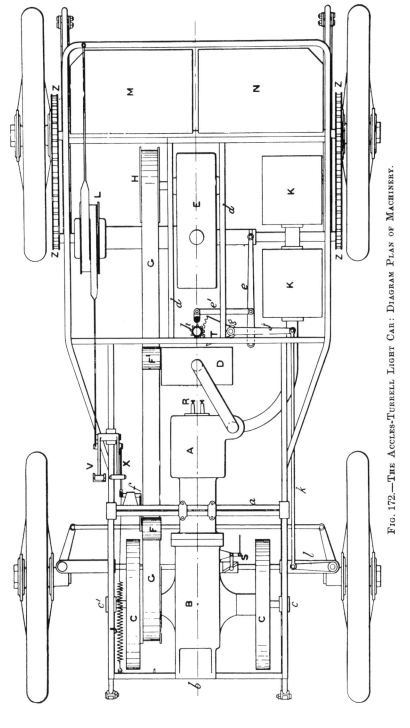

FIG. 172.—THE ACCLES-TURRELL LIGHT CAR: DIAGRAM PLAN OF MACHINERY.

241

16

MOTOR VEHICLES AND MOTORS

The crankshaft is carried in bearings at its extremities c and c', as well as by those in the crank case, in the usual way. Two fly wheels c c are used, so that the inertia stresses on the shaft are equal on the two parts.

Ignition apparatus

Ignition is electric, secondary batteries being carried under the seat, and the period of ignition controlled by very simple means. On the valve-motion shaft is an insulated cylinder at s, spirally embedded in the surface of which is a piece of brass. Bearing upon this cylinder is a conducting finger, which may be moved angularly in plan. Thus by placing this finger so that it bears upon the centre of the cylinder as shown, a mean time of ignition will obtain. By placing at any angle from this position an earlier or later time of ignition will obviously result, as fully explained hereafter.

The ignition spark is made at the two-pole ignition plugs placed in the the end of the cylinder at R. In principle the electrical arrangement is the same as in the Benz car, but the details and the method of changing period of ignition are different. This is shown by the diagram, Fig. 173, in which A is the valve-motion spindle carrying a vulcanized fibre cylinder, or bobbin, B, on which is a strip of brass c, and upon which rests the conducting finger D, pivoted at e, and movable to any angular position between the limits shown by the dotted lines by means of connections to an arm, as at a. Thus if the finger were moved over to g, it would be just making contact, while, where it is, it is just leaving it; and if moved over to f, the cylinder B would have to make about one-eighth of a turn before contact commences. Variable period of ignition is thus simply obtained by strong pieces of mechanism.

The connections are, it will be seen, from the battery E to a part of the motor, and to the interior or primary low tension coil winding, and by way of a trembler to the finger D. The strip c being in metallic contact with the spindle A, circuit is completed when the finger D rests upon it. Then, by means of the trembler, a rapidly alternating current is inductively created in the high tension coil G, the two ends of which are coupled to the two sparking plugs H H. These are placed in one large plug easily removed. The high tension circuit, being thus confined to the coil and the plugs, makes insulation simple and easy, and removes high-tension circuit possibilities of trouble away from the more complicated connections.

Exhaust

The exhaust passes into a double box silencer K K. The carburettor D is of the surface type, but of novel arrangement. The carburetted air passes by the pipe from D to the admission valve, the mixture and the ignition period both being controlled by handles on the steering pillar.

Transmission gear

The transmission of the power from the motor is by a single belt G to a pulley at H on the end of a spindle in the gear box E. At F F′ are jockey pulleys, F being mounted on a stud spindle carried by an arm pivoted at f and pulled upon the belt by a spiral spring at J, the jockey pulleys increasing the part embraced of the periphery of the pulleys. The tightness of the belt when running may be lessened to any extent by the pedal x, pressure on which lessens the tension on the spring J. The bandbrake at L is operated by the pedal v, which at the same time lessens the tension on the spring U and allows the belt to run loose. The band of the brake L runs on the outside of the differential gear box.

The speed gear in the box Ė, which is carried by the bars *d d*, is shown by the diagram, Fig. 174. On the spindle which carries the pulley H are keyed three pinions of different diameters. These gear alternatively with three spur wheels on the hollow spindle w, to which end on movement may be given by the forked lever *e* operated by the link *e'*, to which motion is given by a short arm on the bottom of a spindle within the steering pillar, the lower end of which carries the pinion H. By moving the arm *e* operating the short hollow spindle w, see Fig. 174, the key 1 is caused to

Speed change gear

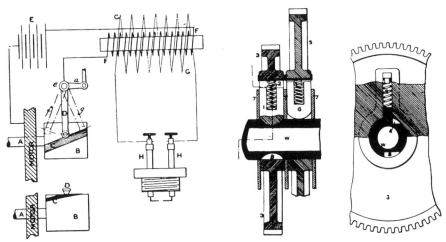

FIG. 173.—ACCLES-TURRELL LIGHT CAR ELECTRIC IGNITION ARRANGEMENT.

FIG. 174.—SPEED CHANGE GEAR OF ACCLES-TURRELL LIGHT CAR.

slide up in the slot in the wheel 3, riding up the curved bottom of the key seat 4 in which it sits when driving. The key will then ride on the plain part of the spindle and the wheel cease to drive. By pulling the spindle w towards the second wheel 5, until the key seat is under the key 6, that wheel will become a driver of the spindle instead of wheel 3, and a speed will then be obtained proportional to the diameters of the wheels in gear. A third speed is obtained in the same way. Thin flat rings 7 are carried on the spindle w to keep the wheels in their proper positions, and spring 2 force the keys 1, 6 into their seats. The wheels 3, 5 are always running with those on the pulley spindle which drive them, but they only drive the spindle w when either of the keys is in the position of key 1.

For a reverse motion the gear shown by Fig. 174A is employed. In it the first motion shaft K is that which carries the pinions for driving the

Reversing

243

several wheels mentioned with reference to Fig 174. One of these, A, Fig. 174A, is of double length, so that it may not only drive the slow speed spur wheel c, but also, through the intermediate pinion B, may drive spur wheel D in the opposite direction. All the wheels are always in mesh, but are only driving when the key of either of the wheels is in the key-seat.

The belt necessarily runs at a high speed, so that the tension is moderate for the power transmitted; but the great variation in the tension, due to power transmission variations, makes it desirable that a wider belt should be used.

When a change of speed has to be made the belt is slacked by the pedal x, so that the spindle w may be easily moved, and as little jerk

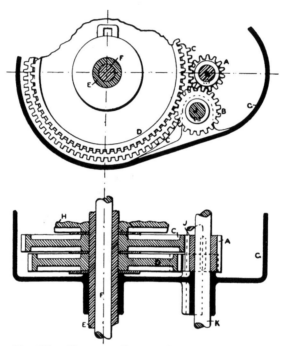

Fig. 174A.—Reversing Gear of Accles-Turrell Car.

thrown on the keys and gear as possible. There is but one key to each wheel, and hence a whole revolution will sometimes have to be made between release of one wheel and engagement of the other. How far the wear on the ends of the keys and of the edge of the seat may give trouble, experience only will show, but all these parts being well hardened the wear may be very slight with careful usage.

The steering is effected by the pinion at h gearing with a quadrant T pivoted at g, and carrying an arm j, connected by a rod k to the arm l on the top of the near steering-axle pivot, the two stud axles being connected in the usual way.

From the chain pinion shaft, driven by the differential gear, motion is

244

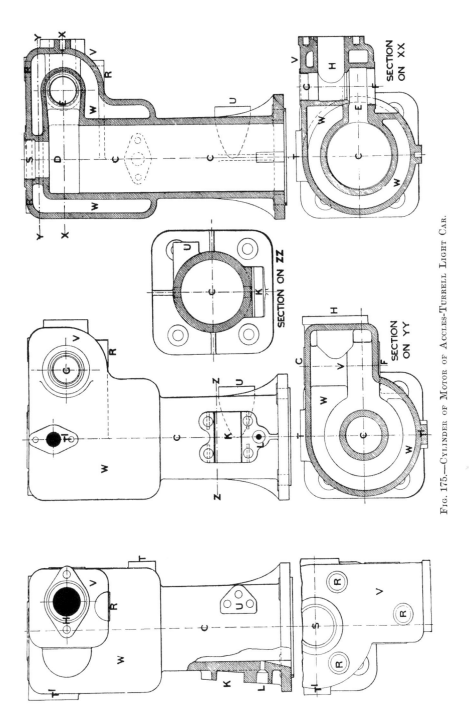

SECTION ON XX

SECTION ON ZZ

SECTION ON YY

FIG. 175.—CYLINDER OF MOTOR OF ACCLES-TURRELL LIGHT CAR.

245

conveyed by the chains z to the rear wheels, which like the steering wheels are 30 in. in diameter.

For taking up slack in the chain, due to wear, and for maintaining the proper position of the main axle, a pair of adjustable radius rods are employed.

The engravings of the motor cylinder, of which six views are given in Fig. 175, may be considered self-explanatory. It will be seen from the vertical section, and from the section upon it at x x, that the seatings for both valves at v are surrounded by jacket water, the space w for the water being everywhere large. At the head end of the cylinder at s is the seating for the two-pole ignition plug. At T is the entrance for the jacket water, and at T', the outlet. At K is the seat for the transverse supporting bar a, Fig. 172, L being a lubricator seat. c and F are respectively inlet and exhaust valve openings, and R are plug-holes for removing the jacket core. The thicknesses are apparently greater than necessary.

At M, Fig. 172, is an iron water tank, and at N a sheet-copper petrol tank, holding 4 and 3½ gallons respectively, and sufficient for 100 miles and for 60 miles respectively, on good roads.

The tool basket in front, and the motor cover, are easily removed, and the gear is easily accessible at the back as well as under the foot board. The weight of the car is about 1,036 lb. = 9¼ cwt.

The gear speeds with normal speed of motor are for 3½, 10 and 20 miles per hour, with reverse at 4 miles per hour. Extended experience has yet to be gained with this promising car.

THE PEUGEOT LIGHT TWO-SEAT CAR, 3½ HP.

Another attractive French light car is that made by MM. Peugeot, Frères. It is a two-seat car, in which the motor and gear are almost identical with those of the larger car, as already described at length in Chapter IX. The general appearance of the car is shown by Fig. 175A.

The front and driving wheels are of the same diameter, the motor is of 3·5 HP., and the total weight is about 850 lb. Either tube or electric ignition are used, and a fixed front axle with short stud steering axles as in the larger cars. The gearing, which is for three speeds and reversing, is easily accessible under the footboard, and those parts of the motor that most often require attention are easily reached by opening small doors at the back.

In the Peugeot engine it will be remembered there is a tongue which engages with a groove in the periphery of a disc between the two cylinders, for operating the exhaust valves. This is the only working piece within the crank case, which is in excess in number over those in the Daimler motor crank case of the Panhard. It is, however, a very important piece, and although it may run for a long time without attention, it is one that might give trouble. Access to it is provided for by making a small door in the upper part of the crank case.

MODERN LIGHT PETROL MOTOR VEHICLES

In the larger cars the jacket water circulates in the tubing of the frame, but this has not been followed in the small cars, and the gilled tube cooler which is used is sufficiently effective to make but a small quantity of water reserve requisite.

Cooling water

The petrol tank is carried in sight of the riders behind the dashboard at a sufficient elevation to provide for a gravity feed. It is estimated to carry enough for about 120 miles on good roads. The speed when carrying two passengers and a child is said to reach 18 miles per hour.

The steering gear is similar to that of the larger cars, with a handle of bicycle type. It is fitted with a bandbrake, operated by a pedal and a more powerful hand lever bandbrake on a pulley at one side of the

Steering gear and brakes

FIG. 175A.—THE PEUGEOT LIGHT TWO-SEAT MOTOR CAR, 3½ HP.

differential gear, the single driving chain sprocket wheel being on the other. The wheels are of the suspension type, with tangent spokes.

THE PANHARD & LEVASSOR LIGHT TWO-SEAT CAR, 4 HP.

Another French car of attractive appearance and make is that of MM. Panhard & Levassor, which is illustrated by Figs. 176–180A, the first two being from photographs.[1]

This is not a small model of the Panhard larger car with the Phœnix Daimler vertical motor in front, but is a differently arranged vehicle with single cylinder nearly horizontal motor behind, and with a straight

General

[1] Figs. 176 and 177 are from *The Autocar*, No. 184, p. 371.

247

central pivoted front axle. The motor is of the Daimler-Dion & Bouton type, designed by M. Krebs, and its arrangement and that of the machinery is shown in Figs. 178 and 178A.

FIG. 176.—THE PANHARD & LEVASSOR LIGHT TWO-SEAT PETROL MOTOR CAR, 4 HP.

The underframe is a combination of the strengthened wood structure carried out in front by four tubes, which in graceful form converge at a steering head, taking the place of the older forms of locking plate used with carriages, which have, as this has, an ordinary centre pivoting axle, not the two short stud axles at the end of a fixed axle so generally used with motor vehicles.

The machinery is enclosed in a backward extension of the body, and is carried by an interior enclosing frame, supported in part by the bearing castings of the main and second shaft bearings. These castings extend vertically up and down the flanges, which are connected at top and bottom, forming a rigid frame carrying the motor and mechanism independently of the frame, as is shown by Figs. 178 and 178A, and also in the perspectives,[1] Figs. 179 and 180.

The motor is nominally of 4 HP. Its cylinder N is ribbed and air cooled, and the head is water-jacketed round about the exhaust valve at **The Krebs motor**

FIG. 177.—THE PANHARD & LEVASSOR LIGHT TWO-SEAT CAR: FRONT VIEW.

P a small water tank being carried above, so as to secure natural circulation. It is fitted with a govenor Q, carried by the spur wheel R, which is driven by a pinion R′ on the crankshaft at half the rotary speed of the latter. On the inner boss of the wheel R is the cam which gives motion to the exhaust valve by means of the rod O. The governor actuates an **Governor** arm pivoted at 31 on a short spindle on the crank case, as seen in Figs. 178 and 179, a small arm at the end of this spindle giving motion by a horizontal rod to a throttle valve at r.

The carburettor h is of the float feed kind already described. It is supplied from the tank G by the pipe g. Ignition is by tube at n, heated

[1] Figs. 179 and 180 are from *The Automotor and Horseless Vehicle Journal*, No. 30, p. 286

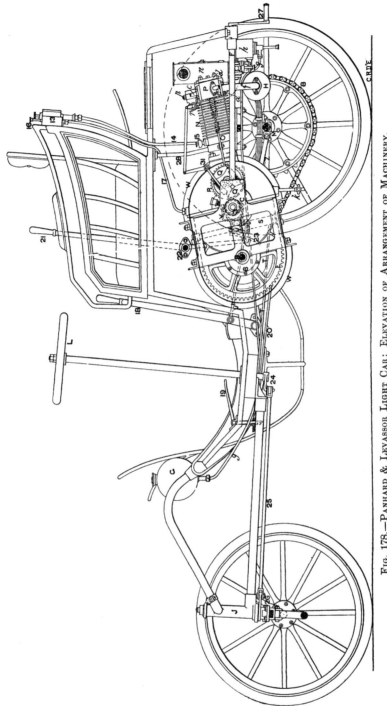

Fig. 178.—Panhard & Levassor Light Car: Elevation of Arrangement of Machinery.

MODERN LIGHT PETROL MOTOR VEHICLES

by a burner g'. The pipe from the carburettor to the admission valve is seen at a, Figs. 178 and 180, in which f is the exhaust connection, d the

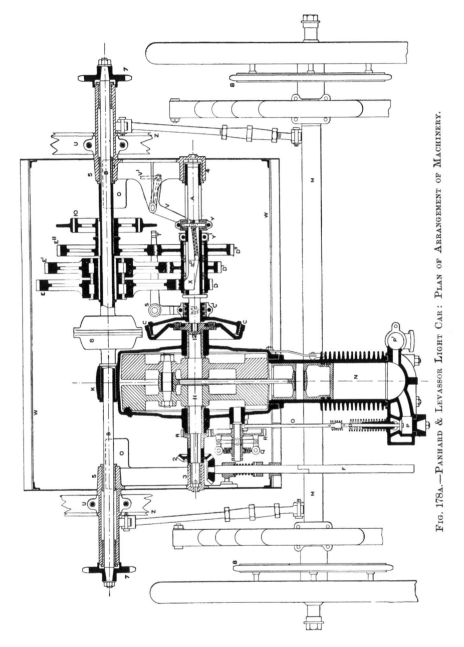

FIG. 178A.—PANHARD & LEVASSOR LIGHT CAR: PLAN OF ARRANGEMENT OF MACHINERY.

exhaust valve cover, e water inlet, c water outlet, h' the friction brake, s the sliding collar and key for moving the key x, Fig. 178A, of the speed-changing gear, and A, B are the first and second motion shafts respectively

251

MOTOR VEHICLES AND MOTORS

In Fig. 179, *e*, *c* are the water inlet and outlet respectively, *f* the exhaust-pipe connection, *m* spur wheel carrying governor and cam for exhaust valve, and *k* dirty oil outlet plug in the crank case. The parts lettered have been mentioned.

Transmission gear — Turning to Fig. 178A it will be seen that the end 12 of the off half of the crankshaft is bored and fitted with a bush, in which the inner end of

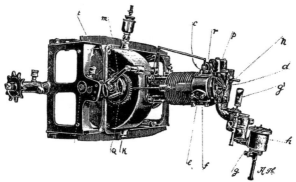

FIG. 179.—KREB'S MOTOR AND GEAR FRAME, PANHARD LIGHT CAR.

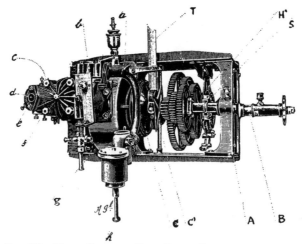

FIG. 180.—MOTOR GEAR AND GEAR FRAME, PANHARD LIGHT CAR.

the first motion shaft A fits and runs. Bolted to a flange formed on the same part of the crankshaft is the outer part of the main cone clutch c, which gives motion to the inner cone, which fits and slides upon the squared part of the shaft A (see Fig. 180A). The boss of this part of the cone is turned and grooved to form a hold for a thrust collar c′ for putting the cone clutch in and out of contact. The thrust collar c′ is pivoted to one arm of the bell-crank lever, which is pivoted at s, and operated by the

252

pedal 19, and rod T, Fig. 178, which passes through the eye at T, Fig. 178A. The end of the crankshaft 11 is housed in a closed bearing at 3, Fig. 178A, and the corresponding end of the first motion shaft is similarly housed at 4. At each end is a collar and thrust adjustment. The interior part of the shaft A is surrounded by a short tubular spindle carrying the three spur pinions D D' D", the bosses of which are internally toothed for the reception of the movable key x, which moves in a long slot in the tubular shaft A. The bore of these spur pinions is partly bushed on either side, so that they may run on the slotted spindle at the difference in speeds due to their diameters. This is necessary because all the pinions, being permanently in mesh with the wheels E E' E", are sometimes running on the shaft even when they are doing no work. The key x finds its way into the inside of one or other of the pinion bosses, moved by the collar Y, which is pivoted on one end of the bell-crank lever V, and actuated by a lever 18 through the rod v' from the driver's seat. The sliding collar Y (see also Fig. 180A), carries a pin, which passes through the slot of the tubular part of the spindle A, and takes hold of the tail of the key x, which is pivoted upon it at x. When the collar pulls the key in the direction of A, the inclined edge of the part entering the pinion bosses allows it to be pulled out of the notch in the boss, and correspondingly the rounded nose of the extremity enables it to be pushed from one boss to the other, a spring on the key tail tending always to make it enter the notches in the bosses. The spindle A is solid at the square part where the inner part of the cone clutch c fits upon it, and it is plugged at the other end. A collar Y' tightly embraces it to keep the pinions D in their place.

Speed gear change-device

Thus either of the pinions D may be made driver of either of the wheels E. Only the wheel E" is keyed to the hollow shaft B, which it drives. The others, E E', are mounted upon bushes which run upon an enlarged part of the tubular spindle surrounding the second motion shaft B, which receives its motion from the differential gear at 6. The two wheels E E' have enlarged bosses inside, which are rachet or pawl gears, to permit of the difference of rotation in the differently sized wheels which are always in mesh. (The English Patent relating to this gear is No. 2960 of 1899.)

The solid part of the shaft B, which carries the sprocket pinions 7, has an excellent bearing 5 at each end, the casing being cast with, and therefore rigid with, the frame ends, carrying the whole of the machinery. These bearings also rest in plummer blocks U U on the main frame. It is further carried by a central bearing at K.

The proper relative positions and the adjustment of the distance between the sprocket pinions and sprocket wheels 8 are made and maintained by the radius rods Z Z.

The brake pulley H', Fig. 180A, is grasped by strong semicircular clamping pieces 10, equally effective with the carriage running in either direction, and actuated by a pedal and rod 20 attached to the lever and

Brakes

253

link arrangement 29, 30. This brake, it will be noticed, is only effective through the chains. Spoon brakes 23, carried by arms on a cross shaft 22, and attached by a hand lever 21 on the off side, are provided.

Long leaf springs are used on the hind axle, and a single transverse spring on the front axle, which is curved downwards. The spring rests upon the straight parts of the axle near the wheels, and in the centre of its length it is grasped by the clamp at the lower end of the vertical pivot pin J in the steering head. The lower part of the pivot pin is fitted with an arm 26, connected by a rod 25 and arm 24 to the steering pillar, which is operated by a wheel L. The steering is thus done through the front spring.

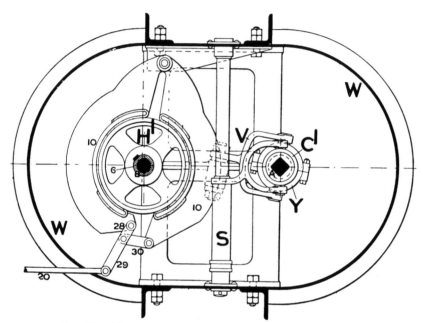

FIG. 180A.—FRICTION BRAKE OF PANHARD & LEVASSOR LIGHT CAR.

Steering axle   It is worth while noting that with all their experience, MM. Panhard & Levassor do not consider the extra stability on the front wheels, obtained by a fixed axle and differential angularity of the steering axles of the short stud or pivoted kind, of sufficient importance to be adhered to where structural reasons do not require it, and where there is plenty of room for the front wheels to lock round with a straight axle.

The car is started with a handle 27 on the spindle F and bevel gear 2.

Lubrication   The lubrication of the parts within the crank case is by oil from the lubricator at 16, through a pipe 17, Fig. 178. The lubricating oil for the piston is carried by the lubricator 13, and distributed by the pipe 14. At 15 is a small cup squirt for forcing kerosene into the cylinder for cleaning purposes.

MODERN LIGHT PETROL MOTOR VEHICLES

Experience is required to show the practical sufficiency and the effect of wear of the always-running gearing, and of the shifting key method of fixing the one or other pinion. There can be no doubt as to the meritorious points of the car, but as very few have been made nothing can be said of the modifications which may prove necessary in the gear details.

THE DE DION & BOUTON VOITURETTE.

Another light car to which reference must be made is the de Dion & Bouton Voiturette, the external features of which are shown by Fig. 181. This is a three-seated car with a tubular underframe, supported on leaf springs over the front steering axle, and by springs over the rear driving

FIG. 181.—THE DION & BOUTON THREE-SEAT VOITURETTE, 3 HP.

axle. The motor is of the Dion-Bouton-Daimler type, but provided with a water jacket, in which circulation is maintained by a pump. It is said to be capable of working up to 3 HP. The arrangement of the motor is much like that of the quadricycle, but it is provided with two speeds, 12 and 30 kiloms., or, say, 7 and 18 miles, and a clutch by which either or both may be thrown out of gear by a pedal. This gear is enclosed in an oil bath, and is shown by Fig. 181A, in which A is the motor shaft carrying pinions B and C tight thereon, driving the wheels D and E respectively, which are loose upon the bushes carrying them. At G and G' are friction blocks, which may be alternatively made tight in the drums F or F' by the right and left handscrews on the spindles of the pinions H and H', moved by means of the sliding spindle K. The blocks G and G′ are carried upon a shaped central spindle J. The whole apparatus must be taken to pieces to get at the friction adjusting screws.

A vapouriser of the atomiser type is used, and electric ignition. Two bandbrakes are provided, one on the differential box periphery on the main axle, and one on a drum on the second motion shaft, both actuated by pedal connections. A cooling coil is placed in the front of the car.

The mixture and period of ignition, and the changes of speed are all controlled by the lower wheel seen on the seat pillar. The car weighs, with water and spirit, nearly 7¼ cwt., and is only occasionally fitted with the reversing gear shown below.

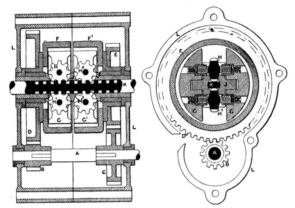

FIG 181A.—THE DE DION & BOUTON TWO-SPEED GEAR.

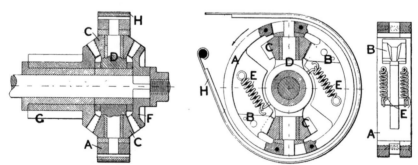

FIG. 181B.—THE DE DION & BOUTON REVERSING GEAR.

Within the pulley A, Fig. 181B, is a pair of stops B B, into which a tooth of each of the bevel pinions C C engage the spider D, being pulled in their direction by the springs E E. Wheel F thus drives pinion G, when pulley A is gripped by bandbrake H, the teeth of pinions C C come out of B, and D, being fixed, G is driven in the reverse direction at half speed.

256

Chapter XV

PETROL MOTOR CYCLES

THE variety of the applications of the light, high-speed petrol motor has been to a considerable extent exemplified by the typical varieties of vehicles which have been illustrated in the preceding chapters. The almost unlimited adaptability is further seen in the applications to motor cycles of all kinds, namely, bicycles, tricycles, and quadricycles, some of the last being, in older terminology, simply light, four-wheeled carriages. As a rule, however, they have the frame, and one or both of their seats of the cycle-saddle kind, which places them in a class distinct from the carriage class.

MOTOR TRICYCLES.

At an early stage in the development of the Daimler motor, of the smallest sizes after the Levassor application to a motor vehicle, MM. de Dion & Bouton turned their attention to it in very small sizes for propelling tricycles. After numerous experiments, and building numerous tricycles, the Dion & Bouton motor cycle reached a stage of practical success, which placed it amongst those that had to be reckoned with in high-speed and long-distance competitions. The de Dion and the Bollée

In 1895 M. Leon Bollée made the more powerful form of rear-driven motor tricycle, which has since been known as a Bollée. It was quickly designed—a flash of the Bollée genius ; it differed from any other motor vehicle, and comprised a number of novel and useful devices which have since been a good deal used. One of the first of these Bollées was exhibited at the Imperial Institute Motor Car Exhibition, in May, 1896, and in the hands of the inventor astonished every beholder of its evolutions.

In 1896, the Dion tricycle ran in the Paris-Marseilles race, making an average speed over the whole distance of 14·8 miles per hour, and in the Paris-Bordeaux race, 1899, another of the same make accomplished 28·1 miles per hour, the first being fitted with a ¾ HP. Dion-Bouton, air-cooled Daimler motor, and the second with a 1¾ HP. motor of the same kind. Subsequently these motor tricycles were fitted with 2·25 HP. motors, and some with a two-speed gear. They have become extremely popular machines, and very large numbers are in use, many of their owners accomplishing long journeys with regularity upon them. The motor tricycle in 1899

MOTOR VEHICLES AND MOTORS

The motors employed are all what may be called de Dion-Daimler motors, the carburettors are of very simple and even crude design, and all are fitted with electric ignition.

THE DE DION-BOUTON TRICYCLE.

Figs. 182 and 183 shows a side and back elevation of a de Dion-Bouton tricycle of the 1898 type, fitted with a 1¾ nominal HP. motor.

Frames

The construction and design of the frame is similar to that of an ordinary tricycle, with the exception of the trussed form of fork, seen in Fig. 182, which gives strength to what would be a weak structure for the purpose if built of simple tubes of the same weight. From Fig. 182 it will be seen that the motor is attached to the rear of the main axle, which it drives by single reduction gear. The motor is supplied with carburetted air from a vessel placed between the main pillar of the frame and the back.

The mixture is fired in the motor cylinder electrically, by spark at the points of an ignition plug, which receives current from a battery suspended from the top frame tube, and an induction coil carried on the bridge of the main axle.

The carburettor

The carburettor c, Figs. 182 and 184, is so shaped as to fit between the saddle-post and back tubes of the diamond frame, to which it is attached by means of light clips. The details of construction are shown in Fig. 184. It consists of a light box c of sheet brass, the lower portion of which forms the petrol reservoir, the spirit being introduced through the aperture 1. Inside the box at the top is brazed a tube 2, which forms the guide for a sliding tube 3, which is open at both ends. At the lower end of the tube 3 is soldered a flat metal plate 4, and through its length passes the wire 5 attached to a float 6, by means of which the level of the petrol is indicated.

Air and vapour valves

To the top of the carburettor box are attached the air and carburetted air controlling valves. These are combined and form a cylindrical double valve 7, in the shell of which are three apertures 8, 9, 10, as seen in the details with Fig. 184. This double valve contains independent cylindrical valves 11, 12, which are operated by means of the lever handles A, B, Fig. 182, mounted on the top stay tube of the frame. These levers are connected by light rods to the lever arms 13, 14, on the valve spindles. The two valves are of the same size, and nearly meet in the middle of the containing shell 7. A copper pipe 15 passes inside the box c from the aperture 10 in the double valve to the cone-shaped end of a chamber 16, in which is a copper cylinder 17 containing discs of wire gauze. To the end of the chamber 16 is connected the pipe 18, Figs. 182, 183, 184, 187 and 188, which leads to the admission chamber of the motor. At the bottom of the vessel, below the level of the petrol, is a pipe 19, to one end of which is connected a branch pipe 20, from the exhaust pipe 31 of the motor, while the other end has attached to it a downwardly projecting

258

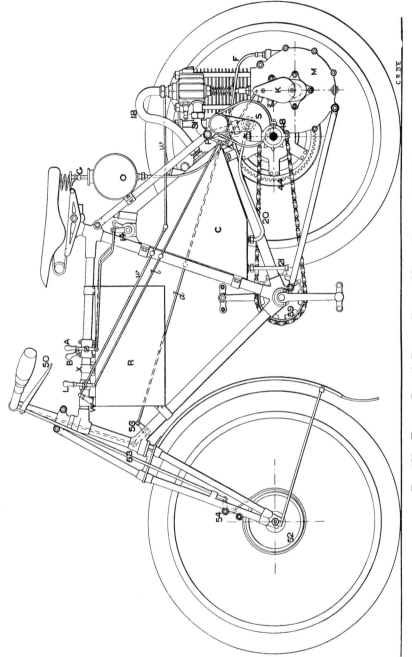

Fig. 182.—The de Dion & Bouton Petrol Motor Tricycle, 1898-99.

259

pipe 21, out of which the heated gases escape after passing round the pipe 19 within the carburettor.

Action of carburettor

The action of the apparatus is as follows: On each suction stroke of the motor a partial vacuum is formed within the box c, and consequently air rushes down the pipe 3, the height of which, above the surface of the spirit, is regulated by means of the float, and indicated by the wire 5, 6. Sweeping over the spirit, below the plate 4 and the wetted surfaces,

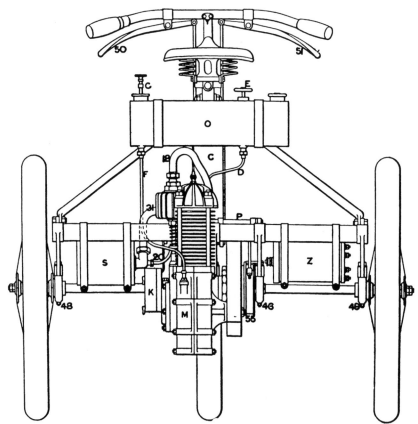

Fig. 183.—The de Dion & Bouton Motor Tricycle, 1·75 HP.

the air becomes richly impregnated with hydro-carbon vapour, and the mixture thus formed passes out of the carburettor box through the aperture 9, regulated by the valve 11. At this point it mixes with pure air, entering at 8, the diluted mixture passing thence through the aperture 10, controlled by valve 12, and down the pipe 15 into the chamber 16. Within this chamber is the gauze cylinder 17, which prevents the possibility of the vapour within the carburettor becoming ignited by a back fire in the motor. The mixture passes out of the chamber 16 by way of a pipe 18, which connects it to the admission pipe of the motor.

PETROL MOTOR CYCLES

The pipe 19 at the bottom of the carburettor becomes heated by the passage through it of the exhaust gases, and by raising the temperature of the petrol causes it to be more readily vaporised. In very hot weather it is usual to disconnect the pipe 20 altogether, as the petrol vaporises sufficiently readily without it. A small cock in the pipe would give a ready means of adjusting the supply of hot gases to the pipe within the carburettor, and save the trouble of disconnecting and leaving the pipe open.

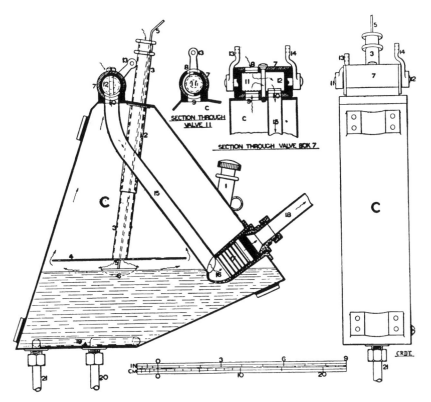

FIG. 184.—CARBURETTOR AND MIXING AND THROTTLING VALVE OF DE DION MOTOR TRICYCLE.

The action of valve 11 may be described with reference to the diagram, Fig. 184, which shows the relative positions of the parts 8, 9 in the valve and in the valve case. It will be seen that as the lever 13 is moved towards the right, more air will be admitted at 8, and less carburetted air at 9. Movement of the arm in the opposite direction produces a contrary effect, so that with this valve and the valve 12 the quality and the quantity of the charge entering the motor cylinder are completely under control.

Thus, to recapitulate, it will have been seen that the suction by the motor piston, acting through the pipe 18, draws a supply of diluted carburetted air through pipe 15, the pure air having entered at 8, and the air

Action of mixing and controlling valve

261

which has been carburetted having entered at the top of the pipe 3, on its way to the surface of the petrol, and thence by the port 9 to the port 10.

The de Dion & Bouton Motor

The motor M of the tricycle, which is seen in Figs. 182 and 183, and in detail in Figs. 185–188, is of the vertical, single cylinder, Daimler type, originated by Messrs. de Dion & Bouton in 1896, and subsequently modified by many improvements in detail. Figs. 185 and 186 show a vertical section and an elevation of the latest type, developing 1¾ nominal HP.; the actual B. HP. which the motor can give continuously being from 1·3 to 1·4. It is constructed in three principal parts: the cylinder head 1 and cylinder body 2, both of which are of cast iron, cooled by means of radiating ribs; and the crank case 3, which is an aluminium casting, divided along the central vertical plane, and bolted together, as seen clearly in Figs. 186–188. The joint between the cylinder head and body, and that between the body and crank case, are accurately turned and faced, and held by means of four long vertical studs 5, which are screwed into large bosses in the aluminium crank case.

The bore of the cylinder is 2⅝ in. = 67 mm., by 2¾ in. = 70 mm. stroke. The piston 6 is of very light construction, and has three $\frac{3}{16}$-in. piston rings. The cross-head pin 7 is held in bosses in the piston casting by means of the set screw 8. The connecting rod 9 is a steel forging of light H section, with brass bushed eyes at either end. The rod passes through a rectangular slot 10 in the crank case, through which oil from the lower part of the case is thrown up by the crank and connecting rod, and thus lubricates the cylinder. The slot 10 is not made of larger dimensions than those shown, in order that an excess of oil shall not reach the cylinder and piston. The crank on the driving shaft 11 is formed by means of a fixed pin joining the two fly-wheel discs 12, 12, which are enclosed within the crank case.

Cylinder head and valve box

The valve chamber 13, which contains the admission and exhaust valves, is cast with the cylinder head as close as possible to the cylinder wall, so as to minimise the port space leading to the combustion chamber. The valves are placed in a vertical line: the admission valve 14 at the top, and the exhaust valve 15 below. The admission valve is of the usual spring-held type. The seat and valve-stem guide are formed by a separate removable piece, into which fits the admission chamber pipe 16, the whole being held together by pressure of the set screw 17 at the top of the bell-shaped cover 18. This cover is held in place by a form of bayonet joint, the cover having at its lower end three projections, which engage with an interrupted annular groove 19, so that when the set screw 17 is tightened, the cover is held firmly in place.

Exhaust valve

The exhaust valve 15 is held on its seat above the chamber 20 by pressure of the spring 21, the lower end of which is carried by a cup 22 held by a small cotter in the valve spindle, which is guided by a long bush screwed into the bottom of the valve chamber.

The valve is made of steel, which seems to be the best material for

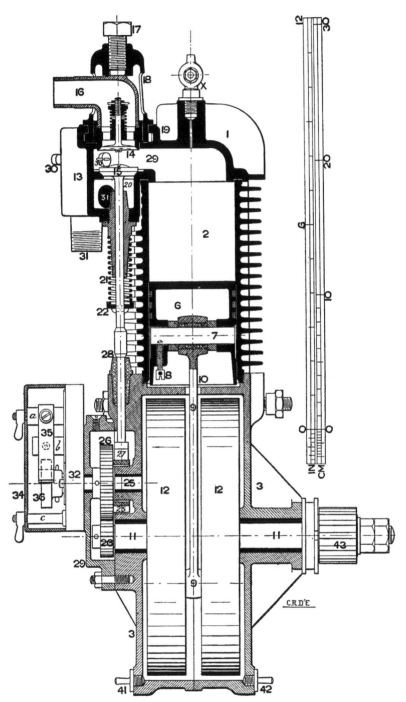

FIG. 185.—THE DE DION & BOUTON MOTOR, 1·75 HP., 1899.

263

these small things, and works well on the cast-iron seat. No real trouble attends these valves, but possibly nickel steel would best suit the work. The valve is lifted in the usual way by means of an annular cam 33, projecting from the inside of a spur wheel 26 on the spindle 25. This wheel gears with a pinion on one end of the crank shaft, the cam being thus revolved at half the speed of the motor. The cam bears against a sector-shaped head 27, at the lower end of the vertical rod 28, the upper end of which is enlarged where it meets the exhaust-valve spindle. The whole of the cam gear is enclosed in a box filled with oil, formed by one side of the crank case, and the cover 29 held by studs, as seen in Figs. 185 and 186.

The cycle of the motor The motor works on the usual four-stroke cycles. On the first down stroke the admission valve opens and allows the charge to pass from the carburettor to the admission-chamber pipe 16, Figs. 185 and 186. On the following up stroke the admission valve closes, and the charge is compressed into the combustion chamber and port space 29, and fired by the passage of an electric spark between the points of the sparking plug 30. The piston next performs the working stroke, at the completion of which the exhaust valve is lifted by means of the cam 23, and the products of combustion pass into the chamber 20, and thence by way of pipe 31 to the exhaust silencer s, Figs. 182 and 183, a very small portion being diverted to aid the vaporisation of the petrol in the carburettor, as previously described.

Ignition of charge The variation of the point in the stroke, at which the charge is fired in the motor cylinder, is effected by means of the adjustable contact maker κ, which is mounted at the end of the second motion spindle 25, and controlled by means of the small hand lever L, connected by the rod b, Figs. 182, 188, to a lever arm on the sleeve L', Fig. 188, which is free to turn on one end of the spindle P, carried in bearings brazed on the axle-bridge tube. The motion of the sleeve L is transmitted to the contact device κ by means of the lever arm and rod k.

Electric ignition apparatus The contact maker consists of an adjustable vulcanite disc 32, which is held against the cover 29 of the cam chamber, by means of the plate and the studs 33, which pass through slot holes in the vulcanite. On the vulcanite disc are mounted three brass standards a, b, c, the upper and lower of which a, c have their ends turned down and screwed. Over these ends is slipped the aluminium cover 34, enclosing the contact making device κ. At one side of the standard a is fixed a steel spring 35, on the end of which is a projecting head which bears against a steel disc 36 fixed to the extremity of the spindle 25, and on the circumference is cut a notch d. The rotation of the disc 36 causes the head of the spring once every two revolutions of the motor to fall into the notch d, thus allowing the spring to press against the platinum tip of a set screw 37 in the standard b. The standards a, b are in electrical connection with the binding screws 38, 39 on the side of the vulcanite base 32, there being connected one to a battery terminal and one to an end of the primary circuit of the induc-

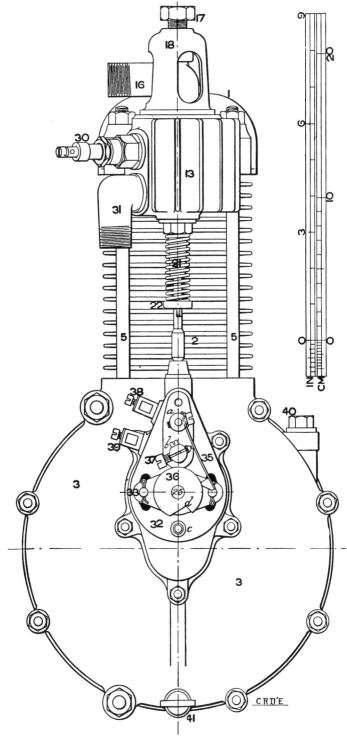

FIG. 186.—THE DE DION & BOUTON MOTOR, 1·75 H.P., 1899 PATTERN.

tion coil z, Figs. 183 and 189, so that whenever the head of the spring 35 falls into the notch d, the primary circuit is completed through the spring and screw 37, and the induced secondary current is discharged between the points of the sparking plug 30.

Lubrication The lubricant is supplied to the crank case of the motor, through the aperture 40, closed by a screw plug, and withdrawn through the apertures 41, 42, closed by thumbscrew plugs, as seen in Figs. 185 and 186. A special form of cylindrical oil tank o, Figs. 182 and 183, is usually attached to the back stays of the tricycle frame; it is divided internally into two compartments, the larger of which contains a reserve supply of petrol, which may be fed to the carburettor box c, through the pipe d, Fig. 183, the supply being controlled by a valve operated by the hand wheel e.

At the other end of the tank o is a smaller compartment, containing a supply of lubricating oil for the crank case, to which it is connected by the pipe f, the oil being injected by the driver, by means of small pump operated by the plunger g, without the necessity of his leaving his seat.

Transmission gear The motor drives the rear axle of the tricycle by means of the machine-cut steel pinion 43, which gears with the bronze spur wheel 44, bolted to the case of the differential gear, Figs. 187 and 188, on the driving axle 45, which runs in ball bearings 46, 47, 48, 49, attached to brackets brazed to the bridge tube of the tricycle frame. The engravings, Figs. 182, 183, 187 and 188, show the gear as made in 1899, but the partial gear protection shown is that of the 1898 pattern. This gearing is now completely covered in by an aluminium case similar to that of the crank. To some recent tricycles a two-speed gear of one or other of several patterns is fitted, as a means of hill climbing with less pedal work, and also as a means of preventing the stoppage of the motor when it falls below a certain minimum speed. Reference to these gears will be made hereafter.

Electric ignition apparatus The electrical outfit for firing the charge in the motor cylinder is shown diagrammatically by Fig. 189. It consists of a four-cell primary battery carried in the case r, Fig. 182, which is slung from the top stay of the tricycle frame. This battery is of the dry cell type, and has a working life of about 300 hours. One terminal of the battery is connected by wire a, Fig. 189, to the terminal b at one end of the induction coil z, Figs. 183 and 189. This terminal is cross connected to a similar one c, which is connected through the contact-making device k, by wires d, e, to one end of the primary circuit of the induction coil, at terminal f; the other end of this circuit at terminal g is connected by wire h to one of two binding screws, mounted upon the vulcanite base y at the top of the tricycle head, Fig. 183. These terminals are connected by wires j, k, which pass through the tubular handle bar to the handle switch shown in Fig. 189. This switch operates in the following manner: the horn handle a is free on the end of the handle bar, and has at its inner end a metallic sleeve and index, which may be set to positions marked "start" and "stop" on the handle bar. Inside the handle bar at its outer end, and attached to the

266

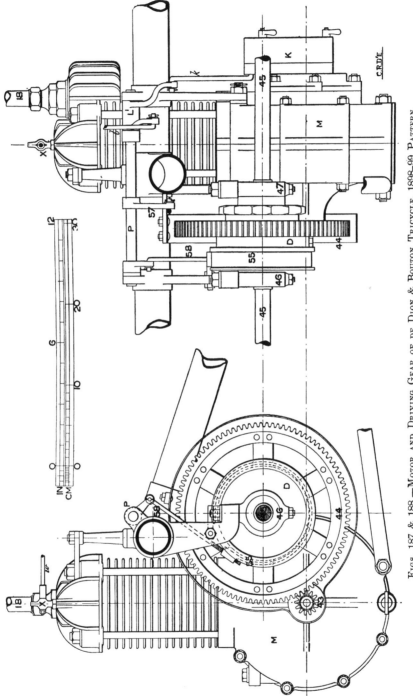

FIGS. 187 & 188.—MOTOR AND DRIVING GEAR OF DE DION & BOUTON TRICYCLE, 1898-99 PATTERN.

267

MOTOR VEHICLES AND MOTORS

handle A, is a vulcanite piece B, upon which is fixed a metal strip C. This strip, when the handle is turned to the "start" position, makes contact between the ends of the spring-controlled rods r r, which pass through the vulcanite piece D, and to which are connected the wires j, k from the terminals on the block y. When the handle is revolved to the "stop" position, the metal strip C takes up a position centrally between the ends of the rods r r, so that the circuit is broken. From this switch, when the circuit is completed, the current flows by way of wire l through the plug switch x, Figs. 182 and 189, which is mounted on the top stay of the tricycle frame, and thence by wire m to the negative terminal of the battery. The plug switch x enables the rider, when the machine is left unattended, to effectually break the electrical circuit, so that the motor cannot be started accidentally or by interference. The positive secondary terminal

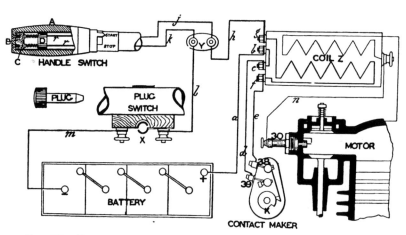

FIG. 189.—DIAGRAM OF ELECTRICAL IGNITION APPARATUS AND CONNECTIONS.

of the coil z is connected by wire n to the insulated conductor within the sparking plug 30. The other or negative end of the high tension part of the coil z is connected, as a convenient place, to one of the coil carriers, and thus the circuit is completed.

Brakes The tricycle is controlled by means of two bandbrakes which are operated by the levers 50, 51 pivoted below the handle-bar on either side of the head. The lever 50 operates the bandbrake 52, Fig. 182, through the rod 53 and bell crank 54. The lever 51 operates the bandbrake 55, Figs. 183, 187 and 188, on the exterior of the differential gear box D on the driving axle. The movement of the hand lever 51 is communicated through the bell crank 56 and rod d to the lever 57, Fig. 188, on the spindle P, a second lever arm, on the end of which is connected by the adjustable rod 58, to one end of the band 55 of the brake, the other end of which is attached to the bracket of the axle bearing 46, Figs. 187, and 188.

268

The relief pressure cock x at the top of the motor cylinder is con- **Compression relief** trolled by the driver by means of the handle w, Fig. 182, actuating the jointed rotative rod *w*. To prevent the noise made by the escape from this cock when opened from frighting horses, it should be coupled up in a small pipe to the silencer s. This, as will be seen from Fig. 183, could be very easily done.

The motor is started by the rider, by pedalling in the usual manner, **Starting** the relief pressure cock being first opened. When the motor is working properly, the cock x is closed, and pedalling stopped ; the motion imparted to the chain by the sprocket pinion on the driving axle causing the silent pawl clutch 59 on the crank spindle to be put out of gear.

THE PHÉBUS MOTOR TRICYCLE.

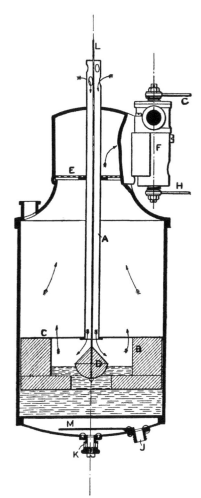

FIG. 189A.—THE CARBURETTOR OF THE PHÉBUS-ASTER TRICYCLE.

A motor tricycle of the same type as that last described is the Phébus, which is fitted with the "Aster" motor. Of this motor it is not necessary to say more than that it differs only in minor points from those of several other makers, and in the use of copper radiating gills tightly fixed on the cylinder and radially corrugated. The advantage of this modification is not very apparent.

The carburettor used on this tri- **Phébus carburettor** cycle may be described with reference to Fig. 189A, although it is only a mere modification of the Daimler surface carburettor. On the suction stroke of the motor, which is connected to the opening at the upper part of the air and vapour valve F, air is drawn into the pipe A through holes at its upper end, these holes being usually covered with gauze. The air so drawn in is delivered at the bottom of this pipe, impinges upon the conical cork float D, from which extends a wire, which passes through the top of pipe A, at L, and acts as a petrol indicator. By impinging upon the cone D the entering air is supposed to be deflected upon the surface of the petrol within the large float B, thus causing a disturbance, which gives it opportunity of absorbing petrol vapour. At all events, the petrol does vaporise,

269

and with the air that takes it up, it passes away through the perforated plate c into the storage part of the vessel, and out through a gauze-covered plate E to the valve F. The handle G controls the quantity of carburetted air allowed to go to the motor, and the handle H controls the quantity of air and the quantity of carburetted air together going to the motor. At the bottom of the carburettor vessel is a space into which a little of the exhaust products are admitted at J, and pass away at K, a baffle-plate M being inserted to cause the cases to impinge on the bottom of the carburettor to heat it and encourage vaporisation.

THE ARIEL MOTOR TRICYCLE.

General arrangement

A modification of the de Dion tricycle is that known as the "Ariel," as shown by Fig. 190. The motor is of the de Dion Daimler type but not of de Dion and Bouton make, and in several details it differs from that of the French firm. The disposition of several of the other parts is also different. The motor, the top of the cylinder of which is seen at c, is of 2·25 nominal HP., and is placed in front of the main axle and between the two tubes of a lower back stay fork. The whole of the gear is covered in by a case which is in front of the axle instead of behind it, as in the more usual arrangement.

The whole of the space within the tubes of the main frame is filled in with a vessel which forms carburettor, battery case, and petrol supply. This combination secures neatness and compactness, and avoids the necessity for the separate tank behind for all ordinary distances. The space B is allotted to the battery Air is admitted to the carburettor by a valve with a lever o, and after carburetion it passes off through the pipe K to the cylinder, the quantity passing being regulated by the valve lever N. The carburettor is heated by exhaust products carried by the pipe z. A petrol level indicator wire is seen at M, the petrol inlet is at T. At Q is a cap on a tube, within which is a long needle valve which regulates the flow of petrol from the spare oil space into the carburettor space. At I is the battery switch, and at L is the induction coil, the circuit being completed by a loose plug seen at Y.

The compression tap H on the top of the cylinder, or tap for relieving the compression pressure when starting, is operated by a handle at I and connecting rod T.

Carburettor

The construction of the carburettor and petrol and battery carrier will be understood from an inspection of the sections given in Fig. 190A. In these A is the exterior of the case, which is of the shape of the ordinary tourists' bicycle carrier. It is divided into three compartments: B for

Battery

carrying spare petrol, C the carburettor space, and S space for battery R R of

Supply tank

primary or secondary cells. The spare petrol space is filled by removing the large screw cap Q, which carries a test rod Q' for ascertaining the quantity of spare petrol. From the space B, petrol is admitted to the

270

Fig. 190.—The Ariel Petrol Motor Tricycle, 2·25 H.P.

271

carburettor space by raising the needle valve E which is enclosed in the tube F and is carried by the thumbscrew G. When this is raised, and the screw cap Q loosened to prevent the formation of a vacuum, the petrol flows through the pipe D.

Between the top of the carburettor space and the top of the case is a connecting tube H, within which is an indicator rod or wire K attached to, and rising or falling with, the float J, and thus showing the level of petrol in the carburettor. When the machine is not in use, the rod K is pushed down, and the sliding cover T moved so as to cover top of the air inlet tube H. The carburetted air passes away to the motor through the pipes N, O,

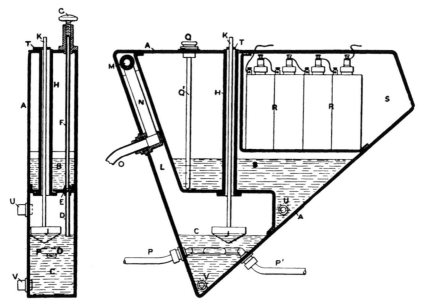

FIG. 190A.—CARBURETTOR SPARE PETROL TANK AND BATTERY CARRIER OF
ARIEL MOTOR TRICYCLE.

admitted thereto by the adjusting valve M at the top of the vapour space L.

To encourage the vaporisation of the petrol in the colder weather, a small branch pipe P is taken from the exhaust pipe and led through the carburettor in zig-zag form to an outlet P', which is led to the silencer.

At U, V are draw-off plug caps for emptying petrol from the chambers B, C.

It will be seen that the carburettor is of the simplest possible form, nothing in fact but a box in which the petroleum spirit is allowed, and sometimes encouraged, to evaporate in the presence of, and be taken up by, air which enters the tube H in obedience to withdrawal of carburetted air from O. No doubt the constant moving of the petrol and splashing over all the surfaces resulting from jumping and jolting over the roads keeps them wetted with a film of petrol in the best condition for evaporation encouraged by the moving air.

Although the general arrangement and scheme of connections in this tricycle may be gathered from the description and engravings herein given, and from the explanations given with reference to the de Dion tricycle, Figs. 182–188, the diagram, Fig. 190B, may be conveniently used for further explanation. It will of course be understood that the parts

Pipe and electric connections

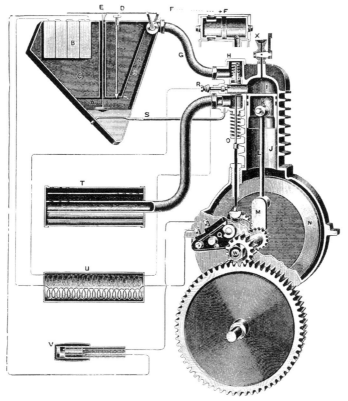

FIG. 190B.—ARIEL MOTOR TRICYCLE: DIAGRAM SHOWING SCHEME OF MECHANICAL AND ELECTRICAL CONNECTIONS.

here brought together are not shown in their relative positions, but are so placed as to show the general scheme of pipe and electrical connections.

The carburettor A is seen connected by the pipe G to the admission valve box H, the passage from the carburettor to the pipe G being controlled by the vapour and air valves in the mixing valve chamber F, shown to an enlarged scale in section, and seen in side view with its actuating levers and rods at the top of the carburettor. The exhaust is shown as passing to a silencer T, shown in section, the small branch from the ex-

273   18

haust for heating the petrol in the carburettor being shown at s, the battery connections to the coil u, sparking plug r, and handle switch v are indicated by fine lines. The primary and secondary circuit arrangement is shown complete with the handle switch at v, but the removable plug switch previously mentioned is not indicated. Other parts shown in the engraving and lettered are: b, battery.; c, spare oil tank; d, needle valve for regulating petrol into carburettor; e, indicator wire from float in carburettor; i is the combustion chamber of the cylinder j, above the piston k; m is the crank connected to the piston by the rod l; p is the cam for lifting the rod for opening the exhaust valve o; and q the contact breaker for production of spark at the plug r; and x is the compression relief cock.

Starting   Cranks, pedal and free wheel are provided for starting by pedalling, and once started the compression tap is closed and the carburetted air and air valves respectively regulated by the handles n, o. The ignition plug is at r, and the starting and stopping switch in the handle at x. A large crank-case lubricator is seen at e with a cap at u. In general it may be said that the working of the motor and connected parts is the same or will be understood by the description given of the de Dion.

These tricycles are in extensive use and have done some very heavy long-distance running.

THE RENAUX TRICYCLE.

Figs. 191 and 192 illustrate the Renaux tricycle fitted with a motor of 3 nominal HP., and is remarkable as being the most powerful vehicle of its weight ever made.

The motor   The position and outline of the horizontal motor are clearly seen in the engravings. Electric ignition is used, and only the battery, in a case, is carried on the top tube of the frame. The petrol is carried in a cylinder, of the usual ugly after-thought appearance, while the greater part of the frame space is unoccupied. The cylinder contains a compartment for lubricating oil and for the induction coil. A float-feed carburettor of the Longuemarre type is used. Differential gearing of the usual kind is dispensed with by making the big end of the connecting rod large enough to embrace the carrier of a bevel pinion, which with two bevel wheels on the crank discs forms a differential gear in the centre of the case. The two parts of the motor shaft, which is excentric to the main axle, are fitted at their exterior ends with pinions which gear into internal spur wheels on the axle.

One of these tricycles, ridden by M. Renaux, ran the Paris-Malo race in August, 1899, a distance of 231 miles (372 kilom.), in 7 hours 11 minutes, or at the mean rate of 32·2 miles per hour over the whole distance.

Very complete drawings of this tricycle and its machinery except that the crank shaft is shown as running in atmospheric bearings, are given in the Patent Specification, No. 13136 of 1898.

FIG. 191.—THE RENAUX PETROL MOTOR TRICYCLE, 3 HP.

FIG. 192.—THE RENAUX MOTOR TRICYCLE.

275

MOTOR VEHICLES AND MOTORS

The Turrell-Bollée Voiturette.

We come now to an entirely different form of tricycle, namely, one of the Bollée type referred to at the commencement of this chapter. Figs. 193–201 show the form of this car as modified and made in England from the designs of Mr. C. McR. Turrell. The modifications do not affect the principles of the design, so that a description of this car will be in all essentials a description of the earlier Bollée car. The differences are chiefly in details of the motor, the arrangement and size of the seats, and in several small parts to which reference will be made.

Figs. 193 and 194 show a side elevation and plan of this English-built three-wheeled voiturette.

General arrangement It is constructed to carry three persons, two in front and the driver behind, and weighs complete about 6 cwt. The motor is carried on the outside of the frame on the left-hand, the cooling of the cylinder and combustion chamber being effected by radiating ribs only, no water being carried. It drives, by means of one of three sets of spur wheels and pinions, a countershaft which in turn drives the rear road wheel by means of a belt, one of the chief features of Bollée's invention being that the driving wheel axle is carried by levers pivoted on the frame of the car, so that by pushing the axle backwards or forwards the belt tension may be increased or decreased, and the motor mechanism thus put into or out of gear with the road driving wheel. The spur driving gear enables speeds of about 6, 12, and 18 miles per hour to be given to the car. There is no reverse motion, this being considered unnecessary for so light a vehicle, except in the eyes of the British law.

Frame The underframe is constructed of steel tubes, and consists of a rectangular structure F, the front tube of which is prolonged on either side and carries at its ends the vertical tubes 1, which form heads for the vertical pivots of the steering wheel axles. To the top of the wheel pivot on the driver's right hand is fixed the steering bar z, the wheel pivot on the opposite side being operated by means of the crosslink 2 connecting the arms 3. At about the centre of the rectangular frame F is a cross stay tube 4, and the strut and trussing rods 5, one of which on each side give additional strength to the longitudinal tubes of the frame.

Motor The motor M, seen in Figs. 193 and 194, and in detail in Figs. 195, 196, and 197, has a single horizontal cylinder without a water jacket, the cooling, as before mentioned, being effected by longitudinal radiating ribs which are about $\frac{1}{8}$ in. thick and $\frac{7}{8}$ in. deep, and cast with the cylinder and cylinder head. It develops $2\frac{1}{2}$ HP. at about 750 rev. per minute. For long continuous runs experience has shown that cooling by radiating ribs as used by Bollée is not effectual for a motor of this power, the portion of the head around the exhaust valve passage especially sometimes reaching red heat. When this happens, it is impossible to prevent the exhaust valve from leaking, which causes loss of power. More recent developments in this direction, as

276

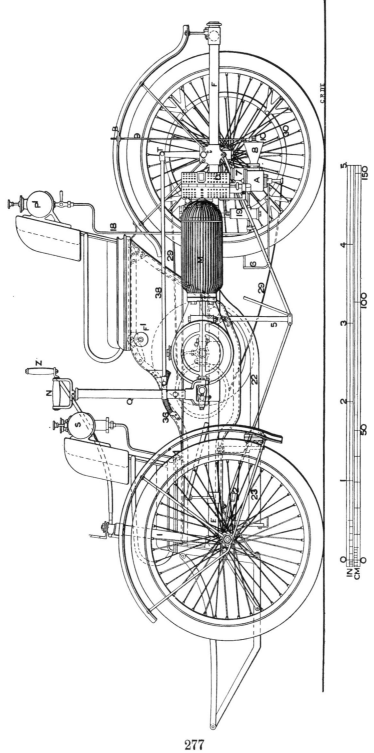

Fig. 198.—The Turrell-Bollée Tricycle Voiturette, 2·5 H.P.

277

in the Darracq-Bollée car, have increased the efficiency of this cooling, but generally it may be said that water cooling is preferable for this size of motor.

The supply of petrol for the motor is carried in the tank P, Fig. 194, on the right-hand side of the frame. It feeds the float vessel A of the carburettor, which is of the Phœnix-Daimler type, by means of the pipe 6, Figs. 193 and 194. As before described in connection with this form of carburettor, the float vessel A feeds the petrol as required into the carburettor 7, into which air is drawn through the conical pipe 8, Figs. 193 and 196, the openings in the end of which are seen in Fig. 196. By means of the small hand wheel B, at the upper end of the vertical rod 9, the plate 10, in which are holes corresponding to those in the mouth of the air tube, may be revolved for regulating the quantity of air admitted to the carburettor. On each suction stroke of the motor the admission valve 13 is lifted against the resistance of the spring 14, and the charge is drawn from the carburettor by way of pipe 11 to the chamber 12, and through the port 15 into the cylinder. On the next in-stroke it is compressed into the combustion chamber and port space 15, and fired by the ignition tube E, Fig. 196, at the back of the combustion chamber.

Ignition The tube is heated by a burner 17, both being enclosed in the box D. The oil supply for the burner is carried in the tank P', at the back of the driver's seat. It is fed by pipe 18 through the cleaning vessel 19, and thence to the burner, which is of similar construction to that used on the Panhard and Levassor car before described. On the completion of the working stroke of the piston, the exhaust valve 20, Fig. 195, is lifted, and the products of combustion pass into the chamber 21, and thence by pipe 22 to the silencer 23, which consists of four short tubes placed side by side horizontally. These tubes have their back ends connected by a chamber to which is also attached the exhaust pipe 22. The front ends of the tubes are closed by a plate, which is held by means of bolts connecting it with the flanges of the chamber. The gases escape from the silencer by means of holes formed in the bottom part of the tubes at their front ends.

Exhaust valve The exhaust valve 20 is operated by means of the cam K, Figs. 197, 198 and 201, at the outer end of the spindle 24, which is driven at a speed of one to two from the engine shaft 25 by means of the pinion a and wheel g. The cam K runs on a roller k at the end of the lever G¹, the other end of which lifts the exhaust valve rod with which it is kept in contact by the spring arrangement seen in Figs. 195 and 197. The large spring 45 acts in the usual way in keeping the exhaust valve on its seat. The small springs 46 and 47, prevent the end of the lever G¹, with its adjusting screw 48, from leaving elastic contact with the bottom of the exhaust valve stem. Rattle is thus prevented, and adjustment provided for. When the speed of the motor exceeds the normal, the governor H, Figs. 198 and 201, comes into action, and by displacing the cam K prevents the exhaust valve from being lifted.

Governor The governor consists of the cylindrical weights w, connected by means of

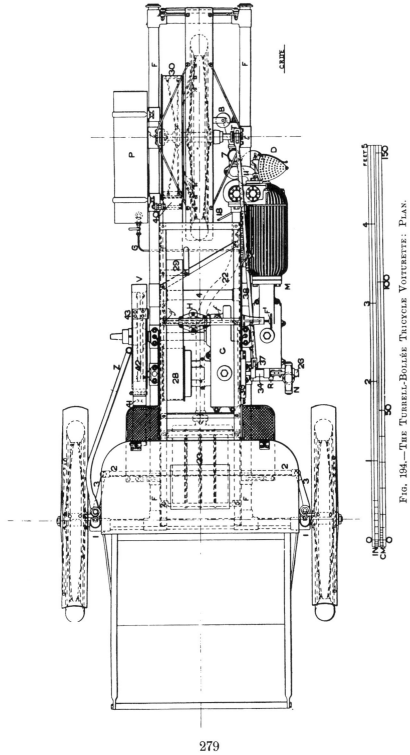

FIG. 194.—THE TURRELL-BOLLÉE TRICYCLE VOITURETTE: PLAN.

279

links l to the joint-collar m, fixed on the spindle 24, and to corresponding joint projections n at the end of the tubular spindle h, which is driven by the gear wheel g. When the normal speed is exceeded, the governor acts in the usual way, and the spindle 24 is pulled in the direction of the arrow, Fig. 198, causing the cam к to miss the roller k at the end of the lever G^1, and the exhaust valve is left closed until the speed is again reduced to the normal, and the cam brought back into place by means of the spring z on the spindle 24, when the exhaust valve is again lifted.

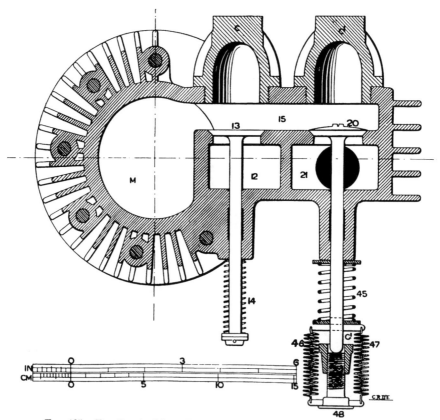

FIG. 195.—THE BOLLÉE MOTOR-SECTION OF CYLINDER HEAD AND VALVE BOX.

Control   The motor is stopped by screwing in the milled head F^1 at the left-hand side of the driver's seat. This operates the horizontal spindle f, Fig. 201, the inner end of which is pressed against the upper end of the vertical bar g^1, seen in section in plan, Fig. 198. The bar is pivoted from the seat frame, so that when F^1 is turned the lower end of the bar g^1 is pressed against the end of the spindle 24, and the cam к pushed out of contact with the roller k so that the exhaust valve cannot be lifted.

Transmission gear   On the engine shaft 25 are keyed three spur pinions a, b, c, which

280

gear with corresponding spur wheels *d, e, f,* carried on the countershaft 26 in such a manner that any one of them may be keyed to the sleeve 27 while the others run loose. The sleeve 27 carries at its opposite end the belt pulley 28, which drives the drum 30 by means of the open belt 29, the drum being attached to the driving wheel of the vehicle. Three changes of speed are obtained by means of the spur gear which is enclosed in the gear case G.

The speed changing is effected by means of the handle N, which also controls the mechanism for putting the motor into or out of gear with the road wheel, and for applying the principal brake. These operations are effected by means of the parts now to be described.

The handle N is mounted at the upper end of a vertical rod 31, Figs. 197, 198 and 201, which passes through the hollow standard Q, and carries at its lower end the pinion *j*, Fig. 198, which gears with a rack cut on the end of the countershaft 26. The handle N and rod 31 may be fixed in one of three positions by means of the key *l*, Fig. 197. By lifting the handle against the resistance of the spring *p* at the lower end of the rod 31, it may be rotated, and the key *l* dropped into one of the three notches at the upper end of the standard Q, and held in place by the pressure of spring *p*. The

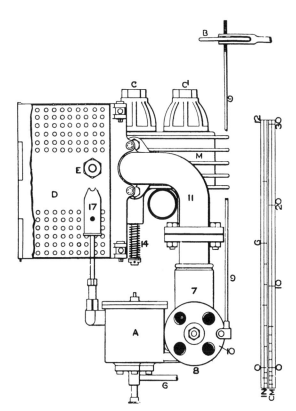

FIG. 196.—THE BOLLÉE MOTOR : ARRANGEMENT OF FLOAT FEED CARBURETTOR AND IGNITION TUBE AND LAMP.

lower end of the standard Q is forked, as seen in Figs. 193 and 197, and pivoted transversely on the longitudinal pivots x, Fig. 197, at the end of the sleeve R, at one side of which is a groove *r*, in which the rack on shaft 26 is free to slide, so that the standard Q may be moved backwards or forwards, and with it the sleeve R. At the position Q¹ on the standard Q is a projecting piece connected by links 32 to a short spindle 33 Fig. 201, which may be moved in the direction of its axis in the bearing 34 carried

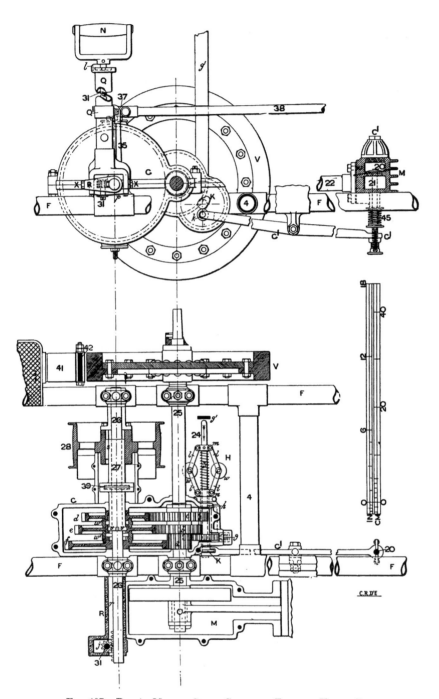

FIG. 197.—BOLLÉE MOTOR: SPEED GEAR AND EXHAUST VALVE GEAR.
FIG. 198.—PLAN OF SPEED GEAR AND GOVERNOR.

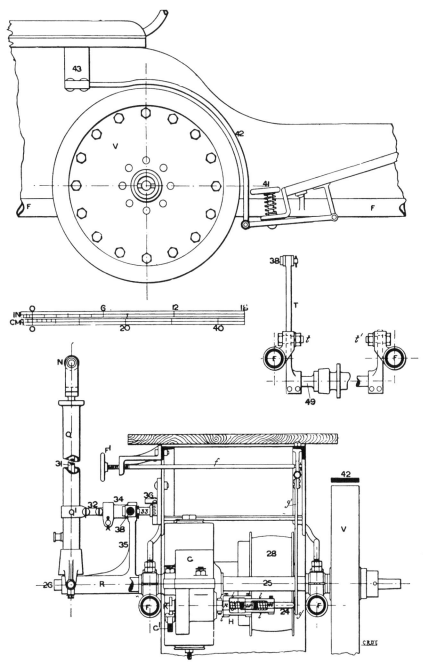

FIG. 199.—BOLLÉE VOITURETTE: FLY-WHEEL BRAKE.
FIG. 200.—DRIVING AXLE SUSPENSION AND CONNECTIONS.
FIG. 201.—BACK VIEW OF SPEED CHANGE LEVER, GOVERNOR, AND MECHANISM.

283

by the arm 35 from the sleeve ʀ. At the inner end of the spindle 33 is a key which may be engaged with the toothed sector 36, Figs. 193, 194 and 201, which is attached to the frame, so that the standard ǫ may be held at any required angular position on either side of the vertical. On the bearing 34 in which the spindle 33 slides, and at its inner end, is an ear 37, connected by the bar 38 to the upper end of the lever ᴛ, Figs. 193 and 200, which is pivoted at *t*. Between the downwardly projecting arm of lever ᴛ and a similar arm pivotted at t^1 on the opposite side of the frame is carried the axle 49, on which the driving road wheel and belt drum 30 revolve, so that the distance between this axle and the countershaft 26 carrying the belt pulley 28 is controlled by the angular fore and aft movement of the standard ǫ.

As above described, by turning the handle ɴ the pinion ᴊ, gearing with the rack on the end of the countershaft 26, causes it to move transversely in its bearings. On the shaft 26, between the belt pulley 28 and the gear case ɢ, is a collar 39, which is free to revolve in a closely-fitting box on the sleeve 27, so that the transverse movement of the shaft carries the sleeve with it. At one end of the sleeve 27 is a feather *s*, which slides in a slot in the boss of the belt pulley 28, so that the pulley is always driven by the sleeve; at the other end of the sleeve are four short feathers *v*, one in each quadrant, which fit into key ways in the bosses of the spur wheels *d, e, f*, spaces being left between the middle and the two outer wheel bosses, so that when the feathers *v* are in these spaces all three wheels are loose on the countershaft 26. The middle wheel *e* is carried between the outer wheels *d, f* by means of the collars *w, w¹* which are fixed to the wheels *d, f*, but which form bearings carrying the portions of the boss of wheel *e*, which project on either side of the wheel centre. The three notches at the top of the standard ǫ in which the key *l* may be fixed correspond to the positions of the sleeve 27 at which one or other of the three wheels *d, e, f* are keyed to the sleeve.

Belt tension In order to put the motor into or out of gear with the road wheel, the standard ǫ must first be pushed outwards away from the driver. This disengages the key on the end of the spindle 33 from the toothed sector 36, and so leaves the standard free. In order to put the motor into gear, or in other words, in order to tighten the belt 29, the standard ǫ is then pushed forward, this movement being transmitted to the lever ᴛ by means of the bar 38 as previously described, so that its upper arm moves in the same direction as the standard ǫ, and consequently the axle 49 of the road wheel is pushed backwards, and thus the distance between the centres of the belt pulleys 28 and 30 increased and the belt 29 tightened. When the standard ǫ is held in the vertical position the belt 29 is slack and the motor out of gear, and when the standard is moved backwards from this position the front portion of the surface of the belt drum 30 is pressed against a wooden block 40, Fig. 194. This is carried by the tubular frame, and when the pulley is pulled against it it forms a powerful brake. There is also an auxiliary brake operated by depressing the foot-plate 41, Figs. 194 and 199, **Brakes** thus pressing the leather band 42 on the surface of the flywheel v, the upper end of the band being fixed to the framework at 43.

Some of the Bollée voiturettes have made long runs at high speeds, including the Paris-Dieppe race in 1897, when, what was then the astonishing mean speed of 24·5 miles per hour over the whole distance was reached.

Large numbers of these voiturettes have been made and used in **General** France. They have not, however, been very favourably received in this country, although their power and handiness, and in some respects simplicity, are freely admitted.

FIG. 202.—SIMMS PETROL MOTOR TRICYCLE VEHICLE. (*See next page.*)

There is, perhaps, no car in which belt driving is so successfully and advantageously employed; and the arrangement which enables the one movement to slacken the belt and apply a powerful brake is an excellent one, and has been adopted by Vallée in his recent racing car and by others. One cause of the failure of the car in this country was the low position of the rider, a position which placed him in the dust of the road in summer and in the mud in winter: this objection, coupled with the noise made by the motor and the frequent small troubles with its unprotected valves after running any considerable distance, may be taken as explanations of the little encouragement which this type of vehicle has met with in this country.

MOTOR VEHICLES AND MOTORS

Simms "Motor Wheel."

Another tricycle, but of quite a different kind, is the Simms "motor wheel" illustrated by Fig. 202. It is arranged to carry a seat, not a saddle, for the driver as well as for the passenger, and is fitted with a Simms motor of the de Dion Daimler type of two sizes, either a 1½ B.HP. or a 2¾ brake HP. The motor is fitted with magneto-electric ignition apparatus known as the Simms-Bosch. The single wheel is used as a rear steering wheel, the two driving wheels on a through axle being in front. The motor and gear are in front under the front seat. The driver starts the motor by crank pedals and chain in the usual way, and a rest is pro-

FIG. 203.—SIMMS PETROL MOTOR, 2¾ HP., WITH FAN

vided for both feet. The general arrangement can be understood from the engraving, which shows that the steering wheel is to the right hand, and the brake and ignition timing device on the left. A second brake is fitted to a drum on the main axle, and this is applied by foot.

Motor   The 2¾ HP. motor is illustrated by Fig. 203, which is from a photograph of one fitted with a fan for giving a forced-air cooling. The same engine is shown generally by Figs. 204-206, and in detail by Figs. 207-210. The motors, it will be seen, are horizontal, with very wide radiating cooling ribs, the valves being placed at the end of the cylinder. A is the automatic air and vapour admission valve, E the exhaust valve actuated by the cam L, connecting rod L′ and arms b, c on the short spindle a, Figs. 204 and 210, the cam L being on the spindle I driven by the pinion G on the crankshaft B and wheel H (see also Fig. 208). The rod L′ is kept in place by the radius link l. The cycle of the motor is that usual with the single cylinder Daimler, the piston P being connected by a channelled rod to the crank pin K of the crankshaft B, which is balanced

286

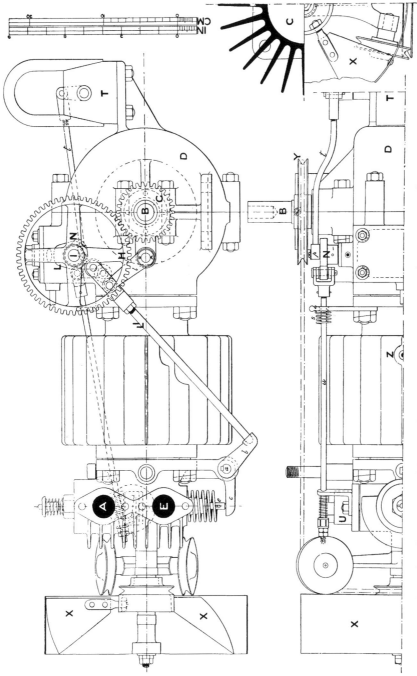

FIG. 204.—Simms Petrol Motor, 2¾ HP, with Magneto-Electric Apparatus and Cooling Fan: Elevation.
FIG. 206.—End Elevation and Section of Cylinder.
FIG. 205.—Part Plan.

287

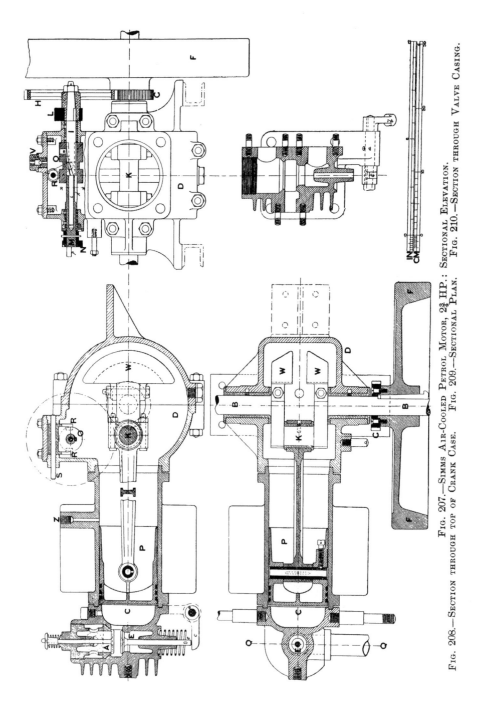

FIG. 207.—SIMMS AIR-COOLED PETROL MOTOR, 2¼ H.P.: SECTIONAL ELEVATION.
FIG. 208.—SECTION THROUGH TOP OF CRANK CASE.   FIG. 209.—SECTIONAL PLAN.
FIG. 210.—SECTION THROUGH VALVE CASING.

by the applied weights w w. At the end of the cylinder is the fan with the blades x x, driven by cord belt running over idle pulleys from the pulleys y on the crankshaft, a method of assisted air cooling which is not desirable. (See Chapter XVIII.) The magneto-electric generator т is placed on a bracket cast with the crank case D, seen in section in Fig. 207. In Fig. 208 the upper part of this case, where it carries the half-speed spindle I, is shown in section, the lower part showing the flange by which it is

Magneto-
electric
ignition ap-
paratus

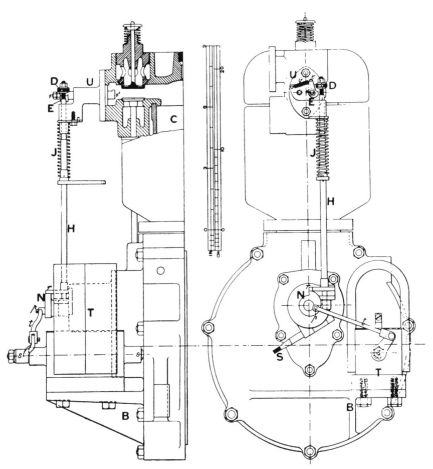

FIGS. 211 & 212.—MOTOR FITTED WITH SIMMS-BOSCH MAGNETO-ELECTRIC IGNITION APPARATUS.

attached to the cylinder end. At one end of the spindle I is the steel cam L, on which runs the roller of the rod U. At the outer end of this rod is an adjustable striker nut, which operates the contact-breaker in the space near the two cylinder valves. For this apparatus reference will be made to the engravings Figs. 211–213. These show the apparatus as attached to a vertical motor; they equally permit an explanation of the apparatus in the horizontal motor, Figs. 204, 205 and 208. In Figs. 211 and

212 the rod operated by the sharp stepped cam N is marked H. At its upper end is the adjustable striking nut D, which on the descent of the rod H strikes the little projection on the short arm E on the spindle F in the boss U. On the inner end of the spindle F is the contact-making finger F'. The rod H moves under the influence of the spring J as the lower end follows the cam N. When the piece r falls into the step of the cam N, the nut D hits the shoulder on the arm E a sharp blow, and the end of the finger F' in the valve space is struck off the contact stud K (see Fig. 213), and a spark results. The oscillating shield, which moves between the fixed armature and the permanent magnets of the magneto - electric generator T, receives its motion from a rod t working on a little crank pin p in the side of the cam N. Whatever the position, therefore, or change of position of the cam N rotatively with regard to the spindle I, the relative positions of the generator shield and the contact-breaker finger F' will remain unaltered. This is important, because the advance or retardation of the period of ignition is obtained by moving the part M of the spindle which carries the cam N relatively to the part I. The part M is hollow at

Control of ignition period

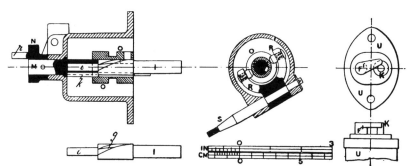

FIG. 213.—IGNITION ADVANCE AND RETARDING DEVICE AND INTERIOR CONTACT-PIECE OF SIMMS-BOSCH APPARATUS.

t, Fig. 213, and the part i of the spindle I is turned down to fit in it (see detail). On one side of the tubular part of M is a keyway k, over which the grooved sleeve nut o slides when moved by the fork R R on the spindle S (see also Fig. 208). If the nut o be by this means moved towards the left, though both parts of the spindle be rotating, the one will be moved angularly relatively to the other by the amount of the pitch of the screw groove g. The sleeve o cannot rotate on M, but the part I can in obedience to the spiral or screw groove. Thus the position of the cam N with relation to the spindle I and wheel H and crankshaft pinion G may be varied so as to get an early or a late period of ignition. In Fig. 213 the parts shown in black section are steel. One terminal of the generator is earthed to the motor frame by direct contact, and the other by a wire to the outside end of the insulated stud K, the inner end of which is seen in Fig. 213, and the outer end shown below the spring v in Fig. 212 but not marked.

There is thus only one wire employed. This apparatus will be referred to hereafter in dealing with different forms of electric and electro-magnetic ignition devices. There appears to be no reason why a rotary generator should not be used instead of a reciprocation or oscillating generator, if the sudden separation of the contact piece and finger be made by the sharp step cam and spring-moved hammer rod. The advantages of the mechanical generator in place of a primary or secondary battery are undoubted, but there are objections to the use of internal contact-breaker. These, how-

Fig. 214.—Columbia Petrol Motor Carrier Tricycle. (*See next page.*)

ever, are not necessarily serious, and may, if the contact-breaker be suitably placed, prove not to be insuperable. A rotary movement of the generator might be expected to introduce more trustworthy and less noisy parts, and it is desirable that some less noisy method of striking the contact-breaker should be devised.

Assuming now that the motor tricycles described are sufficiently typical of the design generally of to-day, a description may now be given of a motor tricycle constructed and arranged as a light goods carrier.

291

MOTOR VEHICLES AND MOTORS

The Columbia Motor Carrier Tricycle.

General arrange-ment

Figs. 214 and 215 are perspective views, and Figs. 216–221 show general arrangement and details of a three-wheeled vehicle for the delivery of light goods, constructed by the Pope Manufacturing Company, of Hartford, U.S., and known as the Columbia Motor Carrier. It consists of a double box-shaped body, the construction of which is seen in the photograph views, Figs. 214 and 215. The body is carried by a tricycle frame of steel tube, and insulated therefrom by means of light leaf springs, Fig. 214, which, together with the pneumatic tyres of the road wheels,

FIG. 215.—THE COLUMBIA MOTOR CARRIER TRICYCLE.

minimise the danger of injury to fragile goods when running over rough road surfaces.

The driver sits on an ordinary bicycle saddle carried upon an extension of the central frame stays at the back of the body.

Attached direct to the frame tubes on either side of the driver's seat are the motor and transmission gear.

The rear road wheels are the drivers, and are carried by a through axle, upon which is mounted the differential gear. The power is transmitted from the motor shaft through one of two sets of spur gears, giving speeds of five and ten miles per hour to the vehicle.

The steering is effected by the single front wheel, which is controlled

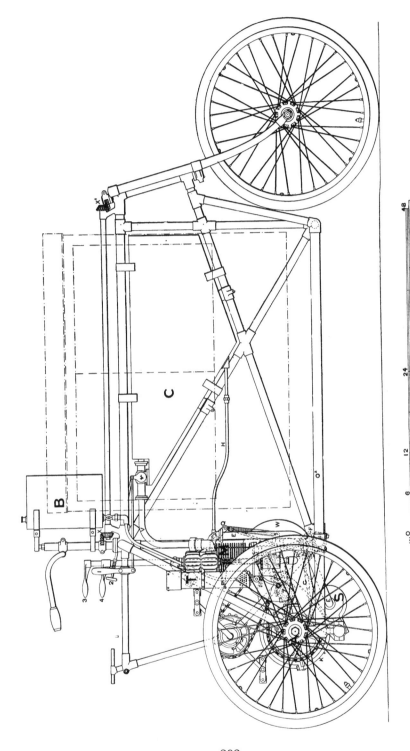

FIG. 216.—THE COLUMBIA PETROL MOTOR CARRIER TRICYCLE : ELEVATION.

293

from the steering bar in front of the driver, through the bevel transmission gear x x', Fig. 216.

The motor is started by pedalling an independent crank-spindle, which drives the motor-shaft through a pitch chain gear and claw clutch. The foot power may also be employed to propel the tricycle without working either the motor or change speed gear.

The supply of petrol carried is sufficient for a run of about 100 miles at average speed.

Motor The motor M, Figs. 216, 219, is of the vertical single cylinder air-cooled type, constructed closely on the lines of the early tricycle motor, 1¼ HP., of Messrs. de Dion & Bouton, the chief points of difference being that the admission valve, as well as the exhaust valve, is mechanically operated; the ignition is by hot tube instead of by the electric spark, and the fly-wheel is placed outside the crank case instead of inside.

Figs. 217 and 218 show a cross section, and a part sectional elevation of the motor. The vertical air-cooled cylinder c is mounted upon a crank case D, which is supported from the cross tube o of the frame by means of brackets and clips L L'. Enclosed by the crank case D are the balanced crank 13 on the crankshaft 4, which runs in bearings at either end of the case, and the cam shaft 6 which is driven at half the speed of the motor by means of a spur pinion 3, which gears with the spur wheel 2 on the countershaft 6. Upon this shaft are cams 1, 5, which operate respectively the admission and exhaust valves v, E, which are normally held upon their seats by springs 11, 12.

At R R' are screw plugs above the valves E, v, by the removal of which the valves are easily accessible. The valve boxes are, it will be seen, given very large radiating surfaces.

The chamber below the admission valve v, Fig. 217, is connected by a pipe v, Figs. 216, 218 and 219, through a valve v, Fig. 216, to the upper portion of a surface carburettor c, Fig. 216, which is in its essential features similar to that of Messrs. de Dion & Bouton. The lower portion of the carburettor box forms the principal petrol reservoir for the motor, the spirit being rendered more easily vaporised by heat from a small portion of the exhaust gases which pass to a chamber at the bottom of the carburettor by pipe H, Figs. 216 and 219. By means of the pipe v the carburetted vapour is drawn from the top of the carburettor box, and is regulated as to quantity and quality by rotary valves on the chamber v, the quantity valve being controlled by the lever 1, Fig. 204, and the quality valve by the hand wheel 2. This regulating device will be seen to be similar to that used by de Dion & Bouton on their tricycles. The other petrol reservoir, B, supplies the ignition tube burner, which is enclosed in the case T, Figs. 216 and 219.

The small end of the connecting rod 8 is pivoted on a hollow pin 9 in the piston body. The crank pin 7 is also hollow, and is formed with one of the crank discs, which are balanced by attached weights as shown. A

294

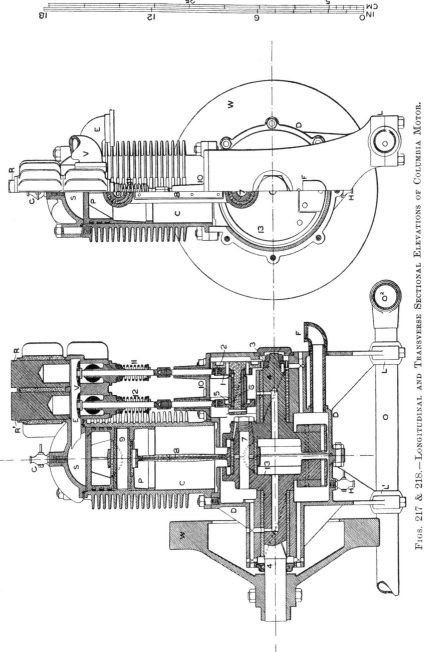

FIGS. 217 & 218.—LONGITUDINAL AND TRANSVERSE SECTIONAL ELEVATIONS OF COLUMBIA MOTOR.

295

great deal of attention has been given, it will be seen by an inspection of the engravings, to the means of lubrication all over the motor. For the crank case oil enters at F, and is drawn off at H.

The cycle of operations within the motor cylinder is the same as that of all small motors of this type, namely, suction, compression, explosion, exhaust, suction, etc.

Ignition

On the first down stroke of the piston P, the admission valve v is lifted by the cam 1, and the charge is drawn into the cylinder through the pipe v from the carburettor c. On the following up stroke the charge is compressed into the combustion chamber s and the part above the valves E, v, Figs. 205 and 206, and fired by means of the ignition tube enclosed in the box T. The piston then performs its working stroke, on the completion of which the exhaust valve E is lifted by means of the cam 5, and the products of combustion are expelled through pipe E, Figs. 216, 218 and 219, into the silencer s, from which they escape into the atmosphere.

Transmission gear

The crankshaft of the motor is prolonged outside the crank case at one end, and to it is keyed the flywheel w, Figs. 217 and 218, the boss of which forms part of a fixed coupling connecting the crankshaft to the prolongation shaft 5, Fig. 219. By means of a friction clutch 25 the power is transmitted from shaft 5 to shaft 6, upon which is a pinion gearing with the spur wheel 8 on the second motion shaft 7 of the speed-changing gear.

Mounted loose upon shaft 7 are two spur pinions, one of which is marked 9, and either of which may be locked to the shaft by means of a sliding claw clutch 10, which with the main clutch 25 is operated by the hand lever 3, Figs. 216 and 220, mounted at the top of a vertical spindle at the driver's left hand. This spindle has fixed at its lower extremity a disc 37, upon the upper face of which is a cam groove 38, 44, Fig. 221, in which run the ends of rods 40, 43, which control respectively the change gear clutch 10 and the main clutch 25. The clutch 10 is operated from the bar 40 by means of the lever arm 39 on a short vertical spindle, which carries a forked arm, Fig. 220, the extremities of which engage with pins on a collar running in a turned groove in the clutch piece 10.

The bar 43 controlling the main clutch carries at its outer end a forked arm controlling a collar running in a groove on the piece 25, keyed to shaft 6, Figs. 219 and 221, by the longitudinal movement of which the friction clutch is operated.

The groove on the cam disc 37 is so shaped that when the speed gears are to be interchanged the clutch 25 is first disengaged.

Gearing with the free spur pinions on shaft 7 are spur wheels 11, 12, Fig. 219, of which 12 is mounted loose upon the solid through axle A, which carries at its far end one of the driving wheel hubs 22. The wheel 11 is mounted loose upon a sleeve 17, which is itself free to rotate on the axle A, and carries the other wheel hub 21. Keyed to the solid axle A and to the sleeve 17 respectively are two spur wheels 14, 15, Fig. 219, which gear with spur pinions running on pins 13, 13, fixed in bosses formed

296

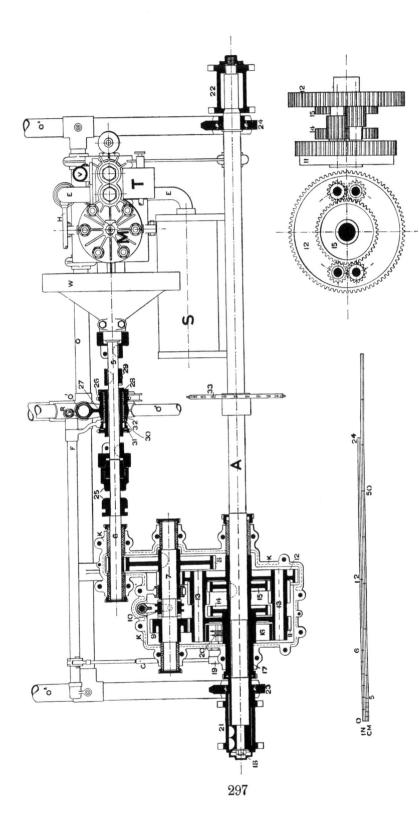

Fig. 219.—Sectional Plan and Elevations of Mechanism of the Columbia Motor Carrier Tricycle.

297

in the webs of, and connecting, the spur wheels 11, 12. It will be seen that this arrangement constitutes a differential gear, which permits of the axle A and sleeve 17 revolving at different speeds relative to one another, while both are receiving power from the motor.

The whole of the gear which transmits the power from shaft 6 to the hind axle is enclosed in an oil-tight gear case K suspended from the frame tubes.

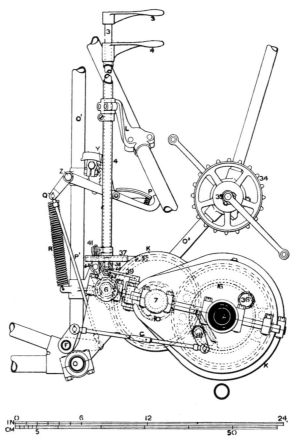

FIG. 220.—SIDE VIEW OF STARTING AND MANIPULATING GEAR, COLUMBIA CARRIER TRICYCLE.

Starting   As previously stated arrangements are made so that by the pedals and crank spindle 35 the driver may start the motor without moving the tricycle, or may propel the tricycle independently of the motor and transmission gear.

The operation of starting the motor is effected through the pitch chain gear 31, 34, Figs. 216 and 221, which transmits the power from the crank spindle 36 to a sleeve 32, Fig. 219, running freely upon the shaft 5, and which carries at the opposite end, from the sprocket pinion 31, a clutch

298

piece 28. The sleeve 32 has some freedom to slide longitudinally on the shaft 5, and is controlled by the hand lever 4, Figs. 216 and 220, which is mounted at the top of a sleeve free to turn on the spindle 3 of the lever controlling the speed-changing gear. At the lower end of this sleeve 4, Figs. 220 and 221, is a short arm 41 connected by a link to a sliding piece 42, projecting downwards, from which is a short forked arm pressing against the end of the sleeve 32.

When the hand lever 4 is moved so that the piece 42 moves towards the motor, the forked arm presses the sleeve 32 in the same direction against the resistance of the spring 30, so that the outer claw face of 28 at

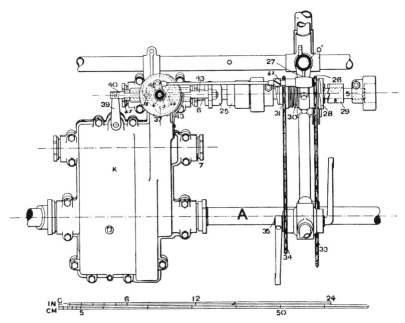

FIG. 221.—PLAN OF STARTING AND MANIPULATING GEAR, COLUMBIA CARRIER TRICYCLE.

the end of the sleeve 32 engages with a claw piece 29 keyed to the shaft 5, which, it will be remembered, is a prolongation of the crankshaft of the motor. When the claw pieces 28, 29 are in engagement, the motor crankshaft may be driven by the pedal and crank 35, and the motor started.

In order to transmit the power direct from the pedal to the rear axle, a reverse movement is given to the hand lever 4, which allows the spring 30 to press the sleeve 32 away from the motor, so that the internal claw face of 28 engages with corresponding claws on the hub of the sprocket pinion 26. This pinion is mounted loose upon the brush surrounding the sleeve 32. The sprocket pinion drives the axle A by means of a pitch chain and sprocket wheel 33, Figs. 219 and 221, so that the foot power is now conveyed to the axle direct without driving either the motor or gear.

Brake

The motion of the tricycle is checked by means of an expanding band-brake 16, Figs. 219 and 220, inside the spur wheel 11 on the driving axle. The brake is operated by depressing the pedal P, Fig. 220, which is pivoted at z on the frame tube o'. The pedal lever is bell cranked, and the end of the short arm Q is connected by a tie-rod P' to a lever arm on a hollow spindle F. A second lever arm on this spindle is connected by an adjustable rod G to a short arm at the outer end of a spindle 19, which passes into the gear case K. At the inner end of the spindle 19 is mounted a double cam, which, when angular movement is given to the spindle, pushes apart the ends of the band of the brake 16, causing the band to press against the

FIG. 222.—THE MOTOR MANUFACTURING CO.'S DE DION MOTOR QUADRICYCLE.

inside of the drum with the same effect as on an external band. When the cam is returned to the normal position between the ends of the band, the spring 20, Figs. 219 and 220, holds the band free of the drum.

By means of a ratchet Y, Fig. 220, the brake pedal P may be held in any desired position.

The wheels are of the cycle type, with tangent spokes and steel rims, holding single tube pneumatic tyres.

This vehicle is made under the American patent of Colonel Pope, H. P. Maxim, and H. W. Alden, No. 621532, of March, 1899, and it illustrates a type which must in the early future come into extensive use, as it will perform all the work at present done by the man-power tricycles, and what is more important, will with one man do a large part of all that work which at present requires, for the longer-distance deliveries, a horse and cart and man, costing for upkeeps at least 40 per cent. more than the motor. (See Chapter on Popular and Commercial Uses and Cost of Working.)

MOTOR QUADRICYCLES.

Small motor carriages, known as quadricycles, are now made in considerable numbers by several manufacturers at home and abroad, most of them being driven by de Dion motors, applied in precisely the same way as they are in the de Dion tricycles already described. Some indeed of

FIG. 223.—THE ARIEL MOTOR QUADRICYCLE. (Weight 4 cwt., ready for the road.)

these quadricycles are simply the tricycle with the front wheel removed, and the fork connected up to the steering arms of the stud axles of a light two-wheel fore carriage, usually with a comfortable seat for one passenger.

One of these quadricycles is shown by Fig. 222, as made by the Motor Manufacturing Company of Coventry, and Fig. 223 shows a similar quadricycle, but with differently arranged carburettor and battery, and differently placed motor and mechanism, made by the "Ariel" Company.

One of these ran the 100 miles and hill trial on 1·687 gallons of petrol, or 60 miles on a gallon.

Since the motor tricycles have been fitted with 1¾ nominal HP. motors, the power has been found to be ample for carrying two persons on all ordinary roads at good speeds, either with the tricycle and a little two-wheeled trailer, which is common, or with the quadricycle illustrated by Figs. 222 and 223. It will be noticed that the two side tubes forming the longitudinal rear extension of the little two-wheeled fore carriage of either of these quadricycles are attached to the axle tube or bridge of the tricycle, and a front fixture is obtained by an attachment to the upwardly inclined tube of the tricycle frame near the head.

Motor Tricycle and Quadricycle Axles and Gear.

The engravings, Figs. 223A and 223B, illustrate two of the principal varieties of driving axles as used for tricycles and quadricycles.

A well-known form of bridge axle (as made for French users by MM. Malicit & Blin, of Aubervilliers) is shown by Fig. 223A, in which the two parts of the axle A meet at the centre in a block 3, which carries the pins on which the pinions 2, gearing with the bevel wheel 1 on the inner ends of the axles, run. The outer ends of the two pins for the pinions 2 are clamped between the boss of the main spur wheel 4 and that of the sprocket wheel 6, the main spur wheel being prevented from wobbling on the pins of the pinions 2 2, by a ring of balls not shown, between the back of the 5 and a ball race ring on the boss of that side of the case. The wheels 1 are screwed upon the two parts of the axle, and a pin, not shown, is put through for security. The differential gear box is centrally divided, and the two inner ends of the axle tube are screwed into its side at the centre, and the bridge tubes F screwed into bosses on the upper part. The bridge struts are brazed on to the bridge tube and clipped to the axle tubes. A brake band of rather small dimensions is seen at 5, and the sprocket teeth for pedalling at 6. Access to the differential is obtained by taking out the eight bolts round its case, and taking off the two angle frame stays, when the whole will come asunder.

With this arrangement the chain is always running, the free wheel being on the pedal axle.

Another and a much used arrangement of axle and gear is the bridgeless axle, as shown by Fig. 223B. It is the form used on the Empress tricycles, and the differential gear is similar to that used on the de Dion & Bouton machines, and on the Columbia machines, as illustrated by Fig. 207, p. 288. In this the back horizontal fork F is attached to the exterior of the differential gear case, and the angle stays are clipped upon the axle tube near the cone cup for the ball bearings, in which the axle A A runs, the ball cone being carried by an inner extension of the wheel hubs H H. The axle is in two parts, the inner end of the off half of the axle being turned down at 1, and entering a cup 2, in which it is free to rotate. On the bosses of the pinions 2 2, forming enlarged inner ends of the two parts of the axle, runs the large box boss 4 of the main gear wheel 5. This is driven by the pinion on the end of the shaft of the motor M. Closing the box boss 4 is the sprocket wheel 8. Carried on four pins in the box boss

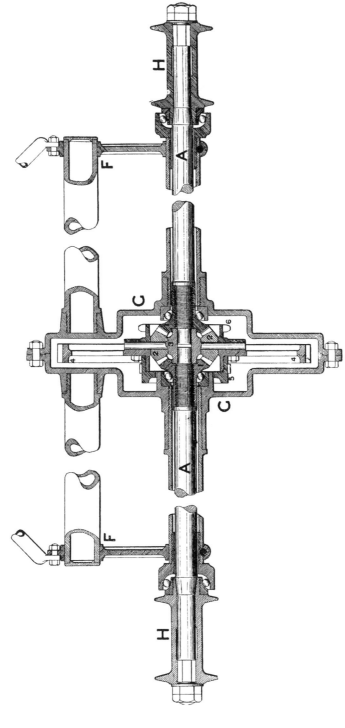

FIG. 228A.—BRIDGE AXLE, DRIVING GEAR, AND DIFFERENTIAL MOTION OF TRICYCLE.

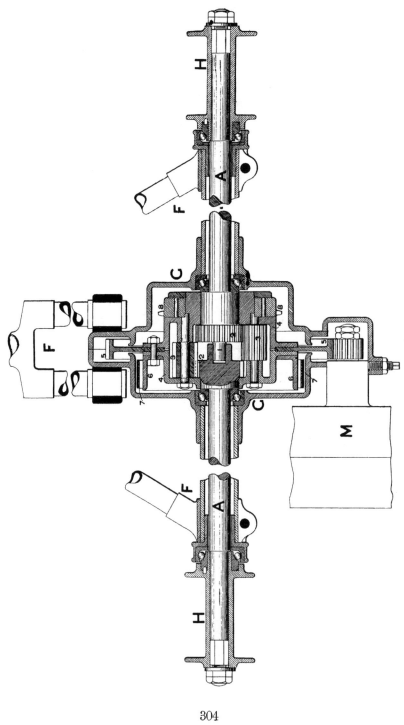

FIG. 223B.—AXLE AND DIFFERENTIAL GEAR OF EMPRESS MOTOR TRICYCLE.

304

are four pinions, two of which 3 3 are shown. One of these gears with the pinion 2 on the near side, and the other with that on the off side, and the two pairs of pinions 3 3 gear with each other. Hence all the gear is locked until the resistance to motion of the two parts of the axle varies, then the lock is destroyed, and the pair of pinions 3 3 rotate on their pins in proportion to the differential resistance experienced by the two road wheels. The action of the gear was fully explained with reference to the same form of gear as used on the Columbia carrier, Fig. 207.

MOTOR TRICYCLE, TWO-SPEED GEAR.

For several reasons a variable ratio between the speed of a motor and the axle of the tricycle it drives is very desirable, not only for climbing hills and for slow-speed travelling, but as an addition to variation of speed by varying mixture and ignition, always a wasteful process.

Although the advantages are obvious, the speed gears yet offered for attaining them have not been unattended with more or less disadvantages, and are not much used. One of these disadvantages results from a defect of those air-cooled motors which, although of a given nominal power, are quite incapable of reaching that power for more than a very short time. Hence a gear that would enable a rider to ascend a long hill of given grade would also show that the motor gets too hot to work at anything like full power on a long climb. Some of the machines that can just ascend a hill with the single reduction gear, perhaps with some help from the rider, would, of course, with speed gear ascend that hill without assistance if the motor would work equally well at near its full load and the higher speed ratio which the gear would permit. This, however, the undersized and insufficiently cooled motors will not do, because a long spell of high speed of motor under moderately high resistance causes the cylinder to get so hot that it stops. The stop is caused, among other things, by the reduction in the quantity of the charge that can be got into the cylinder when so hot.

With the larger and more powerful motors this difficulty is practically overcome; and at the same time the necessity for the two-speed gear is lessened, because the larger power enables the experienced rider to do the long climb at moderate speed by careful attention to mixture and ignition, the number of strokes per minute not being sufficient to raise the temperature of the cylinder, in the time, beyond the working limit.

This method, however, requires a motor much more powerful than is required by most riders for average running; and for moderate speed machines a smaller motor, if provided with ample cooling, would with two-speed gear meet all requirements. The two-speed gear has therefore a place, and a few types may be usefully described.

The form best known is the Didier gear, which is sufficiently shown by Fig. 223c. For this the crankshaft of the motor is extended, and the usual fixed pinion is replaced by a loose pinion with a claw-clutch face. A clutch B fixes this pinion when moved into position by the fork C, lever D, and connection E. When this pinion is fixed, the motor drives the main wheel F in the usual way. For obtaining a second and lower speed the clutch B is thrown over to and fixes the pinion G, which, by the intermediate wheel H, drives the wheel J, on the spindle of which is a pinion K, which is fixed by the clutch L; the speed is by this means reduced in proportion to the relation between the dimensions of the pinion G and wheel J.

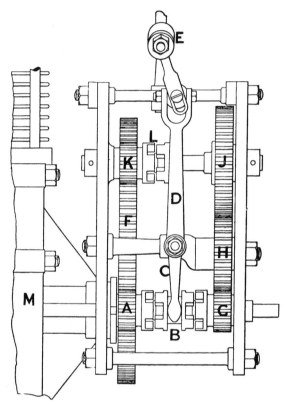

Fig. 223c.—Didier Two-Speed Gear for Motor Cycles.

As shown in Fig. 223c, the clutches are both out of gear, and the motor then runs free with its shaft loose in the pinion A. It will be seen that when one pinion is put in gear the other must be out of gear; but both pinions A and K are always in mesh with the wheel F, and hence one or other pinion must always be rotating upon its spindle. The frame for carrying this gear is very simply made from flat bar iron.

Fig. 223D illustrates a form of two-speed gear made by M. A. Eldin, of Lyons. In this A is the extended crankshaft on which the usual pinion gearing with the main wheel K is replaced by a loose pinion J, which has upon it a flanged disc O, and a long sleeve boss carrying a pinion G. At the extremity of the crankshaft is a pinion B, which gears with the pinion C on the pin E, and formed in one with the smaller pinion D, which in turn meshes with the wheel G, the pin E being firmly fixed in the gear case F, which by its boss L runs freely upon the sleeve H of the pinion J. The method of operation of this gear is as follows : at P is a strong spring ring which ordinarily presses tightly outwards into the flange O of the disc on the pinion J, and by means of the pawl or wedge N, is forced, as soon as relative motion between L and O take place, into strong contact therewith, and in this way the whole of the apparatus rotates with the crankshaft, and the machine is driven at the normal speed. To drive at the lower

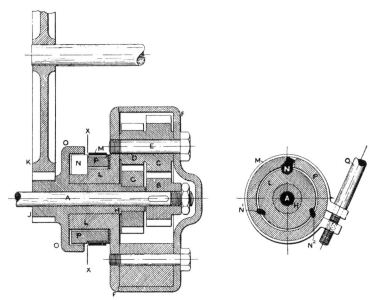

FIG. 223D.—ELDIN TWO-SPEED GEAR FOR MOTOR CYCLES.

speed the ring P is forcibly closed inwards by the brake ring M (see separate figure, which is a section on the line x—x) and the screw rod Q ; by this means the disc O on the pinion J is released, and the ring P grips and holds the boss L of the case, the pinion J is no longer driver, except through the gearing within the case, which works as follows : the small pinion B drives the wheel C, and thereby the pinion which gives motion to the wheel G. on the end of the sleeve boss at H, of the pinion J, which is thus driven at the low speed ratio of the gear described. This apparatus is very compact, and the most ingenious part of it is the device by means of which the strong ring P is alternately the driver of the pinion J and holder of the gear case, the hold of the ring P within the flange being automatically

increased beyond its own spring power by the wedge tumbler pawl N.
It will be seen that the whole of the working parts run within an oil-
tight case.

Another form of two-speed gear is that illustrated by Fig. 223E, known
as the "Cherrier," and exhibited at the last Automobile Exhibition in
Paris by Messrs. Dalifol & Thomas. This is so designed that the existing
pinion on the shaft of the motor A is replaced by the external part of the
cone clutch B. Into this fits the cone B', forming a clutch by means of
which the motor may be in gear or run free, the cone being put into gear
by means of a fork at L. On the spindle O driven by B', are pinions F and
G, which have sleeve bosses carrying the outer parts of two clutches C and
D, the inner cones C',D' being put into or out of gear by the levers K K'.
When the clutch C is in gear, as shown in the engraving, the pinion F

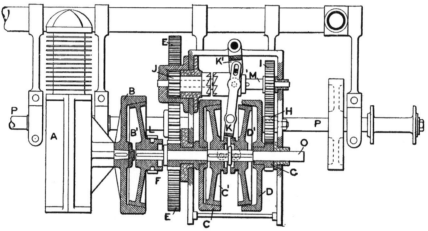

FIG. 223E.—"CHERRIER" TWO-SPEED GEAR FOR MOTOR CYCLES.

drives the main gear wheel E, on the cycle axle P, at normal speed. When
the clutch D is in gear, the pinion G drives the wheel I through the wheel
H; and thereby the pinion J, gearing with the main wheel E, thus imparting a slow speed, due to the difference between the diameters of the pinion
G and wheel I. On the spindle M of the latter is a saw-tooth claw clutch,
which is thrown into gear when the clutch D' is put in; but which would
automatically go out of gear, even if not quite thrown out by the lever K'.
It will be seen that this pinion runs always, even when the normal speed
pinion is in gear. The Eldin gear is preferable, and one of the best made.

MOTOR BICYCLES.

In 1885 Daimler made one of the motor bicycles described with reference to Fig. 29, p. 61; but from that time only spasmodic efforts were made

PETROL MOTOR CYCLES

in this direction until about 1895. Daimler had provided the whole of the general outlines and chief essentials, including a motor and carburettor. A few attempts were afterwards made to produce a steam bicycle, but nothing came of these. The first petrol motor bicycles that attracted attention were the Wolfmüller, made under the patents of Wolfmüller & Geisenhof, and exhibited in the Imperial Institute Exhibition. (English Patent No. 7542 of 1894.)

Motor bicycles have not, however, been much favoured by the public, although they have gone through numerous stages of improvement, and are now again receiving the attention of several makers. It is impossible to predict the future of these vehicles, or the extent to which the objections which are urged against them may be overcome by expert riders, and those to whom their comparative low cost is an advantage. They offer, of course, the other several advantages of the ordinary bicycle, as compared with the ordinary tricycles, of lightness, easy steering, single or narrow wheel track, and small space occupied. It is, however, felt that considerable skill is required to start most of them; and that in case of side slip and in traffic they are heavy to deal with in emergencies, and in the disadvantageous position of a rider dismounting in unpremeditated ways and suddenness.

Origin and development

The Wolfmüller bicycle was fitted with a double-cylinder motor, direct coupled, locomotive fashion, to the overhanging cranks of the hind wheel axle. The design contained numerous well thought out details as to valve motion and control; but the motor being limited in speed to that of the rear driving wheel, it was necessarily heavier than a motor of the same kind would be running at four times the speed or more, and driving through reduction gear. The ignition was by tube, and there were no means of speed regulation except by varying the mixture. This, as much experience with other vehicles has proved, is almost impracticable for any purpose, and for regulating a slow-speed motor of a bicycle, which requires all the rider's attention for other matters, quite impossible. Electric ignition has, therefore, come into use as the only suitable means of ignition for such vehicles. This bicycle was and is of interest, but the reader is referred to the specification above mentioned. After its failure but little was done for some time in this direction; but although the motor bicycle has not attracted very much attention, or received much public support, a description of the typical arrangements of frame motor and gear may be of some service.

The diagrams on page 311 show that simple as the motor bicycle problem might off hand appear to be, there has been and is great variety of opinion even on the first point that presents itself; namely, the best position for the motor.

The diagram, Fig. 224, represents the Wolfmüller drop frame machine with nearly horizontal cylinders A, with connecting rods B on overhanging cranks c on the ends of the axle of the hind wheel D, the two air admis-

The Wolf-müller

309

sion, two exhaust, and two ignition turning valves being arranged in the valve box E. At F was a box containing a downward flame Bunsen burner for heating the ignition tubes, the valve box and burners being between the two frame tubes G. The carburettor and petrol supply vessel J was carried between the four front inclined frame tubes, and by fixture to the two lower tubes G, which acted as air tubes to supply the burners from the strainers L. A mixing valve was placed at H, actuated by the thumb lever and cord K.

Dion-Bouton   Fig. 225 shows the arrangement of the de Dion & Bouton bicycle. The frame is of the usual top tube Humber pattern, but with lengthened lower back stays and the trussed fork. The motor A in this case is of the de Dion Daimler form 1¾ HP., placed within the back fork in the front of the rear wheel, which it drives by a belt from a pulley C on the motor shaft to a similar pulley on E, the outer end of a spindle, which carries at its inner end a pinion gearing with a spur wheel F on the driving-wheel axle. Nearly all the reduction in speed between motor and driving wheel is done by this gearing, the rear belt pulley being only a little larger than that on the motor shaft. The ordinary cranks, shaft and pedals are represented by fixed arms carrying foot rests G. The battery H is suspended from the top tube in front of the carburettor J, the sparking coil K and condenser being fixed to the back stays. A silencer is placed at L. There is a brake on the front wheel, and a bandbrake on the rear axle. The wheels are 26 in. diameter. The carburettor and petrol reservoir holds about six pints. The machine is only recommended for fine weather or dry roads, and for path pacing purposes, and it is only made to order.

Fig. 226 shows an arrangement by Messrs. Shaw & Sons, and is in some respects the same as the Dion-Bouton, the first motion being by chain instead of belt. The motor A is between the lower back forks, and it drives a short second motion spindle carrying a sprocket wheel E, by a chain. The sprocket pinion on the motor shaft is smaller than E, and with it makes a speed reduction. A further reduction is made by the sprocket pinion on the spindle of E, and sprocket wheel F on the rear road wheel which is driven by chain. The coil K carburettor J and battery H are carried in the same places.

Pernoo   Fig. 227 shows the arrangement of M. Pernoo's motor bicycle. The frame is ordinary Humber pattern, except that the lower back stays are extended backward from the rear wheel axle, and at their extremities the little high-speed motor A of 1·25 HP. is carried, supported in part by the rods, to the top of the back stays. A carburettor J is placed behind the main pillar, and the battery H and coil K are carried on the top tube of the frame. The machine is started by pedalling in the ordinary way, and the motor drives by chain with single reduction as seen. This is the only bicycle carrying the motor at the extreme end of the back of the frame.

Blessing   Figs. 228 and 229 show the similar arrangements respectively of the

310

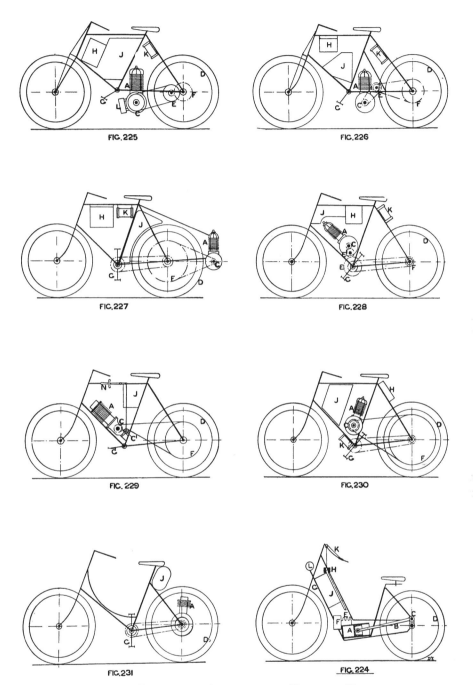

FIG. 225

FIG. 226

FIG. 227

FIG. 228

FIG. 229

FIG. 230

FIG. 231

FIG. 224

Figs. 224–231.—Diagrams of Arrangement of Motor and Mechanism of Motor Bicycles.

311

"Sanciome" and of the machine of Messrs. Blessing & Co. Both have the motor fixed to the lower main forward member of ordinary Humber-type frames. In Fig. 228 the speed reduction is made by a spur pinion c on the motor shaft gearing into the wheel E' on a second-motion shaft; the wheel E' in turn gears with the spur wheel E on the crankshaft, on the other end of which is a sprocket wheel carrying a chain to a sprocket wheel F of similar size on the rear wheel axle. The speed reduction is thus the same as though c geared into E. The carburettor is above the motor carried by the top tube, and the coil is behind the saddle.

In Fig. 229 the similarly-placed motor drives the rear wheel by belt direct with a single speed reduction, the slack in the belt being taken up by means of a jockey pulley c', controlled by the hand lever N, pivoted on the top frame bar.

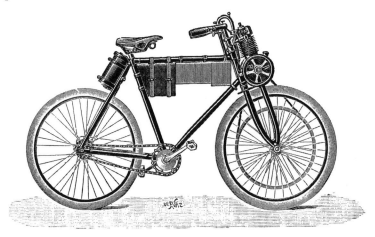

FIG. 232.—THE WERNER MOTOR BICYCLE.

Carcano Fig. 230 shows the arrangement of a machine made in Italy and called the "Carcano," and in it the motor A is placed in the most obvious position, namely, in the open space above the bottom bracket of an ordinary machine, and attached to the main diagonal pillar. It drives by belt with single reduction in speed. The carburettor J is in the usual place of a carrier, and the battery H where the saddle-bag usually is placed. The coil K is rather badly situated, unless these machines are always used in fine weather only.

Lawson Fig. 231 shows the arrangement of H. J. Lawson's drop-frame machine with the motor of 1¼ HP. placed with its crank coincident with the driving wheel axle, and connected to it by internal hub reduction gear, the motor shaft passing through the wheel and carrying a second flywheel there. Ignition is by tube, or may be electrical, and petrol is carried in two egg-shaped vessels below the saddle. The machine is started by pedal and chain gear in the ordinary way.

312

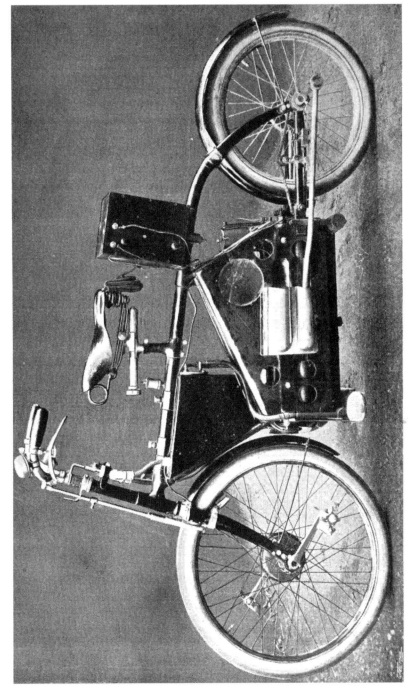

Fig. 233.—Holden's Four-Cylinder Petrol Motor Bicycle.

313

Werner
Fig. 232 shows the arrangement of the Werner bicycle, which again differs from all the others in that its little 1 nominal HP. motor is carried upon an extension of the front fork head, and drives the front wheel with a round belt and single reduction, thus converting an ordinary bicycle into a front driver. The front fork is trussed to secure strength. The bicycle is complete for use in the ordinary way, which is certainly desirable, and the motor affords the occasional help regained on long runs and for helping up hills. It is questionable, however, whether it would be of much help up stiff hills on which the speed of the machine would be very low. The carburettor J, battery H, and coil K are carried in the best positions, and as a large number of these machines have been made it must be assumed that they afford bicyclists the help many require, and that the advantages are greater than the disadvantages attending the use of a machine weighing 65 lb. These machines are being introduced into England by the Motor Manufacturing Company.

Holden
One of the most noteworthy of the motor-bicycles is that made by Major Holden, R.E., F.R.S., as this contains several valuable departures from the details adopted by previous designs. It has a four-cylinder motor, direct coupled locomotive fashion to overhanging cranks on the driving-wheel axle, the driving wheel being of small diameter. The motor, although it has four cylinders and the necessary valves, is very simple in construction.

Figs. 233–239 illustrate this bicycle and its parts, Fig. 233 being from a photograph of the off side.

Holden's Motor
Figs. 234, 235 and 236 show the construction of the motor, which, having four opposite cylinders, gives a push and pull on the cranks at every revolution. From the elevation, Fig. 234, which is partly in section, it will be seen that the cylinders A A' are thin steel tubes fitting tightly into chases turned in the valve-box castings at either end B B'. In the centre of these tubes, rather more than the length of the stroke of the pistons C C', are slots D through which passes the cross-head pin E, which is fixed in the boxes F F' midway between the pairs of pistons. On the end of the cross-head pin E are pivoted the connecting rods G G', as seen in the plan, Fig. 235; the pistons all move simultaneously, but the pair at either end alternatively receives an impulse. Major Holden has used two different forms of valve gear, with one of which he has also fitted a very simple device for simultaneously opening the exhaust valves for starting or for control.

Valves and valve-gear
In the longitudinal section, Fig. 234, the arrangement of a pair of valves in the casting B is seen, the exhaust valve H and the carburetted air admission valve J being opposite each other, so that the removal of the valve seat K makes the exhaust valve removal or introduction easy. On the tops of both the valves are projections which meet each other within a distance which is equal to the range of the movement of either valve, no other check being required. The valve seat K is held in place by the bush

314

L, the centre of which is tapped at M with a thread suitable for the pipe conveying the combustible charge of air and carburetted air. The exhaust valves receive motion from tappet pieces N N' on rocking spindles P P', which are actuated by fork pieces Q Q. These forks are moved by cams

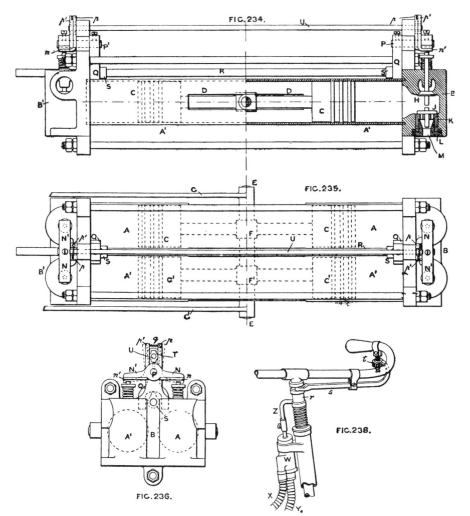

FIGS. 234-236.—ELEVATION, PLAN, AND END VIEW OF HOLDEN'S FOUR-CYLINDER PETROL TRICYCLE MOTOR.
FIG. 238.—MIXING CONTROL VALVE ON HANDLE BAR.

ss on the ends of the shaft R. This shaft is rotated by a worm and wheel driven by a chain from the rear axle, as seen in Fig. 233.

Under the ends of the tappet pieces N N' are subsidiary pieces $n\,n'$, which are pivoted upon the rocking spindles P P' and form parts of belt crank levers, the upper arms of which are seen at $p\,p'$ where they are

Compression, release and control

315

pulled towards each other by a small spring q (see Fig. 236). Between the arms $p\,p'$ is a double cam T, at each end of the spindle U. This spindle is adjustable as to position by the rider. In the position shown the two arms $p\,p'$, Fig. 236, are in their position of nearest approach, but if the spindle U be turned through 90° the arms $p\,p'$ will be pushed farther apart and the pieces $n\,n'$ will be deflected angularly downwards, and the

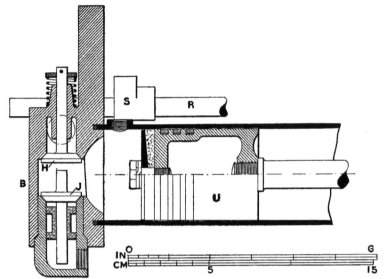

FIG. 237. –HOLDEN'S PETROL MOTOR: DETAILS OF SECOND FORM OF VALVE ARRANGEMENT.

FIG. 239.—PERSPECTIVE VIEW OF HOLDEN'S BICYCLE MOTOR.

exhaust valves will be opened or prevented from closing. By these means not only is starting made easy but control as to speed is obtained.

The arrangement of the valves and cylinder-head, as used in the most recent bicycle, is seen in Fig. 237, which also shows the cam for operating the valves at one end, and the piston. The valves and their operation have already been described. A thin ring of asbestos cord is put into the bottom of the chase, into which the ends of the cylinders fit.

PETROL MOTOR CYCLES

The piston u is shown half in section and half in elevation. The **Piston**
piston has no rings, although three are shown in the engraving, but at
the back of it is a steel disc of slightly conical form, kept pressed against
a packing of asbestos cord by a screw and stiff spiral lock washer. If any
wear of the asbestos takes place, the disc, by its own elasticity, will,
through a small range, follow it up, and it remains tight.

Fig. 239 is from a photograph of the cylinders and valve gear of one **Valve motion, early form**
of the first of Major Holden's motors. In this a rocking shaft, extending
from end to end of the motor and fitted with tappet arms at its extremities
for actuating the exhaust valves, is used. The rocking shaft has upon it
two short collars cut to a cam form. Sliding on the rocking shaft is a
piece which receives motion from the cross-head. Its ends are so shaped that
as they alternately push against the inclined or spiral-cut cam collars they
give the rocking shaft a partial rotation alternately in opposite directions.
In this motor the subsidiary tappet pieces for keeping the exhaust valves
open were not used.

The pipe used by Major Holden to connect his carburettor and ad- **Vapour tube**
mission valves is the flexible metallic tubing, which he finds perfectly
satisfactory; one end is provided with a nipple for screwing into the valve
case bushes M, and the other with a nipple to screw into the mixing valve
w, Fig. 238. By reference to Fig. 233 it will be seen that the carburetted
air from the carburettor enters the mixing-valve by the pipe x, where it **Mixing-valve**
meets the supply of pure air entering the valve, and passes to the motor
through the pipe Y. The quantity of air, and therefore the quality and
quantity of the charge which is allowed to go to either of the four cylinders
alternately, is controlled by the piston in the mixing-valve, the position of
which is regulated by the stem z, fastened to the sliding collar r, on the
stem of the handle bar; the position of this collar is regulated by the
forked lever s, pivoted on the handle bar and adjusted by the milled nut t.
This adjustment is very easy by thumb or finger, without taking the hand
off the steering bar.

Air enters the mixing-valve w through a self-closing valve, and it **Self-closing valves to air inlets**
may be remarked that Major Holden puts a similar valve on the inlet to
his carburettor, so that under no circumstances can petrol vapour escape.
The result is that he does not know what stale petrol is.

The ignition is by electric spark, a secondary battery and coil being
carried behind the saddle, as seen in Fig. 233. The wire connections to
the coil have in circuit a switch on the handle bar. One contact make
and break is used and the secondary current distributed to the four
ignition plugs. For these plugs, see Chapter XX.

For lubricating the various parts Major Holden uses a very ingenious **Lubricator**
and very simple mechanical lubricator, this is shown in Fig. 240. In this
a spindle A, in the upper part of the oil cup, has in it a wide groove in
which runs a wire chain B freely suspended to the bottom of the cup. A
light bicycle chain from the cam shaft gives motion to the spindle A, and

317

MOTOR VEHICLES AND MOTORS

as it slowly revolves, carrying with it the chain, the latter carries up oil which is partly given up in the groove. Within the centre of the broad groove is a small one in which hangs a wire c, the end of which nearly touches the bottom of a tube D in which a large hole is made as seen below c. The wire cannot swing far, and so it delivers a regular capillary supply of oil to the tube D, which delivers it by pipe F to the cylinders. The lubricator can be made with almost any number of separate deliveries in the same or different quantities, and is regular and certain in action.

The bicycle is built rather low, so that it is easily mounted while standing on the ground, and yet a comfortable position of leg is obtained with the feet on the ample foot-rests seen in front of the cylinder case in Fig. 233.

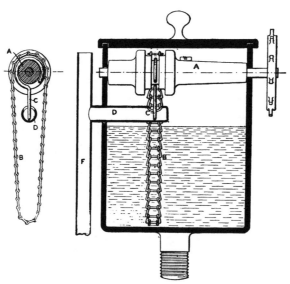

Fig. 240.—Holden's Mechanical-Feed Lubricator.

The ends of the cross-head pin are covered by cover-plates on the sides of the motor case; these plates, as will be seen from Fig. 233, have two corrugations, the upper one being intended merely as a resting-place for the ends of the connecting-rods, should it become necessary to disconnect them so as to run the machine by the pedals, in case of accident to any part of the motor.

General　　Although directly connected to the pistons, the starting is easy and the running of the bicycle is quite comfortable, and without any noticeable jerkiness of motion. It may be said that the working of the motor and bicycle generally is perfectly satisfactory. It will, of course, have been observed that the system involves the use of a very small driving wheel, which very much limits speed on poor roads.

318

PETROL MOTOR CYCLES

One point to which special attention should be drawn is the simplicity **Adaptability of
the motor** of construction, particularly from the workshop point of view, of this four cylinder, or rather four piston, two cylinder, impulse every stroke, motor. The end pieces forming the valve boxes involve only simple machine work, the cylinders are merely pieces of thin steel tube cut off and the ends trimmed and pushed tight into the chases made in the end pieces by three bolts which hold the whole motor together. The motor of the bicycle illustrated has cylinders 2·125 in. diameter, and 4·5 in. stroke, the total weight of the bicycle 104 lb., and the compression pressure used is about 80 lb. per square inch.

There are other bicycles of which particulars might be given, but those described are assumed to be sufficiently typical of the directions in which inventors have worked.

Chapter XVI

CARBURETTORS

Early forms

MOST of the surface carburettors that have been made in the last few years follow the ideas which underlie the construction of the early carburettors of Daimler, Benz, and Blackburn. The first is illustrated on page 71, the second on page 77, and Blackburn's, which was made in 1878, and therefore preceded the others, consisted of a chamber, the bottom of which carried petroleum spirit, upon which jets of air from a number of tubes played, a similar number of tubes carrying the air off with its contained spirit vapour.

Theory of action of surface carburettors

In all these carburettors, as in some of the earlier so-called gas-making apparatus, the one idea is to cause air to blow upon or bubble through the spirit, or to put the air into intimate contact with it, and by these means to disturb the spirit so that in bubbles or spray it should present as much surface as possible to the rapidly-moving absorbent air. The action sought is about the same as that which takes place when the wind is said to "dry up" the puddles in a road in so much less time than the sun would do it with no wind.

Some of the carburettors made with this object must be looked upon as crude, including those of the de Dion tricycles ; and if the petrol grows stale in them, there is no cause for surprise, in that selective evaporation must occur under such circumstances.

It is obvious that what is required is the presentation of the maximum of petrol surface with the minimum weight to the passing air, or, in other words, to present petrol surface without thickness. Many ways of effecting this may be devised, including such mechanical methods as those used in the Mueller and other petroleum spirit-lighting, gas-making apparatus.

Simple forms possible

The utmost simplicity is, however, required, and one of the most perfect would be the application of capillarity and of other simple means of securing automatically petrol-wetted surfaces. A carburettor box might, for instance, be filled with plates so placed that they are occasionally awash, and are kept moist always at the lowest part by capillarity, the air being forced to traverse the greater part of their surface. It is obvious that if these plates were perforated they would present larger surfaces, and

320

that if wire netting, like that used for sieves, with, say, 300–400 holes per square inch, were employed, the wetted surface presented to the moving air in quite a short path might be very large. A carburettor on this system would be of small dimensions as compared with most of those in use, and petrol in them would be uniformly evaporated if there were no outlet for the carburetted air except that to the motor.

One point with regard to carburettors which is often overlooked is the fact that the carburation of the air means evaporation of the spirit, and that this evaporation is attended with degradation of temperature. During a large part of the year, therefore, artificial heat has to be supplied to the spirit or to the air to make up for the refrigerative effect of evaporation.

The spray-making or atomising vaporiser is simply another means of causing a small quantity of petrol to present the largest possible surface to the moving air.

Spray-
making
vaporisers

The surface carburettor, as generally made, is supposed to offer one advantage, namely, some storage space which will contain enough vapour for starting before the carburettor is working. This, however, is not necessarily the case, as the tube between the atomising carburettor and the motor valve may be quite sufficient for this purpose. Moreover, experience with the latter kind of carburettor in several forms, including the Maybach and the more complicated forms of Longuemare and the Maybach-Panhard, the Lepape, and others, show that the large chamber surface carburettors are not a necessity.

The atomiser carburettor in various forms has existed for many years, and several of those brought out recently are mere variations on those of Abel (Otto Co.), Smyer, Schiltz, and others, although some of the early inventions were vaporisers for kerosene.

Early forms

The Smyer's vaporiser (Patent No. 11290 of 1895) is shown in section by Fig. 241. It would work perfectly as a carburettor for motor purposes. The oil or spirit, once it enters the apparatus, meets with a most perfect disintegrating treatment in presence of, and with, the rushing streams of air, so that the latter quickly becomes saturated.

In Fig. 241, A is a conical annular space formed between a conical case C and an inner cone B. At the bottom is a suction pipe D, connected to the motor cylinder. At the top of the case is an injector atomiser, to which the spirit is led by the pipe H. The spirit finds its way into the thin annular space between the cone-ended pipe M and the fine-pointed, screw-topped needle valve G. The pipe M is placed within a cone-ended pipe formed with the piece P, to which the head Q of the atomiser is fixed. The part between P and the part Q is cylindrical, and is perforated with air holes O. It is surrounded by a hit-and-miss air admission ring I, with corresponding holes. When the suction stroke of the motor piston occurs, a partial vacuum is formed in the pipe D and space A, and air is drawn in at the holes O and down the annular space to the restricted nozzle at G. This causes an indraught of the spirit, adjusted in quantity by the needle-valve, the in-

Smyer's
vaporiser
as
carburettor

draught of air being regulated by the ring I. The entering spirit is atomised at the nozzle, and being entrained by in-rushing air, passes over the surface of the inner cone and down to the restricted annular space at E. By these means, if the spirit is not in excess, it will all be taken up by the air which is thus carburetted.

For increasing the certainty of action of the carburettor in cold weather, a branch from the exhaust pipe enters at F and up the pipe S, heating the inner cone and passing off by the pipe T.

It is obvious that this would be an easily adjusted and simple carburettor, and it is fifteen years old.

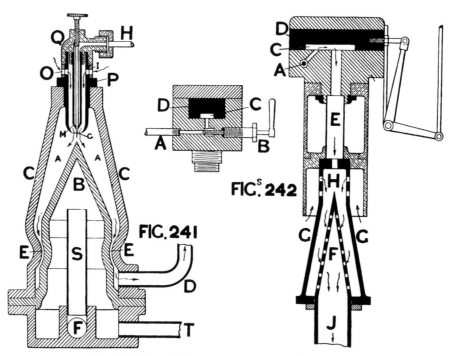

FIG. 241.—SMYER'S VAPORISER PROPOSED AS CARBURETTOR.
FIG. 242.—WORDSWORTH'S VAPORISER.

Words-worth's carburettor

One other form may be described, namely, that of Wordsworth & Wolstenholme (Patent No. 15507 of 1886). This is shown by Figs. 242, and will be seen to include a measured petrol feed.

The petrol enters at A, adjusted by a conical screw valve B, and enters a groove C in a reciprocating piece D. The charge of spirit taken into this groove is dropped into the fall tube E, and finds its way into the annular space between the inner and outer cones F, G. At the top H of G is a short perforated cylindrical part, which admits air when drawn in by the engine through the pipe J, the spirit being broken up at H, and further rubbed and taken up in the rush through the fine holes in F.

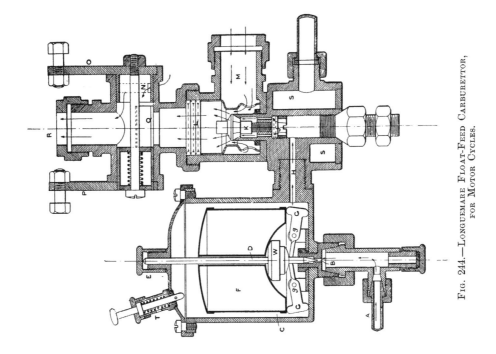

FIG. 244.—LONGUEMARE FLOAT-FEED CARBURETTOR,
FOR MOTOR CYCLES.

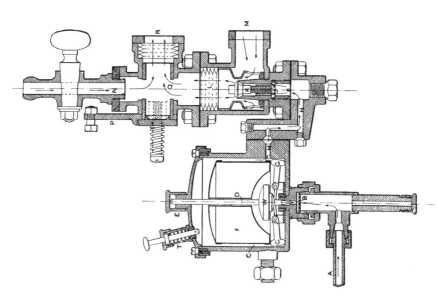

FIG. 243.—LONGUEMARE FLOAT-FEED ATOMISER
CARBURETTOR, FOR MOTOR CARS.

323

It will be seen that this carburettor could be easily improved by making the slide D round instead of rectangular, making it longer in proportion to the length of the groove and forming a small hole at the end which would admit a very small quantity of air with the entering petrol. The part H could be fitted with a hit-and-miss ring, and the area of the perforations there should be as great as that of those in the inner cone F. The part H could be connected direct to the slide case. The grooved slide could be worked by pusher rod and spring return, and, in fact, by making it of two diameters, each end in holes in the case, might be automatic and adjustable by stroke-limiting set-screw or other equivalent means.

These examples are sufficient to show what kind of carburettor invention is already open to users and improvers.

I may now describe some of those carburettors now in use other than the Maybach, already described at length with reference to the Daimler and other motors and carriages.

Longue-mare's carburettors Fig. 243 shows in section Longuemare's float-feed atomiser carburettor, as used for motor cars. The float feed is similar to that known as the Daimler-Maybach, a float F sliding freely upon a needle valve o, which it lifts by its descent with the fall in the level of the petrol. This it does through the medium of the pivoted weighted arms G, the fingers on their inner ends lifting the collar w on the valve. The lift of the valve is determined by the length of the part projecting into the screw cap E. At T is a spring-held valve. On the top of the valve stem is a button which may be pressed to admit air to permit the out-flow of the petrol at H into the carburettor. Petrol enters at A, and passes through a gauze strainer at B. The attachment to the motor supply pipe is at R, suction by the motor piston drawing in air, which should be at a temperature of from about 60° to 85° F., at M. The air is directed into the lantern round the petrol admission valve-shaped spraying nozzle or plug K. Round the screwed stem of this plug are small holes, which admit petrol from H. After leaving the nozzle at K in spray form, it passes with the air carrying it through the layers of gauze at L, similar layers being fixed at R.

The quantity of air admitted to the carburettor at M is controlled by the cylindrical valve Q and lever P, any further admixture of pure air being determined by the cock N.

The quantity of petrol admitted is determined by the spraying plug K, the conical part of which is finely grooved on its surface.

The carburettor used for motor cycles is similar in most details, but the cylindrical valve at Q controls, by the lever P, the quantity of air drawn through the carburettor, and therefore the quantity of vapour used; while the supplementary pure air admixture is controlled by the similar valve N and lever O.

For encouraging the evaporation of the petrol in cold weather a branch from the exhaust pipe delivers hot gas into the space S, the exit being by a small pipe at the side not shown.

CARBURETTORS

At the air entrance M there should be a gauze air strainer, and care must be taken that neither exhaust gases nor products of combustion from the lamp of tube ignition apparatus can enter. To prevent this a pipe is sometimes attached to the union nut or flange at M, the orifice of the pipe being placed where pure air can be best obtained.

These carburettors have been designed to suit the arrangement common with motor cycles and French cars, and are largely used, the surface carburettor being in less favour.

A carburettor used in France and known as the " Universal " is shown **Huzelstein's carburettor**

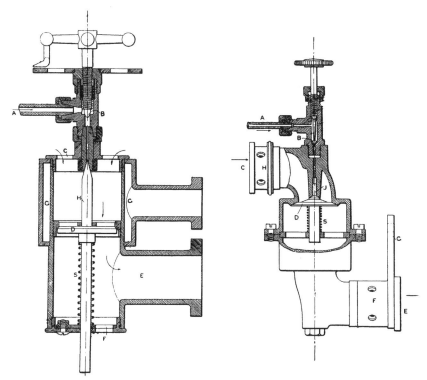

FIG. 245.—THE HUZELSTEIN CARBURETTOR.   FIG. 246.—THE ABEILLE CARBURETTOR.

in section by Fig. 245. In this the petrol enters at A, adjusted in quantity by screw valve B, and descends a small pipe and breaks up on the point of the valve stem H. Thus broken up it is carried and taken up by the air entering at C, the complete formation of the air petrol mist being aided by heat from the inner chamber walls and by the rush and rubbing through the fine narrow space between the valve D and its seat, the valve being kept normally on this seat by a spring S. The air from without enters holes c, which are covered more or less by a hit-and-miss plate not shown. Between the periodic suctions of the motor piston there is a slight dwell of the air in the air chamber, which is heated by exhaust gases in the space G. It will be seen that the valve D, which is drawn down by the motor piston suction, will have a variable opening. This, however, is not disadvantageous, because the carburation, by violent rubbing and mixing, will be equally efficient with the wider or narrower opening, inasmuch as when it is open widest, at the time of greatest velocity of the motor piston, the rush of air will be most rapid. The relation between quantity passing, and size of orifice, thus remains unaltered, and the velocity of air and the rubbing at the valve edges are also unaltered. A supplementary supply of air is admitted at F, the quantity being determined by a sliding plate.

Abeille carburettor Another form of carburettor is that shown in section by Fig. 246. It is known as the Abeille carburettor, and in some respects resembles the last described, but for the completion of the process of atomising and carburation, the spreading of the spirit over the broken conical surface of the valve D, and the commotion and rubbing between it and its seat, are relied upon. Petrol enters at A, adjusted as to quantity by the valve B, from which it passes downwards to the grooved grinder stem J of the valve D, over the surface of which it spreads, the petrol being drawn in by the rush of the air entering the holes in the hit-and-miss adjustable head H, which is similar to those described with reference to the Daimler and Panhard carriages. After passing the valve D, which is kept normally upon its seat by the spring s, it finds its way as carburetted air to the outlet to the motor at E, pure air being admitted, by the holes in the head F, as required for making the mixture burn with explosive combustion. The pure air admitted is controlled by lever G, which opens or closes the holes under those in F. The relative positions of the inlets and outlets may be those shown, or may be, as the connections require, by unscrewing the screws which hold the two parts of the body together, and putting them in holes at 90° or 180° from those they are now in. No means are provided for heating the air, so that this carburettor is not suitable for very cold weather work, and is not suitable for use in any country where the air frequently carries much moisture in cool weather when the temperature is considerably above freezing. **Formation of snow at carburettor openings** The refrigeration due to evaporation will cause formation of snow at the holes in the cap H, just as it does in those of Panhard cars, for instance, with gravity feed carburettors, when the heater pipes are not on, even though under the protection of the bonnet. This has happened often in England, but seldom in France, where the heater pipes for the carburettors are not now always used.

Some improvement is wanted in all these float-regulated valves, inasmuch as when either ascending or descending a steep hill they are liable to stick in consequence of their inclined position, and the float vessel fills up and overflows.

Chapter XVII

MOTOR-CARS WITH PETROLEUM (HEAVY OIL) MOTORS

ALTHOUGH petroleum as kerosene is less than half the cost of petroleum spirit, the difficulties which attend the perfect formation and combustion of petroleum vapour under the varying conditions of motor vehicle working, have hitherto proved sufficient to prevent the heavy oils from coming into use to any noteworthy extent. Messrs. Roots & Venables, however, have persevered in their efforts, and are making vehicles fitted with the Roots oil engine arranged for this application. Of the lower cost of fuel per mile run there can be no doubt, nor can there be of the working of the motor from the mechanical point of view ; but satisfactory as is the combustion and nature of the exhaust products from these engines when used for ordinary purposes, it cannot be denied that the bad exhaust from similar engines working under the extreme variations of power required on a motor car, is often exceedingly objectionable, even to those disposed to overlook many defects in motor vehicles.

Use of the heavier oils

Concerning Vaporisers and Vapour Control.

Improvement in the oil engine in this respect has been made during the past two years, and further improvements are possible or conceivable, but before the heavy oil engine is acceptable as the motor for vehicles for towns, where the traffic causes frequent stoppages, much improvement in the method of preparation of the combustible charge, the control of its admission, and in its combustion, will have to be made. It is necessary that all that enters the cylinder shall be completely burned, and that combustion shall be uniformly good under all variations of conditions and circumstances of working.

Vaporisers

Here the difficulty lies which has puzzled inventors for at least sixteen years.

The methods and means of vaporization of heavy oil, as contrasted with the conburation of spirit, constitute the first and greatest part of the difficulty, and one of the earliest to see the nature of this difficulty and point a way of surmounting it, so far as to meet the conditions of ordinary fixed engines, was M. V. Schiltz, a doctor of medicine at Cologne, in 1885. Much disappointing experimental work has been done since then in the

Vaporiser difficulties and forms

327

attempt to enable the petroleum engine to work under the conditions of a gas engine. To this day, however, the petroleum engine works with a charge which is not a mixture of oil gas, and air, but, as Schiltz suggested, a mixture of heated air heavily charged with petroleum mist and of pure air.

The vaporiser problem

Vaporisers are of many forms and kinds, and all, as shown by the author's articles and reports,[1] on trials made by himself and by the Royal Agricultural Society, have received various modifications. They range from designs which exemplify the elementary idea of stewing the oil in a hot pot to the well-thought-out combination of an atomising process with air sufficiently heated to counteract the refrigerating effect of evaporation, and with the necessary conditions for vaporising the mist produced and maintaining it to the last moment in that condition. Between these two extremes there have been many stages, including attempts to gasify without a proper comprehension of the physical conditions involved in the process and in the use of the product. In nearly all those which are most successful and simple, the guiding principle is the breaking up of the oil by spraying or trituration in presence of warmed air which moves rapidly over heated surfaces on its way to the engine cylinder. The atomising action is an important part of the process, though it may be sufficiently carried out by very simple means, such as admitting the oil with air into a narrow and more or less tortuous passage, or past a mushroom valve into a more or less obstructed passage. The air should be warmed before the point of oil admission is reached, and it must move at high velocity so that it and the oil may be rubbed and knocked sufficiently to produce an oil mist or fog. This mist must then be further heated and not allowed to condense. Whatever the methods employed, uniformity of the conditions met with by the oil must be maintained. Hence vaporisers heated by the exhaust gases are unsatisfactory, except for uniform loads, and quite impracticable for a very variable load, and varying quantity of exhaust gases.

Heating vaporisers by exhaust gases

Equally useless for extremely variable loads is a vaporiser which, though heated uniformly by other means, depends chiefly on that latter part of the process of vaporisation, which is done in the hottest part of the vaporiser immediately before entering the cylinder. Such vaporisers may work with sufficient perfection to suit fixed engines, and others working where variations of power requirements are not extreme, and where a little imperfect exhaust is of no importance; but where the conditions of working require a vapour quality of precisely the same value whenever a charge is taken, whether regularly or with extreme irregularity, the construction of vaporiser must be such as will secure the complete conversion (and use) of the oil admitted at each suction stroke into uniform quality of oil fog. A large part of this conversion must be done

[1] Among these may be mentioned those in *The Engineer*, vol. lxxiii. p. 539, and vol. lxxviii. pp. 7–11, 25–28, 63, 173.

by mechanical abrasion in a passage small enough in section or of such a form as to cause the air to sweep through at a very high speed, or to be violently eddied on its way to the hottest part of the passage, where the fog is slightly superheated as the last part of the process when it is entering the cylinder.

The oil vapour which can be successfully used is not in the condition of an oil gas. It is made by such means as those referred to in vaporisers, the passages of which have a mean temperature which does not heat the air to more than from 350° to 450° F. Such a vapour is much more readily ignited, when mixed with sufficient air, than oil gas, and it is an unstable vapour. Hence it must be made only as wanted, and then used completely, if it is to be satisfactorily used. Incomplete use leads to partial condensation and the formation of secondary products. It must not be allowed to come to rest, but must be kept in violent commotion from the time its formation commences to the time ignition takes place. It cannot be stored. *Vaporiser product not an oil gas*

From these considerations some idea of the vaporiser difficulties for motors for vehicles may be gathered; but it will be seen, that to have made the vaporiser is only a part of the problem, though the most important.

The difficult problem, and the next thing essential for satisfactory working under motor-car conditions, is means for completely emptying such a vaporiser every time oil is admitted to it. This involves something in the nature of an anticipating governor, which, for the work required, seems an impossibility. Whether the motor be governed by controlling the admission of oil to the vaporiser, or by closing the exhaust or by keeping it open, there must occasionally, in the manipulation of a motor vehicle in difficult street traffic, be a charge of oil delivered to the vaporiser just as a dead-stop has to be made, and that may be immediately followed by another stop and start. Under these circumstances a double charge of oil goes next stroke into the cylinder, partly as oil and partly as vapour, and incomplete combustion follows for two or more strokes. *Use and control of petroleum vapour*

Thus exceedingly close governing is required from this point of view, and yet for running purposes it is desirable to be able to work the motor at different speeds within a considerable range. Close governing when this is a requirement is, to say the least, not easy. Moreover, the ordinary arrangement of governor-controlled oil admission, gives an oil charge for every working stroke when the speed is below the normal. The governor only cuts out the oil supply when the speed is above the normal, or when the engine stops. This is not sufficient, because the full charge of oil is given at every working stroke, whether slow speed is due to full load or to the use of the brake for manipulating in traffic. This being the case, the occasional exhaust of unburned oil vapours seems inevitable by means at present available.

329

Possible gas-making vaporiser It is conceivable that an oil feed of minute accuracy, and a gasifying vaporiser, with gas-receiver and valve-controlled gas admission to cylinder, may yet be made, which under all ordinary exigencies of interrupted working will make the use of kerosene sufficiently unobjectionable. The difficulties must not, however, be lightly estimated. The gasifying vaporiser would require very powerful Bunsen burners to raise and keep it at the strong red heat that would be required to make a fixed gas, and any incompleteness in the formation which gave a mixed gas and fog would result in a tarry product of combustion and the stoppage of the motor. The heating of the vaporiser must therefore be powerful and certain. This would consume a good deal of oil, and probably so much that the advantage of kerosene on the score of economy might almost vanish. The oil at present burned by the vaporiser lamps is from 0·083 to 0·125 of the whole of the oil an engine uses. Assuming it took only

four times as much as this, or, say, 0·4 of the whole, to make a fixed gas, and allowing for the losses in the process and cooling of the gas from its high temperature, if that were necessary, the oil consumed would perhaps reach 0·5 of the whole. Thus a 10-HP. engine requiring 6 lb. of oil per hour and 0·6 per hour for lamp would, if converting the oil into gas, require at least $6·6 + (6·6 \times ·5) = 9·33$ of oil per hour, equivalent to raising the price of the oil from 6*d.* to 9*d.* per gallon. This does not consider the weight and space occupied by the apparatus, but if it be a sufficient estimate of the cost, there would still be a considerable margin to pay for these extras, and in favour of the heavy oil.

Forms of vaporisers One form of vaporiser will be shown in connection with the Roots & Venables motor about to be described. Another **Crossleys'** form is that of Messrs. Crossley Bros., as shown in diagram form by Fig. 247. In this A is the chimney for a lamp placed at B. This chimney has cast upon it a number of ribs *c*, in each of which is one notch alternately at one side and the other. Over this chimney is placed an exterior tube fitting it. Below this chimney is the vaporiser proper D D',

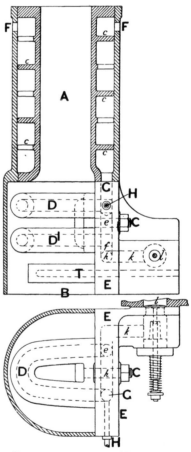

Fig. 247.—Crossleys' Vaporiser.

made up of two simple cored castings both held in place by one **T**-headed bolt c, the holes in these castings meet holes drilled in the casting E which supports the chimney. Air to be heated enters at F and thence, by the circuitous course formed by the notched ribs, reaches the passage G in a heated state, and passes downwards, where it meets the incoming jet of oil from the pipe H. (In the plan the hole c should be shown by a full line circle.) This oil becomes spray, and is violently disintegrated in its passage with the air through D, whence it reaches a passage e, shown in dotted lines, and enters the passage in D′, and arrives at a passage f immediately under G and e. From this it passes into the horizontal hole h, which leads to the port k and the vapour valve l, and enters the cylinder An ignition tube T, covered with a cast-iron protector, is placed below the vaporiser tubes.

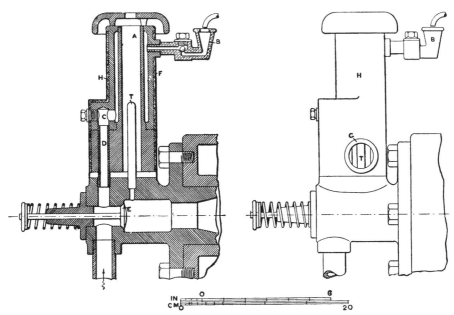

FIGS. 248 & 249.—BLACKSTONE'S VAPORISER.

Other forms of vaporiser, involving the same general principle and features, are used by Messrs. Fielding & Platt in their oil engines, and by Messrs. J. & F. Howard. A very simple form is that of Messrs. Blackstone & Co. This is shown by Figs. 248 and 249. In these, the chimney A forms the interior walls of a narrow annular chamber within the cylinder H, into which oil dropped in at B is drawn with air entering at the same place. The air and oil are rushed round the chimney and through the annular space F at high speed, and the oil is violently disintegrated by the time it reaches the passage c and passes down the tube D on its way to the cylinder by way of the supplementary air-admission valve E. The ignition tube T and the vaporiser are heated by a lamp flame entering the hole G. The vaporiser is lagged with thin asbestos H and sheet iron F.

Fielding's and Howard's

Blackstone's

These vaporisers are sufficient to show the character and diversity of design of vaporisers used for stationary-engine purposes. They would not be sufficient for motor-carriage purposes unless supplemented by other control devices for accomplishing the objects described on page 327, 329.

The objection at present made to the heavy-oil motors, that they give off a great deal of offensive oil-vapour smell, in addition to the bad exhaust, will in the future, no doubt, be very much lessened, difficult as it is to prevent the small leakages and creepings of kerosene. Better pipe jointing and other joints are required, and much better and tighter lamp and lamp fittings for heating the ignition tube and vaporiser. It is the capillary

FIG. 250.—ROOTS & VENABLES' HEAVY-OIL MOTOR CARRIAGE.

spreading of kerosene leakages that occasionally makes this oil dangerous where powerful lamps are used, and where the general belief that this oil is safe causes disregard of proper precautions. That the difficulties at present attending the use of the heavier oils will ultimately be overcome there cannot be much doubt, but foreign experimenters are at present far behind the results obtained in this country, not only for internal combustion engines, but for powerful burners for steam-raising purposes.

ROOTS & VENABLES' PETROLEUM MOTOR CAR.

A two-seated car, worked by a heavy-oil motor of 2½ brake HP. as made by Messrs. Roots & Venables, is shown in perspective by Fig. 250, and Fig. 251 is a reproduction of a photograph of the car with the body removed so as to show the arrangement of the motor and mechanism. Figs. 252 and 253 are respectively sectional elevation, sectional plan of the motor; Fig. 254 is a separate view of the vaporiser.

There are several points of departure from general practice in the construction of this car apart from the difference in the motor. These will be referred to as they occur in the following description.

The side members and two transverse parts of the carriage frame are of light channel and angle iron respectively, the tubular water-cooling frame formerly used having been given up. The frame is carried

Roots' oil motor

FIG. 251.—ROOTS & VENABLES' MOTOR CAR: FRAME, MOTOR AND MACHINERY.

on fore and aft double springs, and on it is carried the whole of the working parts, practically independent of the body.

The motor has a single cylinder 4⅜ in. diameter and of 5-in. stroke, the piston being coupled to a **T**-ended connecting-rod, which engages with a balanced crank. The balance weight balances the crank and half the weight of the connecting rod. The big end of the connecting-rod carries a brass, held in position by a strap-bolt z. The air and vapour admission valve E is automatic in its action, but the exhaust valve F is operated by a hollow sliding piece G moved by an excentric H, the valve being pulled down upon its seat by a volute spring J. At K is the vaporiser, which is of the warm air-current, mist-forming and vaporising type, to which oil is supplied by a " Roots " oil feeder R, seen in section in the plan Fig. 253. The vaporiser itself is seen in section in Fig. 254, and partly so in Fig. 252.

Vaporiser and oil feed

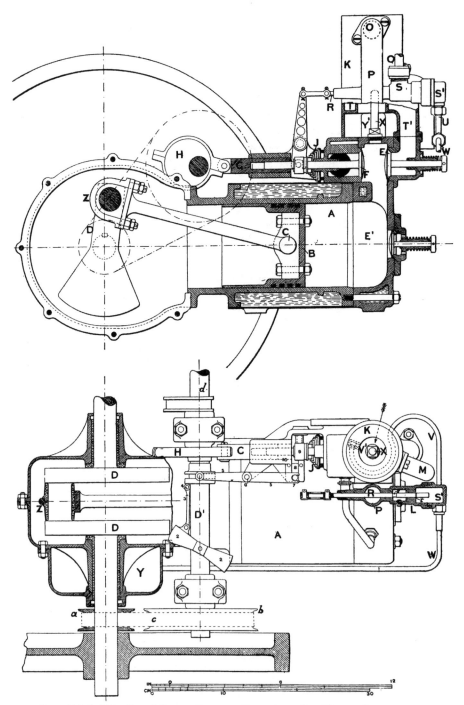

Figs. 252 & 253.—Roots' Single-Cylinder Horizontal Oil Motor: Elevation and
Plan, partly in Section.

334

From these it will be seen that when the motor piston makes a suction stroke the valve E opens and air is drawn up a tube M′ into the vaporiser at M; from this point the path of the air is as shown by the arrows in Fig. 254. The diaphragms N N cause the air after admission to traverse a circuitous path around the central chimney of the vaporiser, and emerging at O, the warmed air passes down the passage P, where it sweeps past the oil-feed valve R at a time when the groove s, Fig. 253, has taken oil from the receptacle s, in which it is shown, and carried it into the pipe P. From this point the warmed air with its charge of oil passes into the passage T round the vaporiser, and thence down the passage T′ past the inlet valve E, and into the combustion chamber E′. The oil enters at Q into the receptacle s. The feed rod R reciprocates, and the groove s enters s, and on the return stroke carries a grooveful into the passage P. At the same time the very much smaller groove t enters the oil receptacle s, and on the return stroke it carries a small quantity of oil into the receptacle s′. This oil trickles down the pipe U and thence along the pipe V, accompanied by air under pressure from W to a burner nozzle V′ at the bottom of the vaporiser chimney, where it heats the ignition tube X, and the products of combustion heat the vaporiser.

The supply of air for this, at a pressure of about 10 lb., is obtained from the chamber Y. This is kept supplied from the crank chamber, which, except for an air inlet valve, is closed. Air

**Heating ig-
nition tube**

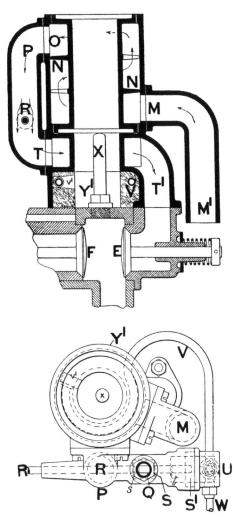

FIG. 254.—VAPORISER OF ROOTS' OIL MOTOR.

enters on the instroke of the piston, and passes, at each outstroke, through an outlet valve into the chamber Y, in which there is a safety or blow-off valve. A large part of the space below the vaporiser and surrounding the ignition tube X is filled in with asbestos paste and fibre Z. By means of the supply of oil down the pipes U, and air from pipe W, the

Bunsen burner v′ makes any other ignition tube heater unnecessary once the motor has started, but the tube and vaporiser are at first heated by a separate lamp placed so that its flame enters the opening at the bottom of the vaporiser at Y′.

Governor The governor is seen at 2, 2, Fig. 253. It is pivoted upon the valve motion shaft D′, and when rotated with it, tends to a position normal to the crankshaft, instead of at an angle therewith which it takes when the motor is at rest. By this resolved centrifugal effort it pulls the link 3 connected to the grooved collar at 4, and thereby alters the position of the lever 5 pivoted at 6, which by the fork at 7 moves the sliding piece 8. When the speed of the motor rises, the piece 8 is pushed towards the collar 9 on the exhaust valve spindle, and when the latter is moved by the pusher piece G, the piece 8 is interposed, and the exhaust valve consequently cannot return to its seat. When the motor has returned to the normal speed, the next push by G liberates 8, the exhaust valve closes, and a new charge is admitted by the valve E, and a new stroke is made by the feeder valve rod R.

Supplementary air valve A large part of the whole of the air used in the motor passes through the vaporiser, but a supplementary air supply is admitted by the short-stroke valve at the end of the combustion space.

Water cooling The cylinder and valve box are, as will be seen, water jacketed, water being carried in two side tanks, and circulated round the fly wheel in a tube four times coiled round it with sufficient clearance from its periphery. This coil, under the influence of the air driven by the fly wheel past and through it, acts as a cooler, and circulation is maintained by a pump operated by the excentric 15. Oil is carried in one tank on the near side. A cover not seen in the detail drawings is placed over the vaporiser, the lamp space, and air inlet, as seen in Fig. 251.

Transmission gear The valve motion shaft D receives its motion at half the speed of the crankshaft from a toothed pinion a driving a wheel b by means of a Renold silent chain c. The shaft is made much larger than is required for the valve motion, as it is used as a second motion shaft for the transmission gear, a chain pinion and clutch being carried on the end d. Upon this pinion runs a chain which drives the main axle by a wheel towards the centre of the axle, as seen in Fig. 251. This provides the slow speed controlled by a conical friction clutch. For the higher speed, a similar chain pinion and clutch are carried on the end of the extended crankshaft, and upon this chain pinion runs a chain which drives the main axle at the high speed. There are thus only two speeds given by the chain gearing, any intermediate speed between the high and low being obtained by lost motion or slip in the clutch on the crankshaft. Similarly, any driving below the speed of the second motion chain is obtained by a regular clutch slip, or as regular as clutches will permit.

Driving clutches A good deal of trouble has been taken to arrive at suitable dimensions and angle of the cone clutches, which run metal to metal. They must be small

336

enough and the angle large enough to cause them to slip more or less when permitted, and not large enough to be able to master their load until well pressed together, otherwise manipulative slipping is exchanged for suddenness of action, there being nothing between holding and not holding. Experience has shown that it is extremely difficult to get clutches, and cone clutches as much or more than others, to transmit motion in a definite and uniform manner from a constant driver speed, but Messrs. Roots & Venables appear to be satisfied with the action of the clutches they use, and in the manner used, and they report that they neither seize nor slip when not required to do so. The low speed clutch is wider than the other.

Two spoon brakes are used, and one band brake on the crankshaft actuated by a pedal.

Brakes

The car weighs under 5 cwt., without oil, water, or passengers, and has therefore 0·5 B.HP. per net cwt. of car. It was exhibited at the Richmond Show in 1899, and ran very well, except for the objectionable smell of the exhaust at starting, and occasionally during the running, and the smell from the powerful lamp employed for the ignition tube and vaporiser. The smell of burning or heated petroleum from slight leakages was also very noticeable, but this could of course be stopped.

Weight and power

There are several other makers who are turning their attention to heavy oil cars in this country, but no noticeable advance has been made. The Koch car made in France was exhibited, and to some extent run in this country last year, but although the carriage was good in design and appearance, its motor manners were not satisfactory, the difficulties referred to in the beginning of this chapter not having been met, and apparently not appreciated. The motor is of 6 B.HP., and has two pistons in one cylinder, the combustion space being between them, and the pistons working in opposite directions coupled by short connecting rods to rocking arms, and thence by connecting rods to a double crank beneath. The English Patents for it are Nos. 3771 and 6711 of 1899. When the engine has been at work a short time on a load which is well within the range of its best full load and its least light load, it will work without much that would be objected to on country roads; but when called upon to work in streets with a maximum variation of load, and at all re-starts, the characteristics of incomplete combustion of kerosene make themselves unpleasantly conspicuous.

Other heavy oil cars

Chapter XVIII

CYLINDER COOLING AND WATER COOLERS

Air cooling

THE selection of the method and apparatus employed for limiting the rise of temperature of the cylinders of internal combustion motors has a much more important bearing on the power required and the speed of a given car than is generally supposed. The limit to the sufficiency of simple atmospheric radiation is reached with very small sizes of motors, and, practically, it is also soon reached even when this is aided by forced air circulation.

Very small cylinders may be sufficiently cooled without any special provision, because the masses and surfaces of the metals of, and metallically connected with, the cylinder are large in proportion to the cylinder volume. Larger sizes may be air cooled by radiation, by increasing the heat emissive surfaces by forming ribs or gills upon the cylinder and valve cases. The limit to this, however, is soon reached, because no matter how large the emissive surfaces may be, they can only radiate the heat transmitted to them by the limited receptive surface of the roots of these ribs. When the motor is moving quickly through the air at the rate of the carriage, the radiation is increased by an uncertain amount of increase in temperature difference, but the specific heat of air being low, natural air currents are insufficient. Recourse has been made to mechanical means of augmenting the weight of air passed over the surfaces per unit of time, and this method is being used; but an adequate appreciation of the physical and mechanical conditions involved will show that this must be a wasteful method of cooling cylinders of motors of more than 1 or 2 HP., and an uneconomical substitute for the water-cooling jacket.

Limits to air cooling

Water cooling the most economical

It is very much cheaper in power and money to evaporate water by the heat that must be dissipated from the cylinder, than it is to use fans to obtain a forced air cooling; and this equally applies when the fan is used to cool the water used in jackets. It is still more uneconomical to use fan and water-circulating pump too, in order to prevent water loss by evaporation, which would otherwise occur through absence of circulation.

It is far cheaper to effect the cylinder cooling by means of water which circulates by convection through pipes having surface enough to carry off

338

CYLINDER COOLING AND WATER COOLERS

all the heat, provided that this surface can be so disposed as to effect its object without materially increasing the wind-resistance surface. It is worth while to enquire into the thermodynamic and economic aspects of this question.

The thermodynamic results of tests of oil and gas engines give some instructive figures. For instance, the ratio of the heat carried away in the jacket water to the heat in thermal units, supplied in the oil or gas when working on full load, is from 0·266 to 0·37. With certain sizes of oil engines tested by the author, these quantities represented thermodynamically no less than 1·65 HP. per effective or B.HP. In the vehicle motor, however, it may be taken that considerable radiation and conduction losses occur which may reduce the quantity to be artificially carried away, especially as it is common to run the motors with much higher jacket temperatures than are permitted in fixed engines. For this reason the total heat to be carried away artificially may not reach more than about 2,500 units per B.HP. hour. Even this, however, is so formidable a quantity that it is clear that, for motors of any size, cooling by air must remain quite impracticable even with a multiplication of cylinders by means of which small dimensions would be secured. The efficiency of any fan or air propeller that could be used would not be more than 0·6, so that it is obvious that ordinary motors depending on surface radiation cannot be economically cooled by air currents. Water must therefore be used.

Some evidence of what is required may be gathered from the 1899 trials of vehicles on the fifty-mile test of the Automobile Club. Most of these vehicles employed coolers of the tube and gill or radiator kind, already described, with a small circulating pump. Others, however, had simply water-tanks, with inefficient water circulation. The result was that a good deal of water was evaporated, one 4-HP. car having used 6·75 gallons in the time occupied by the run, namely about four hours, and another used 5·6 gallons. It may thus be taken the heat given off by the cylinders of these motors was sufficient to evaporate about 1·5 gallons per hour, or 15 lb. This is equivalent to at least 220 B.T. units per minute. If we take it as 200 units, we then have as the mechanical equivalent $\frac{200 \times 772}{33,000} = 4\cdot66$ HP., and this is the net power that would be required if the water cooling were replaced by some mechanical means of cooling having an efficiency of 1·0.

Other cars, with radiators and small pumps for maintaining a good circulation through them, and through the tanks and the cylinders, evaporated smaller quantities of water. Of three of these, of an average of 5 declared HP., the mean quantity used was 1·6 gallons, or 4 lb. per hour. The lowest quantity used by any such car fitted with radiator and circulating pump was 2·5 lb. per hour, the cooling surface being large. This quantity would, of course, have been very much larger if, with the tank and pipe connections used, the water had not been kept

339

in circulation by a pump. The evaporation of this 2·5 lb. per hour may therefore be taken as the least that, with the cooling method employed, would keep the water below the maximum temperature permissible. It may also be taken that the mechanical work done in driving the circulating pump was a necessary help to the cooling by surfaces which were made more efficient by its use.

Therefore a water cooler with which there is no special means of circulation, and in which the effective surfaces for cooling by natural draught and radiation are no more than in the car instanced, will, if the water be kept cool by a mechanically forced air current through tubes or otherwise, require, as a minimum, the mechanical equivalent of the thermal work, which would otherwise be done in evaporating that water.

Power equivalent of the cooling by evaporation

Now, the evaporation of 2·5 lb. of water per hour, even at the low temperature of evaporation which obtains, would require about 2,750 B.T.U. The mechanical equivalent of this is—$\dfrac{2,750 \times 772}{60 \times 3,300} = 1·07$ HP. To this must be added the difference between absolute efficiency and the real efficiency of the apparatus, which will make the actual power required to do the equivalent cooling by fan draught not less than $\dfrac{1·07 \times 1}{0·6} = 1·7$ E.HP.

This is a very large demand upon a motor of about 5½-HP. and it is possible that the actual natural radiation helps in this work more than has been here estimated. It is a question on which experimental evidence is wanting, but there can be no doubt of the large power required for this method of cooling.

Fan-cooling most wasteful

From this it is clear that the most wasteful method of cooling is one which depends on the refrigerative effects of mechanically provided air-blast or currents. It is further obvious that the evaporation of water by the waste heat is the cheapest method of cooling, and that the best cooler will be that which presents very large surfaces in such a way as not to increase the unavoidable wind surface of the carriage, and at the same time in such a form and arrangement as will secure a good, rapid circulation of the water without even a circulating pump. To put a gallon of water per day into a cooler is little trouble and no expense.

The best cooler

The cooler that uses least water may be the least economical

From this little inquiry it will also be seen that from the economical and machine-efficiency point of view, the form of air-tube, fan-forced draught cooler used in the Cannstadt car is very unsatisfactory. At the end of about four hours it had lost only about 2 oz. of water, and the two gallons it contained was, at the end of the run, at a temperature of not much, if any, above 105° F. This was pointed to with pride, the pointer not seeing that it was a reason for condemnation, inasmuch as the low temperature was obtained at the loss of a great deal of power, sometimes very much wanted at the peripheries of the driving-wheels, and a corresponding loss of petrol, which would otherwise be available for extending the distance the car could run.

CYLINDER COOLING AND WATER COOLERS

Several forms of jacket water coolers have been described in connection with the carriages to which they are attached. With most of them a small pump is used to circulate the water, and in one, the Cannstadt-Daimler, a fan is also used to draw air through a large number of tubes in a water vessel after the manner of a tubulous surface condenser.

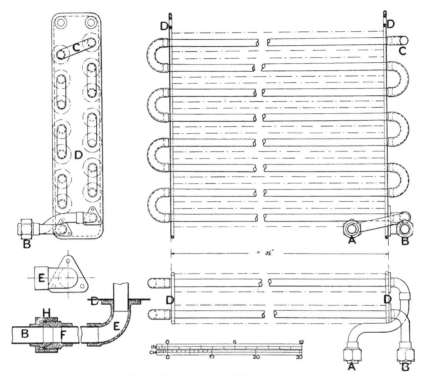

FIG. 255.—CLARKSON'S WATER COOLER.

Fig. 255 illustrates a form of cooler made by Messrs. Clarkson & Capel, **Clarkson's cooler** with thin copper tube covered with spirals of spirally coiled wire, as shown by Fig. 256. The tube is spirally corrugated, and the spirals of wire sit in these corrugations. After winding, the tube with its wire covering is tinned, so that tube and wire have continuity of metallic contact, and the whole is held together and preserved from corrosion. For a motor of 5 B.HP. about 40 feet of thin $\frac{5}{8}$-in. tube of this kind is put together, as

FIG. 256.—CLARKSON'S COOLER TUBE.

seen in Fig. 255, the straight parts of the tube being carried by sheet-iron end supports D, and connected by bends. The water is circulated by a

341

pump, enters at A, and after traversing both tiers of tubes emerges at B. The end connecting flange and bend is shown at E, and the form of metallic union joint is separately shown at H.

Estcourt's water cooler   A powerful natural circulation cooler, made up in part of this tube, has been designed by Mr. E. Estcourt, and made for him by Messrs. Clarkson & Capel for use on his Daimler carriage, which has now been

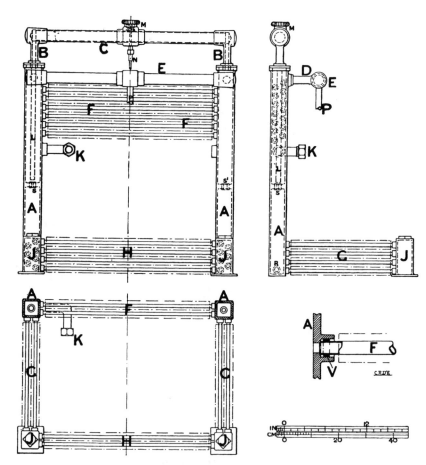

FIGS. 257.—ESTCOURT'S JACKET WATER COOLER.

in almost constant use for over two years. This cooler is illustrated by Figs. 257 and 258, the latter also showing Mr. Estcourt's starting device. The car is one of the early standard Coventry cars, with motor which formerly, that is, two years ago, was known as of 4½ HP. nominal. The same size of motor now, however, gives 6 B.HP., and is easily capable of giving 5½ B.HP. for considerable stretches.

Natural circulation   The cooler is designed to work with natural circulation only, so as to dispense with the usual pump. It consists of three lower sets or tiers of tubes, connecting two short and two tall water columns, and one upper set, connecting the upper part of the tall columns. These are all seen in the scale drawings, Fig. 257, and in perspective in Fig. 258. The columns are aluminium castings provided with recessed bosses, as shown in the detail in Fig. 257, in which F is part of a tube, A a piece of one of the columns, and V an indiarubber gauge-glass ring which makes a water-tight, easily-made joint between tube and socket. The short columns J J are simply connectors. The high columns A A are stopped at s s', but a hole is made in stop s', and a funnel-topped pipe L passes down the left column, and through the stop s, to receive and deliver any water that may be carried up with (or resulting from condensation of vapour in the separate pipe c), down to the cool water at the bottom of the column. Three of the water-pipes from the top of the motor cylinder are connected at K, and another from the other side at P. The first deliver into columns at L, and the second, P, delivers into the same column by the pipe E and the connection D. At M is an inlet for water, and N is an overflow pipe for any excess water, M being always open or easily uncovered to permit the escape of vapour. By insertion of the stop at s (through which the pipe L passes instead of being short as in the engraving) the circulation is determined in the directions shown by the arrows on Fig. 258. The hot water from all the cylinder top connections is delivered into the upper part of the tall left-hand column. It thence passes through the double rows of top pipes F, above the motor bonnet, down the right-hand column, through the bottom layers of pipes to the bottom of the left-hand column, from which a branch is taken to the bottom of the cylinder jacket.

Taking the place, as this cooler does, of a large part of the dashboard, it does not add to the wind resistance surface, and it saves 2 cwt. as compared with the pump and tank systems. In this cooler there are about 170 feet of thin, wire-covered copper tube 0·5 in. outside diameter, and there is beside the cooling surface of the columns and the cross tubes E, C. With it Mr. Estcourt has run 160 miles before make-up water became really necessary. The circulation being good an equable temperature is maintained, and this keeps the valves in good order.

Numerous other coolers might be here described, but under the name of radiator they have now become more or less standardised as to construction. Most of them are similar to the forms described with reference to the Panhard and other cars.

A radiator, in the form of an apron, or other forms, is made by M. Apprin, of Rue de l'Abondance, Lyons, of tubes deeply corrugated circumferentially, the corrugations forming thin hollow fins or ribs close together. This would probably make a very effective cooler, but perhaps no better than one made with the Rowe tube, as used for condensers and other purposes. See Figs. 419, 439. The Apprin tube is used by Marshall & Co.

Other coolers

Chapter XIX

SOME MOTOR VEHICLE COMPONENTS

Estcourt's starter

FOR starting the motor from the seat Mr. Estcourt has devised the apparatus shown by Fig. 258, in which B is a coarse, saw-toothed thin wheel on the end of the motor shaft, and running in part in a slot in a flat bar beneath it. At A is a ring surrounding a strong ribbon clock spring, and carrying a long pitch bicycle chain attached to a cord which runs under a grooved pulley on the left hand, and round pulleys to the seat. By pulling a handle on the end of this cord the cycle chain is pulled off A, it gears with the wheel B, and starts the motor. When the motor starts, the teeth of B automatically disengage themselves from the chain, and the cord being let go the chain is again taken up by the clock spring. The apparatus, although made up of parts that came to hand, is, like the cooler, quite a success.

Other forms of starter acting on similar general lines, but with the friction pawl of the old silent feed motion, have been since made for these cars, and Faugue and Dangleterre used another form on the few cars they made last year. Messrs. Hope & Co., of Water Street, Liverpool, make another.

STEERING AND DRIVING-WHEEL AXLES.

Although the motor vehicles manufactured in this country have reached quite a considerable number, the demand for standard component parts, such as axles, has not attained sufficient importance to lead axle-makers to devote the special attention required to the production of axles really suitable for motor-car work. Some of the axles used in England are made here, but a good many have been imported from France, whence come also a good many of the springs. It is obvious that this is only a question of expediency, inasmuch as no better carriage forgings and springs are made in the world than can be obtained in the United Kingdom, but special patterns are difficult to get when all makers are busy on standard things. This, however, cannot be so generally the case with springs, such as many old coach-making firms have for generations made for carriages of various patterns.

SOME MOTOR VEHICLE COMPONENTS

A few illustrations of French-made axles, as used on the fast petrol motor cars of different sizes, may, however, be of interest. Fig. 259 is the well-known English "Patent" axle of Collinge, of 1787. It is here inserted as an example of successful results which follow the determination to use the best workmanship and the best materials for carrying out the

Collinge
axle

FIG. 258.—ESTCOURT'S WATER COOLER AND STARTING GEAR ON HIS DAIMLER MOTOR CARRIAGE.

best design. The idea of high quality or exactness of fitting, which was followed in making these axles, was for many years unknown in any other class of common machinery, and hence the Collinge axle is still the axle best known more than a century after its introduction. The right and left-handed nuts at G were not used in all, but the oil-cup cap H, to keep the well-ground-in parallel axle A' in the "box" C, and the oil collar K, dust ring and recessed collar F, were always used, and the good selected faggoted scrap for the axle A was a point of honour.

Lemoine's steering axles

In the Ackerman steering axle, Fig. 260, one of Lemoine's, there is no screw end, and the oil cup is formed by the end of the wheel boss, or hub, as we now most often call it. The spoke roots are held at c between the two flanges, one of which, after the manner of Hancock's wheels, Figs. 9 and 10, p. 12, is pulled up towards the other, and on the spokes by connecting belts, no nave being employed. The axle A has forged upon it seats B for springs, and a fork through which passes, at right angles to it, the pin M upon which pivots the boss L of the Ackerman or stud axle A'. This axle, it will be seen, has a downward inclination, so that the periphery of the wheel where it touches the ground will approach the point of incidence on the ground of the centre line of the pin M produced. This causes the impact on collision with obstructions to be delivered more or less nearly on that line instead of out somewhere towards K or C, and hence the length of the couple against which steerage resistance has to work is lessened. Some English makers, when using wheels not dished, as in the case of Fig. 260, give the inclination to the pin M, and leave the axle A' nearly or quite horizontal. It will be seen that in Fig. 260 the wheel is kept on the axle by the leather and metal rings G and bolts. N is the arm for the steering lever connections.

Fig. 261 shows another of Lemoine's steering axles. Here, again, the pivot L, which runs in a socket instead of a fork, is vertical, and the axle inclined from it. The axle A' is forged with the pivot pin, which fits inside the bottom of the socket and the outside of the internal pin M, the arrangement securing maintained lubrication for top and for the bottom at F'. The axle A', with the pivot pin L, is prevented from coming out of the socket when A lifts, by the clip bolt P, and the adjusting screw pin M is locked where put by a spring catch N.

In a modification of this form of axle, M. Lemoine inserts, in place of the check leather, recess and collar at F', a pair of steel cups with ball-ring chases, and this no doubt turns more easily, and in practice should be satisfactory, as it has been even with the heavy axles and vehicles made in this country by the Lifu Company from the designs of Mr. H. A. House.

The steering arm is attached at any angle required upon the downward extension L' of the pivot pin. In the hub of Fig. 261 the outer part C is carried by a sleeve which fits over the ordinary "box," and makes an excellent firm job, while the enlarged end covers and protects the oil-cup cap, which is similar to that of Fig. 259. M. Lemoine makes several of the forms of pivoted steering axles.

Undesirable forms of steering axle

Fig. 261A illustrates a form of Ackerman axle advertised in the United States. It is here inserted so as to call attention to a form that might easily be a dangerous one. The axle pivot is formed by two sets of balls carried between the cup and the coned points of the set screws.

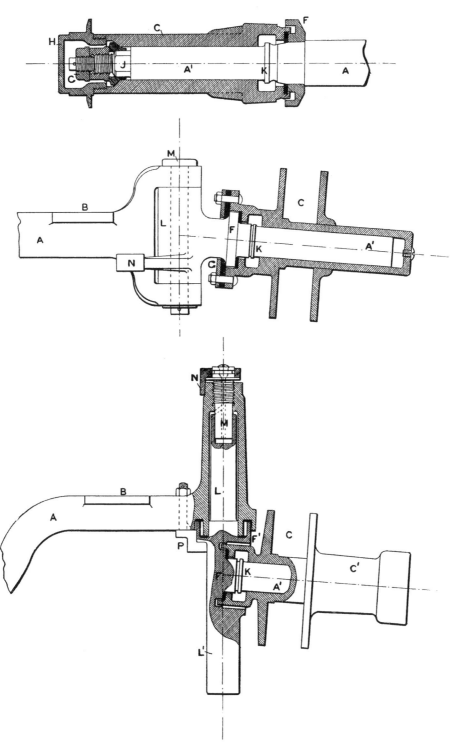

Figs. 259-261.—Types of Collinge and of French Steering Axles.

347

The breakage of steering axles is not a common thing, but it is not rare, and several have broken in this country, not in the fork, but near the root of its junction with the axle. With the form most commonly used in this country, namely, Fig. 260, this is where the breakage would be expected, because the pivot pin sufficiently connects and makes one of the short axle arm and jaw of the fork, so far as jaw-spreading stresses are concerned.

In Fig. 261A, however, it will be seen that there is nothing to

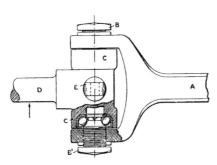

FIG. 261A.—BALL AND CONE PIVOTED STEERING AXLE ARM.

prevent the spreading of the jaws of the forks, so that unless of very great strength at the bends the vertical resultant of the horizontal pressure on the conical setscrews under the load, and the heavy shocks often visited upon these axles, will sufficiently spread the jaws to allow the cone setscrews to be wrenched out of the cups, and the axle will carry away. For this reason this form of axle should be very fully tested before trusting it for high speeds, or avoided.

Lemoine's fixed axles

Figs. 262 and 263 show two forms of Lemoine's main driving-wheel axles with the two different forms of hubs shown by Figs. 260 and 261, but with the inner end of the hub formed with flanges E to carry the sprocket wheel D, and in one case carry also the brake ring. In Fig. 262, the axle A has a spring seat at B, and is kept in the hub by the leather and iron check rings G, which run against the collar F. The hole below B is for the attachment of the radius rod. In Fig. 263 the wheel is kept in place on Collinge's system, and near the spring seat B is forged on the axle a bearing for a link pivot as used in one of Serpollet's cars.

The foregoing are good examples of fixed axles which will probably remain in favour although the live axle will probably be more used in the future than it has been in the past.

Strength of axles and springs

With regard to the strength and deflection of axles and springs, it is not thought desirable to occupy space with any theoretic investigation, as the variety of unassignable stresses to which they are subject in practice makes appeal to practical experience necessary in any case, and to this it is thought best to go direct, especially as English makes are not generally likely to err on the side of insufficiency. For those, however, who wish to enter upon this side of spring and axle questions, reference may be made to MM. Vigreux, Milandre & Bouquet's little treatise on details of *Voitures Automobiles*, Part I., and to Budd's book on *The Underwork of Carriages*.

As a further example of the extent to which parts for motor vehicles

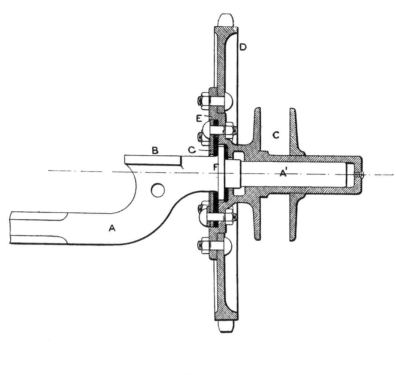

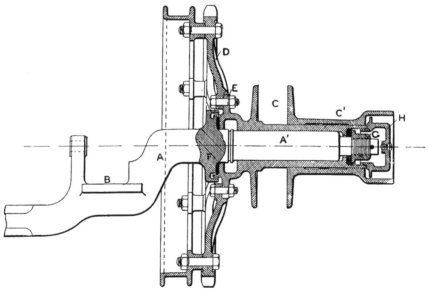

FIGS. 262 & 263.—LEMOINE'S DRIVING-WHEEL AXLES, SPROCKET WHEEL, AND
BRAKE RING ATTACHMENTS.

349

are made as stock productions in France, reference may be made to the
many forms of gearing manufactured by MM. Malicet & Blin, including
A stock various sizes of differential gear, such as that shown in parts by Fig. 264.
differential In this the box is made in two parts, A A', bored to fit the two parts of an
gear axle the inner ends of which are screwed to fit the nuts o o', which
fit in the spider which carries the pins E E upon which the pinions
G G' revolve, and gear with the wheels c c'. The pins are clamped
in grooves between the two flanges F F', and the wheels c c' are fixed
upon the axle ends. It may not be long before such fittings are obtainable
as stock goods in England.

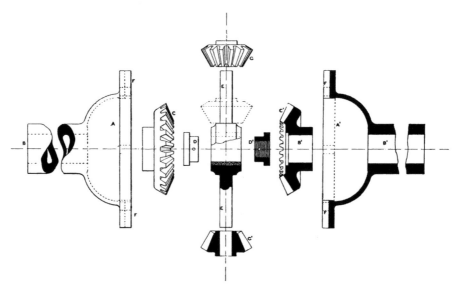

FIG. 264.—FRENCH DIFFERENTIAL GEAR PARTS FOR MOTOR CARS.

SPOON OR TYRE BRAKE SHOES.

As a rule, the spoon brake or tyre brake shoe has been made much
too small, often a mere finger, which deforms the part of the tyre it bears
upon and causes a good deal of destruction. Lately, however, a great
improvement has been made in this respect, and two good forms of shoe
are shown by Figs. 265 and 266.

Price's Fig. 265 shows Price's shoe brake. The shoe A presents a cover for
shoe brake a considerable arc of the wheel tyre T, and is so carried upon the arm
c that it can, under the resistance of the springs s s', accommodate
itself to the wheel, though the brake itself may move up and down in
obedience to the changing deflection of the springs supporting the body
of the car. The shoe is attached to a piece B by two bolts, and is easily
renewable. The piece B is adjustable on the bolt c', and the piece F is
adjustable so that the tension on the springs s s' shall hold the shoe in

350

the position which makes it touch the tyre with approximately equal pressure all over.

Sampson's shoe brake

Fig. 266 is another form of shoe brake, designed and made by Mr. J. Lyons Sampson, in which uniform touch is obtained with freedom for the shoe to accommodate itself to the tyre during the vertical movement of the carriage body on its springs. On the centre of the shoe A is cast a box B, within which is a bush F with set screw E, by which the bush is fixed adjustably on the end c' of the brake arm c. Round the bush is a spiral spring s, bearing at one end against the shoe, and at the other against the setscrew boss E. By these means the shoe is free to move

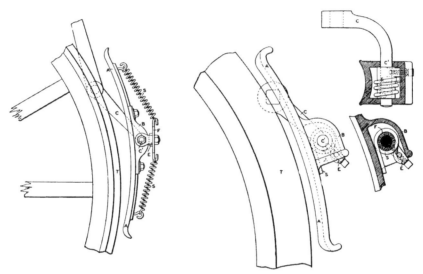

FIG. 265.—PRICE'S BRAKE SHOE.   FIG. 266.—SAMPSON'S BRAKE SHOE.

through a limited range on c', under the restriction of the spring s, the adjustment of which secures parallelism of approach of the shoe to the tyre.

THE BENZ SLOW-SPEED GEAR.

Fig. 267 illustrates a form of slow-speed gear, known as " Crypto " gear, used on some of the Benz cars of 1899 and 1900 for hill climbing. In this A is the boss of the fast pulley F, fixed on a sleeve B and to the differential gear case M. D is the boss of the loose pulley L. It is fixed on the one end of the sleeve E, and carries at its outer end the pinion G formed upon it. In ordinary running L and F run at the same speed upon the shaft N, and the gear does not work although always in mesh. When it is desired to mount a hill, the belt is put on the loose pulley L, and a band Q is caused to grip the gear-case periphery. The loose pulley then drives the fast pulley at a very slow speed by the pinion G gearing

351

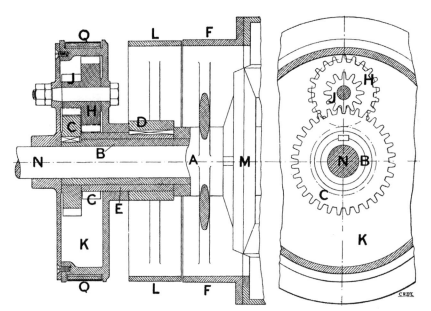

FIG. 267.—THE BENZ SLOW-SPEED HILL-CLIMBING GEAR.

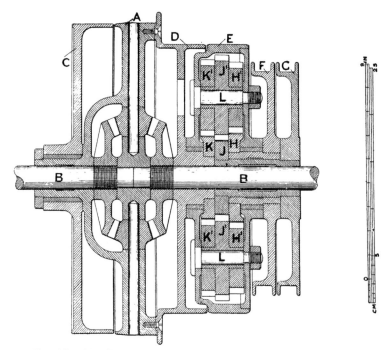

FIG. 267.—THE MARSHALL COMBINED SLOW-SPEED AND REVERSING GEAR.

SOME MOTOR VEHICLE COMPONENTS

with the wheel H, pinion J, and finally the wheel C, which is fast on the sleeve of the fast pulley.

THE MARSHALL COMBINED REVERSING AND HILL-CLIMBING GEAR

The Marshall hill-climbing and reversing gear, which is similar to that used on the Benz car, is illustrated in section by Fig. 267A, in which the central pulley A carries the differential gear upon the two parts of the solid spindle B B. At G is the loose pulley for the low speed belt which runs either upon A or upon C. At D is the fast pulley for the high speed drive, the belt for which may run either upon it or upon E as a loose pulley. The pulley D is fastened to the differential gear pulley A. The pulley E runs loose upon bushes. By the side of the pulley E are brake pulleys F and G upon sleeves, which at their inner ends carry respectively the pinions H and J. Upon the boss of the pulley D is a pinion K. Carried by the sides of the hollow pulley E is a pin L, upon which run the three pinions fixed together as one H', J' and K'. By fixing the brake pulley G, and thereby the pinion J, and then driving the pulley E, the pinion J' rolls round on J, and carries forward with it the pinion K driven by K', and thereby the pulley D at a slower forward speed, because the epicycloidal path of J' in one rotation round J is greater than that of K' round K. By fixing the pulley F, and thereby the pinion H, the pinion H' rolls round H, and the pinion K' gives to K and thereby to D a slow motion backward, but at a slightly higher speed than the forward motion, resulting from the fixture of the pulley G, because the epicycloidal path of H' in one rotation round H is smaller than that of K' round K. When F is fixed, the pulley G is also driven at a slow speed backwards.

Chapter XX

ELECTRIC IGNITION AND IGNITION APPARATUS

Growing use of electrical methods

THE many advantages secured by the use of electrical methods of ignition for light petrol motors are now generally recognised.

The fact that the electric spark has not completely displaced the hot tube, and that tube ignition still has its advocates amongst motor car constructors and users, may be traced to the following causes:—

Comparison of merits of tube and of electrical

(1) The vital principle of the tube ignition, namely, the petrol required for the burners, is the same as that upon which the motor depends, and must therefore be obtainable wherever it is possible to use vehicles propelled by petrol motors.

Tube ignition

(2) The use of the tube ignition apparatus does not require any special knowledge on the part of the driver, other than that necessary for the general management of the mechanism of the car.

(3) Where the motor is required to work continuously at full power, as in races, tube ignition is the simplest, and gives practically everything that may be desired.

Other reasons which sustain the use of tube ignition, especially for racing cars, are that the weight of the complete apparatus is small, and that in the hands of skilled drivers it is one of the least likely parts of the mechanism to cause trouble or delay.

Against the more important points mentioned above as secured by the use of tube ignition, the following disadvantages may be contrasted:—

(1) In the hands of the every-day user it introduces an element of danger from fire in the case of accidents which but for this would be nothing more than temporarily unpleasant.

(2) From the point of view of the designer, the use of tube ignition constitutes the petrol motor a practically inflexible source of power, thus putting upon the speed-changing gear and friction clutch the greater part if not the whole work of obtaining variable power and speed at the driving-rod wheels.

Electrical ignition

Electrical methods of ignition, while introducing into the petrol motor car a principle involving knowledge apart from that necessary for

354

the management of other parts of the mechanism, will, when their working is understood, be found to secure many advantages over tube ignition without its disadvantages. The following important features attending the use of electric ignition may be mentioned :—

(1) The danger of fire in case of accidents is entirely removed.

(2) The instant at which the charge within the motor cylinder is fired may be varied by the driver from his seat, so as to secure the greatest obtainable power and efficiency from the motor under all conditions of load.

(3) The time at which the electric spark is formed within the motor cylinder is regulated by a mechanical contrivance operated by the motor, and although the quality and quantity of the charge and its pressure may be varied through a considerable range, ignition will always take place at the assigned instant, as the electric spark is capable of igniting a mixture not only weaker but of greater variation in strength than can be fired by a uniformly heated tube. It will thus be understood that, when electric ignition is employed, the power and speed of the motor may be varied through wide limits, either by varying the time at which the charge is fired, or by throttling the charge drawn in on each suction stroke of the piston, and so varying both the quantity of the charge in the cylinder and the final pressure of compression.

The flexibility thus secured will permit the use of a less complicated speed-changing gear, and reduce the wear and tear on the transmission mechanism, as the speed of the car may be varied considerably without interchanging the gear wheels or disengaging the friction clutch.

Before describing in detail the apparatus employed for the purpose of electric ignition, mention must be made of the various systems commonly met with in motor vehicles, and these may be divided into two classes, namely,— **Various systems of electric apparatus**

(1) That in which the igniting spark is formed by the discharge of a current of high tension but very small in quantity.

(2) That in which the spark is formed by breaking a circuit round which flows a current of low tension and larger in quantity.

Each class is represented by two combinations of the several parts of the apparatus, which secure the same end by slightly different means.

The high tension method of electric ignition is most frequently met with in practice in connection with motor tricycles and light vehicles, such as those of de Dion & Bouton, and in the carriages of Benz. It involves the use of a primary or secondary battery, from which an electric current may be obtained, which is powerful in quantity but low in electromotive force. An induction coil or transformer, by means of which a current small in quantity, but of high electromotive force or tension, is obtained by the influence of the primary current. One or more automatically operated switches for alternately interrupting and completing the primary circuit, and a sparking plug or plugs at which the high tension spark is formed within the combustion chamber of the motor. **High tension system**

355

MOTOR VEHICLES AND MOTORS

Low tension system The apparatus employed in the low tension systems comprises a magneto or dynamo electric machine driven by the motor, and capable of furnishing a current of moderate quantity and tension. A conductive circuit which includes a mechanically operated switch by means of which the circuit may be broken at any desired instant within the combustion chamber of the motor, and in some cases a coil having a large self induction for the purpose of intensifying the primary current at the instant of breaking the circuit. In practice the low tension system is best known in connection with the Cannstadt Daimler vehicles and the carriages of M. Mors.

THE DE DION & BOUTON HIGH-TENSION SYSTEM

The de Dion & Bouton apparatus The electric ignition apparatus employed by de Dion & Bouton on their tricycles and light cars comprises a primary battery and an induction coil. In the primary circuit there is one mechanically operated switch for alternately making and breaking the circuit; one terminal of the secondary or high-tension circuit is connected to any part of the motor vehicle, while the current is led from the other terminal, through an insulating sparking plug, into the combustion chamber of the motor.

Fig. 268 shows diagrammatically the arrangement of the various parts of the apparatus.

The battery The primary battery is composed of four elements of the type known as dry cells. Each of the cells consists of an outer box of sheet zinc. Inside the zinc box is a carbon plate around which is compressed a mixture of graphite manganese di-oxide and syrup, so as to form a symmetrical block. This block is bound round with several wrappings of cloth, and between it and the sides of the box is a packing of cork dust moistened with a solution which is one of several known to electricians, and is probably a saturated aqueous solution of ammonium chloride with a powerful oxidizing salt and glycerine. The top of the cell is finally sealed over with marine glue. One binding screw is attached to the carbon plate which projects above the sealed top of the cell, and the other to the zinc box. The complete cell weighs about $3\frac{1}{2}$ lb., and is capable of furnishing **Duration and running down** an average current of about 3 ampères at 1·25 volts for from 300 to 400 hours on the intermittent work of the electric ignition apparatus. After this period, owing to the chemical changes which take place, the internal resistance of the cell increases to such an extent that sufficient current cannot be supplied for the satisfactory working of the induction coil. A new battery must then be obtained, as, once exhausted, the cells can only be revived by renewal of most of the constituent elements.

E.M.F. of the combination It will be noted that the current furnished by one cell has only a very low electromotive force, and as 5 volts are required to send sufficient current round the primary circuit of the apparatus, it becomes necessary to use four cells joined in series, namely, the zinc terminal of the first cell to the carbon of the next, and so on. This arrangement leaves a free carbon terminal on the first cell and a free zinc terminal on the last cell, and the

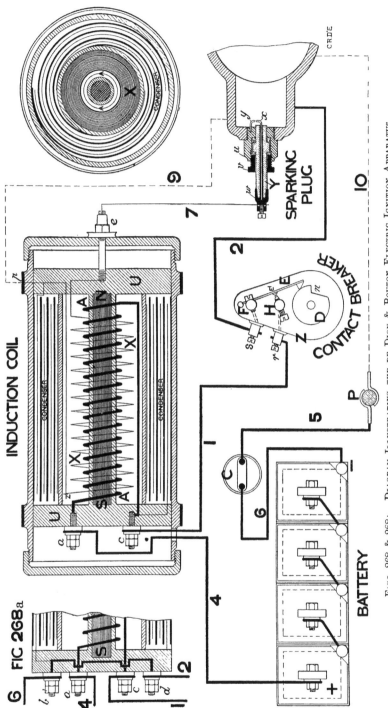

INDUCTION COIL

CONDENSER

FIC 268a

SPARKING PLUG

CONTACT BREAKER

BATTERY

Figs. 268 & 268a.—Diagram Illustrating the de Dion & Bouton Electric Ignition Apparatus.

357

MOTOR VEHICLES AND MOTORS

battery will be capable of supplying an average current of 3 ampères at a pressure of 5 volts = the sum of the four electromotive forces, each of 1·25 volts ; the current, however, will remain the same as that given by one cell, as, although flowing at four times the pressure, it is opposed by the internal resistance of the four cells added together.

Testing To avoid the inconvenience of unexpected failure of the battery, it is advisable to test its current from time to time by connecting up an ampère meter in the primary circuit, and setting the contact breaker on the motor so that the circuit is temporarily completed. Contact should, however, be only made for as short a time as possible, otherwise considerable waste of current will result. To produce an effective spark at the ignition plug, a current of 3 ampères should flow in the primary circuit. If obtainable, a sensitive voltmeter should be used in preference to the ampère meter, as during its use very little energy is taken from the battery ; the terminals of the voltmeter must be connected across the free terminals of the battery. For reliable working 5 volts should be indicated.

The induction coil The induction coil, the construction of which is shown diagrammatically in Fig. 268, is the next important part of the ignition apparatus which will be described in detail. As previously mentioned, its function is to generate, by the influence of the battery current, a current of high tension but small in quantity, which will pass at definite intervals in the form of a spark or series of sparks between the joints of the ignition plug within the combustion chamber of the motor, generating at the same time sufficient heat to ignite the compressed carburetted vapour.

Principles of construction and working For a clear understanding of the main principles which underlie the construction of the induction coil reference must first be made to the diagrams Figs. 269 and 270, of which Fig. 269 shows a simple electromagnetic apparatus consisting of a soft iron bar N, S upon which is wound a coil A A of insulated copper wire, the ends of the coil being connected, one to the positive terminal of a battery B through the switch C, and the other to the negative terminal direct. When the switch C is open, no current can flow round the coil A A, and the bar N S does not possess any magnetic properties ; but if the switch C be closed, a current will flow round the coil, and the bar becomes possessed of all the properties of a permanent magnet.

When a steady current is flowing round the coil A A, the bar and the space around it are permeated by magnetic lines of force distributed approximately as shown in Fig. 269. These lines of force constitute a stable magnetic field the directive action in which lies along the lines in the direction from N to S outside the bar. When the flow of current from the battery is stopped by opening the switch C the magnetic properties of the bar disappear.

Induction coil phenomena Phenomena of especial importance in their bearing on the induction coil are exhibited at the period of generation and destruction of the magnetic field, *i.e.* of making and breaking the electric current.

At the instant of closing the switch the flow of current does not rise

358

at once to its full value, but is delayed during a period which increases with an increase in the number of convolutions in the coil and the maximum strength of the current. This delay is due to the fact that when, as in the case now under consideration, a closed electric circuit is so disposed as to be cut by the lines of force of a magnetic field of increasing

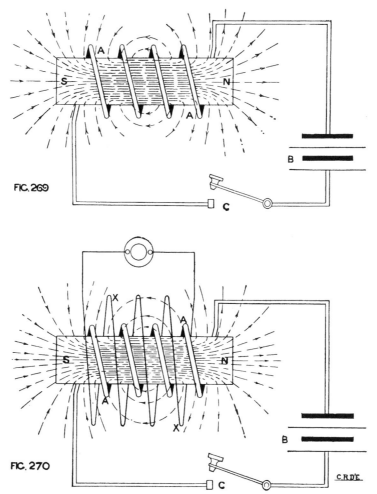

FIG. 269.—DIAGRAM OF MAGNETIC FIELD AND PRIMARY COIL.
FIG. 270.—MAGNETIC FIELD AND PRIMARY AND SECONDARY COILS.

strength, an induced current is created in it flowing in the opposite direction to the battery current and tending to oppose the changes in the electromagnetic state of the system.

The existence of this self-induced current, as it is termed, depends on **Self-induced** the unsettled state of the magnetic field, and it decreases in intensity as **currents** the battery current establishes the steady magnetic field, step by step as it were, suppressing the opposing induced current, which ceases altogether when the full flow of battery current has been established.

Corresponding to this self-induced current generated on make of circuit, a second induced current is created on break of circuit. This current follows as a result of the cutting of the convolutions of the coil A A by the lines of force of a magnetic field of decreasing intensity, and it flows in the circuit in the same direction as the battery current, and in the effort to sustain the flow it sparks across the gap between the contact points of the switch at the instant of breaking the connection. The spark formed by the passage of the self-induced current increases in intensity with an increase in the number of convolutions in the coil, the intensity of the magnetic field, and the quickness of breaking the circuit.

The magnetic field and primary and secondary coils
Referring now to the diagram Fig. 270, this shows the same apparatus as Fig. 269, with the exception that a second coil of insulated wire x x has been wound on the iron core over the coil A A, and completely insulated from this coil, the ends of the secondary coil x x may be assumed to be connected up through an instrument capable of indicating the flow of electric currents of momentary duration.

If now the battery circuit is completed by closing the switch c, at the instant the current starts flowing in the coil A A a current will be generated in the coil x x due to the fact that the planes of its convolutions are intersected by the magnetic lines of force of the field created by the flow of current in the coil A A, and as the magnetic field is increasing in intensity the flow of induced current in x x is in the opposite direction to the flow of current in A A, while its existence is only of very short duration, coinciding with the time occupied by the battery current in attaining its full value.

Corresponding to the induced current flowing in x x on make of the battery circuit, there will be a similar momentary flow of induced current on break of circuit, but owing to the fact that in this case the magnetic field is of decreasing intensity, the induced current flows in the coil x x in the same direction as the battery current in A A.

The intensity of these induced currents in an independent circuit, just as in the case of the self-induced currents, depends on certain laws which may be stated as follows :—

Intensity of induced currents
Whenever a closed conductive circuit is cut by magnetic lines of force, a current is generated in the circuit, the electromotive force of the current being proportional to—

(1) The number of convolutions in the circuit which lie in planes cut by the magnetic lines of force.

(2) The total number of magnetic lines of force which cut the convolutions at any instant.

(3) The rate of change in the total number of lines of force which cut the convolutions at any instant.

This last law is of especial importance in its bearing on the induction coil, showing that, all other conditions remaining the same, the electro-motive force of the secondary current, and therefore also the quantity of the current flowing in a circuit of given resistance, will increase propor-tionately with increased quickness in creating or annulling the magnetic field, *i.e.* in establishing or stopping the flow of battery current.

The induction coil, which is shown diagrammatically in Fig. 268, is based on the principles above stated.

The induction coil

It consists of a core N, S, made up of a bundle of soft iron wires cut to the required length and inserted into the bore of the vulcanite bobbin U. The wire of the primary coil A A is first wound upon the bobbin, the ends being connected to the terminals A, C. The primary wire is of consider-able size, usually No. 16 S.W.G., insulated by a double cotton covering, and the coil as a rule consists of only two layers.

Primary winding

When the primary winding is completed the outer layer is wrapped round with paraffined paper, so as to insulate it thoroughly, and upon this is wound the secondary coil, which consists of many layers of fine, silk-covered copper wire, usually No. 36 S.W.G., each layer being insulated from the next following by means of paraffined paper. The inner end, or origin, of the secondary coil at z' is connected to a metal band p, near one end of the outer cylindrical vulcanite case D of the coil, and from here con-nection is made to earth through one of the metal straps by means of which the coil is carried from the bridge tube of the tricycle frame. Since the motor is in metallic contact with the frame tubes it is also in electrical connection with the metal band p and with the origin of the secondary coil. In the diagram this connection is indicated by line 9. It will be noted, also, that the negative terminal of the battery is earthed to the motor as indicated by line 10. In this manner the electrical potential of the origin of the secondary circuit, and therefore that of the earthed point y of the sparking plug, are reduced to the lowest value in the system ; the negative terminal of the battery affording a constant dead " earth " at a much lower potential than that of the metal parts of the machine as a whole may sometimes be. The outer end of the secondary coil is con-nected to the terminal e, and thence by wire 7 to the insulated sparking point x of the plug.

Secondary winding

Connections

Point of zero potential

When the secondary coil is completed, it is surrounded by layers of insulating material, and upon this is wrapped the condenser, which con-sists of two long strips of tin foil separated throughout by a layer of paraffined paper, one of the tin-foil sheets has its outer extremity con-nected to the terminal c, and the other to earth through the metal band p. Outside the induction coil wire 1 connects the terminal c to the terminal r, on the vulcanite base of the mechanically-operated contact breaker, by means of which an intermittent current is caused to flow round the primary winding of the coil. The other terminal s of the contact breaker is earthed to the motor by wire 2, and consequently it is also connected to one of the tin-foil sheets of the condenser.

The condenser

Mechanical contact breaker

The contact breaker consists of a steel disc D, mounted on the extremity of the cam shaft of the motor, which revolves at half the speed of the crankshaft. Cut in the disc D is a notch n, and against the circumference of the disc bears the head of a steel spring E, the other end of which is fixed by a setscrew to the brass standard F, which is connected by a wire in the vulcanite base to the binding screw s. At the point E′, on the spring E, is rivetted a small platinum button, and opposite to this is a platinum-tipped screw H, carried by a brass standard, connected to the binding screw r. The contact screw H is so adjusted as not to be in contact with the platinum button E′ during the period when the head of the spring rests on the circumference of the disc D; but when in virtue of the rotation of the disc the head of the spring falls into the notch n, the platinum button E′ is caused momentarily to tremble on the point of the screw H.

Battery connections

Referring now to the battery connections, it will be seen that the free positive terminal of the four cells is connected by wire 4 to the binding screw a, on the coil bobbin, while the negative is connected through the handle switch C, and the plug switch P, to earth, and so to the terminal s on the contact breaker.

The arrangement of connections between battery and coil shown by Fig. 268 is that used on the most recently constructed tricycles of de Dion & Bouton. In the earlier form of the induction coil, many of which are still in use, the connections are made as shown by Fig. 268A. There are four binding screws at one end of the coil bobbin instead of two as in the recent coils. The positive terminal of the battery is connected by wire 4 to the binding screw a, which is connected to one end of the primary winding of the coil. The other end of the primary goes to binding screw c, which is connected by wire 1 to the binding screw r on the contact breaker; the other binding screw s on the contact breaker, instead of being earthed, as in the most recent arrangement, is connected by wire 2 to the binding screw d on the coil bobbin, this is cross-connected inside the coil to binding screw b, which is connected through the plug switch and handle switch to the negative terminal of the battery. The coatings of the condenser are connected to binding screws c and d, and so across r and s of the contact breaker. These induction coils are manufactured by the firm of Bassée & Michel.

Action of complete apparatus

The action of the complete apparatus, shown in Fig. 268, is as follows: Once every two revolutions of the crankshaft of the motor, at a point when the piston is nearing the end of the compression stroke, the head at the extremity of the contact spring E falls into the notch n on the disc D, at this instant, the primary circuit being completed, a current flows from the battery, starting from the positive terminal and passing along wire 4 to binding screw a, round the primary winding A A of the coil to c, and from here, by wire 1, to the terminal r on the vulcanite base of the contact breaker. From here the current flows to H, and at the instant when E′ touches H, the current passes through the spring E and pillar F, to the

362

terminal s, and thence along wire 2 and through the metal parts of the motor and tricycle frame to the plug switch P. The circuit is completed from here by wire 5, the handle switch c and wire 6 to the negative terminal of the battery.

On the completion of the primary circuit through the contact spring and screw, the flow of current on the primary coil A rapidly rises to its full value against the opposing self-induced current generated in the coil, and establishes a powerful magnetic field, the lines of force in which intersect the planes of the convolutions of the secondary circuit, creating therein, during the period occupied by the battery current in rising to its full value, a difference of electric pressure between the end z′, which is earthed, and the end connected to the terminal e.

This difference of electric pressure resulting from the increasing density of the magnetic field is, however, not great enough to cause a spark discharge between the points x, y of the plug, owing to the fact that the rate of change of the density of the magnetic field is retarded by the opposition of the self-induction of the primary to the rapid flow of the battery current. When, however, the battery circuit is broken, either as a result of the trembling of the contact spring E, or when the head of the spring is lifted out of the notch n by the rotation of the disc D, the magnetic field is almost instantly destroyed through the agency of the condenser, as will be hereafter explained. Resulting from the very rapid rate of cutting of the secondary convolutions by the magnetic lines of force set in motion at the instant of breaking the battery circuit, there will be set up between the ends of the secondary coil a difference of potential amply sufficient to spark across between the points of the plug.

Retardation of magnetic flux

If the screw H is so adjusted that the platinum button E′ on the spring E trembles on the point of the screw during the period when the head of the spring is free in the notch of the cam disc, a series of sparks will pass between the points of the plug corresponding to the fluctuations in the magnetic field, thus insuring the ignition of the charge in the motor cylinder.

The trembler

From what has been already stated, it will be understood that one of the most important factors in the production of induced currents is the time rate of change of the intensity of the magnetic field in which the secondary coil is immersed. A high time rate of cutting of the secondary convolutions on make of circuit cannot be secured owing to the retarding effect of the self-induction of the primary coil on the battery current. Correspondingly, a sufficiently high time rate would not be secured on break of circuit if the primary self-induced current, flowing then in the same direction as the battery current, were allowed to sustain the flow by sparking across the gap between H and E′ as contact is being broken. It is for the purpose of suppressing the self-induced current in the primary on break of circuit that the condenser is employed in conjunction with the induction coil. As will be seen from Fig. 268, its metallic surfaces are

Rate of magnetic intensity alternations

363

connected independently, one to the binding screw *c*, and so to binding screw *r* on the contact breaker, and the other to earth, and so to binding screw *s*

Effect of the condenser
on the contact breaker. The result of the addition of the condenser is, that at the instant when the circuit is broken by E′ parting contact with H, the self-induced current flowing from the primary coil at *c*, instead of passing in the form of a spark and re-establishing the magnetic field, is discharged into one coating of the condenser through the wire 8, the other coating being earthed. This discharge of current into the condenser produces as it were a heaping up of electric pressure at the end of wire 8, and finding no outlet, the wave of the electric pressure sweeps back through the primary coil A, instantly demagnetizing the core, since its direction of flow is the reverse of that of the original self-induced current. This oscillatory discharge, due to the condenser, continues in alternating waves of rapidly decreasing amplitude, the complete suppression of the self-induced current being effected in a very small fraction of a second.

Merits of de Dion system
The de Dion system of electric ignition has the merit that the combination of apparatus employed is one of the simplest by means of which high tension electric currents can be generated. There can, however, be no doubt that a large primary current is required in proportion to the useful work performed by the secondary current, with consequent shortening of the life of the battery. The weight of the complete apparatus is also very considerable, being about 20 lb. From the point of view of efficiency, the fault lies with the mechanically operated contact breaker, which, although probably one of the best of its kind, does not do that which is essential for the efficient working of an induction coil, namely, produce a very rapid succession of complete breaks and makes of primary circuit. True that the contact breaker is adjustable so that the platinum button E′ does not tremble on the point of the screw H, but since the actual period during which the head of the contact spring is free to vibrate in the notch *n* of the cam disc, when the motor is running at normal speed of 1,200 revolutions per minute, is only about 0·01 second, it is doubtful whether, even with the very stiff spring employed, the period of vibration is such as admits of even one additional break and make of the circuit during the period when the spring head is free to vibrate. Even if vibrations do take place, their amplitude must be so exceedingly small that the circuit is not completely broken at all, as the flow would be sustained by a minute spark across the gap. The greatest effect would therefore be to cause a fluctuation of the battery current and of the magnetic field, which would never rise to the maximum nor fall to the minimum possible.

THE BENZ ELECTRIC-IGNITION APPARATUS

The Benz system and apparatus
The failings of the mechanically operated contact breaker are to some extent remedied by the use of two automatically operated switches, as in the Benz system, the arrangement of which is shown diagrammatically in Fig. 271. The apparatus employed is in general form similar to that of

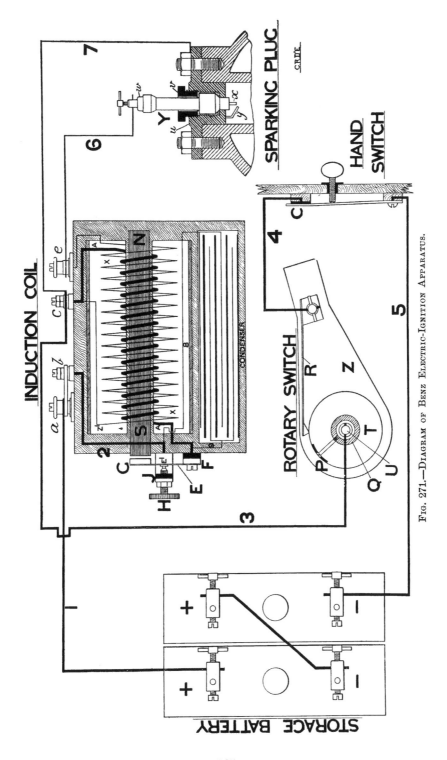

INDUCTION COIL

SPARKING PLUG

HAND SWITCH

ROTARY SWITCH

STORAGE BATTERY

CONDENSER

Fig. 271.—Diagram of Benz Electric-Ignition Apparatus.

365

de Dion & Bouton. The primary current is supplied by a set of two storage cells connected in series, these being capable of delivering the necessary current of from 1·5 to 2 ampères, at a pressure of 4 volts, for about 20 to 25 working hours, and if properly constructed will retain their charge for a month or more when intermittently employed. When one charge is exhausted, the cells may be re-charged from any source capable of supplying a continuous current at from 5 to 6 volts pressure.

Magnetic contact breaker The induction coil is identical in principle and of similar construction to the de Dion coil, but it includes a magnetically operated contact breaker in the primary circuit. This contact breaker consists of a steel or brass spring E, carried by a brass base F, at one end of the coil bobbin. On the end of the spring E, opposite the iron core of the coil, is a soft iron head G, and on the spring at E′ is rivetted a platinum button. Opposite E′, and in contact with it, when the current is not allowed to flow round the primary circuit, is a platinum-tipped screw H, carried by a bridge piece J.

Rotating switch and connections In addition to the magnetically operated contact breaker, there is also in the primary circuit a mechanically operated switch for completing the primary circuit during a definite period each time the piston of the motor approaches the end of the compression stroke. This switch consists of a vulcanite disc T, mounted at one end of the sleeve Q, which carries the exhaust valve cam, and is driven at half the speed of the crankshaft of the motor. On the periphery of the disc T is a small brass plate P, which is connected to the sleeve by a countersunk screw, sleeve Q being in electrical connection with the negative terminal of the battery through the spindle U and wire 3. Upon the periphery of the disc T rests the button at the end of a spring R, which is carried from a vulcanite arm Z mounted on the stud shaft U, so that the position of the head of the spring R may be adjusted relatively to P.

The spring R is connected by wire 4 to the hand switch C, and from here connection is made by wire 5 to the negative terminal of the storage battery.

Condenser The condenser with this form of induction coil consists of alternate layers of tin foil and paraffined paper packed close together in the lower part of the box containing the coil. The ends of alternate sheets of the tin foil are connected, and one set is joined by wire 8 to the bridge piece J of the contact breaker, and the other set by wire 9 to the base F carrying the contact spring.

Action of the apparatus The action of the apparatus is as follows : Once every two revolutions of the motor, the metal plate P of the rotary contact switch comes into contact with the head of spring R. When switch C is closed, this causes the primary circuit to be completed, and allows the current to flow from the battery along wire 1 to the terminal *a* on the coil box, and across by a fixed metal plate to terminal *b*. From here the current passes along wire 2 to the platinum-tipped screw H, which is in contact with the button E′ on the spring E, so that the current is free to flow round the

366

primary winding A of the coil to terminal c, and by wire 3 to the rotary switch. The circuit is completed from here by wire 4 to the hand switch, and thence by wire 5 to the negative terminal of the battery.

The flow of current will, however, only be instantaneous, because the moment the core N, s of the coil becomes magnetized it attracts the head c on the spring E, and, bending the spring, causes the platinum button at E′ to part contact with the tip of the screw H, so breaking the primary circuit, and stopping the flow of current the instant after it has been established; the action of the condenser being, as in the case of the coil used in the de Dion apparatus, to suppress the spark discharge of the primary self-induced current, which otherwise would take place on break of circuit, and to increase the rate of demagnetization of the core of the coil.

With the cessation of the current in the primary, the spring E, being released, flies back, and again completes the circuit by causing E′ to strike H, so allowing a second rush of current to take place. This re-magnetizes the core, and the head G being again attracted a second break of circuit follows, and so on, until the rotation of the disc T causes the plate P to part contact with the head of spring R, so completely breaking the primary circuit until the time arrives for a fresh charge to be ignited.

One of the chief advantages gained by using the magnetically operated contact breaker lies in the fact that the makes and breaks of the primary circuit are more perfect, owing to the greater amplitude of the vibrations of the contact spring, continually reinforced by the impulses given to the head by the magnetic attractions. In practice it is necessary, in order to secure the maximum obtainable number of breaks and makes of the circuit during the period when P and R are in contact, to adjust the screw H, so that the platinum tip is caused to press lightly on the button E′ against the resistance of the spring. This effectively makes the circuit, and allows the magnetic field to be fully established. Then, under the powerful attraction of the core on the head G, the spring is further deflected, causing an effectual break of circuit, while at the same instant the condenser steps in and destroys the magnetic field. The stress on the spring being relieved, it flies back with considerable force, and dwells for an instant on the tip of the contact screw, so allowing the battery current to re-establish the magnetic field, while the rebound of the spring greatly quickens the following break of circuit. In this way the vibrations are rapidly set up, and judging by the pitch of the note sounded by the spring when the coil is at work, there are about four complete breaks and makes of the circuit between the make and break by the plate P and spring R of the rotary contact switch. As a result of these rapid fluctuations of the magnetic field a practically continuous stream of sparks, of the hot and flaming type, pass between the points of the plug during the period when contact is made by the rotating switch, and the effectual ignition of the charge is thus ensured.

Magnetically operated contact breakers

MOTOR VEHICLES AND MOTORS

Ignition Plugs

Sparking or ignition plugs

The various forms of ignition plugs employed for high tension systems, of which the best-known forms are shown by Figs. 268 and 271, and Figs. 272, 273, 274 and 275, are all essentially similar, consisting of a metal base u, which is screwed into the wall of the combustion chamber, and which carries by means of a screw gland v and asbestos packing an insulating stick of porcelain y, through which passes the wire of the high potential sparking point x within the combustion chamber. At the outer end of the porcelain stem this wire is attached to a fixed binding screw w, which

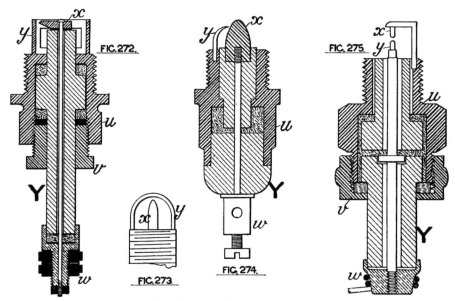

Fig. 272.—Bassée & Michel Sparking Plug.
Fig. 273.—Major Holden's Sparking Plug.
Fig. 274.—Reclus Plug.   Fig. 275.—Bisson & Berger Plug.

is connected by a carefully insulated wire, Figs. 268 and 271, to the terminal e of the secondary circuit. Opposite the sparking point x is a similar point y, brazed into the metal base of the plug, and so electrically con-

The de Dion plug

nected to the earthed end of the secondary. In the de Dion plug, Fig. 268, the sparking points consist of similar wires approached within about $\frac{1}{32}$ in.,

Benz plug

while in the Benz plug, Fig. 271, the wires are of rather larger gauge set $\frac{1}{16}$ in. apart. The plug shown by Fig. 272 is an improved form made

Bassée plug

by Bassée & Michel. In this the high potential wire carries a finely pointed piece x, closely approaching a ring y, connected to earth. This is a very correct arrangement of the sparking points. The discharge takes place more readily when a fine point at high potential approaches a plane surface connected to earth; and when this surface is in the form of a

368

cylinder surrounding the high potential point the spark spreads in a flame-like form on the surface of the cylinder opposite the point. The arrangement of the sparking points shown by Fig. 273 is that employed by Major Holden on his motor bicycle; as in the previously mentioned plug, the spark spreads in flame-like form along the arched wire y connected to earth. Fig. 274 shows the "Reclus" unbreakable plug. It is very strongly constructed but the arrangement of a mass of metal at the end x of the high potential wire does not appear to be a good one, and the length of insulating surface between u and w is not sufficient to prevent leakage of the high potential secondary current if the atmosphere is at all damp. The plug shown in Fig. 275, made by Bisson, Berges & Co., has the merit that the porcelain stick being in two parts is much less liable to fracture, due to the difference of temperature of the ends.

Owing to the high pressure of the charge in the motor cylinder at the time when the igniting spark passes, the difference of potential between the sparking points, even when these are only $\frac{1}{32}$ in. apart, probably rises to nearly 10,000 volts; it will therefore be obvious that in damp weather considerable leakage will take place all along the wire connecting the terminal e of the secondary with the sparking plug, particularly as a flexible insulated wire is commonly employed, and allowed to touch metal parts connected to earth. When the insulation becomes covered with a film of moisture, leakage from the uninsulated extremities along the damp exterior to points in contact with metal parts may take place so freely as to completely prevent the passage of the igniting spark at the plug. It would therefore appear very advisable to substitute for the flexible wire a thin stiff insulated rod, bent as required, so as not to come in contact with metal parts at any point.

Considerable loss of pressure will also result in damp weather if the porcelain insulator in the sparking plug is allowed to become dirty, while a fracture of the porcelain within the metal base generally causes failure of the ignition, as, owing to the greater resistance to the passage of a given length of spark through the compressed carburetted vapour, the discharge will more readily take place through the fracture between the wire which passes to the high potential sparking point and the metal base of the plug.

The form of ignition plug employed by Messrs. Accles & Turrell for their light petrol motors is worthy of notice. In this plug both wires which form the sparking points are led into the combustion chamber through separate insulated porcelain stems carried by a common metal base. Insulated wires connect the terminals of the secondary of the induction coil to the terminals of the sparking wires at the extremities of the porcelain insulators. By this arrangement the secondary circuit is everywhere insulated from earth, with resulting diminution of trouble and loss owing to leakage.

The combinations of apparatus which have now been described may

Holden plug

Reclus plug

Bisson plug

E.M.F. of spark

Leakage

Insulation

Accles-Turrell plugs

Separate complete secondary circuit

be said to cover in principle the whole field of high tension electric ignition devices, as any others which may be met with in practice are only modifications in detail of the systems employed by de Dion & Bouton and of Benz.

The general working of high tension apparatus, when proper precautions as to insulation are observed, may be said to give every satisfaction in practice; but the supply of current to the induction coil by means of primary or secondary batteries is often a troublesome and expensive item, while the sparking plug is a delicate piece of apparatus, easily broken, and, if of defective construction, it may cause much loss of current, and irregular firing of charge in the motor cylinder, with accompanying loss of power and very objectionable smell of the exhaust due to the escape into the atmosphere of unburned hydrocarbon vapour.

Low-Tension Ignition Systems.

By the use of the low-tension electric currents for purposes of ignition practically all the difficulties of the high tension systems are avoided, and there appears to be little doubt that ignition by low tension electric currents will eventually supersede all other methods.

Principle of low tension apparatus As previously mentioned, all successful low-tension ignition systems depend for their action on the spark produced on breaking a circuit round which flows a current of moderate quantity and pressure, which is usually generated by some form of dynamo-electric machine driven from the motor, one spark being relied upon to effect the ignition of each charge.

The Bosch system The low-tension system, which combines the patents of R. Bosch and F. R. Simms, is one which is finding favour amongst motor car constructors, owing to its simplicity and reliable working. The apparatus employed consists of a magneto-electric generator of the inductor type, patented by Bosch, No. 15411 of 1897, and a simple low tension circuit, in which is a special form of switch operated mechanically outside the motor cylinder and breaking circuit within the combustion chamber. Mr. Simms' patents, Nos. 7195, 7196, 12433 and 24859 of 1898; and Bosch, No. 26907 of 1898. The arrangement of the apparatus is shown diagrammatically by Fig. 276, and Figs. 277–280 show the principle on which the action of the magneto-electric generator depends.

The magneto generator The magneto generator consists of a powerful compound permanent magnet A A of horse-shoe form, fixed to a bed-plate B of cast bronze, the pole-pieces of the magnet being so formed as to partially surround the armature c, which is fixed centrally in the gap between them. The armature consists of a soft iron core c of the Siemens' H type, in which are two longitudinal channels D D, filled by a coil of insulated copper wire, the outer end of which is connected to the insulated terminal a, while the inner end is connected to the metal core of the armature, and is so electrically connected to the motor through other metal parts of the motor vehicle. In the gap between the fixed armature and the pole-pieces is a

370

magnetic shield, consisting of two longitudinal soft iron segments E E, mounted at either end on discs F of bronze, which carry short stud spindles working in bearings in bronze end-pieces G fixed across the field magnet limbs. The shield is thus centrally supported, so as to be just clear of both armature and field-poles, and oscillatory motion is imparted

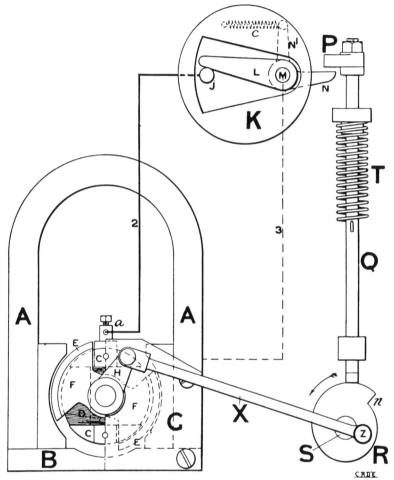

FIG. 276.—DIAGRAM OF LOW-TENSION MAGNETO INDUCTOR GENERATOR AND
IGNITION APPARATUS, BOSCH & SIMMS' PATENTS.

to it by means of a lever arm H and connecting rod X to the excentric pin Z on the cam disc R, on the second motion shaft of the motor.

The principle on which the action of the Bosch magneto generator depends will be readily understood by reference to Figs. 277–280, of which Fig. 277 shows the manner in which the magnetic lines of force are disposed between the pole-pieces of the field magnet when both armature and shield are removed.

Principle of the magneto generator

371

Fig. 278 shows the result of interposing the armature between the field-poles.

Fig. 279 shows the manner in which the magnetic lines of force are disturbed when the shield is interposed in the angular position occupied when the oscillating crank is at the middle of its stroke, in which position it will be seen that a considerable proportion of the magnetic lines leak round from pole to pole through the shield and armature faces without cutting the convolutions of the coil D D.

Finally Fig. 280 shows the change which takes place in the disposition of the lines of force when the oscillating crank operating the shield is at one end of its throw, in which case the density of the lines of force cutting the armature coil is much greater, owing to the resistance to magnetic leakage of the air gap, which is now interposed at x x, in the path of lines which endeavour to leak across without cutting the armature coil.

It will thus be obvious that when the magnetic shield is caused to rotate from the position shown in Fig. 279 to that of Fig. 280, or to a corresponding position on the other side of the horizontal axis, the magnetic lines of force will alternately be allowed to rush through the armature, cutting the convolutions of the coil D D, and the next moment to leak across from pole to pole through the shield and armature faces without passing through the coil. Just as in the case of the induction coil, the magnetic lines of force of the fluctuating field, intersecting the convolutions of the coil D D, generate in it a current, the electromotive force of which is proportional to the number of convolutions in the coil, the difference in the intensities of the magnetic fields when the shield is in the positions shown in Figs. 299, 280, and the rate of oscillation of the shield, *i.e.* the rate at which the lines of force cut the convolutions of the coil.

Application of current and apparatus employed

The current generated in the armature coil of the magneto machines is led by wire 2, Fig 276, to the contact-breaking switch. This consists of a metal base K bolted to the wall of the combustion chamber of the motor, and carrying a fixed insulated spindle J connected at its outer end by wire 2 to the insulated terminal a on the magneto generator. At its inner end, rod J projects into the combustion chamber, and on it rests the arm L, fixed at the inner end of the spindle M, which is in metallic contact with the base K of the switch, and so in electrical connection with the inner end of the armature coil, as indicated by the line 3, Fig. 276. At the outer end of the spindle M is fixed a bell-crank lever N N′.

Sparking make and break apparatus

By means of a spring C and the arm N, the lever arm L inside the combustion chamber is normally held against the rod J, and the electric circuit so completed. The other arm N of the bell crank is under control of the hammer-head P, on the vertical rod Q, which is normally lifted by means of a cam disc R on the second motion shaft s of the motor, so that the hammer-head P is not in contact with N. Once every two revolutions of the crankshaft the lower end of the rod Q is forced into the notch n on the cam disc R by means of the spring T, and the hammer head P, striking the arm N, imparts angular

372

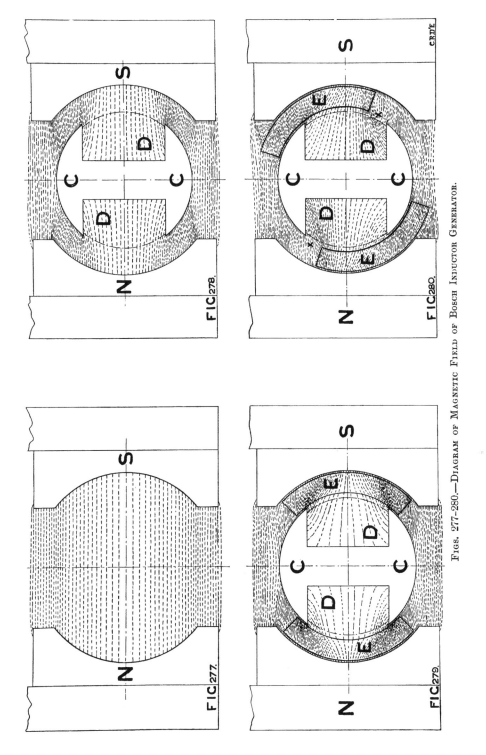

FIG. 277.

FIG. 278.

FIG. 279.

FIG. 280.

FIGS. 277-280.—DIAGRAM OF MAGNETIC FIELD OF BOSCH INDUCTOR GENERATOR.

373

motion to the arm L, causing a sudden break of the electric circuit. This causes a powerful spark to pass between the fixed rod and retreating arm, due to the effort of the self-induced current in the armature coil, superimposed on the current generated in the coil by the oscillation of the shield contemporary with the break of circuit, to sustain the flow of current. To **Varying the period of ignition** vary the period of igniting the charge in the motor cylinder, a mechanical device (Simms' Patent, 1898) is employed, by means of which the position of the notch on the cam disc R may be varied relatively to the position of the piston in the cylinder.

This ignition apparatus works well in practice. The magneto generator, though rather heavy, is reliable in working, and gives a good flaming spark at the contact breaker, on break of circuit, amply sufficient for the ignition of the charge. One magneto generator may be used for any number of cylinders, each having a contact breaker. No more difficulty is experienced in starting a motor with this ignition gear than with a hot tube, and it possesses the great advantage that the motor may instantly be started at any time.

The Mors Electric-Ignition Apparatus.

The Mors dynamo electric apparatus The low-tension system which is employed by Mors on his vehicles (Patent No. 10141 of 1896), is one which has proved reliable in practice. It is identical in principle with the system above described, but instead of using a magneto inductor generator, a small shunt-wound dynamo, giving a continuous current of low tension, is employed in connection with an electromagnetic coil having large self-induction and one or more mechanically operated switches for breaking circuit inside the cylinder or cylinders depending on the type of motor employed. Fig. 281, shows diagrammatically the arrangement of the apparatus employed by Mors for a two-cylinder motor.

The dynamo and connection The dynamo is driven from the crankshaft of the motor, and is capable of supplying a current of several ampères at about 15 volts. The current generated in the armature divides at the brushes $x\,x'$, part flowing round the shunt-winding to magnetize the field, while the main flow passes from the positive brush x, by wire 1, to contact 6 on the hand switch at the top of the steering pillar, and when the pointer on the moving arm of the switch is set to the position marked "Dynamo," the current passes to 8, and thence by wire 11 to the coil.

This is a simple electromagnet, consisting of a soft iron core, upon which is wound several layers of insulated copper wire A, the current after passing round the coil flows by wire 2 to the insulated rod J of the contact switch inside the combustion chamber of the cylinder c. The rod J is connected by wire 3 to the corresponding rod J′ of the switch in the other cylinder c′. The spindles L L′ of the switches are earthed as indicated by line 4, and so connected to the negative brush of the dynamo through con-

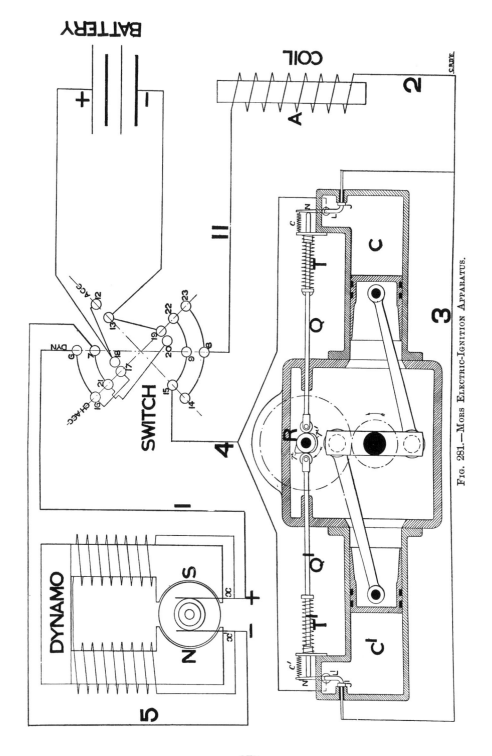

Fig. 281.—Mors Electric-Ignition Apparatus.

375

The sparking make and break

tacts 9 and 7 of the switch and wire 5. The contact spark-making switches are similar to that employed in connection with the previously described system. The spindles L L′ are bent at right angles at their inner ends, arms being thus formed, which are held in contact with the extremities of rods J J′, by springs c c′, which control the spindles through the external arms N N′. The sparking switches are operated by spring-controlled rods Q Q′, the inner ends of which carry rollers bearing on the surface of a double cam disc R, on the second motion shaft of the motor.

Action of the combined apparatus

The action of the apparatus is as follows. In the position shown in Fig. 281 the electric current is completed through both the sparking switches, but as the disc R rotates, the cam r′, bearing against the roller at the end of rod Q′, causes the other end to strike the arm N′, thus causing L′ to part contact with J′. This, however, does not cause a spark to pass, as the circuit is still completed through the switch of the cylinder c. Immediately following the break of circuit at J′, L′, the other cam r, on the disc R, comes into contact with the roller at the end of rod Q, and the switch of cylinder c is operated, causing L to be disengaged from J, and thus completely breaking the electric circuit. When this occurs the self-induced current in the coil A sparks across the gap formed, as L parts contact with J and ignites the charge within the cylinder c. The continued rotation of the cam disc R causes the same cycle of operations to be repeated, but in the reverse order, so that the spark is next formed at J′, L′, and so on.

The storage battery

In the Mors ignition apparatus a storage battery is employed for effecting the ignition when the motor is being started, as the rate at which the crankshaft can be turned by hand does not give a sufficient speed to the dynamo to enable it to excite its own field magnets. Under these circumstances the moving contact of the hand switch is moved so that the pointer is on the line "Acc." Then contact 12, which is in connection with the positive terminal of the storage battery, is connected to 14, and so to 8, and by wire 11 to the coil, while the negative terminal is connected to 13, and so to 15, and to the earthed contacts of the switches at the motor cylinders. The cells are switched out of circuit as soon as the motor starts, and may be charged *en route* by the dynamo, which is capable of supplying more current than is necessary for purposes of ignition. To effect this the hand switch is operated so that the contact arm points in the direction marked "Ch. Acc." Then part of the dynamo current goes by way of contacts 16, 23 to the coil, and by 21, 22 to earth, while part flows from 17 (which is permanently connected to 6 and to the positive dynamo brush) to 18, and to the positive terminal of the battery, and leaves the battery at the negative terminal flowing to 13, and thence to 19, across to 20, and thence to 21 and to 7, and by wire 5 to the negative dynamo brush.

The low-tension systems employed by Lanchester and by Duryea are closely similar to that of M. Mors above described. Lanchester's system (Patent No. 18829 of 1896), is noteworthy by reason of the ingenious design of the contact-breaking switch, which is arranged so that the circuit is normally broken outside the cylinder as well as inside, the first operation being to close the circuit outside the cylinder and then inside, after which the circuit is broken outside the cylinder, and at the same instant the spring-controlled spindle of the switch is released, which causes a very quick break of circuit inside the combustion chamber with an accompanying self-induction spark which ignites the charge. The advantage secured by this arrangement lies in the fact that the current is only supplied to the circuit just as the piston approaches the end of the compression stroke, so that only very little power is required to drive the dynamo or magneto machine which is normally running unloaded. The Lanchester and the Duryea systems
Lanchester's

Advantages of the system

The system of Duryea (Patent No. 20341 of 1898), describes special means of operating the contact switch so that the circuit is normally broken inside the cylinder, and is only completed by the moving arm of the switch coming into contact with the fixed pole inside the cylinder, a fraction of the revolution before the contact is again broken and the igniting spark formed. Duryea

All the systems described above, except that of Mors, employ mechanical means whereby the time at which the igniting spark is formed can be varied relatively to the position of the piston. This enables the best point of firing the charge to be determined for all conditions of working.

Low-tension systems employing a magneto generator in connection with one or more contact breaking switches undoubtedly offer at present the best solution of the ignition problem. Given intelligent construction of the comparatively few and simple parts involved, the apparatus may be expected to have as long a life as any other working parts of the mechanism of a motor vehicle, and it possesses the great advantage that the electric supply is always at hand, and cannot fail unexpectedly, as may happen when primary or secondary cells are employed. A certain amount of difficulty is experienced in designing small electric generators of light weight capable of supplying sufficient current for ignition at the comparatively low speed, at which the motor shaft can be rotated when being started by hand, and for this reason a storage battery is sometimes employed to supply current until the motor attains its normal speed. Conclusions

The use of permanent field magnets obviates, to a great extent, the starting difficulty, but there can be no doubt that from this cause the various types of generator at present employed are all heavier and absorb more power than is necessary for the production of the requisite igniting spark at normal speed of the motor. Generators should be of less weight and power

In most of the systems employed in practice the importance of obtaining a very quick break of circuit by means of the contact switch has been recognised. The intensity of the igniting spark is mainly dependent on Importance of quick break the effort of the self-induced current generated on break of circuit to maintain the flow between the retreating contact points, and it will be remembered that, other conditions remaining the same, the electromotive force of the induced current, and therefore also its total quantity, depends on the destruction of the magnetic field generated by the primary flow of current in the circuit, that is to say, on the quickness with which the flow is stopped by a complete break of circuit at the contact switch. It is not improbable that a rotating make and break revolving in mercury may be hereafter employed for this service.

Chapter XXI

PETROL MOTOR VEHICLE PERFORMANCE AND MECHANICAL EFFICIENCY

THERE is not very much upon which can be founded any conclusions as to the power and efficiency of the transmission gear of even the most-used vehicles.

That the combined effciency of the various cars differs through a very wide range is obvious from an examination of the results of their performances, not only during the 50-mile and hill-climbing trials of the Automobile Club before the Richmond Show, but during the remarkable performances in some of the numerous French races.

There were no competitions and scarcely any recorded performances of modern motor vehicles until 1894. At this time the successful, as it was then considered, running of the Panhard & Levassor, the Serpollet, the Peugeot-Frères, Roger, and other cars, attracted a great deal of attention in France, and particularly in Paris. This led the proprietors of the *Petit*

The "Petit Journal" 1894 competition

TABLE XIII.

"Petit Journal" Competition, July, 1894.

Final Trial, Paris to Rouen, 80 miles.

Constructor of Vehicle.	Persons carried.	Horse-power stated by maker.	Type of Motor.	Type of Boiler.	Prize.
Panhard & Levassor . .	2	3½	Daimler	—	⎫
,, ,, . .	4	,,	,,	—	⎬ 1st
,, ,, . .	4	,,	,,	—·	
,, ,, . .	4	,,	,,	—	⎭
Peugeot	2	,,	,,	—	⎫
,, ...	3	,,	,·	—	⎬ 1st
,, ...	4	,,	·,	—	
,, ...	4	,,	,,	—	⎭
,, ...	5	,,	·,	—	
De Dion & Bouton . . .	6	—	De Dion	De Dion	2nd
Le Blant	9	—	Le Brun	Serpollet	3rd
Vacheron	2	—	Vacheron	—	4th
Le Brun	4	—	Le Blant	—	4th
Roger	4	—	Benz	—	5th
Scotte	—	—	Scotte	Scotte	6th

379

TABLE XIV.
FRENCH MOTOR-CARRIAGE RACES.

Date.	Race.	Constructor of Vehicle.	Persons carried.	Horse power stated by maker.	Distance in miles.	Time. (hrs. mins.)	Average speed in miles per hour.	Type of motor.
June 11, 1895	Paris to Bordeaux and back	Panhard & Levassor	2	3½	744	48 47	15·25	Daimler Petrol
		"	4	3½		72 14	10·3	"
		" Peugeot	5	3½		78 7	9·52	"
		"	2	3½		54 35	13·61	"
		"	4	3½		59 49	12·45	"
		Roger	4	—		59 48	12·45	Benz
		"	4	—		64 30	11·52	"
		Bollée	4	—		82 48	9·0	"
		"	6	—		90 3	8·23	Bollée Steam
September 24, 1896	Paris to Marseilles and back	Panhard & Levassor	4	4	1,077	67 42	15·9	Daimler Petrol
		"	4	4		68 11	15·8	"
		" Peugeot	2	4		71 23	15·1	Peugeot
		"	—	6		75 26	14·3	"
		Delahaye	2	6		81 23	13·3	Delahaye
		"	4	—		75 29	14·3	"
		De Dion Tricycle	4	—		84 27	12·8	Dion
		"	1	¾		73 30	14·7	"
		"	1	¾		83 6	13·0	"
July 24, 1897	Paris to Dieppe	Panhard & Levassor	2	6	106	4 36	23·1	Daimler Petrol
		"	2	6		4 38	22·9	"
		"	2	6		5 19	19·9	"
		" Peugeot	2	6		5 28	19·5	Peugeot
		"	6	6		6 27	16·5	"
		Delahaye	6	6		5 41	18·7	Delahaye
		"	6	6		5 58	17·8	"
		L. Bollée	1	3		4 14	25·1	L. Bollée
		"	1	3		4 42	22·6	"
		"	1	3		5 44	18·5	"
		A. Bollée	2	—		5 17	20·1	A. Bollée
		Mors	2	5		5 46	18·4	Mors
		De Dion Tricycle	1	1½		4 45	22·2	De Dion
		"	1	1½		4 45	22·2	"
		"	1	1½		5 5	20·4	"
		De Dion-Bouton	4	6		4 19	24·6	De Dion, Bouton Steam

July 7, 1898	Paris to Amsterdam and back	Panhard & Levassor	2	8	895	33	4	27·0	Daimler Petrol
		" "	2	8		33	25	26·3	" "
		" "	2	8		34	58	25·7	" "
		Peugeot	2	8		36	20	24·6	Peugeot "
		"	2	8		38	26	23·2	" "
		"	2	6		39	30	22·6	" "
		A. Bollée	2	8		34	8	26·2	A. Bollée "
		"	2	8		35	19	25·3	" "
		Mors	2	6		38	41	23·1	Mors "
		"	2	6		43	58	20·4	" "
		G. Richard	2	4		57	27	15·6	G. Richard "
		L. Bollée	1	4		54	3	16·6	L. Bollée "
		Decauville	1	2½		50	14	17·8	Decauville "
		De Dion Tricycle	1	1¾		41	20	21·7	De Dion "
		" "	1	1¾		54	19	16·5	" "
		" "	1	1¾		58	51	15·2	" "
		"Phébus"	1	1¾		39	36	22·6	" "
May 24, 1899	Paris to Bordeaux	Panhard & Levassor	2	12	351	11	43	29·9	Daimler Petrol
		" "	2	12		11	51	29·6	" "
		" "	2	12		12	32	28·0	" "
		Peugeot	2	10		15	23	22·8	Peugeot "
		A. Bollée		9½		15	52	22·1	A. Bollée "
		Mors	2	15		13	18	26·4	Mors "
		"	2	13		15	20	22·9	" "
		"	2	10		15	51	22·2	" "
		De Dion Tricycle	1	2¼		13	23	26·2	De Dion "
		" "	1	2¼		13	25	26·1	" "
		" "	1	2¼		13	45	25·5	" "
		"Phébus"	1	1¾		15	6	23·2	Aster "
July 16, 1899	Tour de France	Panhard & Levassor	2	16	1,440	44	59	31 9	Daimler Petrol
		" "	2	12		49	37	29·1	" "
		" "	2	12		50	12	28·75	" "
		A. Bollée	2	10		53	30	27·0	A. Bollée "
		Mors	2	12		55	28	26·0	Mors "
		Decauville	1	2½		66	39	21·3	Decauville "
		"	1	2½		76	33	18·85	" "
		De Dion Tricycle	1	2¼		50	58	28·3	De Dion "
		" "	1	2¼		51	29	28·0	" "
		" "	1	2¼		53	37	26·9	" "
		"Phébus"	1	2¼		56	30	25·5	Aster "

MOTOR VEHICLES AND MOTORS

Journal to organise a series of trials on several routes round Paris, and to offer substantial prizes for the vehicles which proved themselves the best under the conditions named. The final trial of this series took place in July, 1894, and was a run from Paris to Rouen, a distance of 80 miles.

In Table XIII. are the leading results of this trial or race, so far as they were chronicled, but only the names of those who went through and completed the runs, are included. This may seem a little unfair to some of the other makers, who went partly through the trials; but the object of these trials was to weed out those cars which could be depended upon from those which could not, and inasmuch as it would be difficult now to discriminate between the most trustworthy car and the moderately good car most skilfully driven, and further, as the records are incomplete, it is best, as well as necessary, to accept the adage that, "nothing succeeds like success," and to give only the names of the successful cars. This course may moreover be taken with the less compunction because the cars which were best in 1894, were constructed by those who are still amongst the makers of the best cars. The first prize was divided between MM. Panhard & Levassor and MM. Peugeot; the 2nd prize went to the de Dion & Bouton steam brake; the 3rd and 6th also went to steam vehicles. It will be noted that the most successful vehicles were only of 3½ HP. The ordinary vehicles of same size and not intended for racing, have now about double that power. Altogether forty-six vehicles were entered for the 1894 competitions, of which twelve were steam.

French competitions, 1895-1899
From 1894, trial races have taken place every year in France and the powers and speeds have grown every year. The spirit motor has been made more powerful, more efficient every year, and the steam carriage has retired.

To show the advance year by year, Tables XIV. and XIVA. have been compiled giving makers' names throughout in the same order, and presenting the most salient facts concerning the cars and their performance in the earlier races, so far as they were recorded, and the results of Nice races in March, 1900.

Growth of power employed
It will be seen that the power employed for cars carrying two or four persons, has grown from 3½ HP. in 1894-95, to 16 HP. in 1899; and already still larger powers are being built, as will be seen from the illustrations and particulars given elsewhere.

Great growth in speed
It will also be seen that the maximum average speed has grown from 15·25 miles per hour, over the run of 744 miles in 1895, to a mean speed, over a distance of 1,440 miles, of no less than 31·9 miles per hour in 1899. These speeds have been much exceeded on short distances, but generally speaking it is the long distance that tells against a racing car when such high averages are maintained.

Recently (February, 1900) in the Pau race, it may be mentioned, Chev. René de Knyff has made a run of 208 miles without a stop, at a speed which, if the time records are correct, gives a mean of 43·5 miles per hour,

382

TABLE XIVa.

Results of the Nice Motor Cycle and Motor Carriage Races, March, 1900.

Date.	Race.	Driver of Vehicle.	Vehicle.	Persons carried.	Declared horse-power.	Distance in miles.	Time.	Average speed in miles per hour.	Type of motor.
							hrs. mins.	miles.	
March 27, 1900	Nice to Marseilles	Réné de Knyft	Panhard & Levassor	2	16		3 25	36·6	Daimler
		Hourgiéres	,, ,,	2	12		3 32	35·8	,,
		Charron	,, ,,	2	16		3 33	35·75	,,
		Pinson	,, ,,	2	12		3 45	33·4	,,
		Girardot	,, ,,	2	16		3 53	32·2	,,
		Levegh	Mors	2	12		3 50	32·5	Mors
		Kœchlin	Peugeot	2	18		3 59	31·4	Peugeot
		Labouré	La Parisienne	–	–		4 51	25·8	—
		Mercedes	Cannstadt Daimler	2	24	125	4 53	25·6	Daimler
		Decauville	Decauville	2	–		6 40	18·6	Decauville
		,,	,,	2	–		7 14	17·3	,,
		Béconnais	Perfecta Tricycle	1	4·5		3 23	37·0	Soncin
		Teste	De Dion Tricycle	1	3		3 44	33·5	De Dion
		Marcellin	Perfecta Quadricycle	2	6		3 46	33·4	—
		Joyeux	Perfecta Tricycle	1	4		3 51	32·4	—
		Bardin	Phébus Tricycle	1	4·5		3 52	32·4	Aster
		Bonnard	Werner Bicycle	1	1·25		5 47	21·6	Werner
							mins. secs.		
March 30, 1900	Nice La Turbie Hill-climbing Competition	Levegh	Mors	2	12		19 2	33·1	Mors
		Charron	Panhard & Levassor	2	16		19 15	32·7	Daimler
		Girardot	,, ,,	2	16		19 59	31·5	,,
		Réné de Knyff	,, ,,	2	16		20 21	31·1	,,
		Hourgiéres	,, ,,	2	16		21 12	29·7	,,
		Pinson	,, ,,	2	12	10·5	23 55	26·3	,,
		Kœchlin	Peugeot	2	18		37 45	16·7	Peugeot
		Labouré	La Parisienne	–	–		39 50	15·8	—
		Boson de Périgord	De Dietrich	3	22		25 19	24·9	De Dietrich
		Théry	Decauville	2	–		34 0	18·5	Decauville
		G. Richard	—	2	–		34 12	18·4	—
		Gasté	Liberateur Tricycle	1	4		20 10	31·2	—
		Marcellin	Perfecta Quadricycle	2	6		21 16	29·6	—
		Clement	Clément Voiturette	2	2·5		22 40	27·7	—
							mins. secs.		
March 31, 1900	Nice La Course du Mille	Levegh	Mors	2	12		1 38·6	36·5	Mors
		Van Berendonck	Voiturette	2	–		2 0·6	29·9	—
		Béconnais	Perfecta Tricycle	1	4·5		1 18	46·1	Soncin
		Joyeux	,, ,,	1	4	1 Standing start	1 27·8	41·0	—
		Gasté	Liberateur Tricycle	1	4		1 27	41·4	—
		De Méaulne	De Dion Tricycle	1	3		1 50·8	33·3	De Dion
		Serpollet	Serpollet Steam Car	–	–		2 0·8	29·8	Serpollet
		d'Oldenbourg	De Dion Steam Car	–	–		2 3·4	29·1	De Dion
		Hamilton	Steam Car	–	–		2 34·8	23·2	—
		Nottbeck	Electric Car	–	–		2 30·4	23·9	—
							mins. secs.		
March 31, 1900	Nice Course du Kilometre	Levegh	Mors	2	12		0 48·2	46·5	Mors
		Van Berendonck	Voiturette	2	–		1 8	32·9	—
		Béconnais	Perfecta Tricycle	1	4·5		0 39·2	57·0	Soncin
		Joyeux	,, ,,	1	4	0·621 Flying start	0 46·2	48·5	—
		Gasté	Liberateur Tricycle	1	4		0 46·4	48·2	—
		De Méaulne	De Dion Tricycle	1	3		0 48·8	45·8	De Dion
		Serpollet	Serpollet Steam Car	–	–		1 5·6	34·0	Serpollet
		d'Oldenbourg	De Dion Steam Car	–	–		1 4·4	34·7	De Dion
		Hamilton	Steam Car	–	–		1 20·2	27·9	—
		Nottbeck	Electric Car	–	–		0 48·8	45·9	—

The material originally positioned here is too large for reproduction in this reissue. A PDF can be downloaded from the web address given on page iv of this book, by clicking on 'Resources Available'.

and is said to have covered 55 kilometers, or 34·1 miles in 33·5 minutes, or at the rate of practically 60 miles an hour. He was driving his 16-HP. Panhard, the cylinders of which have been enlarged to give 25 HP. He is also reported to have covered a distance of 85 miles in exactly two hours, or 42·5 miles per hour. During these races, de Knyff was using both tube and electric ignition. In these runs a strong wind was blowing at the backs of the cars. It will be seen hereafter, that without this wind such a speed would not be possible with the power available.

In June, 1899, the Automobile Club of Great Britain organised and carried out out a somewhat extensive series of trials. The first of these was a hill-climbing trial, carried out on the Petersham Hill, Richmond. The second was a run of 50 miles, comprising a run from Southall to a point beyond Stokenchurch and back ; the route including some long, stiff hills, including one of nearly a mile, with a gradient of about 1 in 11. The Petersham Hill is 1,917 ft. long, with a mean gradient of 1 in 15·8, the maximum being 1 in 9·53. The total rise in 1,917 ft. is 121·5 ft., and if the whole hill were of the maximum gradient the rise would be 201 ft. The third trial was of a few heavy-load vehicles, both of petrol and steam motors on the 25 miles, from Uxbridge to Wycombe and back. *(British Automobile Club trials, 1899)*

One vehicle, the Delahaye, was tested for speed on a private road near Colchester.

With the exception of the speed trial, the salient results are given in Table XV., and from the results of the hill-climbing trial and the distance trial, a number of deductions have been made, giving approximately the mechanical efficiency of the vehicles.

The average actual horse power at the driving road wheels, as given in Table XV., has been calculated from the weight of the car and load, and the speed at which it ascended the hill the inclination of which is given above. The brake horse power of the motors is in most cases known within a very little of accuracy, but there are some motors of which the power is only known as that stated by the maker or his representative. In most cases the power thus stated is more likely to be a little above than below the actual power. *(Analysis results)*

The figures show that the mechanical efficiency of the whole combined transmission gear of those cars, the power of the motors of which are known, does not reach 70 per cent, that most of them are much below this, and that generally the belt-driven cars are at least as high in efficiency as the best gear-driven cars. It is quite likely that some of the cars would have done better if they had had more gradual gradation of speeds, as at the speed it was found expedient to climb the stiffer part of the hill, the motor, in at least one case, was cutting out ; but if the next speed had been put in, it would have slowed down. It is also probable that several of these cars would show a higher mechanical efficiency if tested when running at higher speeds, but there can be no doubt that fully recorded observations on long ascents would be most useful, as showing the relative efficiency of various forms and arrangements of transmission mechanism. *(Mechanical efficiency)*

383

TABLE XV.

Results of Automobile Club Trials of Petrol Motor Vehicles and Deduced Economy and Efficiency.

AUTOMOBILE CLUB 50-MILES TRIAL AND HILL-CLIMBING TRIAL, 1899.

Name.	Total weight, including passengers.	Number of passengers carried.	Horse-power of engine stated by maker.	Transmission gear.	Pints of petrol consumed in 50-miles trial.	Miles run at average speed per 1 gallon of petrol.	Miles run at average speed per gallon of petrol per passenger.	Average speed in 50-miles trial. Miles per hour.	Hill-climbing trial. Mean gradient, 1 in 15·8. Average speed in miles per hour.	Average actual horse-power at driving road-wheels, as given by the hill-climbing trial.	Maximum actual horse-power at driving road-wheels, assuming ¾ average speed maintained on steepest part of hill. Gradient, 1 in 9·6.	Mechanical efficiency of vehicle.
	cwt.											per cent.
Barriere Tricycle	about 3½	1	2	Tooth	6	67	67	11·05	13·926	1·24	1·38 [2]	69
Cannstatt-Daimler Wagonette	,, 24	4	5½	Tooth and chain	20	20	80	11·25	4·00	2·58	2·73	49·6
English Daimler Phæton	,, 23½	4	5½	,, ,,	12	33·3	133·2	12·25	5·387	3·41	3·73	67·9
English Daimler Wagonette	,, 26	4	5½	,, ,,	18	22·2	88·8	11·6	4·41	3·11	3·39	61·7
Critchley-Daimler Car	,, 10	2	4	Belt and tooth	12	33·3	66·6	11·15	5·46	1·34	1·56	39
Benz Dogcart	,, 20	4	5	Belt and chain	18	22·2	88·8	12·35	5·863	3·3	3·47	69·4
Benz "Ideal" Car	,, 9¾	2	3	,, ,,	10	40	80	12·2	5·615	1·475	1·6	53·4
Benz International Phæton	,, 11½	2	3	,, ,,	24	16·7	33·4	8·7	6·99	2·16	2·36	78·5
Lanchester Phæton	,, 16½	2	8	Worm and wheel	14	28·6	57·2	12·1	5·75	2·42	2·67	33·4 [3]
Hertu Phæton	,, [1] 16	2	4	Belt and chain	25	15·7	31·4	6·8	7·42	3·05	3·35	83·6
Delahaye Phæton	,, 25	4	8	,, ,,	12	33·3	133·2		7·91	5·1	5·67	71
English Daimler Car	,, [1] 27	5	5½	Tooth and chain	20	20	100		4·58	3·35	3·66	66·5
Panhard & Levassor Car	,, 20½	2	8	,, ,,	18	22·2	44·4	12·1	8·75	4·65	5·13	64·1 [3]
Vallée Car	,, 13½	2	5	Belt and chain	26	15·4	30·4	11·7	7·8	2·67	2·98	59·6
Mors Car	,, 21½	4	7½	,, ,,	10	40	160	10·5	7·08	3·92	4·33	57·7
Ducroiset Wagonette	,, 31½	6	10	,, ,,	34	11·8	70·8	12·9	6·108	4·9	5·44	54·4
Bergman Orient Car	,, 15½	2	4	,, ,,	28	14·25	28·5	9·5	5·17	2·06	2·28	57·0
Lisgeoise Tourist Car	,, 7	2	3½	Tooth and belt	13	30·8	61·6	11·7	4·95	0·89	0·98	28·0

[1] Estimated. [2] Rider not pedalling. [3] If speed on steepest part of hill was approximately the average speed, the mechanical efficiency would be for Lanchester car, 41·6 per cent.; Panhard & Levassor, 77 per cent.

AUTOMOBILE CLUB HEAVY VEHICLES TRIAL, 1899. 20 MILES.

Name.	Total weight of vehicle and load.	Load.	Nominal horse-power by maker.	Transmission gear.	Pints of petrol consumed on 20-miles trial.	Ton miles total at average speed per 1 gallon of petrol.	Ton miles of load at average speed per 1 gallon of petrol.	Average speed. Miles per hour.
Cannstatt-Daimler Lorry	4·7 tons	2·2 tons	7·8	Tooth	22	34·2	16	5·88
Cannstatt-Daimler Lorry	9·4 ,,	6 ,,	11·8	,,	32	47	30	3·88
English Daimler Van	4·05 ,,	1·8 ,,	11·0	Tooth and chain	24	27	12	5·0

PETROL MOTOR VEHICLE PERFORMANCE

It may be here recorded, although well known to those present at the 50-mile trials, that the average speed in miles per hour was only kept so near the legal 12 miles by slowing down at nearly every milestone, many of the cars doing most of the running at from 40 to 80 per cent. higher speed than this.

The Delahaye car appears exceedingly well under this examination, its efficiency of transmission, taking the B.HP. of the motor at 8, being 0·71.

This was the only car run for speed. The road chosen was a slightly frequented alternate route, under the control of the Colchester authorities, who, with very commendable desire to assist a new engineering industry, allowed the road to be used for the purpose of a high-speed test. The road is nearly level, and at the time was dry and dusty, but not deeply covered with dust. There was no observable wind.

The road being good, hard and dry, it may be taken that the road resistance per ton was approximately 40 lb. per ton. (See Tables IX. and X.) The speed realized was 26·9 miles per hour, or 2,370 ft. per minute. The weight of the vehicle with its load was 23 cwt. or 1·15 tons. The net work done against internal and road resistances was, therefore, in effective HP.

Rolling resistance

$$HP. = \frac{1·15 \times 2,370 \times 40}{33,000} = 3·3$$

The work done against air-resistance at this high speed was necessarily large, and with it the total power exerted at the wheels was greater than on the hill-climbing trials.

Air resistance

Taking the air-resistance area of the car as 15 sq. ft., we have as the work in effective HP. done against the air, as per Chapters V. and VI.—

$$H\text{-}P. = \frac{v^2 A.\ 0·0017 \times V.}{33,000} = H\text{-}P = \frac{1,550 \times 15 \times 0·0017 \times 2,370}{33,000} = 2·85$$

The total net HP. is then $3·3 + 2·85 = 6·15$, and the gross HP. exerted by the Delahaye motor, taking the mechanical efficiency obtained from the hill-climbing trials, was

$$\frac{6·16}{0·71} = 8·17 \text{ gross HP.}$$

This agrees very nearly with the results of the French trials of motors under the auspices of La Locomotion Automobile, which will be dealt with hereafter. The mean of two trials was 9·2 B.HP., or 1 HP. more than is required in the above calculation. If the motor on the occasion of the speed trial did exert more than 8·17 B.HP., it may be easily accounted for by the road resistance being higher than the resistance assumed for the calculation which was near the minimum.

Combined resistances and efficiency of transmission

At present the data available on this subject is only sufficient to show that the machinery of transmission absorbs too much power, and that carefully conducted trials are necessary on this subject.

There is another direction, however, in which the data available from the report of the judges of the Automobile trials may be used to show the power of motors of given dimensions and the relative power of the motors of similar cylinder capacity of different makes.

Table XVI. deals with the motors referred to in Table XV., including those in the heavy trials. Where the horse power of the motor declared by the maker seems open to question, the figures have been modified from other information, have not been employed, or attention is drawn to them. It will be seen that, as in Table XV., an over statement of the B.HP. of the motor, results in an apparent inferiority in the mechanical efficiency of the transmission gear and the whole car. Correct statements by makers are, therefore, best on all grounds.

Deduced power and dimensions coefficients

From Table XVI. the approximate power of any petrol motor, based on usual practice as to design and comparison, may be ascertained, and the table at the same time affords an interesting comparison of the performance of the different motors the particulars of which are given.

Column 12 gives a co-efficient of mean power of the air- and water-cooled motors, which ran in the Automobile Club trials, and of a Dion tricycle not in the trials. By computing the cubic capacity of the cylinder swept through by the piston (col. 6), and the product of this and the revolutions (col. 7) made by engines giving off known power (col. 9), a coefficient (col. 11) may be derived, by which the power of any other engine, not very much larger in dimensions, may be approximately determined, and by which any series of engines from which the coefficient is derived may be compared.

The top line of Table XVI. gives these figures with reference to a De Dion & Bouton tricycle, with a motor of 1·75 HP. These motors are known to be capable of just reaching this power for a minute or two. Col. 6 gives the cubic capacity of the cylinder swept by the piston. Col. 8 gives the developed capacity, that is the cubic capacity multiplied by the number of revolutions per minute (col. 5). In the developed capacity there are 22,350 cubic inches, or 223·5 hundreds of cubic inches. By taking the numbers of hundreds of revolutions, the influence of widely varying speeds or abnormal speeds on the comparison of motors, or on the value of the co-efficient, is rendered negligible, and col. 10, which shows the power per 100 revolutions, shows to what extent the power of any motor is obtained, only by excess in the number of revolutions.

Col. 12 gives a mean coefficient for the whole of each kind of motor, air- or water-cooled, and by using it a correction has been made, which gives in col. 13 a nearer approach to the actual power which might be expected from the motors. Col. 14 shows how far any of the motors are above or below the power that might be expected from them, judging by cubic capacity and revolutions.

Col. 15 gives the mean developed capacity required by the motors per horse power, the mean being very much the same for water-cooled and air-cooled motors, but the latter would only give the power for short periods.

TABLE XVI.

POWER OF MOTORS, DIMENSIONS, DECLARED POWER, AND POWER BASED ON MEAN POWER, CYLINDER CAPACITY AND REVOLUTION.

1	2	3	4	5	6	7	8	9	10	11	12	13	14	15
Name of Motor, Car, or Maker.	Number of cylinders.	D Diam. (in.)	L Stroke. (in.)	N Revs.	Cubic inches capacity col. 4 × area.	Cubic inches capacity per 100 revs.	Cubic inches capacity × normal revs. = A L N.	Declared power of motor H.P.	Declared power per 100 revs.	Declared power per 100 cub. in. capacity and per 100 revs. $\frac{\text{col. } 9 \times 100}{\text{col. } 8}$	K Mean of col. 11.	Power of motor corrected by mean $\frac{\text{col.8}}{12} \times \frac{\text{col.8}}{100}$	Ratio of declared H.P. to H.P. estimated from mean H.P. of col. 12.	K Cub. in. capacity at normal speed per 1 H.P.
De Dion & Bouton	1	2·625	2·75	1,500	14·9	1,490	22,350	1·75	0·1167	0·00784		1·64	1·07	
Barriere	1	2·75	3·0	1,400	17·75	1,775	24,900	2·0	0·143	0·00804	0·00734	1·83	1·09	13,600
Lanchester	2	5·0	4·5	700	176·8	17,680	123,600	8·0	0·8	0·00646		9·07	0·88	
Lisgeoise	1	3·75	5·0	900	55·25	5,525	49,700	3·5	0·39	0·00704		3·64	0·96	
Cannstatt Daimler (wagonette)	2	3·437	4·562	720	84·6	8,460	60,900	5·5	0·765	0·00903		4·38	1·25	
Cannstatt Daimler (lorry)	2	3·81	5·375	660	122·4	12,240	80,800	7·8	1·182	0·00965		5·83	1·33	
Cannstatt Daimler (lorry)	2	4·94	6·125	540	232·6	23,260	125,500	11·8	2·18	0·00940		9·06	1·30	
English Daimler	2	3·562	4·75	720	94·6	9,460	68,100	5·5	0·765	0·00807		4·91	1·12	
English Daimler (postal van)	4	3·562	4·75	800	189·2	18,920	151,500	11·0	1·375	0·00726		10·94	1·01	
English light Daimler														
Benz	2	3·0	5·0	750	67·0	6,700	50,200	4·0	0·534	0·00799		3·62	1·14	
Benz	2	4·75	5·0	500	177	17,700	88,500	5·0	1·00	0·00565	0·00721	6·38	0·785	13,900
Benz	1	4·5	5·25	400	79·4	7,940	31,760	3·0	0·75	0·00945		2·29	1·31	
Hertu	1	5·0	5·5	600	108	10,800	64,800	4·0	0·667	0·00617		4·68	0·87	
Delahaye	2	4·33	6·30	750	186	18,600	140,000	8·0	1·07	0·00572		10·1	0·81	
Panhard & Levassor Daimler	4	3·15	4·7	750	145·5	14,550	109,000	8·0	1·07	0·00734		7·85	1·02	
Mors	4	2·75	4·25	1,200	100·8	10,080	120,960	7·5	0·625	0·00620		8·70	0·86	
Vallée	2	3·25	5·25	610	87·0	8,700	53,100	5·0	0·82	0·00942		3·82	1·31	
Ducroiset	2	4·5	8·0	650	254	25,400	165,000	10·0	1·54	0·00607		11·90	0·84	
Bergman	1	5·0	6·25	800	122·5	12,250	98,000	4·0	0·5	0·00408		7·06	0·525	

387

The Lanchester motor is of different construction to the other air-cooled motors, but it was necessary to put it into this category.

Apart from the comparisons obtained by a glance at the table, the approximate power of a motor of given dimensions and proposed revolutions, or the dimensions of a proposed motor of given power, may be ascertained as follows:—

Let A = area of piston in square inches;
L = length of stroke in inches ;
N = number of revolutions per minute ;
K = co-efficient of B.HP. based on dimensions and revolutions ;
k = co-efficient of developed capacity per HP. ;
H-P = B.HP. of motor.

Then

$$HP. = \frac{A \, L \, N \, K}{100}$$

$$HP. = \frac{A \, L \, N}{k}$$

$$A = \frac{HP. \times 100}{L \, N \, K}$$

$$N = \frac{HP. \times 100}{L \, A \, K}$$

$$L = \frac{HP. \times 100}{A \, N \, K}$$

Formulæ empirical These expressions are, it must be understood, entirely empirical, and are based only on the power practically obtained from motors using petrol vapour mixed with air in the proportions found best by experience to the present time, and also with the compression pressure found best or most expedient for motors as at present used. The compression used varies with different makers and with motors of different sizes. It varies from Major Holden's cycle motor compression pressure of about 80 lb., to that common in the English Daimler motor cars, namely, about 40 lb.[1] With these variations there is difference in power value of cylinder capacity and greater difference in fuel economy, but the mean pressure during the working stroke which has been found by an enormous range of experience to be best for all working conditions, may be considered to be embodied in the practice, the results of which are given in Table XVI., and these expressions. Even had we satisfactory indicated power of these motors, the relative value of the different compression pressures, as well as of all other factors including relation between stroke and area, must be ascertained by dynamometer tests and by extended working. On these grounds, then, the Tables referred to will be found to contain the most useful information. **Very high speed indicator diagrams** Those who wish to base their power calculations on mean pressure during the working stroke, must take indicator diagrams at very high speeds, which is not very easy, nor are such diagrams, when obtained, always very trustworthy. Those taken from a motor which runs normally at 800 to

[1] This has recently been raised as mentioned hereafter.

PETROL MOTOR VEHICLE PERFORMANCE

1,200 revolutions per minute, are of very little value when taken at 400 revolutions per minute, and the break has to be appealed to for actual horse power, and for a mechanical efficiency which only obtains with the particular motor tested.[1]

The mean pressure being obtained, however, the indicated or estimated horse power per cylinder will be for Otto cycle motors :—

$$\text{I.HP.} = \frac{P\,L\,A\,N}{33,000 \times 2}$$

In which P = mean pressure from diagram, L = length of stroke in feet, N = the number of revolutions per minute, and A = area of piston in square inches.

During the past year the proprietors of the French journal, *La Locomotion Automobile*, organised a series of trials of the brake horse power of many of the motors made in France for motor vehicles and motor cycles, and also tests of the brake horse power at the tyres of the road driving wheels driven by these or similar motors. The trials were carried out under professional or professorial guidance at the works of MM. Malicet & Blin.

The brake or dynamometer used for the motor tests was of simple but not good form, a belt embracing half the circumference of fly wheel or pulley, and the motors were tested with surface and with spray-making or atomising carburettors, and with and without silencers. Some were also tested with petroleum spirit of different densities, and a few with kerosene or russolene.

The power delivered to the road wheels of various vehicles was measured by running these wheels upon drums of about 30 in. diameter, on the end of the shaft of which was a dynamometer or friction brake pulley 37·5 in. diameter.

The measurements taken included the power absorbed in running these drums and their shaft, but this was separately ascertained and added to the observed brake power. The apparatus must be assumed to be satisfactory as to the accuracy of its indications, and the power absorbed by it was probably correctly taken to within 2 or 3 per cent. It does not seem, however, that the deformation of the tyres running upon the drums, which would be greater than would be the case on average roads, was taken into consideration. The whole of this work of deformation is not completely returned by the tyre as it leaves the drum, the rubber and canvas tyres being far from perfectly elastic. The loss under this head may not be very large, but it is desirable that it should be ascertained. The mechanical efficiency of the transmission gear, the information most required, cannot be calculated from these trials because the cars were run at the lowest speeds, and the results seem to indicate that the motor was cutting out all through the trial. Otherwise it is impossible to account for

The French motor trials of 1899

Tests of power absorbed by transmission gear

As to accuracy of observation

---

[1] "The Friction Brake Dynamometer," *Proc. Inst., C.E.*, vol. xcv., paper by the author; and *The Steam Engine Indicator and Indicator Diagrams*, by the author. London : The Electrician Publishing Co.

the very low output of the Panhard 8 and 12 HP. cars at 12 miles per hour, *i.e.* an efficiency of only about 49 per cent. Even if a large addition be made to this, the experiments appear to show that the 70 per cent. efficiency taken for Table XII. is higher than is at present reached with the cars of the makers of the highest reputation; even if some addition be made for the difference between running on drums and running on a flat road surface, 70 per cent. is not reached.

Analysis of results

The chief results of the French motor tests are given in Table XVII. with the corresponding cylinder capacity and developed capacities and coefficients. It will be seen that the deduced coefficient giving power per 100 inches capacity per 100 revolutions, is slightly lower than that found from previous trials as given in Table XVI. The Gobron-Brillié motor tested was slightly higher in power than that in the vehicle tested and referred to in Table XVIII., and the Delahaye motor gave more power than that found for the hill climbing and speed test referred to on p. 385. Nearly all the tests of motors were of short duration, namely, 10 to 15 minutes with results recorded for 1 to 3 minutes and not averaged. Most of them were several times repeated. They may be taken as giving the maximum powers which the motors reached for brief periods, often but 1 minute, but they do not really represent the power of the motors. From our point of view they give a very exaggerated notion of the power, and the method of recording the maximum jumps of power readings can only be excused on the ground that these little motors are only called upon to exercise their maximum power for very short periods. Some of them were tested with oils or spirits of different densities, two of them with what are called heavy oils. The specific gravity of these is given as 0·88, which is that of a heavy crude petroleum. The specific gravity of the petrol used is given as 0·68.

Power readings too high

One of the motors so tested, namely, that of the *Touriste* car, gave 3·4 per cent. more power than with petrol, while the Gobron-Brillié motor gave 3·2 per cent. less power. These very short trials, however, afford very little guidance as to the relative power obtainable from motors of given dimensions with the light and heavy oils, nor do they as tests of fixed engines give any indication of the suitability of the motors for using heavy oil under the conditions of frequently varying load of motor car propulsion and manipulation.

Analysis of transmission gear tests

Table XVIII. gives the observed figures and results of *La Locomotion Automobile* trials of the power delivered to the periphery of the road-wheels of vehicles. In some cases the motors of these cars were of precisely the same patterns as those referred to in Table XVII., but as the power absorbed between motor shaft and dynamometer is much higher than that indicated by the practical test of climbing the Petersham Hill, it must be assumed that there were during these French trials some serious losses not accounted for. The figures in other parts of the table are, however, of considerable comparative utility.

TABLE XVII.

RESULTS OF "LA LOCOMOTION AUTOMOBILE" TRIALS OF MOTORS, 1899, WITH DEDUCED VALUES AND COEFFICIENTS.

Motor.	Carburettor.	Diam. of cylinder in inches.	Stroke in inches.	Revolutions per minute.	B.HP.	Cubic inches capacity of cylinders.	Cubic inches capacity × normal revolutions A L N.	Brake H.P. per 100 cubic inches capacity and per 100 revolutions.	K mean B.H.P. per 100 cubic inches capacity and per 100 revolutions.	Remarks.
De Dion & Bouton, 1 Cylinder Vert.	Surface	2·75	2·75	1,473	1·35	16·3	24,010	0·00562		
"	Spraying	"	"	2,057	1·875	"	33,500	0·00560		Highest result.
"	Surface	"	"	2,015	2·12	"	32,820	0·00645		
"	"	"	"	1,444	1·42	"	23,540	0·00604		
La Minerve "	Surface	3·36	3·17	1,672	1·65	28·1	45,500	0·00354		
"	Spraying	"	"	1,342	1·56	"	37,700	0·00414		
De Dion & Bouton "	Surface	2·6	2·75	1,393	1·34	14·6	20,350	0·00658		With silencer.
"	"	"	"	1,360	1·30	"	19,850	0·00655	0·00572	Without silencer.
"	Spraying	2·75	2·75	1,564	1·30	16·3	25,480	0·00510		"
Deckert, 1 Cylinder Vert.	Surface	3·0	3·85	1,650	1·58	"	26,900	0·00587		"
Buchet, 1 " (à culasse)	Spraying	3·17	3·17	1,752	1·97	27·2	47,600	0·00414		{ With silencer. / Petrol 0·680 sp. gr.
" " "	Buchet	3·55	5·12	1,660	2·9	25·1	41,600	0·00696		{ With silencer.
" " "	"	"	"	850	3·36	50·6	43,000	0·00781		{ Hydrocarbon 0·88 sp. gr.
Le Touriste, 2 Cylinders Hor.	T. Bouché	3·55	6·3	633	5·5	124·6	78,900	0·00697		With silencer.
"	"	"	"	622	5·7	"	77,500	0·00735		
Amiot-Péneau, 2 Cylinders Vert.	Longuemare	3·95	6·3	752	7·2	144·0	108,200	0·00665		With silencer.
Hertu, 1 Cylinder Horizontal	Surface	4·92	5·5	1,042	4·15	104·1	108,500	0·00382		
Al Dumas Fils, 2 Cylinders Hor.	Longuemare	4·1	4·95	703	7·15	130·6	91,800	0·00779		Without silencer.
"	"	"	"	1,012	6·0	"	132,200	0·00454	0·00682	With silencer.
Cyclope, 2 Cylinders Hor.	Cyclope	3·35	4·72 combined	700	4·84	83·0	58,100	0·00834		
Gobron-Brillié, 2 Cyl. Vert., two upper pistons having 2·35-in. stroke, and the two lower having 3·15-in. stroke with silencer	Spraying	3·17	5·5	880	5·85	86·9	72,200	0·00812		Hydrocarbon 0·880 sp.gr.
	"	"	"	849	6·05	"	73,800	0·00819		Petrol 0·680 sp. gr.
	"	"	"	1,167	7·08	"	101,500	0·00617		Highest result.
	"	"	"	870	6·1	"	75,600	0·00806		Petrol 0·680 sp. gr.
Delahaye, 2 Cylinders Hor.	—	4·35	6·3	687	10·05	186·8	128,300	0·00817		Highest reading.
"		"	"	893	9·2	"	166,500	0·00553		Mean of two trials.

391

Further trials required
It will be some time before the results of the trials to be carried out this year by the Automobile Club of Great Britain will be available for analysis, but it may be hoped that these will be so conducted as to yield information on the most important questions of mechanical efficiency of different systems of power transmission to the road wheels.

TABLE XVIII.

RESULTS OF "LA LOCOMOTION AUTOMOBILE" MOTOR VEHICLE DYNAMOMETER TRIALS, 1899.

Vehicle.	Approximate weight on driving wheels.		Revolutions per minute of driving wheels.	Corresponding road speed.		Weight supported by brake.		Work absorbed in foot lb. and in kilogram metres per minute.						Brake HP.	Ponce-lets.
								By Brake.		By Rollers.		Total.			
	lb.	Kilogs.		Miles per hour.	Kilometres per hour.	lb.	Kilogs.	Foot lb.	Kilog. metres.	Foot lb.	Kilog. metres.	Foot lb.	Kilog. metres.		
Panhard & Levassor, 12 HP. nominal	1,488	675	113	6·34	10·2	41·6	19·24	159,500	22,041	17,660	2,443	177,160	24,484	5·37	4·08
Ditto, ditto	„	„	201	12·43	20·0	24·78	11·24	163,100	22,592	31,400	4,345	194,500	26,937	5·89	4·49
Panhard & Levassor, 8 HP. nominal	1,323	600	84	4·66	7·5	41·6	19·24	116,800	16,162	12,160	1,681	128,960	17,843	3·90	2·97
Ditto, ditto	„	„	122	6·84	11·0	20·37	9·24	80,500	11,273	17,410	2,413	97,910	13,686	2·96	2·28
Peugeot	1,323	600	85	4·75	7·65	35·88	16·275	99,700	13,794	12,150	1,680	111,850	15,474	3·38	2·58
„	„	„	92	5·15	8·3	32·57	14·775	98,300	13,593	13,210	1,830	111,510	15,423	3·38	2·57
Delahaye	1,764	800	154	8·6	13·9	19·35	8·775	97,900	13,513	22,020	3,049	119,920	16,562	3·63	2·76
„	„	„	82	4·60	7·4	67·04	30·43	181,000	24,953	14,580	2,017	195,580	26,970	5·92	4·49
Gobron-Brillié	1,102	500	224	12·43	20·0	22·98	10·22	166,000	22,993	39,800	5,510	205,800	28,503	6·23	4·75
„	„	„	108	6·04	9·72	45·92	20·83	162,900	22,496	13,420	1,857	176,320	24,353	5·34	4·06
Rochet, 6 HP. nom.	1,268	575	125	6·99	11·25	33·38	15·18	137,200	18,975	15,550	2,150	152,750	21,125	4·62	3·52
„	„	„	88	4·97	8·0	40·21	18·24	116,000	16,051	12,570	1,736	128,570	17,787	3·89	2·96
Klaus „	441	200	164	9·2	14·8	15·96	7·24	85,950	11,873	23,390	3,234	109,340	15,107	3·31	2·52
„ „	„	„	112	6·21	10·0	17·86	8·10	66,250	9,162	8,100	1,120	74,350	10,282	2·25	1·71
Koch (large)	1,764	800	149	8·27	13·41	11·42	5·18	55,760	7,718	10,780	1,490	66,540	9,208	2·02	1·53
„ „	„	„	77	4·35	7·0	42·5	19·285	107,400	14,849	13,700	1,894	121,100	16,743	3·67	2·79
„ „	„	„	85	4·75	7·65	35·9	16·285	101,000	13,842	15,110	2,091	116,110	15,933	3·60	2·74
„ „	„	„	136	7·6	12·24	20·47	9·285	91,500	12,628	24,200	3,345	115,700	15,973	3·52	2·68
Koch (small)	1,433	650	144	8·08	13·0	18·26	8·285	87,000	12,018	24,900	3,442	111,900	15,460	3·39	2·58
„ „	„	„	109	6·09	9·8	27·0	12·27	96,750	13,374	14,400	1,991	111,150	15,365	3·36	2·56
„ „	„	„	119	6·65	10·7	22·6	10·27	89,800	12,401	15,710	2,174	105,510	14,575	3·19	2·43
„Le Touriste	1,323	600	198	11·08	17·82	11·62	5·27	75,500	10,435	26,400	3,650	101,900	14,085	3·08	2·35
„ „	„	„	65	3·73	6·0	31·45	14·27	67,000	9,275	9,580	1,325	76,580	10,600	2·32	1·77
Rouval „	1,543	700	116	6·52	10·5	9·41	4·27	35,800	4,953	16,600	2,294	52,400	7,247	1·59	1·21
Amiot-Péneau	1,653	750	82	4·6	7·4	55·7	25·175	149,800	20,726	13,210	1,828	163,010	22,554	4·94	3·76
„	„	„	79	4·35	7·0	33·66	15·275	87,400	12,067	16,840	2,327	104,240	14,394	3·16	2·4
„	„	„	159	8·88	14·3	13·83	6·275	72,100	9,977	33,850	4,685	105,950	14,662	3·21	2·44

Chapter XXII

ELECTRIC MOTOR VEHICLES

Attractiveness of electric power

GIVEN the electricity supply, nothing would be better than the electric motor for the propulsion of vehicles. This fact has lured many into the expenditure of large sums of money in the attempt to make practical electrically-propelled road motor vehicles. So far the attempts have led to very little success, except for vehicles for carrying light loads short distances, and comparatively with other methods of propulsion, electricity has met with little favour, even for short-distance work, either for private or commercial purposes. The electrically propelled vehicle has, however, made great strides during the past three years, but this advance has been almost entirely with the vehicles for carrying two persons.

Improvements of 1898-99

The improvements have been in the simplification and lightening of the whole of the transmission gear and of the vehicle itself. It may be said that all that is at present required in the mechanical engineering and the carriage building part of the problem has been provided. In these respects the change has been equal to that of the construction of the petrol motor and light steam vehicles. So great is the change that it is unnecessary to dwell upon the history or the construction of the earlier vehicles by Volk, Ward, Clubbe, Immisch, Carli, Blumfield and Garrard, and Jeantaud. All have been surpassed by the vehicles of the last year or two, and the greatest improvement has been made in the United States.

Difficulty as to accumulated supply

The electricity supply is, however, still almost as great a problem as ever, because the weight of secondary battery, which must be carried for even a 25-miles run in any but flat country, is prohibitive. In flat towns or country where roads are good, or, in other words, where the conditions are such as to require very small power, this is not the case, but such districts are few and the limitation does not affect either petrol or steam motors. The past two years have considerably lowered the weight of battery made for vehicle purposes, but even if a battery of theoretically least weight could be used, that weight would still hopelessly handicap the electrical vehicle for long journey work over average country. It is not invention alone that is required to provide the successful electricity accumulator. Discovery must be made of means of employing materials other than lead and its oxides or salts.

394

ELECTRIC MOTOR VEHICLES

This difficulty, however, does not preclude the use of electricity for some forms of light vehicle for short journeys on favourable roads and for owners to whom the cost of working is a matter of little or no importance.

Useful applications

I shall now only describe a few electric vehicles and their gear as representing modern practice, reserving to a future chapter a comparison of the cost of working power and weight of steam, electric, and petrol motor vehicles.

The most extended practical and commercial trial of electric vehicles was that made by the London Electric Cab Co. with a service which was stopped in the latter part of 1899.

Experience in London

This Company started working in the streets of London in 1898, and the service was maintained, with more or less regularity and irregularity, for some months, when the vehicles were withdrawn ostensibly for modifications and repairs. They were subsequently replaced in service for a short time.

The arrangements at first made by the Company for providing current for charging its batteries were foredoomed to commercial failure. Alternate current motor generators were put down. These were supplied with current from Deptford at what was by some electricians considered so low a price that the cost of current to the batteries would, they thought, be as low as if supplied by steam-generating plant on the spot. The Company was advised that this could not possibly be the result, as total cost of the electro-mechanical conversions must inevitably raise the total cost of current to at least 150 per cent. above the cost if generated on the spot.

After running some time on the alternate current motor generator system, the Company was forced to adopt the rejected advice, and put down steam plant.

A satisfactory contract was made for the supply and maintenance of the accumulators, but the stern realities of cost of working or of expenditure and receipts forced the Company to stop working towards the end of 1899, and the whole of the plant and stock to be sold by the order of the Court in March, 1900, together with what in the particulars of sale is humorously called "the benefit of a license" from three syndicates, for which a royalty of £15 per cab and of £4 per cab per year has to be paid.

The cabs are, however, of considerable interest as their construction is that of the first hackney vehicle worked out for and actually used on a public service for altogether about a year. The experience gained with them will no doubt benefit those who are most likely to become possessors of them and the plant and leases.

Table XIX. has been compiled from the records of the trials made by the Automobile Club of France in 1898. The results of similar trials made in 1899 are not yet available for analysis; but, as the outcome of all the work done towards the commerical electrical vehicle, is very little better in France than it is in London, it is obvious that the mere cost of electrical

395

TABLE XIX.

RESULTS OF TRIALS OF HACKNEY ELECTRIC VEHICLES BY THE AUTOMOBILE
CLUB OF FRANCE, JUNE, 1898.

Name of constructor and type of vehicle.	Persons carried, including driver.	Total weight of vehicle and load.		Weight of storage battery.	Weight of load carried, not including driver.	Normal H.P. of motor.	Average speed of the vehicle, miles per hour.	Total cost of running the vehicle while on hire per mile.	Cost of electrical energy consumed while the vehicle is on hire per mile.
		ton.	cwt.	cwt.	cwt.			d.	d.
Jenatzy Coupé .	3	1	15·5	11·1	2·75	4	8·85	7·25	0·41
Jeantaud Cab . .	3	1	7·8	7·85	2·75	3·1	8·9	7·08	0·31
Jeantaud Landau	3	1	12·5	8·85	2·75	3·25	7·66	7·17	0·38
Kriéger Coupé .	5	1	12·3	9·0	5·5	3·4	9·2	7·25	0·33
Kriéger Victoria .	5	1	11·5	9·0	5·5	3·4	8·5	7·3	0·35
Kriéger Coupé à Galerie . . .	6	1	15	9·0	7·9	3·4	8·5	7·2	0·35
Peugeot Petrol Coupé	4	1	5·4		4·14	6	10·0	9·65 [1]	

energy is a small item in the total cost of construction and of working of
such vehicles. The total cost of running, as given in Table XIX., does not
include depreciation, and the cost of the pneumatic tyres for electric vehicles is
taken as 2 francs per day less than for the petrol motor—a difference assumed
in favour of electric vehicles, which the French assign to the more uniform
turning effort of the electric motor. If this be disallowed, then the petrol
motor vehicle will cost, as far as it is indicated by these trials, about the
same as the electric vehicle per mile on hire. If maintenance over a period
of a year be taken into account by reference to experience, then the
apparent superiority of the electrical cab, is converted into a real superiority
of the petrol motor vehicle, especially now that the motor may be started
from the driver's seat. A good deal of the cost per day shown by the table
arises from the running of the motor when the cab is standing, and from
the very high price of petrol in Paris: while, on the other hand, the
charging current is charged at only 1·2d. per kilowatt, or 0·9d. per HP.
hour. Thus every charge has been favourable to the electrically-propelled
cab. The mere figures of the trials are also favourable, but the results of
extended practical working and the known condition of batteries, gearing,
and tyres of these heavy dead-load and small-paying-load vehicles are
enormously in favour of other motors, or even of horses.

Beside the vehicles mentioned, and those referred to in the tables,
there have been those of MM. Bouquet, Garcin, and Schivre, who in 1898
made electric carriages to carry two or three passengers, and weighing
complete about 18·8 cwt. without accumulators, which weighed 7 cwt., and
were said to run the vehicle 68 miles on average roads. In September,

[1] Cost of petrol in Paris, 0·57 franc per litre = 2s. 2d. per gallon. Charging current,
1·2d. per kilowatt.

ELECTRIC MOTOR VEHICLES

1899, two of these vehicles are credited with having run 1,250 miles in the
north of France, and in August of the same year Chasseloup-Laubat is
said to have driven one of the same vehicles from Paris to Rouen, about 95
miles, in 7 hours on one charge. This is a speed of 13·5 miles per hour the
whole journey, and one can only say that it is a pity that vehicles of this
kind behave so much better during these short trials in very expert hands
than they do on a year's trial under ordinary circumstances.

In 1895 Jeantaud entered an electric vehicle, carrying six persons, in
the Paris-Bordeaux race. It covered the first half of the race, but only
after various accidents or breakdowns, and a 100 hours of time for the 370
miles journey. The vehicle, like almost all the others, was heavy. It
weighed with passengers 1 ton 16 cwt., and the batteries 16·75 cwt. The
vehicle ran about 25 miles per charge.

Jeantaud made several vehicles for Chasseloup-Laubat, who competed
with Jenatzy for the kilometre record. The best times he made were:
With a standing start, 1 kilometre in 48·8 seconds; flying start, 38·6
seconds.

Krieger constructed several vehicles in 1897.

A very remarkable vehicle, made for a special purpose, namely, of
accomplishing a record for 1 kilometre, was Jenatzy's electrical four-
wheeled cigar-body *La Jamais Contente*. This vehicle weighed altogether 19·75 cwt. with its batteries. From a standing start it ran the
kilometre in 47·8 seconds, and with a flying start he accomplished it in 34
seconds, or at the rate of 65·8 miles per hour. This is without doubt a
higher speed than any other human being has ever travelled on roads, but
it was only for about three-quarters of a mile that it was maintained.
This vehicle was of no use in any way as a guide for any other class of
vehicle.

For the 1899 trials MM. Mildé & Mondos entered a delivery van in
the Paris trials, the weight of which with load was 2 tons 15·5 cwt., and
of this the load was only 8 cwt.

Altogether the results so far with electrically-propelled vehicles have
not been commercially encouraging, although they may be considered so
from a merely practical point of view in most that relates to the mechanical features of the most recent of the English and American light vehicles.

THE LONDON ELECTRIC CAB.

Figs. 282-284 show a side elevation, plan, and back-end elevation of
one of the vehicles of the London Electric Cab Co. constructed to the
designs of Mr. W. C. Bersey. It is in the form of a closed coupé, to carry
two persons inside and the driver, whose seat is on a raised platform in front,
so that his view of the road is entirely unobstructed. The various parts of
the motive mechanism are carried upon an underframe, which is sprung
from axles by means of leaf-springs. The body of the cab is also indepen-

*General de-
scription of
London
electric cab*

397

dently sprung from the underframe, which almost completely relieves it from shock and vibration due to inequalities of the road, etc. The road wheels are provided with solid rubber tyres.

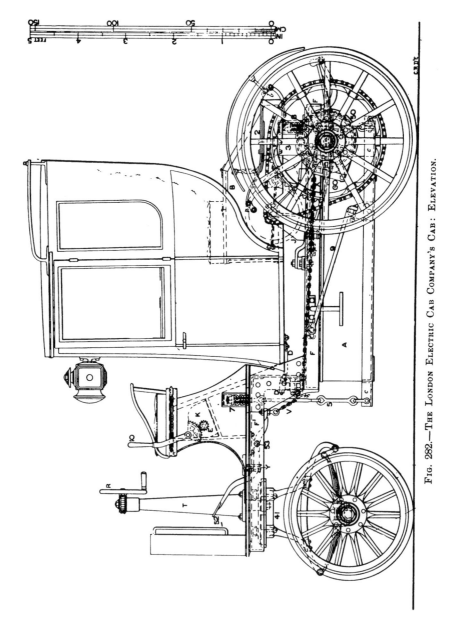

FIG. 282.—THE LONDON ELECTRIC CAB COMPANY'S CAB: ELEVATION.

The storage battery is carried in a box suspended below the underframe, so that it may easily be detached and a freshly charged battery replaced. It consists of 40 E.P.S. cells, supplying current at an average pressure of 80 volts. On the level the average discharge is about 30 ampères, and the capacity of the battery is about 150 ampère hours, so that one charge is sufficient for a run of about 25 to 30 miles, according to the condition and gradients of the road. Various cells were used, and the E.P.S. were finally adopted under a contract for maintenance at 10 per cent.

The motor and driving gear are carried from the underframe at the back of the vehicle, and are closed from view by the rear box of the cab body. The power is transmitted from the motor, by means of spur gear, to a counter-shaft which carries the differential gear, and from the ends of this shaft the power is transmittted to the rear road wheels by means of pitch chain gear.

The controller, by means of which the speed of the motor is varied, is operated by means of a lever at the driver's left hand. Four forward speeds are provided, the maximum being about 9 miles per hour, and one reverse speed of 2 miles per hour. An electric brake, provided by the reaction upon itself of the motor acting as a dynamo, may also be applied by means of this lever. A plug key, by means of which the main circuit between the battery and motor may be broken, is placed in the box below the driver's seat, so that if the driver wishes to leave the vehicle he removes the plug, and so makes it impossible to set the cab in motion. Besides the electric brake above mentioned, two band brakes are also provided, which act on bandbrake pulleys on the rear road wheels, and are applied by means of a pedal in front of the driver. This pedal also automatically breaks the electric circuit, and puts the motor out of action before applying the brakes.

The steering of the vehicle is effected by means of a hand-wheel, which operates through toothed gearing and a locking-plate, to which the front axle is connected.

The weight of the vehicle, complete with the storage battery, approaches 2 tons.

The underframe, which is seen most clearly in side elevation in Fig. 282, and in plan in Fig. 283, is constructed of angle-iron bars, and consists of a rectangular structure F F, at the front end of which, and rigidly connected thereto, is a raised framing F′ F′, upon which is the platform which carries the driver's seat and the several levers, etc., controlling the motion of the vehicle. The underframe is of very strong and somewhat heavy construction, so as to allow an ample margin of strength to withstand rough usage. The angle iron used is 3·5 in. deep, 2·5 in. broad, and 0·375 in. thick. It is sprung from the axles by means of single-plate springs.

The body of the cab is independently sprung from the underframe by means of the plate springs 2, 2 on either at the back ; these springs are bolted at their centres to flat iron standards 3, 3, Figs. 282 and 284, rigidly attached to the underframe, the body being slung from the ends of the springs by means of the links *a, b*.

Under-frame

MOTOR VEHICLES AND MOTORS

In front the body is carried by means of a single cross spring, to the
centre of which it is attached by means of the arm D, Fig. 282, the ends of
the spring being slung by links from the underframe F F.   This front

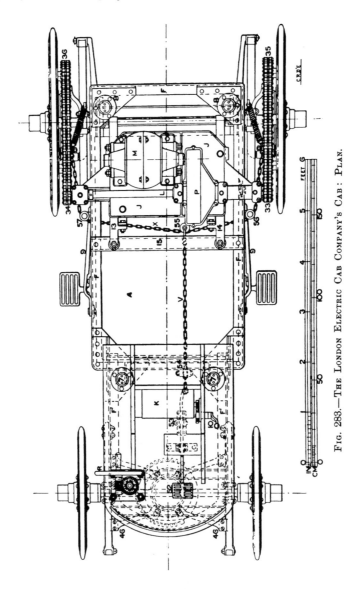

FIG. 283.—THE LONDON ELECTRIC CAB COMPANY'S CAB: PLAN.

spring is behind the vertical part of the angle-iron frame, and does not
show in the drawings, but its form—a flat leaf spring—and position may
be understood from the above and from Fig. 282.

The storage battery is enclosed in the box A suspended below the
underframe by means of the links 5, 5, and 6, 6.   At their lower ends these

400

links are attached to the angle-iron edging *c c* of the battery box A by means of pins, which are easily removable when the battery is required to be changed. At their upper ends the suspension links are arranged so that the weight of the cells is supported by strong helical compression springs within the boxes 7, 7 and 8, 8 which are bolted to the underframe.

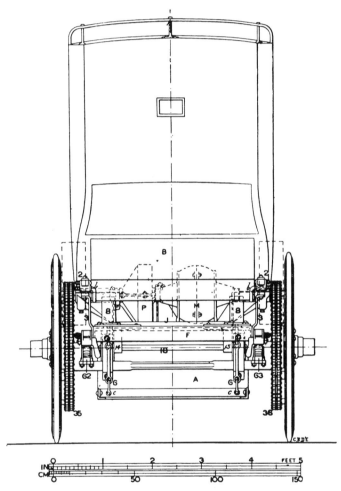

Fig. 284.—London Electric Cab Company's Cab: Back View.

The battery box is restrained from swinging by means of the radius bars 9, 9 on either side of the underframe. These bars were due to the Hon. Reginald Brougham, and supposed to permit the free movement of the box upon its spring supports, but to prevent undue freedom. They are, however, the cause of the delivery to the box of very considerable shocks every time the cab stops and starts under the exigencies of town traffic, and every time the cab runs over a bad crossing or any obstruction. Vertical movement is free, it is true, but the inertia effects are equally and oppositely visited upon the box and body in proportion to the relation between the dip and the horizontal length of the radius bars.

The motor M and the driving gear are carried at the back end of the underframe, and are partially enclosed by the box B which forms part of the cab body.

The motor   The motor, which is seen in Figs. 283–286, and in detail in Figs. 287 and 288, is of the Lundell type, and develops about 3 HP. at normal speed of the vehicle. The armature A is of the toothed laminated drum type, and wound with two sets of windings, which are connected to independent commentators B B, one at each end of the armature spindle c, Figs. 287 and 288. The field magnet D, which will be seen to be of peculiar form, is so designed with the object of reducing the demagnetizing effect of the armature on the field to a minimum, and keeping it as nearly constant as possible while the load is varied through wide limits, so that the brushes once set do not require to be shifted to prevent sparking between no load and full load. It is constructed of mild steel castings, and formed so that it constitutes a cylindrical shell surrounding the armature, the two halves being bolted together by means of the four bolts a, Fig. 288. The ends of the cylindrical body are closed by means of the covers E E, in which are conical projections which surround the commutators, and which carry in their ends the roller bearings F F of the armature spindle. The poles D D of the field magnet are surrounded by the coil H, which like the armature winding is in two independent halves. The brushes are of carbon, and are set symmetrically between the pole tips of the field. They are pressed against the commutator surfaces by means of spring pressure. Fig. 288A shows a detail of one of the brushes. The motor is carried by means of four feet 11 11, 12 12, which are cast with the end covers E E, and fit into pockets on the cast-iron tray J, to which they are bolted as shown.

The design not integral   It will be seen that up to this date the design of a motor cab as a whole, was not reached, the procedure being to obtain and combine the various separate elements as well as these, as completed parts, permitted, not to design so that one part formed part of other parts. Hence a separate motor on a separate bed-plate and separate gear.

Control   As mentioned above, both the armature and field of the motor have two independent windings, and it is by forming various combinations between these windings, a resistance coil, and the storage battery, that the various speeds are given to the vehicle. These several combinations are obtained by means of the controller, which is placed on the box below the driver's seat, and operated by means of the hand lever 10.

The controller   The construction of the controller will be explained with reference to Figs. 289–294, which show a plan and end elevation of the complete apparatus, and diagrams by which the method of forming the various combinations will be explained.

402

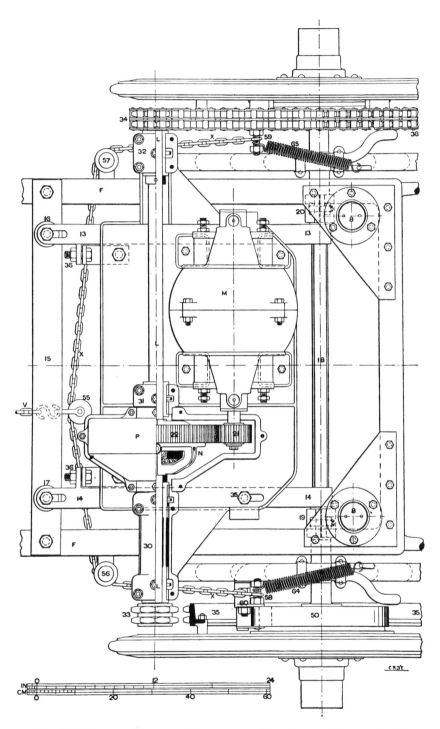

FIG. 285.—PLAN OF MOTOR AND TRANSMISSION GEAR, LONDON ELECTRIC CAB CO.

403

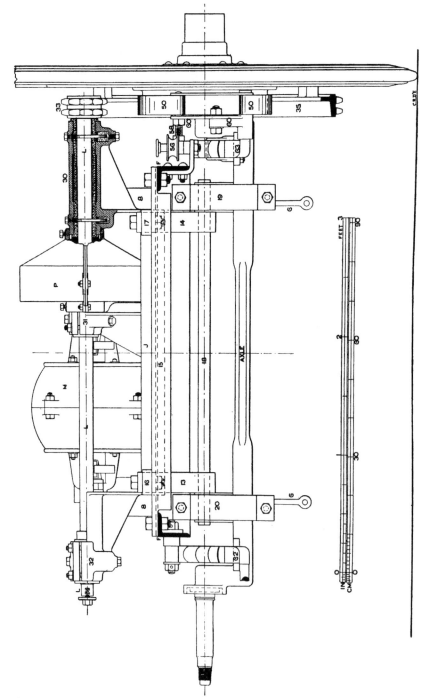

FIG. 286.—London Electric Cab Company's Cab: Back View of Motor, Axle, and Driving Gear.

404

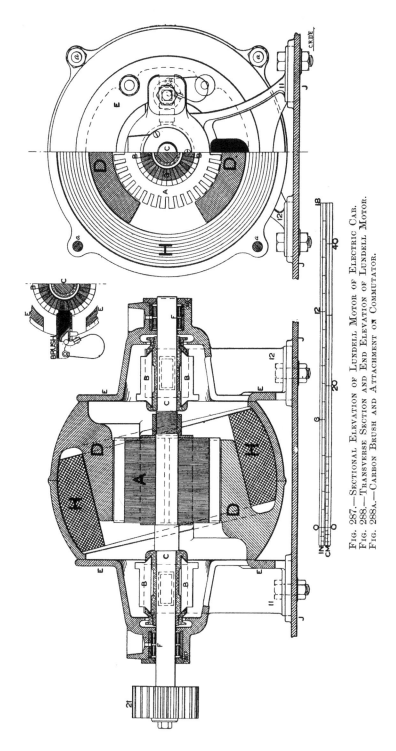

Fig. 287.—Sectional Elevation of Lundell Motor of Electric Cab.
Fig. 288.—Transverse Section and End Elevation of Lundell Motor.
Fig. 288A.—Carbon Brush and Attachment of Commutator.

405

MOTOR VEHICLES AND MOTORS

It consists of a wooden cylinder κ, which is mounted on a spindle A carried in bearings B B, which are cast with the light iron tray c. This tray forms the base upon which the various parts are mounted. The wooden cylinder κ, a development of which cut along the line A, x, Fig. 289, is shown in Fig. 291, is divided into five principal portions by means of the vulcanite rings, a a^1 a^2 a^3, and upon its periphery are mounted metallic contact pieces b b^1, etc., c c^1, etc., d d^1, e e', etc., the surfaces of which are indicated by cross-hatching in Figs. 290 and 291. These contact pieces are of cast brass, and are held to the wooden cylinder by means of counter-sunk screws as shown. They are interconnected in some cases, and in others insulated from one another in the manner which will be subsequently described, their object being to form various combinations of the battery and motor windings by interconnecting the brushes, 1–11, of the controller, Figs. 289 and 290. These are connected to 11 binding screws s mounted upon the cross wooden bar, D. The binding screws s, to which the brushes 1–11 are attached, are connected by means of insulated wires to the storage battery, motor windings, and resistance coil R, in the manner shown diagrammatically in Fig. 291, and also seen in Fig. 294. From the diagram, Fig. 291, it will be seen that the + terminal of the battery is connected to brush 1, and the − pole to brush 11. Brush 2, which it will be noted is the third from the + end, is short-circuited with brush 1 through the resistance coil R, which is carried with the controller in the box below the driver's seat; brush 3 is connected to one end of one of the field-magnet windings, the other end of which is connected to brush 4. The other field winding is connected in like manner to brushes 5, 6; brushes 7, 8 are connected to the commutator brushes v, w at one end of the armature, and therefore to one of the armature windings, and brushes 9, 10 are connected to the commutator brushes v', w' at the other end of the armature, and so to the other armature winding; brush 11 is, as before mentioned, connected to the − pole of the battery.

Motion is imparted by means of the hand lever 10 to the controller drum κ, through a tooth sector gearing with the pinion E mounted at one end of the spindle A. By means of a ratchet wheel F, mounted on the spindle A with which the roller pawl G engages under pressure of a spiral spring H, Fig. 289, there are eight positions corresponding to the eight teeth of the ratchet wheel in which the pawl G will hold the drum κ fixed, unless pressure sufficient to overcome the spring H, holding down the pawl, is applied to the hand lever 10 by the driver. The metallic contact pieces on the drum κ are so arranged that corresponding to these eight positions, in which the drum is held by means of the ratchet wheel F and pawl G, eight combinations of the controller brushes 1–11 are formed, which give four forward speeds to the vehicle, break circuit, give two combinations by means of which the armature windings are disconnected from the battery and are caused to send a current round the field windings, with a resulting powerful braking action on the vehicle through its gear, and one position

406

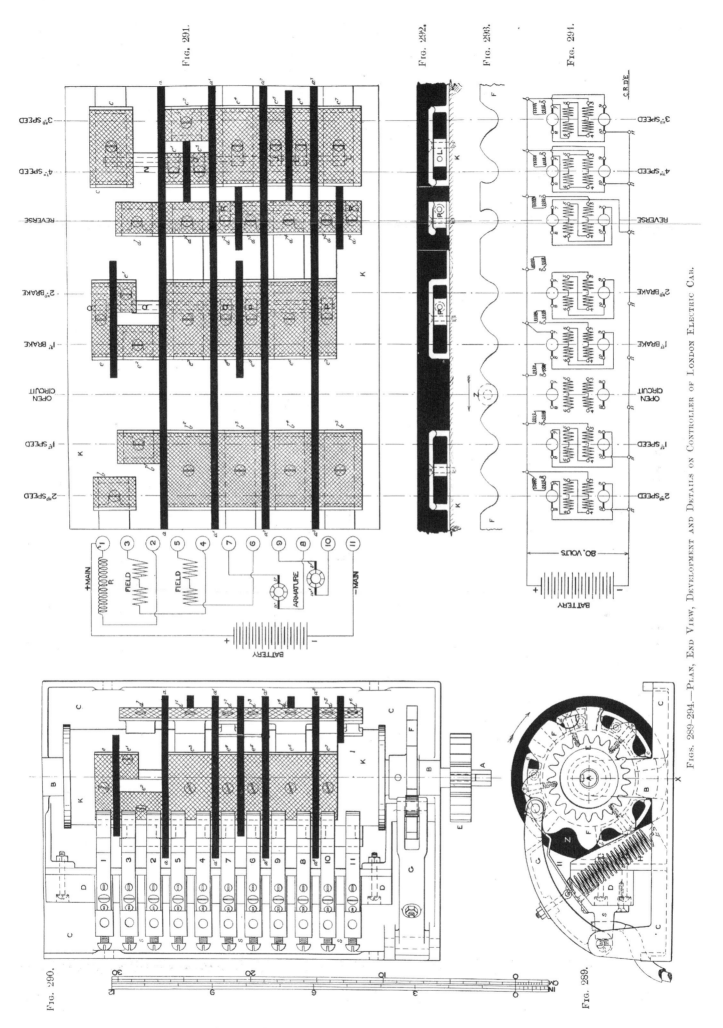

FIGS. 289–294.—PLAN, END VIEW, DEVELOPMENT AND DETAILS ON CONTROLLER OF LONDON ELECTRIC CAB.

The material originally positioned here is too large for reproduction in this reissue. A PDF can be downloaded from the web address given on page iv of this book, by clicking on 'Resources Available'.

ELECTRIC MOTOR VEHICLES

in which the direction of rotation of the armature is reversed, and consequently the movement of the vehicle also.

The manner in which these combinations are effected will best be understood by reference to Figs. 291-294, of which Fig. 291 shows a development of the cylinder K cut at the line A, X, Fig. 289, as before mentioned. Fig. 292 is a developed end elevation of the cylinder. Fig. 293 is a development of the ratchet wheel F cut along the line A, X, Fig. 289, and Fig. 294 shows diagrammatically the combinations formed by the controller in the eight different positions corresponding to the eight teeth of the ratchet wheel F.

Starting with the position of the controller cylinder in which the circuits between battery motor, etc., are all broken, i.e. when the pawl G is engaged with the ratchet F in the position along the line marked "Open circuit," Figs. 291 and 294, which position with reference to the end elevation of the controller, Fig. 289, would be that of the cylinder when it has been rotated so that the pawl G engages with the indentation Z of the ratchet wheel Q. It will be seen that the controller brushes will then have all disengaged from the contact pieces $e\ e'$, etc., and will not yet have become engaged with the contacts $b\ b'$, etc., so that they all lie in the clear space along the line "Open circuit," Figs. 291-294, and the connections are all broken, as indicated by the diagram Fig. 294.

If now the controller cylinder is rotated one tooth of the ratchet wheel in the direction of the arrow, Fig. 289, i.e. to the position along the line marked "1st speed," Figs. 291 and 294, the metallic contact pieces $b^1\ b^2$, etc., will have been moved under the controller brushes, and these will now be connected up in the manner shown by the diagram on the "1st speed" line, Fig. 294. These connections will be followed in detail with reference to Fig. 291. Starting from brush 1, this, it will be seen has not engaged with a contact piece on the cylinder and remains as before connected through the resistance R to brush 2; brush 2 is connected to brush 3 by the contact piece b^1; similarly brushes 4, 5 are connected by contact b^2, brushes 6, 7 by b^3, brushes 8, 9 by b^4, and brushes 10, 11 by b^5. Starting from the + terminal of the battery the path of the current is therefore from brush 1 through resistance R to brush 2, through contact b^1 to brush 3, round one of the field coils to brush 4, through contact b^2 to 5, and round the other field coil to 6; from here the current flows through contact b^3 to brush 7, and round one of the armature windings to 8, through contact b^4 to brush 9 and round the other armature winding to brush 10, and from here through contact b^5 to 11, and thence to the – terminal of the battery; it will thus be seen that the resistance R, both the field windings of the motor, and both the armature windings are in series, the maximum resistance of the circuit is therefore opposing the flow of the current, and the speed of the motor will be a minimum giving the 1st or slowest speed to the vehicle.

The combination for the 2nd speed is obtained by moving the controller

Operations of and by the controller

1st speed

2nd speed

407

drum another tooth of the ratchet in the direction of the arrow, Fig. 289, thus bringing the contacts b b^1 b^2, etc., which are along the line marked " 2nd speed," Fig. 294, under the brushes of the controller, and it will be observed that the only change in the combination from that for the 1st speed is to join brush 1 direct to brush 4, and cut out the resistance R altogether, as brush 2 does not engage with any contact piece on the controller drum. The diagram on the " 2nd speed " line, Fig. 294, shows clearly the change in the combination, and it will be seen that as the resistance R has been cut out of the circuit a proportionately greater current flows through the motor windings, giving a proportionately greater speed to the vehicle.

3rd speed   By a further movement of the controller drum over one more tooth of the ratchet wheel the combination for the third speed is obtained, the contacts along the line marked " 3rd speed," Figs. 291 and 294 being now under the controller brushes. It will be seen that no change in the combination has taken place, so far as concerns the resistance R and the field windings of the motor, but the armature connections have been varied. Starting from brush 6, which is on the contact piece c^4, the current leaving the field windings flows through the contact piece to brush 7, and thence to one of the armature windings at brush v, Fig. 291, but now returning to the contact piece c^4. It will be observed that this is connected by the cross wire J to the contact piece c^5 upon which rests brush 9, which is connected to brush v^1 of the motor; in other words, the brushes v v^1 are connected in parallel, and it will also be seen that as the contact pieces c^6 c^7 are connected by the cross wire L, brushes 8, 10 which rest upon them are in electrical connection, and therefore the brushes w w^1, by which the current emerges from the armature, are also in parallel, and consequently the combination indicated on the line " 3rd speed," Fig. 294, is obtained, namely, resistance R cut out, field windings in series, armature windings in parallel, the resistance to the flow of the current from the battery is consequently reduced by one and a half times the resistance of one of the armature windings with a corresponding increase in the power and speed of the motor and that of the vehicle also.

4th speed   To obtain the fourth and highest speed of the vehicle, the controller cylinder is moved against one tooth of the ratchet in the direction of the arrow, Fig. 289, which brings the contacts along the line " 4th speed," Figs. 291–294, under the brushes of the controller.

It will be observed that no change has taken place in the armature connections, but those of the field have been altered.

As before, brush 2 does not meet a contact piece, and the resistance R is still cut out of the circuit. Brushes 1 and 3 are connected by the contact piece c, and as the contact c is joined with c^1 by the insulated cross wire N, brush 3 is in electrical contact with brush 5. The current therefore enters the field windings simultaneously, and it likewise emerges simultaneously, as brush 4 on contact c^3 is connected with brush 6 on c^4 of the cross wire J. The combination is therefore that shown by Fig. 294 on the " 4th speed " line, *i.e.* resistance ʀ cut out, field windings in parallel, armature windings in parallel. The resistance opposed to the flow of the current has therefore been further reduced by the one and a half times the resistance of one of the field windings, with a corresponding increase in the power and speed of the motor, which gives now the highest speed to the vehicle.

Having now arrived at the limit of movement of the controller cylinder in the direction of the arrow, Fig. 219, to return to the "open circuit" position. If the cylinder is rotated one tooth of the ratchet wheel contrary to the direction of the arrow, Fig. 289, the electric equipment of the vehicle is so connected as to form a brake opposing its forward movement. Electric brake

The contacts along the line "1st brake," Figs. 291 and 294, will now be under the brushes of the controller, and it will first be observed that brush 11 does not engage with any contact piece on the drum, and that it is not connected in any way to brush 10, and consequently that the storage battery no longer supplies current to the motor, which, being driven by the carriage due to its momentum will now become a dynamo. Starting with brushes 8, 9 of the controller, which are connected to brushes v^1 w of the motor, these are connected together through the contact piece e^6. Motor acting as dynamo

The current generated in the armature windings may be considered to start from brush w, it flows through contact e^6 of the controller to brush v^1 of the other armature winding. It emerges from this winding by brush w^1, and flows to brush 10 of the controller, which is on contact e^7, and thence through the insulated cross wire ᴘ to contact e^5, and from brush 6 round one of the field windings to brush 5, which rests upon contact e^3. As brush 4 also rests upon contact e^3, the current now flows round the other field winding to brush 3, and thence through contact e^2 to brush 2, and round resistance ʀ to brush 1, which is on contact e.

As this contact is connected by the insulated cross wire ǫ to contact e^4, the current now flows by way of this wire to contact e^4, on which rests brush 7, which is connected with brush v of the armature winding, from which the current was assumed to start. It will therefore be seen that the motor, acting as a dynamo, is short circuited on itself through the resistance ʀ, and that it therefore tends to check the motion of the vehicle with a force equal to that generated within the closed motor circuit. These various connections are shown diagrammatically in Fig. 294, along the line marked "1st brake."

By moving the controller cylinder another tooth of the ratchet wheel in the same direction as above, a still more powerful breaking action is produced by cutting out the resistance ʀ and short circuiting the motor windings direct, as shown by the diagram along the line marked "2nd brake," Figs. 291–294. The final movement of the controller cylinder brings the contacts along the line marked "Reverse," under the brushes. These contacts connect up the brushes identically with the combination Reversing formed for the first forward speed, except that brush 11, which is connected to the terminal of the battery, and which now rests on contact d^6 of the controller, instead of being connected to brush 10, and so to brush w^1 of the motor, is connected by the cross insulated wire R to d^2, on which rests brush 7, so that the current enters the armature by brush v, and flows round it in the reverse direction to that for the first forward speed of the carriage, which will therefore now be given an equal speed, but in the backward direction. The electrical combination is shown by the diagram on the " reverse line," Fig. 294.

Motor mounting Returning to the mechanical details of the vehicle, it will be seen from Figs. 283, 285 and 286 that the tray bed-plate J, upon which the motor is mounted, is carried upon the underframe of the car by means of two longitudinal iron bars 13, 14, which at their front ends are bolted to the cross channel bar 15, which joins the longitudinal members F F of the underframe, rubber cushions 16, 17 being interposed between the carrying bars of the tray J and the surface of the channel bar 15. At the back ends of the carrying bars 13, 14, bosses are formed, through which passes freely the cross spindle 18, which is carried upon the back axle by means of wood blocks held between the double plates 19, 20. These blocks also clamped to the squared ends of the axle. It will be seen, therefore, that the carrying bars 13, 14 of the motor tray form also radius bars, which preserve a constant distance between the main axle and the countershaft L,

Transmission gear Figs. 283, 285 and 286, from which the power is transmitted to the rear road wheels by means of chain gear. The power is transmitted from the motor M to the countershaft L by means of a cut steel pinion 21, which is mounted at one end of the armature spindle c, Fig. 287, and which gears with a machine-cut bronze wheel 22 upon the differential gear N, which is seen in detail in Figs. 295 and 296. The spur wheel 22 is mounted freely upon the shaft L, and is bolted to a cast-iron ring 23, carried by arms from a boss, also turning freely on the countershaft. Two of these arms have bosses, which carry the fixed spindles 24, 25 upon which the bevel pinions 26, 27 of the differential gear revolve. The pinions 26, 27 gear with the bronze bevel wheels 28, 29, which are keyed, one on each half of the countershaft K. The whole of this gear, and the motor pinion are enclosed by a cast-iron gear case P, which is divided horizontally into halves, bolted together, and with bosses which embrace the exterior of the bearings 30, 31, Figs. 285 and 286, of the countershaft L. The bases of these bearings, and that 32 at the opposite end of the countershaft, are formed with the cast tray upon which the motor is carried. They have plain brass bushes, and are lubricated by means of oiling rings in the manner shown in the section of bearing 30 in Fig. 286. The counter-

Chain adjustment shaft drives by means of double pitch chains from the pinions 33, 34 at its ends, on to the sprocket-wheel rings 35, 36, which are carried from the spokes of the rear road wheels. The double roller chains, which are of Brampton make, are one of many forms of driving chain tried by the

410

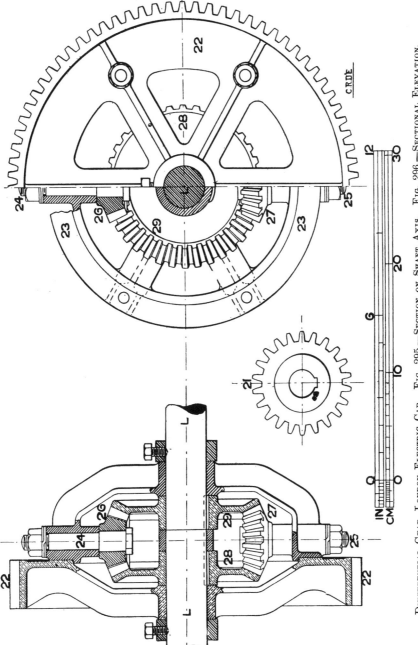

DIFFERENTIAL GEAR, LONDON ELECTRIC CAB.   FIG. 295.—SECTION ON SHAFT AXIS.   FIG. 296.—SECTIONAL ELEVATION.

411

Electric Cab Company. They are said to run silently, and give much less trouble from stretching than single chains. The adjustment for taking up the stretch in the chains is effected by moving the tray J, with the countershaft, along its carrying plate, which is done by loosening the bolts, one of which is seen at 35, Fig. 285, which pass through slot holes in the tray; then tightening the bolts 36 at the front ends of the carrying bars 13, 14, so that the countershaft may be moved parallel with the back axle and the chain adjusted to any desired degree.

Steering gear The steering of the vehicle is controlled by the hand-wheel R at the top of the fixed pillar T, in front of the driver's seat. The details of the stearing gear are shown in Figs. 297 and 298. On the spindle of the hand-wheel R is a worm 37, which gears with a worm wheel 38 at the top of the vertical spindle 39, which passes down the steering pillar T, and is guided at the top and bottom of the pillar by means of bearings bushed with white metal. At its lower end the spindle 39 carries a spur pinion 40, which gears with the toothed wheel 41, which forms part of the locking gear, supported by the front axle of the cab. Concentric with the toothed wheel 41, and carried therefrom by means of four arms (seen in Fig. 298), is the cylindrical boss 42 which guides the turning movement of the wheel 41 about the central fixed cylinder 43, which is formed with the upper fixed half 44 of the locking gear. A ball race is formed in the upper and lower locking rings, and between them are numerous steel balls, seen in Figs. 297 and 298, which reduce the friction of the locking plates to a minimum.

This steering gear, which differs entirely from that of the type usually adopted for motor vehicles, though it has often been used for traction engines, has been found satisfactory for the electrical cabs, in which a high rate of speed is not aimed at, and which require to be easily manipulated in confined spaces. The front end of the carriage is supported through the locking plate gear by single-leaf springs and bar hangers, as shown in Figs. 297 and 298, the locking gear being supported from the bar hangers 45 by means of the arms 46, 47, which at the centre of their length are of rectangular section, as seen in Fig. 297, and are there bolted to the lower locking plate. The upper locking plate is prevented from leaving the lower one by the bar 48, the ends of which project under the ring 42.

Band brakes In addition to the electric brake previously described, there are also two band brakes 50, 51, which act on drums fastened to the spokes of the rear road wheels. These brakes are controlled by the pedal 52, Figs. 282, 283, which is pivoted at 53 on the woodwork of the raised platform above the underframe. To the back end of the pedal lever is connected a chain V, which passes under the roller 54 carried from the underframe. This chain has at its back end a fork carrying a roller 55, over which passes the cross chain X, Figs. 283, 285, which at each side of the underframe passes through a hole in the frame-angle bars F F, and round rollers 56, 57, its ends being connected to the stud pins 58, 59, Figs. 285 and 286, at the upper ends of the levers 60. These levers are pivoted on

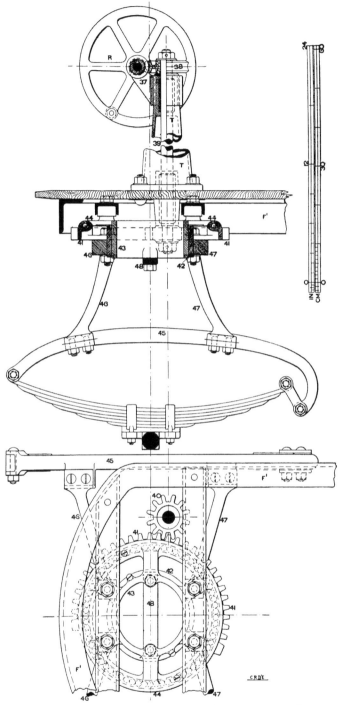

FIGS. 297 & 298.—STEERING AND LOCKING GEAR AND FRONT SPRINGS,
LONDON ELECTRIC CAB COMPANY.

413

the turned ends of the bars 62, 63, which are clamped under the springs on the main axle. It will be seen clearly from Fig. 286 that the fixed end of the band of each of the brakes 50, 51 is carried by the pins 62, 63, upon which the lever 60 are pivoted, while the free ends of the bands are carried by the stud pins 58, 59 at the upper ends of these levers.

Break of circuit on application of brake

Upon depressing the pedal 52 its first action is to break the connection between the battery and motor by opening a switch y, which is in the main circuit, so that the power is cut off before the brakes are applied. When the pedal is released the brake bands are returned to their normal position, free of the drums, by means of the spiral springs 64, 65, Fig. 285. The wheels of the cab are of rather light construction considering the weight of the vehicle. They have wooden spokes, and rims shod with steel bands which carry narrow solid rubber tyres—a narrow tyre being by experience found best for running on greasy pavements.

When these cabs commenced running they were fitted with Renolds' silent chain of insufficient width, and these were removed in favour of the Brampton chains referred to.

The results of trials of a delivery van built on the same system as these cabs, will be found on Table XXI.

Such then are or were the cabs of the London Company, and of which now thirty-six cabs and forty-one incomplete cabs are for sale. They were practical cabs, but not commercial cabs. They were very much too heavy in the frames, the method of suspension both of body and of battery-box involved the use of a heavy frame. They were too heavy, as a result of the separate conception in parts instead of as a whole, and they were heavy in every part. They were under-powered, and were too slow for people accustomed to London hansoms. After the novelty had worn off, public patronage fell off, and adverse criticism as to the rumble of the machinery and the jerking when stopping, starting, and slowing was very common. It cannot be doubted that improvements could be made on these cabs; but even the mystic power of the name " Electrical " will not haul cabs without sufficient foot-pounds, though it may attract the precious metals from what are usually considered safe places.

The " Riker " Electric Motor Carriages.

General description

Figs. 299–302 show a side elevation plan, and back and front elevations of a four-seated electric vehicle constructed by Messrs. Mackenzie & Sons, London.

In its mechanical details and design this vehicle is in all essentials the same as those manufactured in America under the Riker patents.

Fig. 303 shows a plan of the underframe and driving gear of the American vehicle as described in Riker's patent, No. 5570 of 1898, and Figs. 305 and 304 show details of the differential gear and pivoted steering wheel hub.

414

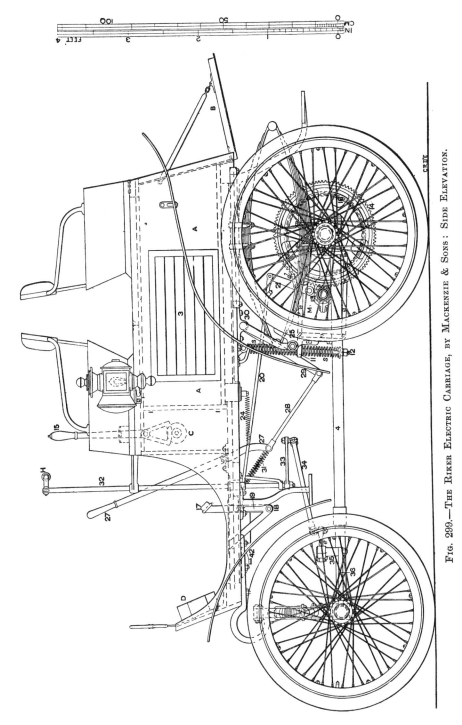

FIG. 299.—THE RIKER ELECTRIC CARRIAGE, BY MACKENZIE & SONS : SIDE ELEVATION.

Referring now to Figs. 299–302, it will be seen that the body of the vehicle is built on the lines of a dogcart, with seats back to back.

The storage battery is carried in the main portion A of the boot of the body, the bottom being completely enclosed, and of especially strong construction to support the weight of the battery. The back end of the boot is closed by a light canvas flap, and by the hinged wooden flap B, which when let down, as shown in Figs. 299 and 300, forms the footboard for the back seats. In the main part of the body under the front seats is a partition 1, which separates the battery-box from the smaller front box containing the controller c, by means of which the speed of the vehicle is varied, and the switch 2 for completely breaking the main circuit between battery and motor. A combined Weston volt and ampère metre, which indicates the discharge of the battery, is placed at D on the dashboard, so as to be always within view of the driver.

Battery

The storage battery consists of forty-four elements, having a capacity of about 130 ampère hours, at a discharge rate of 25 ampères. These are contained in four long boxes, which are inserted or removed through the open back end of the body. The connections between the four sets of cells are all brought to the front ends of the boxes, so as to be easily inspected, the connections of the terminals of the four sets of cells being made by binding screws after the boxes are in place. This arrangement of the storage battery in the body of the vehicle has one serious drawback, namely, the difficulty of removing and replacing the cells, as each of the four boxes is a considerable load for four men to move with safety.

The capacity of the battery is stated to be sufficient for a run of from 25 to 30 miles on level roads at 12 miles per hour.

Underframe

The underframe 4, 5 of the vehicle, from which the body is supported by means of leaf springs, seen clearly in Figs. 299 and 302, is constructed on lines closely resembling those indicated by Riker in his patent (see Fig. 303). The differences will, however, be seen by inspection of the two sets of drawings. The frame consists of two side tubes 4, 5, which are connected at their ends to the front and back axles, and form with them a rectangular frame of diminished breadth towards its front end. In addition to the side tubes 4, 5, there are two stay tubes 6, 7 between the front axle and the side tubes of the frame. The chief feature in the construction covered by Mr. Riker's patent is the method of connecting the side tubes with the axles, so as to allow of a considerable flexibility for the frame to adapt itself to the unequal heights which the wheels may assume when running over rough roads. This is effected in the following manner: The frame tube 4 and stay 6 are rigidly connected to the front axle by means of brazed joints, while those 5, 7 at the opposite side of the frame are loose on the axle and free to turn thereon, the tendency towards side movement being restrained by loose collars c c, Fig. 303, and c, Figs. 300 and 302, at the end of tube 5. At their back ends the side tubes 4, 5 pass through bosses below the cast bearing boxes 8, 9 of the main axle. These

416

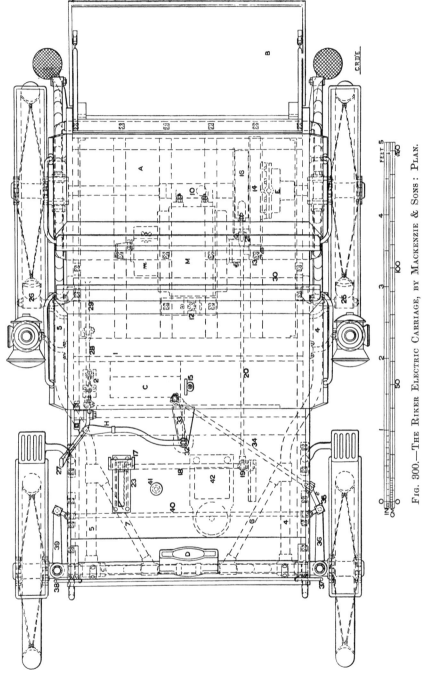

FIG. 300.—THE RIKER ELECTRIC CARRIAGE, BY MACKENZIE & SONS: PLAN.

bosses form bearings in which the ends of the side tubes are free to turn, end on movement of the tubes being prevented by collars in front and collars and screws seen at 4, 5, Fig. 301, at the back. The form, as designed by Mr. Riker, for permitting the same movement of the frame members, consists of a spherical outer bearing *b b*, as seen in Fig. 303.

It will be seen, therefore, that as the side tube 5 is free to turn on the

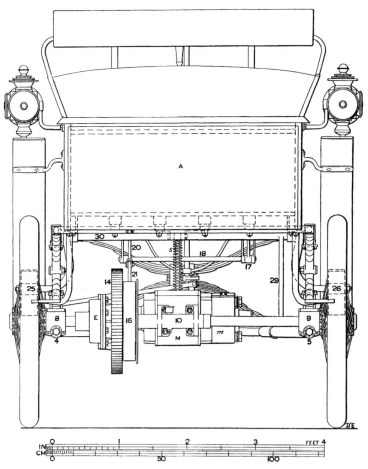

FIG. 301.—RIKER ELECTRIC CARRIAGE: BACK VIEW.

front axle in a longitudinal vertical plane, and in a transverse vertical plane at the bearing box 9, and that the side tube 4 is only rigidly held at its front end, the road wheels may be all at different levels without any dangerous strain being put on the frame tubes, the rigidity of the frame as against deformation in plan being unaffected.

Motor   The motor M is of the Riker two-pole drum armature type, giving about 2·5 B.HP. at 80 volts and 25 ampères, and 1,000 revolutions per minute. The field magnets, which are series wound, are constructed so as to form a box completely enclosing the armature. At one end of the enclosing case there is an extension m which surrounds the commutator and brushes. The bearings of the armature shaft are carried at the ends of the enclosing box, so that all the working parts are completely protected from mud and dust. The weight of the motor is about 175 lb. It is

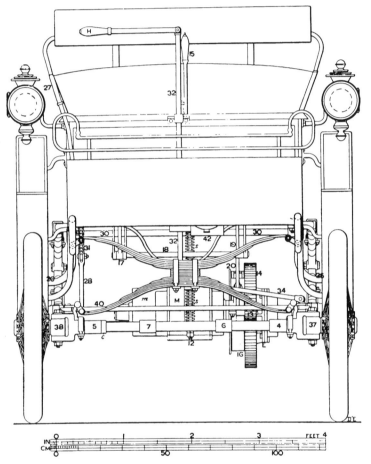

FIG. 302.—RIKER ELECTRIC CARRIAGE: FRONT VIEW.

carried at the back by means of the double bracketed sleeve 10, which is free to turn upon the main axle, and at the front end a projecting lug 11 is bolted to the field casting, and through a slot-hole in the lug passes an eye bolt 12, which is pivoted at its upper end from the body of the vehicle. Upon this bolt above and below the lug 11 are spiral springs s s, which are in compression, a washer and lock nut at the lower end of the bolt holding the springs in place. It will thus be seen that the motor is free to

419

swing radially in a vertical plane about the main axle, but that excessive movement is restrained by the springs s s.

The power is transmitted from the armature shaft to the back axle by means of a machine-cut steel spur pinion 13, Fig. 299, which gears with the bronze spur-wheel 14, carried upon the box E, enclosing the differential gear upon the driving axle. The method of carrying the motor so that it forms its own radius bar, as in the old locomotive practice, ensures the preservation of a constant distance between the axle and armature shaft.

The variations of speed of the vehicle are effected by means of the controller c, which is operated by means of the hand lever 15 at the centre of the front seats. The details of construction are closely similar to those of the controller used on the London electric cabs, described on pp. 406–410. There are three forward speeds, the maximum being about twelve miles per hour, and two reverse speeds. The combinations are obtained by varying the interconnections of the four sets of cells, into which the storage battery is divided, and not, as in the case of the London cab controller, by variously connecting the windings of the motor.

It will be seen that on the Mackenzie-Riker carriage the differential movement of the driving wheels is obtained by means of a compensating gear in the ordinary manner, the main axle being severed within the box E. Riker avoids the severance of the driving axle, and obtains a

through axle by means of the arrangement shown in Figs. 303 and 305. In this the tubular axle A is driven by the motor, the end of this tube carrying the crown B, which carries the short spindles H, C, C of the differential gear, which is enclosed in the hollow hub B of one of the driving wheels. One of the bevel wheels E, Fig. 305, of the differential gear is fixed to the hub D by a key, as shown. The other bevel F is fixed to the solid axle G, to the other end of which the hub corresponding to D is fixed.

The action of this differential gear is as follows : The short spindles c carrying the bevel pinions, being set in motion by the motor, the wheels E, F are driven at speeds corresponding to the relative resistances with which they meet, the wheel having the hub D rotating upon the part H of the solid axle G, which is of reduced diameter.

The hub containing this gear is only 5½ in. in external diameter, so that this differential gear, which in a way constitutes a driving gear, is necessarily of very small dimensions. The advantage of a through axle is thus attended with the disadvantage that the differential gear must be of but small dimensions. The wheel carrying it runs upon the extremity only of the axle, except in so far as it is supported by running on the hollow axle A at K, Fig. 305.

The motion of the vehicle is controlled by means of three brakes. The one used for all ordinary occasions is a band brake 16, which acts upon a drum bolted to the arms of the spur driving wheel 14 on the main axle. This brake is operated by means of a pedal 17 at one end of the cross spindle 18, which is carried by hangers bolted to the woodwork of the

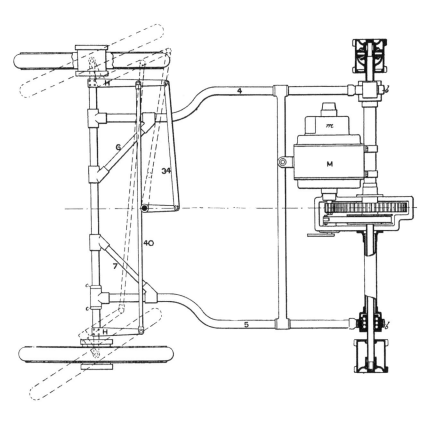

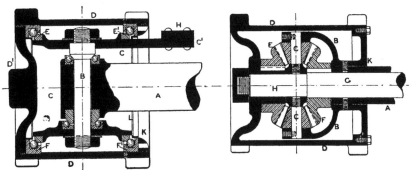

FIG. 303.—THE RIKER ELECTRIC CARRIAGE UNDERFRAME.
FIG. 304.—THE RIKER PIVOTED STEERING AXLE-HUB.
FIG. 305.—THE RIKER HUB ENCLOSED DIFFERENTIAL GEAR.

421

carriage body. At the opposite end of the spindle 18 is a lever arm 19, connected by the tension rod 20 to the long arm of a bell crank lever 21, which is pivoted at 22 upon a pin carried by a collar bolted over the armature shaft bearing at the driving end, as shown in detail in Fig. 306. One end of the band of the brake 16 is carried freely by a pin at the end of the short arm of the bell crank, while the other end is connected by a link d to the long arm. Upon depressing the pedal 17 the end of the long arm of the bell crank is pulled forwards, the rotational movement causing the band to be tightened on the drum at both ends simultaneously, the object being to make a brake which is equally effective when the car is

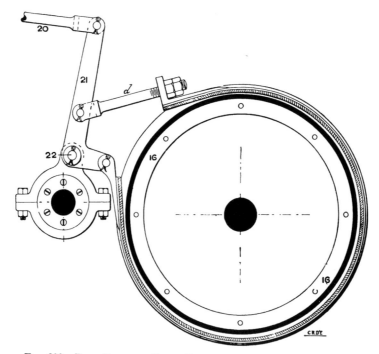

FIG. 306.—BAND BRAKE OF RIKER ELECTRIC CARRIAGE. SCALE, 1 TO 5·5.

running up or down hill. The pedal 17 may be held in any desired position by means of a projection on one side of the pedal lever, which engages with a ratchet cut in one side of the guide plate 23, Fig. 300, in which it moves. When the pedal is released, a spring 24, Fig. 299, pulls back the bar 20, and slackens the band on the brake drum.

The other brakes, which are only used on emergencies, are shoe brakes 25, 26, which are pressed upon the tyres of the driving road wheels. These brakes are operated by the hand lever 27, which is pivoted on the carriage body, and the lower end of which is connected by the link 28 to the end of a lever 29, which is forged on to the cross-bar 30, from the ends of which are carried the arms of the shoes 25, 26. When these brakes are not required, the shoes are held off the tyres by means of the spring 31, Fig. 299.

The steering gear of the vehicle is controlled by means of the hand-bar H pivoted, for transverse movement, at the upper end of the vertical bar 32, which is guided in bearings attached to the carriage body. At its lower end the bar 32 carries a long arm 33, which is connected by the rod 34 to an arm *e*, at the upper end of a short vertical standard 35 fixed to the end of the arm 36, which is welded on to the wheel pivot 37 at one end of the fixed front axle. The purpose of the vertical piece 35 is only to raise the rod 40, which connects the arms 36, 39 on the wheel pivots at either end of the axle, clear of the frame tubes. The steering axle is of the Ackerman type, each wheel being carried on a short axle pivoted on the ends of a fixed axle, forked at the ends. In principle it is the same as that of several of the carriages already described, although mechanically it is inferior because of the short distance between the limbs of the fork and the readiness with which the axle will wear on the pivot bolt.

The arrangement adopted by Riker is of an entirely different character, and may be explained with reference to Figs. 303 and 304, in which A is the fixed front axle, the end of which is enlarged to carry the pivot-pin B, which is fixed in the interior part C of the wheel nave D. On the inner part C are two hardened steel cones E E′, upon which run the balls in the ball races F F′, fixed in the outer part of the wheel nave. The wheel then revolves upon the inner tube C as axle, this tube being pivoted upon the pin B, which itself is carried in ball bearings. Extending from one part of the tube C is an arm C′, to which is fixed the steering lever arm H, seen also in Fig. 304. It will be seen that the ball races F F′ are respectively held in place by the cover D′ and the ring nut K, by means of which the ball-race adjustment is made. The balls being put in place at the end D′, the wheel is inverted, and the race F′ with its balls then put into place and the dust ring L finally inserted. (Patent 5570 of 1898.)

It will be seen that with this arrangement the steering pivot is exactly in the plane of the centre of the wheel, so that the impact of the wheel with any obstacle on the road is received directly by the pin B, and not upon the steering arm H or steering handle. This is a very desirable object, which to some extent is obtained by placing the pivots when outside the plane of the wheel at an angle, so that if projected their axial line would meet the ground at the point of tread of the wheel. The Riker arrangement has, however, the objection that access for the adjustment of the ball race fixings of the axle pivot pin B is difficult.

Warning of the approach of the vehicle is given by depressing the small foot-switch 41, which completes the circuit of the electric bell 42.

The wheels of the vehicle have steel tangent spokes and steel rims, carrying single-tube pneumatic tyres.

Steering gear

The Riker pivoted steering wheel hub

Chapter XXIII

THE ELECTRIC VEHICLE COMPANY'S ELECTRIC CABS AND BROUGHAMS

TWO forms of electrically propelled cabs have been designed and constructed for the Electric Vehicle Company by Messrs. Morris & Salom, of Philadelphia, and some of their cabs have reached London and Paris.

Their general arrangement is quite unlike that of the London Company's cabs, and the details are also different in important particulars.

General arrangement of parts The battery is carried in the main body of the vehicle, and by this means the jolting or knocking, due to the swing of the suspended battery box, checked by diagonal rods and pins, as in the London cab, is avoided.

The driving is by single reduction spur-gear pinions on the ends of the two Westinghouse motors employed, gearing in internal wheels on the driving wheels, which are in front. The motors being independent no differential gear is required.

The cabs weigh about 20 cwt., with their batteries, which weigh 11·1 cwt. **Weight** Pneumatic tyres, 5 in. diameter, were adopted with a pressure of about 60 lb. per sq. in., after trying 3-in. tyres; and wood wheels after **Tyres** trying various forms of cycle pattern and disc or plate wheels.

The battery trays are slid into the cab body from the back, and the contacts are automatically made as they slide into place, the batteries being **Battery contacts** permanently connected to contact pieces on the tray, which are out of circuit when the driver's switch is open.

Figs. 306A and 307 are side elevations of the cab and the brougham respectively of the Electric Vehicle Company of New York. Fig. 308 is a back elevation of the cab, and Fig. 309 a plan of the cab. Fig. 310 shows the positions of battery, controller, switches, and motors. Fig. 311 is a diagram of the several electric connection combinations of motors, battery, by controller.

All the machinery being mounted upon or carried by the axles no separate underframe is required, all necessary attachment being made to **Mounting of the machinery** sills or transoms of the vehicle body. From the plan and the back elevation it will be seen that the motors M^1, M^2 are, by a projection from the field-magnet castings, pivoted upon the fixed front axle A, the radial distance of the armature spindles with the spur pinions on their outer ends, is thus a constant whatever may be the movement of the tail 8 of

Two motors

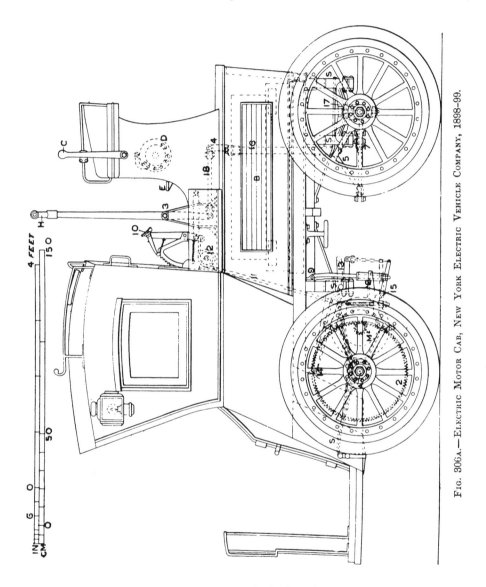

FIG. 306A.—ELECTRIC MOTOR CAB, NEW YORK ELECTRIC VEHICLE COMPANY, 1898–99.

the motor, suspended from the vehicle body by the rods pivoted at 9, upon which are rubber buffers JJ, Fig. 312, above and below the tail. The internal spur wheels, 2, are attached to the rims of the front driving road wheels, the tyres of which are 36 in. diameter, the hubs of these wheels are fitted with roller bearings with end-thrust balls and cones, the rollers being

Roller and ball-bearing axles

425

MOTOR VEHICLES AND MOTORS

about 5·5 in. long, and running upon a steel sleeve on the axle, which is
1·5 in. diameter.

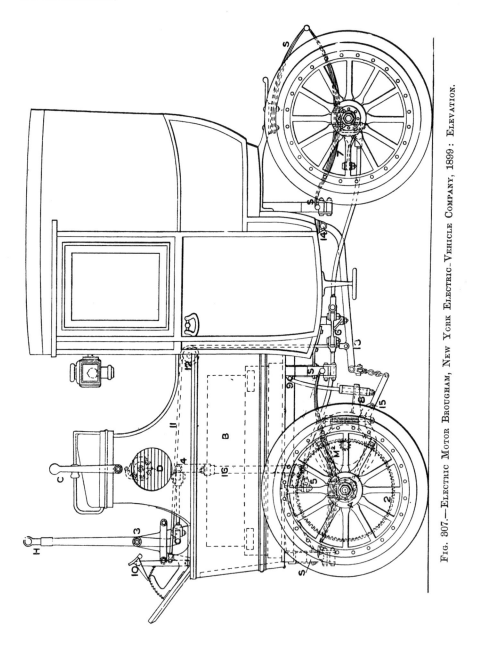

FIG. 307.—ELECTRIC MOTOR BROUGHAM, NEW YORK ELECTRIC-VEHICLE COMPANY, 1899 : ELEVATION.

Rear carriage and steering gear The steering axle, as will be seen from Figs. 308 and 309, carries at its ends steel castings 17 forming the forks for the short pivoted Ackerman axles, the pivots of which 17 are inclined so as to reduce the effect upon

426

the steering leverage of any obstruction met with on the road. These castings also provide bracket extensions, to which are pivoted the hanging links for the transverse springs s s. These transverse springs, as seen in

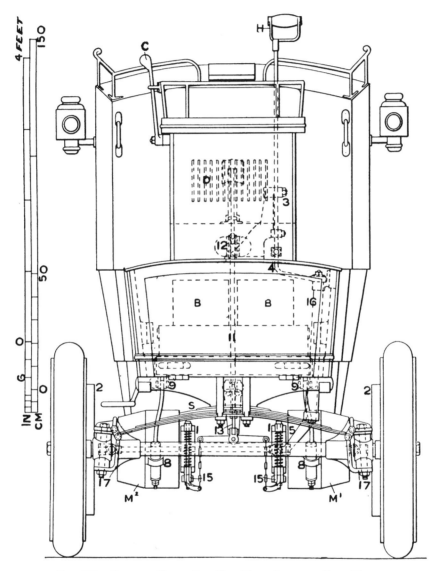

FIG. 308.—ELECTRIC MOTOR CAB, NEW YORK COMPANY : BACK VIEW.

Fig. 308, support the heavy battery-box end of the cab frame by the rear transom; the front springs of the cab are pivoted at their fore end to the sides of the body framework, while their rear ends are slung from the links on the ends of a transverse spring s, clearly seen in Figs. 308 and 309.

427

In the brougham both ends of the main springs are carried by transverse springs in the front of the vehicle, while at the rear end the side springs are partly carried by a transverse spring under the seat, and partly by half springs like the ordinary English broughams.

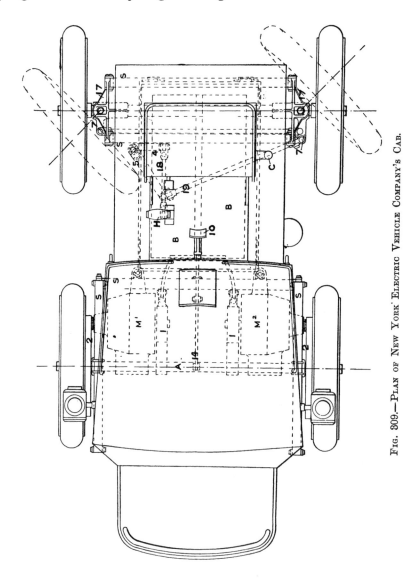

FIG. 309.—PLAN OF NEW YORK ELECTRIC VEHICLE COMPANY'S CAB.

Steering The steering is effected by means of a lever H working in a fore and aft direction, and pivoted at 3. At its lower end is connected a rod 18, coupled to an arm 4, on the top of a vertical spindle 16, on the lower end of which is an arm 5, downwardly inclined to and engaging with a rod 19, coupled at 7 to the steering arm of the near axle. The two steering arms

428

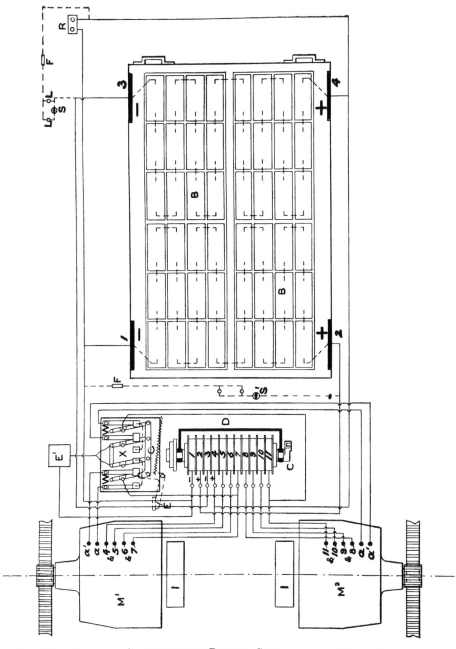

Fig. 310.—Scheme of Arrangement of Battery, Controller, and Motor Connections, New York Electric Vehicle Company.

429

are connected by a rod, as seen in the plan, Fig. 309, the off-axle arm being only of sufficient length for this connection. The arms are parallel when the vehicle is running in a straight course, there being no attempt at differential radiation of these wheels for running round curves.

Brakes The brakes are two in number, seen at 1 1, both being band brakes on drums on the armature shafts, and both worked by means of the pedal 10 on the top of a lever, the lower end of which actuates a wire cord 11, running over a pulley 12, and lifting a bar pivoted at 14, and carrying at its outer end 13 a short whippletree or balance lever, connected by two chains, Figs. 306 and 308, which act upon the brake levers 15 15.

Equal application of brakes These brake levers are pivoted to an extension of one end of the brake bands, and by means of a finger bearing upon the nut on the end of a bolt connecting the brake band ends draws them together. Spiral springs on these bolts release the brake bands when the chains 15 are lowered. Fig. 312 shows this brake-band tackle to a larger scale, as it may be looked upon as one of few good arrangements which are to be found on motor vehicles.

In Fig. 312, the arms E E' are shown as pivoted to the axle at A, and as lined where they rub upon the drum C on the motor spindle B, with wearing pieces, which are pulled

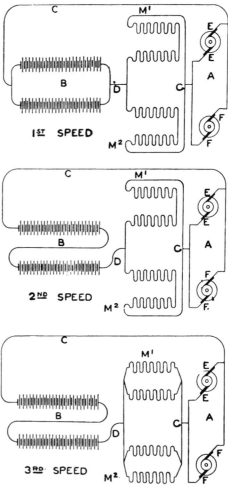

FIG. 311.—DIAGRAM OF ELECTRIC CONNECTION COMBINATIONS FOR THE THREE SPEEDS.

into their place by eye-bolts on pins F F'. One of the two brake levers L pivoted at 15 is shown in its lowest position, the break blocks being free. Upon raising the break connection rod 11, the bar 13 is moved between the guides G, the whippletree with its two chains 20 lifted, and with 15 as a fulcrum the break arm E is lifted, the lever L using the bolt 16 as a support for its short arm 17.

430

The motors are each of 2 HP. at a speed of 700 revolutions per minute.

The battery consists of 48 cells, the plates of which have been of various kinds during the experimental history of the service. A combination much favoured was multiples of two Manchester positive plates to three chloride negative plates, each cell made up of five such plates is said to have weighed 26 lb. and the whole battery is credited with a discharge capacity of 100 ampère hours at 80 volts at normal discharge rate, sufficient it was thought for an average of 25 miles over good town streets and roads.

The battery is divided into two sections, and the series-wound fields of the motors are also in sections, from this it follows that various series and parallel combinations may be made as required for the three speeds forwards or reverse.

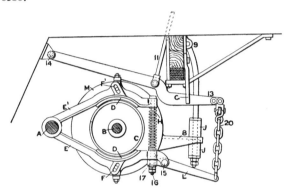

FIG. 312.—BRAKE GEAR OF CAB OF NEW YORK ELECTRIC VEHICLE COMPANY.

The position of the controller D actuated by the lever c is shown in Figs. 306–308, and with reference to its connections is seen in Fig. 310. In the latter figure the battery is seen at B B in its tray making complete connections to the negative contacts 1, 3, and the positive contacts 2, 4. At D the rotary controller is seen, having 11 contact plates coupled up by the leads shown and marked with numbers corresponding with the motor connections. At a, a', are the connections to the reversing and ordinary forward working switch x operated by the pedal E, seen also in Fig. 306. Ordinarily this pedal is maintained in the position for keeping the switches in contact, as shown in the engraving for forward running; for reversing, the depression of the pedal cuts out the right-hand set w of switch connections all of which are pivoted to bar G, and throws in those coupled with the wires a a', of each motor. At E' is indicated the existence of an emergency switch placed in an easily accessible position, so that by the removal of a key the driver is enabled to cut off all current either for stopping purposes or for leaving the vehicle safely. The speeds given by the operation of the controller are 6, 9, 12 to 15 miles per hour. At R, is seen indicated a socket for insertion of a dual plug for charging the

431

MOTOR VEHICLES AND MOTORS

batteries in place, and at s is a switch for two lamps within the battery boot; f indicates a fuse in this circuit. A similar fuse is used in the carriage lamp circuit controlled by the switch s′, under the driver's seat.

Speed variation connections
In Fig. 311, showing the controller connections for the three speeds, b represents the battery; d diagrammatically represents by one wire the whole of the connections shown in their entirety in Fig. 310; M^1 M^2 diagrammatically represent motor windings with the motor fields supposed to exist between them; and g in each case completes the circuit with the armatures supposed to be at a with the brushes e e, f f upon their commutators. In the upper or 1st speed diagram the batteries are in parallel while the field windings are in series, in the 2nd speed diagram the batteries are in series and the field windings in series, and in the 3rd speed diagram the batteries are in series and the field windings in parallel. In each case the motors are coupled together in parallel, with the controller in position for the arrangement shown in diagram, Fig. 311, for 1st speed. The batteries being in two lots in parallel series give a current at only 40 volts through field windings in series and armatures in parallel, so that the starting or running torque is the lowest possible. When in position for the arrangement for 2nd speed the full electromotive force of 80 volts is used with the same conditions as to motor, field and armature connections, while in the arrangement shown for 3rd speed the full electromotive force through field windings in parallel gives the highest discharge at the full voltage.

The charging station
A very complete arrangement of battery-charging and battery-shifting plant was erected in 1898 at No. 1684, Broadway, New York,[1] under Mr. C. H. Condict, by whom I was kindly conducted over the then not quite finished works in the autumn of 1898. The plant includes hydraulic **Charging batteries in cabs** tables on to which the cabs ran when they came into the depôt for recharging, and by which they were raised a few inches to place them in exactly the right position for command of the battery trays by a pair of claw arms which enter the boot and return pulling the battery tray out of the vehicle boot on to a continuous transversely-moving platform. As soon as the battery is on this platform the latter carries it and stops opposite one of several ranges of overhead travelling lifts by which the batteries are taken to position for recharging. By this same movement the travelling platform brought a newly-charged battery opposite the cab boot, and the hydraulic claw bars then operated as pushers and forced the battery tray into the boot, the whole operation taking not more than a minute.

In New York, as in most towns in the United States, any vehicles plying for hire have to contend with extremely adverse conditions as to roadways, and with what is far worse, the abominable system of street tramways everywhere, many of them laid with rails projecting above the paving, hence the use of very large tyres, which, big as they are, were not big

The Horseless Age, September, 1898, p. 9.

ELECTRIC CABS AND BROUGHAMS

enough to prevent the rapid deformation of the wheel rims first used, which acquired the appearance of many-sided polygons. As against this, however, the American company has the advantage of the very high cab fares common in America, especially in New York, while the cost of the batteries and of the current is no more than it is here; it is therefore not very surprising to find that these cabs are credited with having run extremely well under the circumstances of very adverse weather during the past winter.

FIG. 313.—COLUMBIA ELECTRIC MOTOR VAN.

ELECTRIC VEHICLES OF THE COLUMBIA COMPANY.

The Columbia Company of Hartford, Connecticut, one of the combination of manufactures of which Colonel Pope is the able head, has for some time been devoting a great deal of attention to electric vehicle construction. In a previous chapter I have described the carrier tricycle made at these works, and Figs. 313–318 illustrate the prominent features, by means of photographs of exteriors, an Electric parcels delivery van, a Stanhope carriage, and an electrical " Runabout " carriage : these have been designed under Mr. H. P. Maxim.

Fig. 313 is from a photograph of a van of pleasing design and the method of spring suspension, and Fig. 314 is a view of part of the same

Columbia electric vehicles

Columbia electric van

433 28

van looking in between the front and hind wheels, and showing the rear
spring mounting and the laminated spring support for the motor, which
at its other side is pivoted upon the driving-wheel axles.   The motor is en-

Fig. 314.—Columbia Electric Motor-Van: Enlarged View,
showing Spring Suspension of Motor.

Fig. 315.—Columbia Electric Stanhope Carriage.

closed in a conical-ended case, one cone enclosing a spur pinion form of
differential gear of small dimensions.   On the ends of the divided spindle,
which runs from the differential through the hollow armature shaft, are

434

pinions which gear into the internal wheels fixed to the drivers. The arrangement of enclosed motor and gearing involves a good deal of boxed-in mechanism and ball bearings difficult of access. These internal gear wheels are clipped to alternate spokes of the rear wheels by strap staple bolts, wood naves in this case being used. The weight complete is given as 3,900 lb., and the batteries 1,390 lb. The load carried is 1,000 lb. with two occupants. The expenditure in current with full load on good level roads is given as 285 watt hours per mile, and recuperation 440 watt hours. The stated current gives for 8 miles an hour about 3 electrical HP.

FIG. 316.—COLUMBIA ELECTRIC STANHOPE: FRONT VIEW,
SHOWING CENTRE-PIVOTED FRONT AXLE.

The vehicle by which the Columbia Company has attracted most attention to its designs is the Stanhope phaeton, shown by Figs. 315 and 316, one of which is owned by Mr. A. Harmsworth. Fig. 315 is a side view from the front, and Fig. 316 a front view showing how the underframe of the carriage is centrally pivoted upon the front axle, the frame carrying at either front corner a downwardly projecting horn, which take the thrust of the axle in a backward direction but permits it perfectly free play to follow the variations in the roadway. The batteries are carried under the two seats. The carriage weighs complete 2,550 lb. or 1 ton 2·7 cwt., the batteries weighing 1,100 lb. or 9·85 cwt. The expenditure of current on good level roads, with three occupants, is stated to be 144 watt hours per

Columbia Stanhope phaeton

435

mile, and recuperation current 220 watt hours per mile, showing a battery efficiency of about 65 per cent. The expenditure on good level roads represents at 10 miles per hour about 2 HP. The mileage capacity on level roads is said to be about 35 miles.

Fig. 317.—Columbia Electric "Runabout."

Fig. 318.—Columbia Electric Motor Carriage: Back View.

Columbia "Runabout" phaeton   Fig. 317 is a side view of a light popular carriage of the "Runabout" phaeton type, the whole of the body or boot being occupied by the storage

436

batteries. The arrangement of the motor, differential and gearing, differs from that of the Stanhope inasmuch as the rear driving axle is a live axle driven by single reduction gear through a differential. The weight of this car is 1,500 lb., of which the batteries weigh 700 lb. The expenditure in current on good level roads, with two occupants, is given as 125 watt hours per mile, or at 10 miles per hour about 1·7 electrical HP. The recuperator or recharging current is given as 195 watt hours, so that the battery efficiency is taken as about 65 per cent.

Fig. 318 shows the rear view of a four-wheeled dog-cart with the motor and differential carried upon a fixed axle and driven by internal spur wheels on the drivers. **Columbia dog-cart**

Batteries of many forms have been extensively experimented with by the Columbia Company, but their experience only confirms that of the battery users in this country. The Columbia Company do not make batteries, but fit the carriages with any makes which may be preferred. **The batteries**

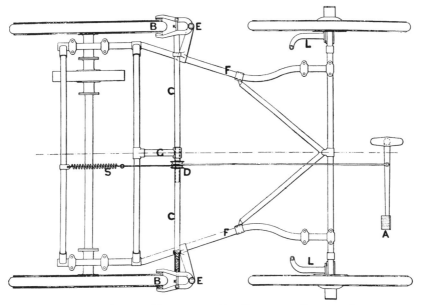

FIG. 319.—DIAGRAM PLAN OF UNDERFRAME, RIM BRAKE, AND STEERING AXLE OF THE ELECTRIC MOTIVE POWER COMPANY'S DOG-CART.

THE ELECTRIC MOTIVE POWER COMPANY'S MOTOR VEHICLES.

This Company exhibited two vehicles at the Richmond show of the Automobile Club in 1899, one of which was a four-wheeled, four-seat dog-cart, operated by a 5 HP. motor of the Mackey type supplied with current by 13·25 cwt. of Crowdus secondary cells, each weighing 24 lb., and giving on a four hours' discharge what was stated to be 264 watt hours. The body was mounted upon an underframe, which in general form and **Dog-cart** **Motor batteries**

437

construction is shown in plan by Fig. 319. The motor was pivoted upon the axle which it drove by single reduction spur gear, the spur wheel being the outside of the differential gear case seen in the diagram, Fig. 319. The front of the motor is suspended from the underframe after the manner of that adopted in the London and in the New York cabs. A feature of the car is the form of rim brake shown in the plan which is made under the patent (No. 360 of 1899) of Messrs. Northey & Dudgeon. Referring to the diagram, it will be seen that two pairs of gripper arms

Gearing

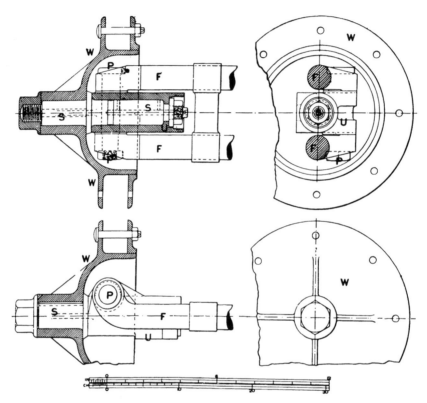

Fig. 320.—Clubbe & Southey's Pivoted Steering Wheel Hub and Steering Axle.

are pivoted at E E, and provided with leather facings where they grip the aluminium rings on the sides of the wheel felloes B B. The operation of the brake is as follows : Within the tubular shaft c c is a solid shaft K K, in the middle of which is a pulley D, and on one end of which is a screw and nut as shown. Over the pulley D runs a wire rope, kept taut by a spring s. By pushing the pedal A the rope operates the pulley D and thereby closes the grippers upon the wheel rims.

Rim brake

Another feature of the car is the Clubbe & Southey pivoted steering wheel hub and steering axle shown by Fig. 320 (Patent No. 17776 of 1897).

Pivoted steering wheel hub

438

This is an ingenious method of arriving at the object afterwards aimed at by Riker in designing the pivot hub wheel described with reference to Fig. 304. The arrangement now illustrated secures the further point that the vertical pivot, on which the steering axle moves, is a little in front of the centre of the axle, so that the wheel runs caster fashion and tends always to run in any position in which the steering gear has for the time placed it. In Fig. 320, F F is the forked part of the front fixed axle, the eyes on the ends of which receive the pivot pin P, which holds the axle bearing U, in which the axle s runs. The short end of the stud axle is fixed tight in the boss of the wheel hub w. At the point U in the sectional plan given in Fig. 320 is fixed by studs the steering arm L seen in Fig. 319. It will be noticed that the stud axle revolves with the wheel in a bearing which has only an horizontal angular movement on the pivot P, and which may be of any reasonable dimensions and easily lubricated. An objection to this arrangement is the shortness of the bearing of the pivot pin in the boss U.

By different combinations of the electrical circuits, four speeds, namely, 3, 6, 9, 12 miles per hour, were obtainable according to the arrangement described in the patent of E. J. Wade (11525 of 1898). The controller used for this purpose is so arranged with the connections of the battery that the circuit is always made on one combination before it is broken on another, so that the sparking across terminals, or the alternative momentary short circuiting of cells, is avoided. One of the combinations enabled the motor to act as a dynamo when running down hill for the purpose of recuperating the battery whenever the opportunity occurred. The total weight is stated to be about 26 cwt.

Electrical circuit combinations

Amongst the few who are working at the electrical vehicle problem is Mr. Carl Opperman, who has recently brought forward an arrangement of motor which drives by means of a worm on the end of its spindle, gearing with an out worm wheel on the outside of a differential motion wheel. The gear runs in oil, and if the motor spindle were fitted with a Mossberg thrust roller bearing the efficiency of the transmission, would probably be higher than several of the much more used forms of gearing.

Opperman's electric vehicles

Chapter XXIV

MODERN AND RECENT STEAM VEHICLES

Renewed
attention to
steam in
France

THE partial success reached a few years ago, by several designers and builders in France, with light and heavy steam vehicles of several kinds, has led to renewed attention in the United Kingdom to the steam engine as the prime mover for motor vehicles, especially for heavy work.   This attention has been the more readily given, because not only is

Advantages
of steam

the steam motor more readily manipulated and more elastic as to power and speed than other motors, but it is better understood and less liable to what were looked upon as uncertain freaks of the oil or spirit motor.   It will work under circumstances of neglect and absence of intelligence in its management that would make most oil motors stop for enquiries, and hence

High pres-
sures, high
duty, and
high-class
boilers,
engines and
fittings now
neccessary

it is a favourite ; but in its most recent forms it has lost much of the worship of its devotees of 1894–98, and is now looked upon with some suspicion because it has taken upon itself the high-bred, high-speed, high-pressure, and high-class material and high-duty attributes, which alone have made possible much that is successful in recent fire engines, firearms, machine tools, bicycles, motor cycles, and racing motor carriages.   Like other systems it has a place among the motive powers for vehicles, and as it is now in use on vehicles of from about 5 cwt. and upwards, that place seems

Wide appli-
cability of
steam

to have but dim confines.   To some enthusiastic engineers there are no limits to its field, but often this feeling interpreted means that most people think they can make a motor car—they do know about a steam engine and about a carriage, and they believe that an oil or spirit engine is an

Limits

inferior and complicated box of tricks.   This notion has caused many to embark on experiments which have proved that a steam carriage has many points in common with those propelled by any other power, many points in which it compares favourably and many unfavourably ; that in brief its comparative simplicity is more apparent than real, that it undoubtedly has an important place among motors, but that, although it was in the field long before the petrol motors, it was not until recently that it rewarded any of those who attempted to tame it and confine it to the dimensions, weight, and simplicity of an acceptable motor vehicle engine.

MODERN AND RECENT STEAM VEHICLES

The Bollée Steam Coach, 1880.

It is now twenty years since Bollée, sen., made the steam coach which The Bollée steam coach is illustrated in Figs. 283 and 284. In 1885 he made an omnibus which, with a few modifications, ran in the 1895 trials between Paris and Bordeaux; but, like all the steam road carriages, it proved to be an exceed-

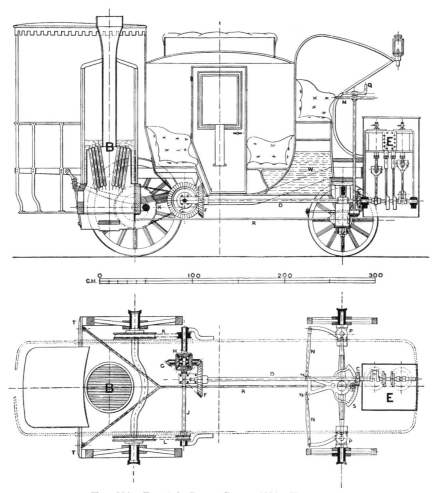

FIG. 321.—BOLLÉE'S STEAM COACH, 1881: ELEVATION.
FIG. 322.—BOLLÉE'S STEAM COACH: PLAN.

ingly troublesome thing to keep in repair. While in order, it would run Paris and Bordeaux race at a high speed; but, like the hare, had to make frequent stops, not for rest but for water. In running it broke many things, and its boiler gave much trouble, but it was the only steam vehicle of the six that started, that completed the journey; that is, only 16 per cent. of the steam vehicles, while

441

Retirement of steam vehicles from French races 53 per cent. of the petrol vehicles completed the tour. In the 1896 race, Paris–Marseilles, all the steam vehicles failed, while the whole of the petrol vehicles, twenty-nine, returned to Paris. In January, 1897, a Dion steam brake in the form of a tractor ran and won the Marseilles-Nice-Turbie race at a mean speed of 18 miles per hour for the whole run of 144 miles, but this was the last appearance of a steam vehicle in any races. This same vehicle had previously been through a great deal of trial and renewal. It weighed about 2 tons, and carried two men, but would haul a heavy carriage.

Arrangement of Bollée coach machinery Figs. 321 and 322 show the coach made by Bollée, of Mans, in 1880–81. It is interesting in several respects. Firstly, it will be seen that the arrangement of the engine E and its transmission shaft D is the same as that afterwards adopted by Panhard & Levassor for the Daimler motor cars, but with steam no manipulating clutch was required, a claw clutch c, not ratchet-toothed as shown, being used to connect the crankshaft and transmission shaft. A mitre wheel F on the end of **Differential gear** this shaft, geared with a similar wheel A on a differential gear H, also with mitre wheels on the transverse shaft J, which carried **Longitudinal main shaft** chain sprocket wheels on its ends. This shaft was carried by a central bearing and by bearings outside the sprocket pinions. The boiler B was of **Boiler** the vertical type, about 30 in. in diameter inside, with a number of vertical field water tubes fixed in the fire-box top plate and descending nearly to the fire. It was of the heavy type so commonly used to that date by all who worked in the intermediate period, no advantage being taken of the experience of Hancock and others. It had a large steam space, and no doubt generally gave dry steam. The chimney went through and just appeared at the top of the cab occupied by the stoker, fuel being carried on the foot-plate and at the sides of the boiler. Water was carried in a forward tank w. The first part of the coach was carried on transverse springs, the ends of which were carried by lugs on the top and bottom of sockets P of the vertical pivots of the Ackerman short-arm axles which Bollée used. These sockets were fixed by central bosses on the ends of the fixed axle. The weight was thus carried by the springs bearing directly upon or very close **Steering axles and fore gear** to the vertical pivots, and not by the cross axle. The steering was done by the wheel M on the top of a tubular spindle within the steering pillar. The bottom of this tubular spindle carried a spur pinion, which geared with an internally toothed sector s pivoted on the cross axle. Formed with the sector was a rear projecting arm to which were pivoted at n n the steering rods N N connected to the arms on the pivoted short-arm axles. Through the tubular steering spindle **Shoe brakes** passed a spindle carrying the brake handle Q at the top, and at the bottom a bevel pinion gearing with a bevel pinion forming a nut on the screwed end of the rod R, which pulled the brake blocks at T T upon the back of the driving-wheel tyres. The wheel base was 7 ft. 4 in. The front wheels were 32 in. in diameter, and the driving wheels 42 in. The gauge centre to centre of wheels was 4 ft. 6 in. for drivers, and 4 ft. 10 in. for front wheels.

The coach was a roomy vehicle, about 5 ft. 3 in. from front to back inside, and the front seat was well protected by a large hood.

It will be noticed that there were in this "caléche" several points in the arrangement of the motor and transmission gear which have since been represented in recent vehicles, but the vertically bent iron bar frame extending all round the vehicle has not been imitated, and probably was not any too effective in maintaining the proper distance between the main clutch and the mitre driving gear. **Anticipating recent practice**

The carriage is said to have run at 18 miles per hour on the level, and to have ascended gradients of 1 in 12 with ease. Its coal consumption per mile is given as 3·6 lb. on the level and water at 60 gallons. This, at 18 miles per hour, represents an evaporation of 9·25 lb. of water per lb. of coal, and a higher evaporative efficiency at lower speeds, which makes the record very doubtful. **Speed** **Fuel consumption**

443

Chapter XXV

SOME SERPOLLET STEAM VEHICLES

The Serpollet instantaneous generator IN 1889 Leon Serpollet invented and made the instantaneous generator or boiler which is so widely known by his name. As at first made, this generator was composed of a number of flat tubes, with only a capillary water space. The tubes were surrounded by a coating of cast iron, cast upon them, and this made them heavy; but the cast iron protected the

FIG. 323.—SERPOLLET STEAM VICTORIA, 6 HP., 1899.

wrought iron or steel tube from rapid corrosion in the high heat of the furnace in and above which they were placed. It also acted as a heat accumulator during the time when the engines were stopped and no water was being pumped through for evaporation. The boiler gave very high pressure steam considerably superheated. Various modifications of form were

444

SOME SERPOLLET STEAM VEHICLES

subsequently given to the tubes and their arrangement and connections, and the water space was increased to $\frac{1}{16}$ in., and afterwards to $\frac{1}{8}$ in. in the smaller boilers, and to about $\frac{1}{4}$ in. in the large boilers, up to 70 HP., used in the very large tramcars in Paris and in the postal and other railway cars used on the Northern of France and other railways.

In 1894 Serpollet made a steam carriage fitted with one of these boilers, and in 1895 one of these carriages was brought to England and tested by Professor A. B. W. Kennedy. Reports of the trials were published in some engineering journals of the time, but as the form and arrangement of the

Serpollet
1894 car

FIG. 324.—SERPOLLET STEAM WAGONETTE, 6 HP., 1899.

boilers, engines and carriages have been all changed since that date (see references on p. 3) I shall here only describe some of the main features of the Serpollet carriages and machinery as now made with boilers fired with kerosene.

Fig. 323 is from a photograph of a Victoria of about 6 HP., with oil tank in front, and Fig. 324 shows a four-seated carriage also of 6 HP., with water tank in front, Fig. 325 being a side elevation of this same carriage, with indications of the positions of the boiler B, engine M, and condenser C.

Serpollet
victoria

The steam generator used in these carriages is illustrated by the engraving Fig. 326, which also shows one of the forms of petroleum burners employed.

Steam
generator

The tubes are enclosed in a double casing packed with asbestos E, the lower part of the casing forming the firebox J, in which is placed the 12-jet burner shown. Immediately above and surrounding the burner,

445

which rests on the angle irons just above the air holes κ, is a coil of round tubes c. These tubes receive the water and pass it on to the zigzag coils B, which nearly fill the case-enclosed space. Finally the heated water and steam pass into zigzag coils of twisted flat tubes A, which deliver super-heated steam. The three sets of coils are connected by bends and unions outside the case, as seen at F F, but in a closed-in space H. Four of the unions are shown, but the bends connecting them are only seen in section by six

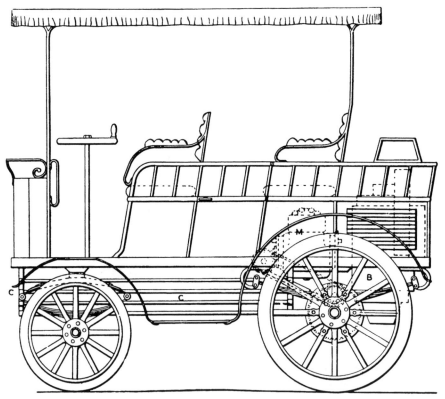

FIG. 325.—SERPOLLET 6-HP. STEAM WAGONETTE: ELEVATION.

rings in the upper Fig. 326A, and in the plan 326B one of those connected to the coils B is seen.

Heating surface   The total length of the tubing of the nominal 6-HP. boiler is about 95 ft., and the heating surface about 25 sq. ft.

Air for supporting combustion enters holes in the lower part of the outer case, as shown by dotted lines, and is then directed downwards by the inner lining so as to be delivered suitably to the burner, and almost un-affected by windy weather.

Oil Burners   The burner, two views of which are shown by Figs. 326D, consists of a

446

cast brass tubular frame carrying twelve Swedish burners, as seen in the end elevation. Oil enters the burner at B, controlled by a small valve on the inner end of the stem C.

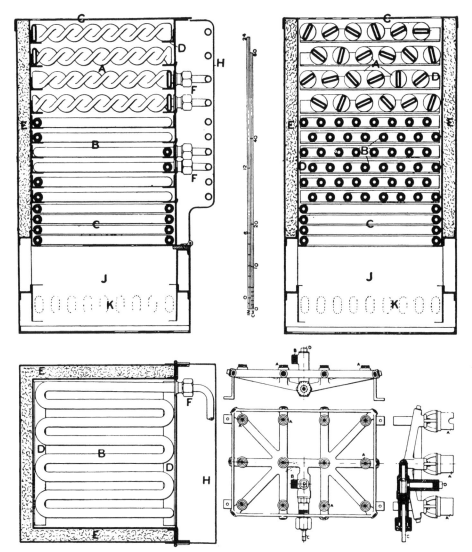

FIGS. 326A, 326B, 326C.—SERPOLLET STEAM GENERATOR, 6 HP.   FIG. 326D.—HEATING BURNER.

Another form of burner used by Serpollet is the Longuemare, shown by Fig. 327. It is provided with two regulating valves H H, one controlling the oil admission from the pipe A, through B, and thence through the vaporising coil C, to B¹, whence it finds its way to the small annular space surrounding the stem K of the smaller conical burner by means of the pipe

Longuemare type

447

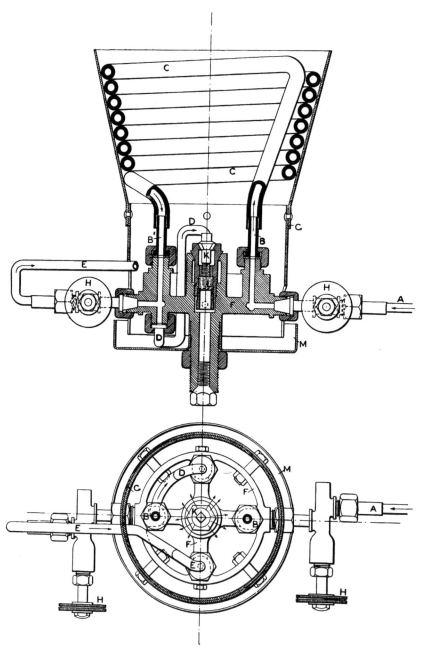

FIG. 327.—LONGUEMARE OIL BURNER USED BY M. SERPOLLET FOR SMALL VEHICLES.

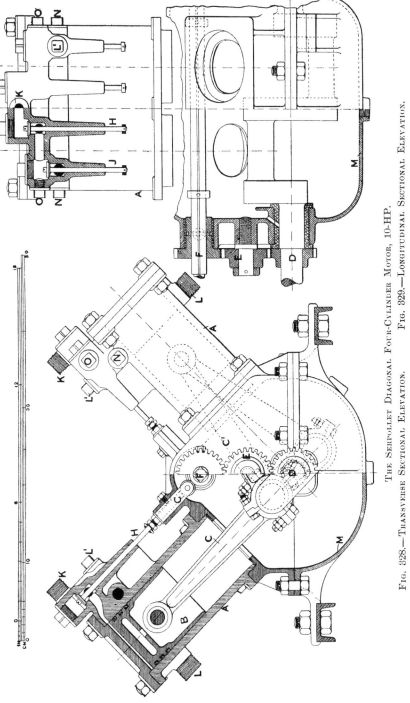

THE SERPOLLET DIAGONAL FOUR-CYLINDER MOTOR, 10-HP.

FIG. 328.—TRANSVERSE SECTIONAL ELEVATION.   FIG. 329.—LONGITUDINAL SECTIONAL ELEVATION.

D, after passing through the gauze at L. By means of the second controlling valve some of the vapour produced, it may be the larger part of it, passes by the pipe E to the annular space round the larger conical burner, which contains the smaller burner K; by closing the last-mentioned valve only the small burner is left at work. Mr. Serpollet has made another form of burner of the vaporiser type for heavy oils.

Four-cylinder diagonal engine

Figs. 328 and 329 illustrate partly in section the four-cylinder diagonal engine of 10-HP by M. Serpollet for the larger wagonette and other cars. This engine is rated as 10-HP. This of course it may easily be with the very high pressure available, say up to 250 lb., in the boiler, and for a heavy hill climb, if necessary, 200 lb. mean pressure in the cylinder. But

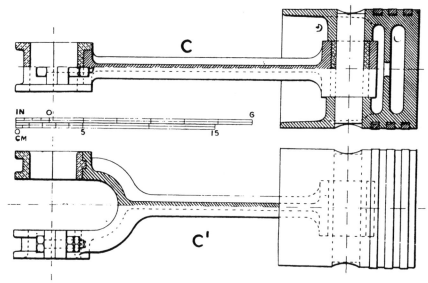

FIG. 330.—CONNECTING RODS AND PISTONS OF SERPOLLET DIAGONAL 4-CYLINDER ENGINE.

Horse-power of engine

with 100 lb. mean pressure the horse-power of the engine at 720 revolutions per minute would be 5·3 HP per cylinder, or say 21 indicated HP. in all, giving at least 17 B.HP. The cylinders A are placed at 45°, and are of 3·35 in. diameter and 3·35-in. stroke. The pairs of cylinders being placed exactly opposite each other, the connecting rod of one of a pair being forked and the other single at the big end as shown in Fig. 330. The

Cam moved valves

steam supply from the boiler enters at K, and is admitted to the cylinder by the mushroom valve on the stem H, actuated by a cam on the spindle F, which moves the sliding piece G, fitted with a roller bearing upon the cam. The camshaft rotates in the same direction as the crankshaft D, driven by spur wheels on either side of them and an intermediate wheel E, by which similar direction of rotation is secured. Similar cams on the same shaft operate the exhaust valves J, which deliver into a pipe surrounding

SOME SERPOLLET STEAM VEHICLES

the cylinders, and which has an outlet at L; N, O, L¹ are plugs closing passages to the several valves. It will be noticed that there is one defect which militates against the economical working of the engine, namely, the large size of the port space and passage between the admission valve

Clearance
loss

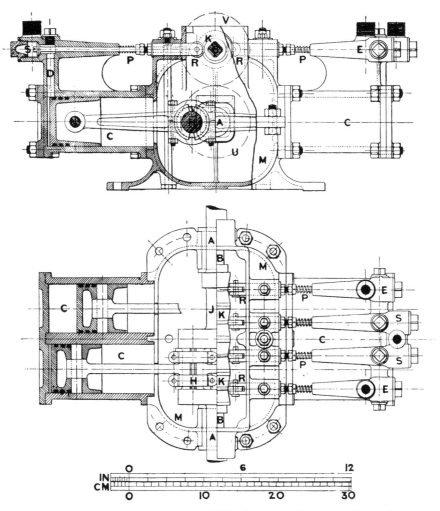

Fig. 331.—Sectional Elevation of 6-HP. Serpollet Horizontal Steam Motor.
Fig. 332.—Sectional Plan of Serpollet 6-HP. Engine.

and the exhaust valve, showing a considerable volume, which is filled with steam at every outward stroke which does no work.

Figs. 331 and 332 illustrate in sectional elevation and sectional plan the 6-HP. engine used in Serpollet vehicles, such as that shown in outline by Fig. 325. In this the crankshaft A is operated by four oppositely-

Four-
cylinder
horizontal
engine

451

Two connecting rods on one crank

Cam valve motion

Power of engine

placed cylinders c, with pistons and connecting rods, the by-ends of which are both of the full width of the crank-dip bearing, but which only embrace about one-third of its circumference. Outside the projecting parts of these big ends are clamp brasses H, Fig. 332, which keep these rods in place. The arrangement of the steam and exhaust valves and ports is in principle the same as that in the previously-described engine, the admission valves s s being placed between the exhaust valves E E and operated by sliding pieces having rollers R R, operated by cams K K on the spindle B, driven by the spur wheels U, V. The cylinders are each of 2·55 in. diameter and of 2·55 in. stroke. The estimated indicated power

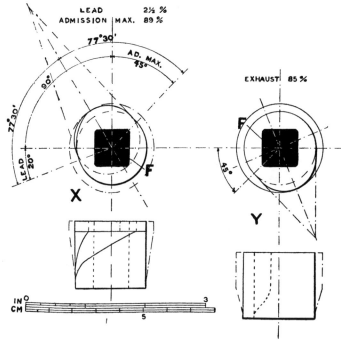

FIG. 333.—FORM AND SETTING OF THE CAMS OPERATING THE VALVES OF THE SERPOLLET STEAM MOTORS.

of this engine with an assumed mean pressure of 75 lb. per square inch will at 700 revolutions per minute, the pistons being single acting—

$$\text{I.HP.} = \frac{4 \text{ P L A N}}{33,000} = \frac{4 \times 57 \times 0\cdot212 \times 5\cdot1 \times 700}{33,000} = 6\cdot9$$

Thus with the high pressure available the power may easily be for a short time as much as 15 HP. actual, while the engine may work economically with the earliest cut-off which the cams can give. The cams just referred to are set out and placed upon the square shaft that carries them,

Reversing and varying cut-off

both steam and exhaust valves, as shown in Fig. 333. It will be seen in

the engraving, Fig. 332, that there are cams along nearly the whole
length of the shaft B, and that the rollers are resting upon only one of
four pairs of cams, one of each pair of cams moves the rollers R and the
valves correctly for going ahead and the other pair for going backwards.

M. Serpollet has devised and used two forms of automatically regulated oil and feed-water supplies to the boiler. Most of the makers who
have adopted an automatic regulation of the oil supply to the boiler
burners have used a diaphragm acted upon by the steam pressure, the feed

Automatic oil and feed-water regulators Type 1

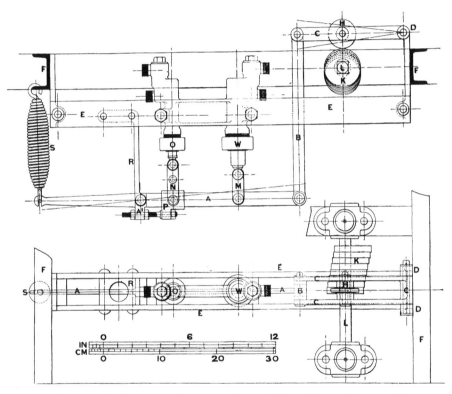

Fig. 334.—Serpollet's Adjustable Automatic Oil and Water Feed to Boilers : Elevation.
Fig. 335.—Plan of Serpollet's Automatic Oil and Water Feed Pumps.

pump being in most cases continuous acting, water delivery to the boiler
being controlled by a bye-pass, which allows the water to return to the
feed-tank instead of going into the boiler. M. Serpollet uses instead one of
the arrangements shown in Figs. 334, 335 and 336. In Figs. 334 and 335,
A is a vibrating lever, to which angular movement is given by the link B
and arms C, pivoted at D D. The arms C are actuated by the cam K, upon
which rests the roller H. The cam K may be moved from the position,
shown in Fig. 335, towards the other end of the spindle L, which carries
it, thereby obtaining for the lever A a stroke which varies from zero with

453

the cam in the position shown, to a maximum when the cam is at its other position. The cam thus gives a variable movement to the lever A, but the relation between the strokes given to the plungers of the oil pump O, and water pump W remains unaltered. For the purpose of varying this relation, the links N, by which the oil-pump plunger is operated, are pivoted upon a sliding block P, the position of which may be varied by means of the screw adjustment at A below the fulcrum of the vibrating lever carried by the bracket R. By thus altering the stroke of one pump the relation between the output of both is varied. The apparatus is mounted in the underframe F, and the roller H is kept always in contact with the cam by means of a spring s of small range.

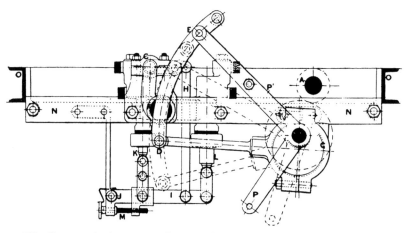

FIG. 336.—SERPOLLET'S ADJUSTABLE OIL AND WATER BOILER FEED PUMPS : ELEVATION. TYPE 2.

Type 2 In the arrangement shown in Fig. 336 the two pumps are worked by the vibrating lever I, pivoted at J. This lever receives its motion from a connecting rod or link H, which is attached to a spindle between bearings C On the visible end of this spindle is a rocking lever carrying a circular block F, within which fits the sector suspended at E from the arm P', the sector working after the manner of a well-known form of link motion. At its lower end is pivoted at D the end of an excentric rod which has a stroke proportional to the excentricity of the sheaf c on the spindle B. In the position shown the angular vibration given to the sector and to the lever F is a maximum, and the greatest stroke is given to the plungers K, L. By dropping the sector and excentric rod to the position indicated by dotted lines, which is done by lowering the bell crank P P', the greater part of the movement due to the excentric takes place in the sector link, **Adjusting** vibrating with the circular block F in its housing whilst imparting very **relation** little motion to it. Thus the rocking shaft carrying the suspending and **between oil** operating link H gives but small motion to the lever I or to the pump **and water**

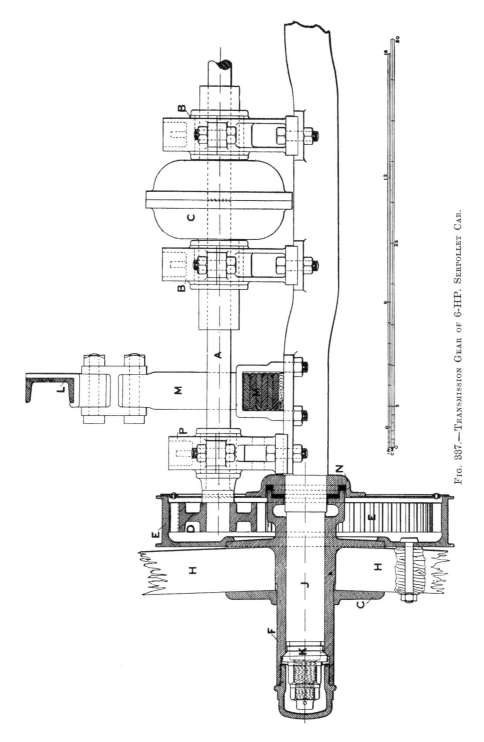

FIG. 337.—TRANSMISSION GEAR OF 6-HP. SERPOLLET CAR.

455

plungers. The stroke of both the latter may be thus varied simultaneously by varying the position of the sector through a very considerable range. To alter the relation between the lengths of the strokes of the oil and water plungers so as to meet differences in evaporative efficiency of oils used, the stroke of the oil pump K is altered by shifting the block adjustable by means of the milled-headed screw M.

Trans-mission gear

During the past year M. Serpollet's arrangements of engine and transmission gear have received several modifications, and I am unable to give complete plans of the steam and transmission connections. Fig. 337, however, shows the arrangement of the main axle with the transmission shaft and spur driving gear. On the axle J are mounted the two bearings B B carrying the hollow sleeve ends of the differential gear in the case C. Within these sleeves run the solid shaft A, in two parts, meeting within the differential gear-box C and at their outer ends supported by bearings P, beyond which they carry the pinion D, gearing with and driving the internal spur wheel E, fixed upon the flange of the hub F, and kept in place through the spokes H of the driving-wheels. Motion is given to the shaft A through the differential gear by means of a strong chain running on a sprocket wheel fixed upon the gear sleeve near the letter A, and upon a similar sprocket wheel on a transverse shaft in front of the engine, as indicated by dotted lines in Fig. 325. The internal spur wheel E is closed against dust by a plate attached to the collar N, the plate being edged with leather.

Only some of the features of the Serpollet cars are thus described, a consolidated or standard pattern not having yet been fixed upon for these small cars. It must remain to a future time to show more completely what the Serpollet car becomes. The very successful running of the Serpollet car in the Nice races (see Table) shows that M. Serpollet has again come to the front to take the position he occupied before he left the road vehicle to devote his attention to the steam tramway cars.

Use of superheated steam and mushroom valves

It may be here mentioned with regard to the engine that the very high temperature and pressure of steam used have been one of the chief reasons for adopting the gas-engine type of admission and exhaust valves. Slide valves had been used until early in 1899, but their use involves the admission of large quantities of lubricating oil, and this rapidly prevents the proper performance of the condenser, which, as well as the feed tank, becomes oil coated.

Engine oil in condenser

To prevent this, however, M. Serpollet adopted a form of separator which considerably modified this objection, but for all this it was thought desirable to keep oil out of the steam as much as possible, and for the very highly superheated steam used to adopt the form of valve shown, and thus to secure the high economy which attends the superheat.

Chapter XXVI

SOME AMERICAN LIGHT STEAM CARRIAGES

FOR other examples of light steam motor vehicles which have assumed anything like practical shape, we have to go to the United States, where, amongst others, there are the cars of C. E. Whitney, now being introduced into this country by Messrs. Brown Bros., and the very similar cars of Stanley Bros., built in the several works of the Locomobile Company of America, and of the Stanley Manufacturing Company. These cars are distinctive in their design throughout. The car itself is distinctly utilitarian, especially that form known as the "Runabout" car.

The design and arrangement of the boiler, engine and transmission gear illustrate well another characteristic of America in the application of machinery, namely, the careful study and development point by point in detail of a thing or machine, the principle and practice in which are already well known on broad lines. It would be difficult to-day for any one to devise an entirely new boiler or a new steam-engine, and so much has been done in the design and construction of both these things that there remains now as conceivable lines of invention only, such as those which have occupied Stanley and Whitney, namely, the invention of specially devised details and arrangements for obtaining a specific object. While several firms in this country and in France have been devoting time and material to the design and construction of entirely new boilers and motors, Whitney and Stanley have both selected well-known types and have carefully developed them for their particular requirements, namely, the propulsion of light road vehicles by what they conceive to be the best means.

While in England and France, for instance, all, perhaps, with the exception of Messrs. Des Vignes & Cloud, have occupied themselves with various forms of water-tube boiler, several of them following the lead of M. Serpollet in the use of either an instantaneous generator or a combination of instantaneous generator with small water reserve, largely with the object of obtaining a *générateur inexplosible*, Whitney and Stanley, on the other hand, have taken extremely simple forms of vertical fire-tube boiler, and, as one may say, treated it to a course of high refinement. They have not

Whitney and Stanley cars

American development of details

Boiler design, old and new

Water-tube boilers

Smoke-tube boilers

457

"High bred" boilers

complicated it, except in so far as they have increased the number of precisely similar parts—namely, the tubes. Then having a high-class boiler of known type, they have assumed that for the purposes of such boilers the owners can afford the use of an expensive fuel, and they have adopted that which will give a hot and perfectly clean flame with very little trouble—namely, American stove gasolene, with which, again, they have had large experience. Though living in the home of the petroleum industries, they have refused to allow the attainment of their object to be delayed by attempts to use the cheaper petroleum products, such as kerosene, the complete combustion of which, under the very varying conditions of motor-vehicle working, they have learned enough to know is extremely difficult of achievement and difficult to maintain in the hands of any but experts.

Expensive fuel necessary

Engines

For their engines they have again made a judicious selection of the available, have put it into the motor vehicle designers' refinery, and with already available transmission gear have produced the vehicles which are now to be described. By using the smoke-tube boiler, they are able to get dry saturated steam, and have no difficulties with the use of ordinary slide valve and link motion. The engines are very small, but the criticism which calls them toys does not unsuit them for their work. They have not yet, however, accomplished that which is looked upon as almost essential by European designers, namely, the condensation of the steam so as to economise water, and so as to use pure water in the boiler and increase the distance that may be run on one supply. Leaving a cloud of steam behind it as it runs may not be a sin on the part of a motor vehicle in the United States, but it is in this our land of critics, and improvement in this respect will have to be made if these cars are to run in most weathers in the towns of England. These little steam cars appear very simple, and they are so in their transmission and motor mechanism ; but when the whole of the boiler parts and connections are considered, the total probably exceeds those of some of the petrol motor cars.

Small, but not "toys"

Condenser required

THE STANLEY STEAM CAR.

Stanley car

Figs. 338 and 339 show respectively a side elevation and a plan, and Figs. 344 and 345 a back and a front elevation of a two-seated Stanley steam car of the "Runabout" phaeton type, manufactured by the Locomobile Company of America, who have adopted this type of vehicle as their standard model, considerable numbers having already been made. Every part of the vehicle is constructed on the lightest lines possible. the total weight with fuel and water being only about 7 cwt. Fig. 338A is from a photograph of one of these cars. In general design the vehicle consists of a light tubular underframe carried direct from the axles, and upon this the body is supported by means of light, double-leaf springs.

General design and arrangement of parts

The carriage body is constructed entirely of wood, and to it are attached all the principal parts of the motive mechanism. The boiler

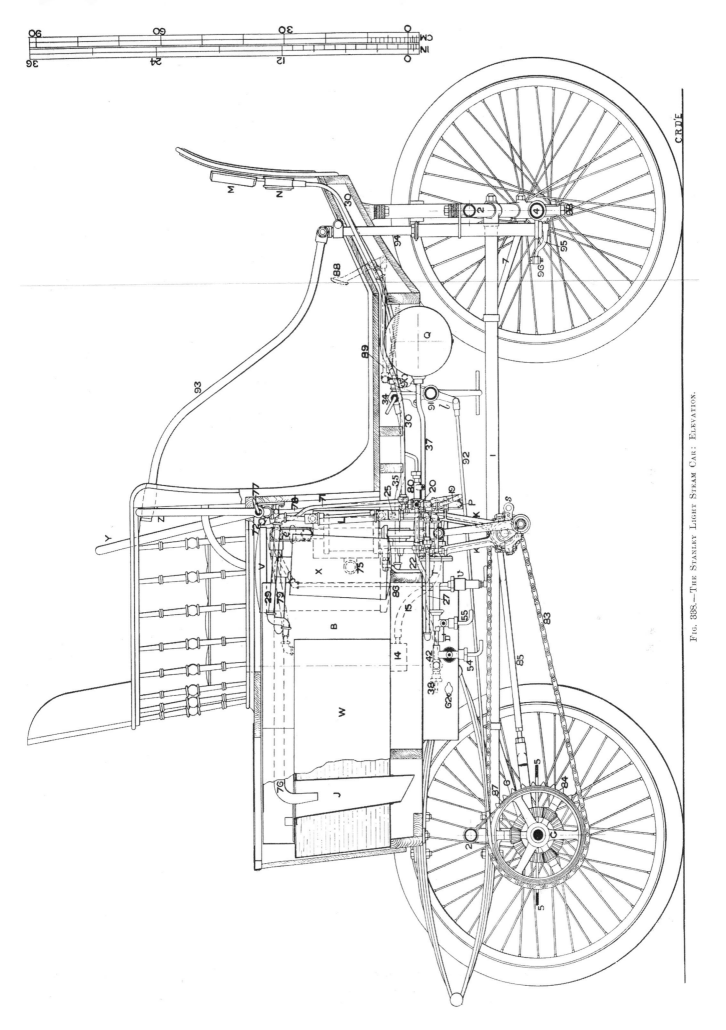

FIG. 338.—THE STANLEY LIGHT STEAM CAR: ELEVATION.

The material originally positioned here is too large for reproduction in this reissue. A PDF can be downloaded from the web address given on page iv of this book, by clicking on 'Resources Available'.

and engine are placed below the seat portion of the body, while the boot is occupied by the feed-water tank and the horizontal extension of the smoke-box, which forms the chimney of the boiler. Three levers, by means of which the engine and boiler are controlled, are placed at the driver's right hand. The steering is controlled by a hand lever, which operates the pivoted front wheels on the Ackerman system. Motion is transmitted from the engine to the rear road wheels by means of a single pitch chain, which drives a sprocket-wheel ring, mounted upon the

Fig. 338a.—The Stanley Steam Car.

differential gear on the divided through axle, at the ends of which the road wheels are keyed. One hand brake acting on the rear axle is employed to check the forward motion of the vehicle. The quantity of water carried is fifteen gallons, sufficient for a run of from 20 to 25 miles on average good roads; and about three gallons of petrol are carried for the boiler furnace, this being sufficient for about 70 miles of level roads.

The underframe is constructed chiefly of steel tubing, and consists of   Underframe two longitudinal tubes 1 1, brazed and bolted at either end to curved cross tubes 2 2, which are themselves brazed to straight cross tubes 3, 4,

459

of which tube 3 is divided at its centre for the reception of the sprocket driving-wheel and differential gear on the rear axle. The inner ends of tube 3 are rigidly connected by the guard plate 5. Additional stiffness is given to the underframe by the stay tubes 6 6, 7 7, and by rods 8 8 at the back.

Boiler The boiler B, which is seen in detail in Figs. 340 and 341, is of the vertical fire-tube type, Patent No. 2844 of 1899. It consists of a cylindrical drum, formed by the upper and lower tube plates 9, 10, and the shell plate 11, to which additional strength against bursting is given by windings 12 of steel piano wire. The tube plates are connected by 298 copper fire tubes A, these being about 13 in. long by 11 mm. or 0·437-in. bore, and No. 16 S.W.G., or 0·56 in. outside diameter. The ends of the cylindrical boiler shell are flanged outwardly and riveted to the tube plates, the inner ends of the rivets passing through and holding upon the flanges steel reinforce and strengthening rings, shown black in the section Fig. 340.

Heating surface The heating surface of the boiler consists of 100 sq. in. of tube plate surface and 4,600 sq. in. of tube surface, or a total of 30·25 sq. ft. This alone gives some idea of the large power which is compressed into the small cubic contents of this boiler. Taking the estimated indicated power of the engine, see p. 464, this gives 8·7 sq. ft. of surface per I.HP. From this it will be readily understood that for short periods the engine may be made capable of considerable hill-climbing power. The weight of the boiler is about 110 lb., empty. Below the lower tube plate is the fire box C, which contains the petrol vapour burner.

Burner This burner consists of a shallow cylindrical chamber D, through which pass 114 short copper pipes E E of 0·437-in. bore, open below to the atmosphere and above to the fire box. In the top plate 13 of chamber D, around the ends of tubes E, are bored numerous small holes H (see plan Fig. 341), which form jets, at which the petrol vapour, which fills the chamber D, escapes under pressure, causing induced currents of air up each of the pipes E. A combustible mixture is thus formed, which burns with an intensely hot, pale-blue flame, the products of combustion passing up the tubes A into the smoke box F, and away into the atmosphere without perceptible smoke or smell. The smoke box and drum of the boiler are thickly lagged with asbestos. The burner is very powerful, and the flame volume large, and it requires careful attention to the rules for its use and for lighting and extinguishing it, or accident might easily occur.

Water supply The water supply for the boiler is contained in the horseshoe-shaped tank W, which holds about fifteen gallons. The supply is drawn from a pocket 14, Fig. 338, at the bottom of the tank on the near side, and flows by way of the rubber pipe 15 past the cock 16 and check valve 17 to the feed pump P, Fig.

Feed pump 342, which is attached to the engine frame, and operated by a rocking lever from one of the crossheads. The water is delivered through valve 18 and along the coil pipe 19 to valve 20, and from here to pipe 21, which is closed at its outer end, so that the water passes along the branch pipe 22,

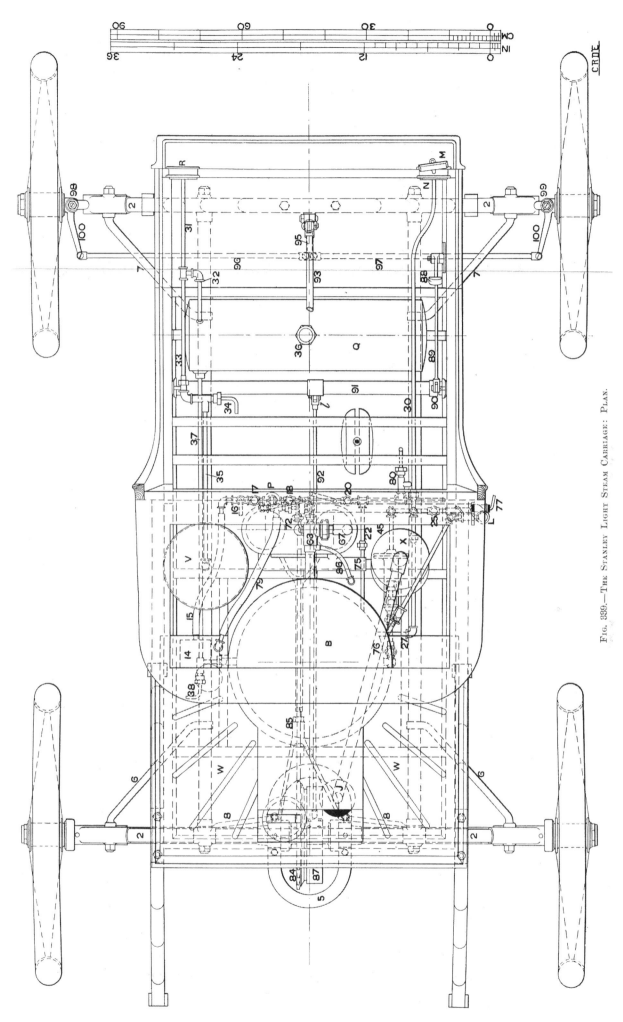

Fig. 339.—The Stanley Light Steam Carriage: Plan.

The material originally positioned here is too large for reproduction in this reissue. A PDF can be downloaded from the web address given on page iv of this book, by clicking on 'Resources Available'.

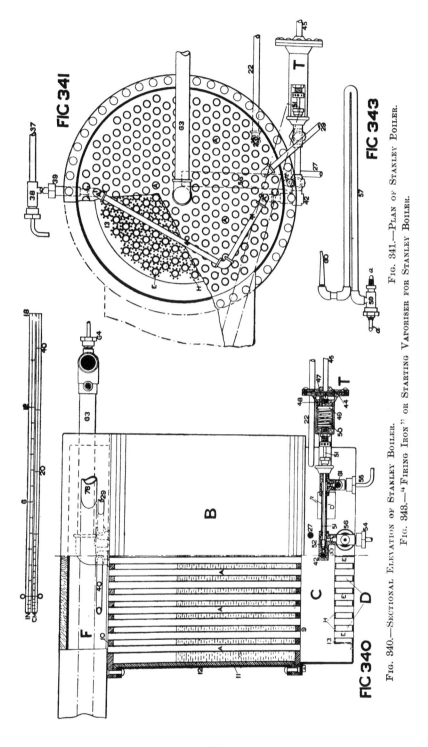

FIG. 341.—PLAN OF STANLEY BOILER.

FIG. 343.—"FIRING IRON" OR STARTING VAPORISER FOR STANLEY BOILER.

FIG. 340.—SECTIONAL ELEVATION OF STANLEY BOILER.

461

MOTOR VEHICLES AND MOTORS

Figs. 340, 341 and 342, entering the boiler at the lower tube plate at 23, Fig. 341. The water level in the boiler is indicated by the gauge glass L, placed at one side of the carriage body, Fig. 342. The gauge glass is connected to the bottom of the water space of the boiler by way of pipe 24, through the check valve 25 and branch pipe 26, to pipe 27, which is screwed and brazed into the lower tube plate. The steam connections to the gauge glass are formed by pipe 28 and pipe 29, which is screwed and brazed into the upper tube plate of the boiler, Fig. 338, with an intervening check valve. Observation of the gauge glass by the driver from his seat is rendered possible by means of the mirror M mounted on the dashboard, and adjusted so that the image of the gauge glass is reflected along the driver's line of vision. There are no cocks or blow-through valves connected with this gauge glass, but the two check valves mentioned automatically close the steam and water connection from the boiler when a gauge glass breaks. This is very simple, but it involves emptying the boiler when a gauge glass breaks, because the check valves otherwise remain upon their seats under the boiler pressure. To remedy this defect, a small bye-pass connecting both sides of one of the check valves, and thus restoring equilibrium of pressure, may be inserted, but in the car owned by Mr. T. W. S. Firth, who has so successfully defended motor-car owners in vexatious police prosecutions, an English set of fittings has been fitted. His car is similar to that shown by Fig. 338A.

The boiler is normally worked at 150 lb. pressure, as indicated by the pressure gauge N placed on the dashboard below the mirror M. The pressure gauge connection is by pipe 30, Fig. 342, which joins pipe 26, and is thus connected to the water space of the boiler.

The petrol supply for the boiler furnace is contained in the cylindrical tank Q suspended from the frame of the carriage body below the footboard, air pressure being employed to maintain the supply. In order that the pressure of air on the surface of the petrol may remain fairly constant while the level of petrol in the tank varies considerably, an air vessel V, placed below the seat on the near side, is employed in connection with the tank Q. The air pressure is indicated by the gauge R on the dashboard; this is connected by pipes 31, 32 to the tank Q, and by pipes 31, 33, cock 34, and pipe 35 to the air vessel V.

When the petrol tank is being filled, cock 34 is closed, and petrol supplied to the tank through the aperture 36. When the tank is nearly full, the aperture 36 is closed by means of an air-tight screw cover, and cock 34 opened and air pressure raised to 20 lb. by means of an ordinary bicycle pump, which is connected to a valve nozzle in pipe 31 below the gauge R.

The spirit flows from the tank by pipe 37, at the back end of which is a cock 38, Fig. 341, by means of which the supply to the furnace is controlled. From here the spirit passes along pipe 39, which is connected by a brazed joint to the lower end of one A¹ of the vertical boiler tubes A.

462

SOME AMERICAN LIGHT STEAM CARRIAGES

At the top, tube A¹ is connected by pipes 40, 41 to another A² of the boiler tubes, while the lower end of this is connected to the chamber 42 of the automatic valve T, which is controlled by the steam pressure, and which regulates the quantity of oil vapour passing into the chamber D of the furnace.

The automatic valve T, Fig. 340, consists of a thin metal diaphragm 44, to the centre of which is fixed a guide plate 47. The edge of the disc is firmly gripped between the end cover and the body T of the valve. One side of the diaphragm is under steam pressure, the connection being made by pipes 45, 46, 26, 27 to the bottom of the water space of the boiler, so that the diaphragm is deflected more or less as the steam pressure varies. When the pressure rises, the diaphragm is forced inwards, and the guide plate 47 pressing upon the three pins 48, one only of which is seen in the section, Fig. 340, pushes inwards the boss 49 on the spindle 51 against the resistance of the spring 50. This causes the cone valve 52 at the back end of the spindle 51 to approach its seat, so reducing the quantity of oil vapour passing as the steam pressure rises, while above a certain pressure, valve 52 is closed, except for a small bye-pass 53, formed by a small groove, which permits the passage of sufficient vapour to keep the burner alight.

Under ordinary conditions of working, cock 54, in the casting 42 containing the valve 52, is open, and cock 55, at the end of pipe p, is closed, so that the petrol vapour is free to pass into the furnace chamber D, through the nozzle N, Fig. 341, within the pipe 56, air being at the same time sucked in at the outer cone-shaped end of this pipe.

In order to start the furnace when the boiler is cold, the apparatus shown in Fig. 343, which is known as the firing iron, is employed. It consists simply of a bent pipe 57, at one end of which is a cock 59, while to the other end is brazed a conical nozzle 60. When the furnace is to be started, the pipe 57 is heated to a dull red, and inserted into the fire box through the door D', so as to rest upon the burner 13 plate of the chamber D. The nozzle a of the cock 59 is then screwed into the aperture 61 in the side of cock 55, by turning the handle d of the cock, and the conical nozzle 60 is inserted into the air pipe 56 leading into the chamber D.

Pressure having been raised in the air vessel V, and air space above the level of petrol in tank Q, the cock 54 is closed, and cock 38 on the other side of the boiler partially opened. A small quantity of petrol in the liquid state is thus allowed to flow up pipe A¹, across the top of the tubes, and down pipe A² in the boiler and fills the pipe p of the automatic regulating valve 52, which, it will be understood, is now full open, as there is no pressure in the boiler. Cock 55 is now opened, and the flow of petrol continues past cock 59, and round the bent pipe 57 of the heated firing iron in which the petrol will be vaporised, emerging at the nozzle 60 under pressure and so filling the chamber D of the furnace. A lighted taper is now inserted into the fire box through the door 62, Fig. 338, and the petrol vapour

463

ignites at the jets round the air pipes E, the temperature of the firing iron being maintained by the flame, while the water in the boiler is being heated.

When a steam pressure of 20 lb. is indicated by the gauge N, the temperature of the water and steam in the boiler will be sufficient to completely vaporise the petrol in the vertical tubes. A^1, A^2 and cock 54 being now opened the vapour passes through the nozzle N, and the furnace flame rises to its full intensity. Cock 55 is then closed and the firing iron removed.

Steam connections Steam is supplied from the boiler to the engine by pipe 63, the quantity being regulated by the valve 64, which is controlled by the hand lever Y at the driver's right hand. This lever operates the valve through the tube 65, near the inner end of which is a short downwardly projecting arm 66, connected by a link to the valve rod 64. The steam, after passing the valve 64, is led into the steam chest of the engine through the bent pipe 67, the connections for which are formed by union nuts locked by the spring plates E, held together by a central bolt.

Engine The engine, which is best seen in Figs. 342 and 342A, is of the vertical coupled type, having two cylinders, each 2·5 in. diameter by 3·5 in. stroke. The normal speed of the engine is 400 revolutions per minute, and the normal working pressure 150 lb. per sq. in. With a cut off which will give a mean pressure of, say, 50 lb., the engine will be of 3·5 estimated I.HP., or of, say, 3 B.HP., which, it will be seen, may be doubled or trebled for a short time by using a later cut off, and the higher pressure at which the boiler may be worked. The engine frame consists of a light casting K, which is attached to the cross bar 68 of the carriage body by means of two plates, one of which is seen at r, Figs. 338 and 342A. The bearings of the crankshaft are formed at the lower end of the arms of the frame, these being strengthened by the cross stay rod s. In another arrangement of this bearing the brasses are cut diagonally to suit the direction of the main pull upon them, and the caps are formed as shown in Figs. 342 and 342A. The cranks, which are at 90°, are keyed to the crankshaft ends outside the bearings, while the excentrics and sprocket driving pinion 82 are mounted on the shaft between the bearings. Ball bearings are employed at the big ends of the connecting rods. The slide valves are operated through simple reversing link motions, the links being connected by light rods to the arm 69 on the spindle 70, pivoted on the engine frame. A second arm on this spindle is connected by rod 71, to an arm 72, at the inner end of the tube 73, which passes through the tube 65 of the main steam valve control gear, and carries at its outer end the lever Z, under control of the driver.

Lubrication Lubrication of the slide valves and cylinders is effected by means of the oil cup and cock 74, Figs. 342 and 342A. The cylinders and steam chest of the engine are thickly lagged with asbestos, not shown in Figs. 342 and 342A.

464

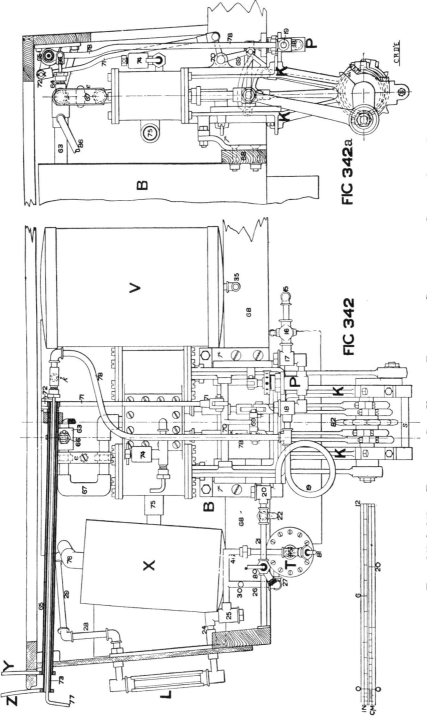

FIG 342a

FIG 342

FIGS. 342 & 342A.—FRONT AND SIDE VIEW OF ENGINE AND CONNECTIONS OF STANLEY STEAM CAR.

465

30

MOTOR VEHICLES AND MOTORS

Exhaust silencer

The exhaust steam passes from the engine by pipe 75 to the silencing drum x, and from here by pipe 76, perforated in the part within the drum. This pipe passes through the smoke box of the boiler, and at the back end is bent down into the coned pipe J. The steam is thus to some extent dried before escaping into the atmosphere, partly by its passage along pipe 76, and partly in the pipe J by its mixture with the hot products of combustion blast. The feed water in tank w is also to some extent heated by the exhaust steam and products of combustion.

Feed water pump

The supply of water to the boiler is controlled by the driver by means of the handle and rod 77, Fig. 342, which passes through the tube 73 of the reversing gear, and at its inner end operates the cock k, which controls the bye passage of water from the pump along pipes 78, 79, Fig. 339, into the top of the feed-water tank w. By adjusting cock k, more or less water may be caused to be simply passed back into the tank by the pump, while the remainder is forced into the boiler. The operation of blowing down the boiler is effected by opening the cock 80 connected to pipe 21, and the gauge glass may be blown through by opening cock 81 in pipe 46 leading to the automatic regulating valve of the boiler furnace. A spring safety valve

Safety valve

v, at the bottom of pipe 86, brazed into the steam pipe 63, allows the boiler to blow off at 200 lb. pressure. At the bottom of the valve is a pivoted handle for holding the valve off its seat when necessary. When the carriage ceases running, and steam has fallen in pressure, this valve may be opened to prevent the formation of a vacuum in the boiler and its consequent filling up by suction through the pump connections of water from the tank. To prevent this, and for cutting off the pump connection with the tank, there is the cock 16, which users have been known to forget on the one hand to open and on the other to close for admitting water to the pump and for preventing suction from the tank by the cooling boiler respectively.

Transmission gear

Power is transmitted from the crankshaft of the engine by means of the sprocket pinion 82 and pitch chain 83, which drives on to the sprocket-wheel ring 84, mounted on the exterior of the differential gear G, on the driving axle of the car. The transmission gear is thus as simple as that of a bicycle.

Differential gear

The differential gear is of the three-pinion type, as seen in the section Fig. 338. The spokes of the sprocket wheel, three in number also, carry the band-brake pulley, and between them are the bosses into which through the brake pulley periphery are inserted the steel pins on which run the three small pinions of the differential gear. The radial distance between the crankshaft and driving axle is preserved by means of the adjustable forked radius rod 85, by means of which also the chain may be tightened, but only through a very small range. The chain is of special make with links arranged so that one may be easily inserted or taken out. It is illustrated by Fig. 342B, which is full size. The side links A have holes E, into which the pins C, which are flattened where they are held, pass. From the hole E the narrowed parts of the pins pass into the

466

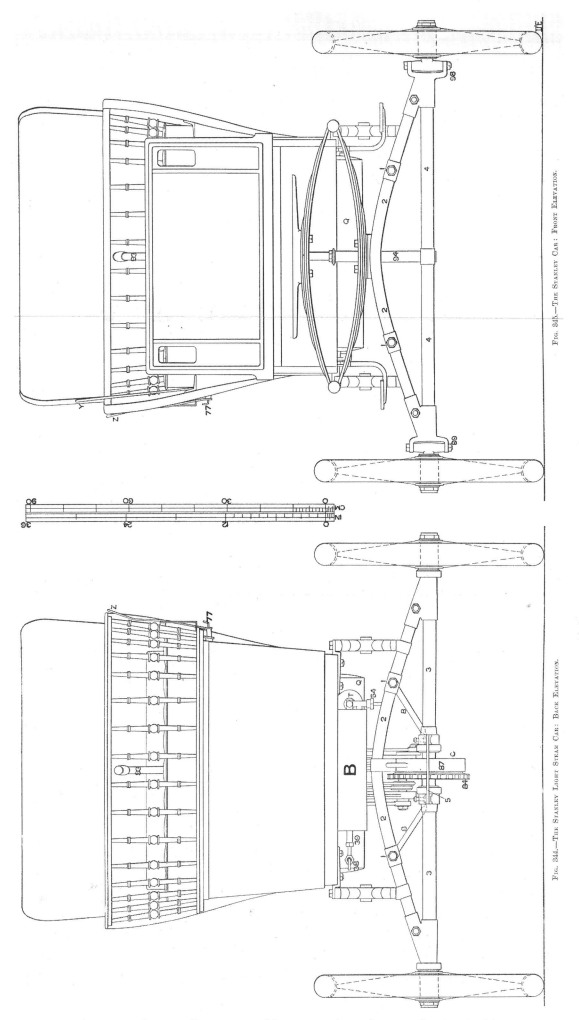

Fig. 84b.—The Stanley Car : Front Elevation.

Fig. 84a.—The Stanley Light Steam Car : Back Elevation.

The material originally positioned here is too large for reproduction in this reissue. A PDF can be downloaded from the web address given on page iv of this book, by clicking on 'Resources Available'.

SOME AMERICAN LIGHT STEAM CARRIAGES

narrower elongated holes in which they fit. The pins are thus fixed without heads or riveting, and one pair of links and one block B may be removed with facility.

The vehicle is fitted with only one brake, this is a hand brake 87, **Brakes**
acting on a drum attached to the sprocket-wheel ring on the exterior of the differential gear. The brake is operated by means of the pedal 88, which is connected by the rod 89 to the lever 90 mounted on the tube 91, which is pivoted from the wood frame of the carriage body. A second lever *l*, on tube 91 is connected by rod 92 to one end of the brake band, the other end of the band being attached to the frame tube 2. It will be seen that the band embraces about five-sixths of the whole circumference of the pulley, and is thus a very powerful brake, but it may be noted that it operates through the differential gear. As a further brake the driver has at his disposal the reversing gear of the engine which acts through the chain, but is very useful for gently running down hills.

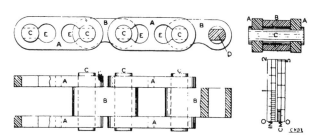

FIG. 342B.—THE BALDWIN DETACHABLE LINK CHAIN OF THE STANLEY STEAM CAR.

The steering of the vehicle is controlled by the lever 93, which operates **Steering gear**
a vertical spindle 94, carried from the tubular wider frame and upon the top of which it is pivoted. At the lower end of this spindle is an arm 95, which is connected by rods 96, 97 to arms 100 on the steering-wheel pivots 98, 99. It will be noticed that the steering arms 100 are placed at a few degrees from a right angle with the steering axle, sufficient to secure the necessary differential angularity of the steering wheels on turning curves of small radius.

The wheels are of the cycle type, with tangent spokes and steel rims, carrying single-tube pneumatic tyres.

The proof of sufficiency of the various parts of a car, especially **Novelty and proof of experience**
when of general design which departs from ordinary practice, from which our experience is obtained, must be gained from the use of that car. Past experience with a different class of thing does not necessarily apply, except with very considerable judgment and liberality. The car just described weighs, as has been stated, only about 7 cwt., while on the other hand it is provided with very considerable power. This power is exceedingly important for hill climbing, when only certain few parts need any very great

467

strength, namely, the driving wheels and the things that transmit the driving power to them. When, then, the greatest power has to be exercised, strength and weight in other parts of the car than those mentioned may be

Strength with lightness

superfluous. When running along the level or down hill, only sufficient strength is required for the transmission of smaller power and for the maintenance of the desired position of the front wheels by the steering gear. Some parts of this steering gear, such as the arms 100 and their pivoted connections, appear to be very light and without doubt an English maker would add to their strength; this would probably be wise, although the effort on ordinarily good roads necessary to the steering movement is with so light a car extremely small. The question seems not so much whether the parts are strong enough for what we may look upon as the chief stresses they have to bear, but whether these parts and their connections, including those with the woodwork of the body, are such as will stand the constant repetition of the vibratory shocks which accompany ordinary running at rather high speed on wheels of small diameter which are so much more affected by bad roads than the larger wheels, such

Slight elastic flexure may be better than rigidity gained by weight

as those of a hansom cab. The demand for greater strength may also be urged, more especially for the front axle and its connections, in view of the very heavy stresses that are undoubtedly thrown upon them when the exigencies of country touring, and avoidance of erratic people and horses, make it necessary to do a little steeplechasing over gutters, curbs, side walks, and ditches or heaps of stones at the roadsides. In reply to this, however, the makers of the " Locomobile," who, it may be mentioned, have a branch at Sussex Place, Kensington, may say that the very light weight will enable their car to do a good deal of this performance without injury, and that if steeplechasing is much to be indulged in they can add half a hundredweight of metal to the parts requiring it and still have a very light car. Weight does not always mean strength, any more than a heavy packing-case is necessarily more proof against railway porters' treatment than a lighter flexible basket trunk. The extent to which flexibility and " bendability " may be actually called upon, and not simply relied upon, is a matter for experience as well as the knowledge and judgment of the designer. If an English engineer were called upon to-day to design for the first time some of the ironwork of an omnibus, the horses would grow tired much sooner than they do.

Chapter XXVII

AMERICAN STEAM VEHICLES (*continued*)

THE WHITNEY LIGHT STEAM CARRIAGE.

A NOTHER light American steam car, equally well known by name as the Stanley, is the Whitney steam car, one of which was exhibited at the Agricultural Hall Exhibition last summer, and has since been taken up in this country by Messrs. Brown Bros. Several of these forms of this car have been made, that which was shown here having a water tank, over the front wheels, and another at the back enclosing the petrol tank. The engine and boiler were in type much like that of the Stanley car already described, but the driving mechanism was entirely different. Instead of chain and sprocket-wheel transmission, a form of differential gear combined with the crank was employed, and the engines, instead of being rotative were reciprocating, and, after the manner of mine pumping engines, operated connecting rods by means of triangular bell cranks. The differential crank axle was driven by these connecting rods directly as in a locomotive.

Other forms of the Whitney car include several operated by chain gear transmission and rotative engines of practically the same design as those which were made to operate the bell cranks. One of these is shown by Fig. 346. The framework, running gear and mechanism of this type of car is shown by Figs. 347–356. From Fig. 346, the general arrangement of the carriage will be gathered, and from what is shown in the accompanying details, the character of the whole design can be gathered without complete general drawings such as are given of the Stanley car.

In the engravings Figs. 347 and 348, A is the boiler which supplies steam, through the $\frac{5}{8}$-in. copper supply pipe B, to the engines E, the exhaust from which passes by the pipe F, to the silencer and feed heater G, and thence by the pipe H, through a part of the tank W. Controlled by the lever K is a valve by which the exhaust may be passed direct from the silencer by the pipe H′, and down through the chimney C, and a blast down this chimney may also be obtained from a live steam jet supplied by the pipe D, and the valve to which it is seen to be connected. The feed water

The Whitney steam car

Differential gear with crank

General arrangement

Steam supply

Exhaust

Feed pump

469

is supplied to the boiler by means of a pump P, which receives its supply from the pipe P′ from the water tank, and sends the water by the pipe L into the combined silencer and feed heater G. From this heater the water passes by a pipe M into the feed-heating coil M′, Figs. 349 and 350, entering at 2, and passing out at N through a check valve N′, and thence into the boiler. Below the valve in the check valve case a small pipe enters from

Exhaust silencer and feed heater

Fig. 346.—The Whitney Steam Car, with Hood.

Hand feed pump

a hand feed pump placed at v, by which the feed may be at any time made up or augmented from the warm feed water of the silencer G.

Petrol supply and regulator

The oil or petrol for heating the boiler is carried in a tank below w, or in most cases immersed in it, and from this tank it passes by the pipe Q, to the burner by way of a pipe passing through two of the smoke tubes and thence down to the steam diaphragm regulator R, which controls the issue of the vapour under pressure at the jets entering the three tubes s s s.

Air pressure on petrol

The pressure of the oil in the tank is maintained by a pump s′, occasionally operated by a pedal 4. At 6 is seen the water gauge, fitted with steam control and water blow-out, and at 7 is a steam oil pump supplying

470

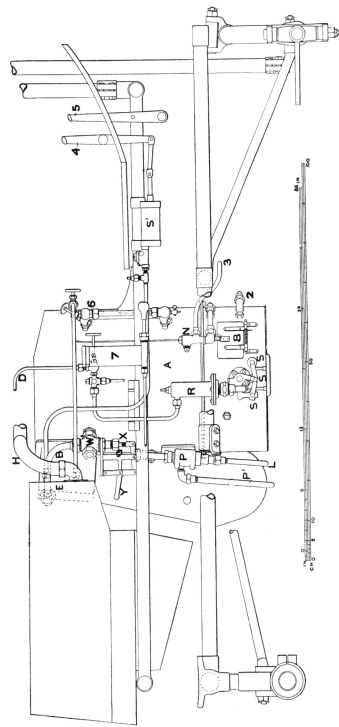

FIG. 347.—WHITNEY STEAM CAR: ELEVATION OF UNDERFRAME AND MACHINERY.

471

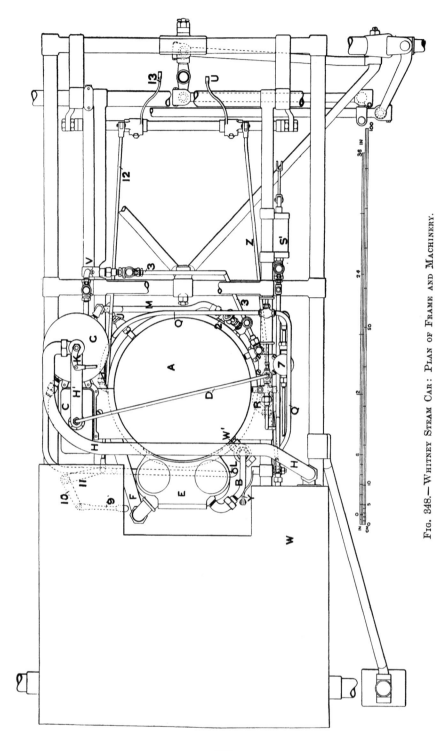

Fig. 348.—Whitney Steam Car: Plan of Frame and Machinery.

472

oil to the engine valves and cylinders. At 8, Fig. 347, is a spring-closed
door for access to the tops of the burners, which are seen more fully in
Figs. 349, 350 and 352.

Fig. 349 is a plan of the burner box in which is shown the disposition
of the air admission and annular vapour Bunsen burners 8, of which there

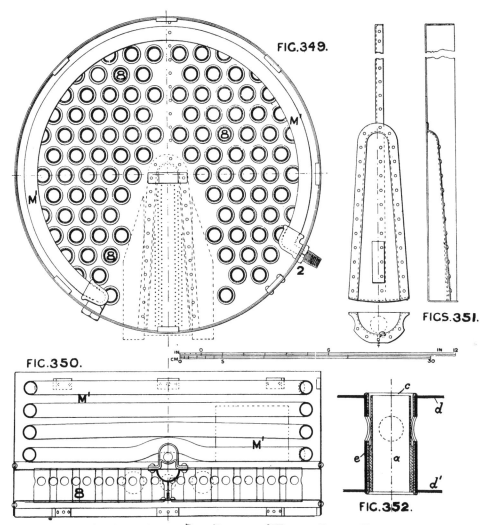

FIGS. 349–352.—PLAN, SECTION, AND DETAILS OF WHITNEY BURNER, FEED HEATER,
AND BURNER DETAILS.

are in the view shown 102. The burners are of very ingenious design for
obtaining at little cost a fine orifice suitable for burning this vapour. One
of the burners is seen in section to a large scale in Fig. 352, and they are
seen in their position in the vertical section Fig. 350. They will be seen
to consist of an internal tube a, turned down to about half thickness at the

473

upper, so that it is separated there from the surrounding enclosing tube c, which is expanded into and riveted over on the surfaces of the two plates $d\ d'$, forming the burner plates, which are kept apart by the loose distance piece tube e. In the last-mentioned tube and through the tube c, are drilled four holes which admit the passage of the vapour from the space between the plates $d\ d'$, into the narrow annular space between the upper part of

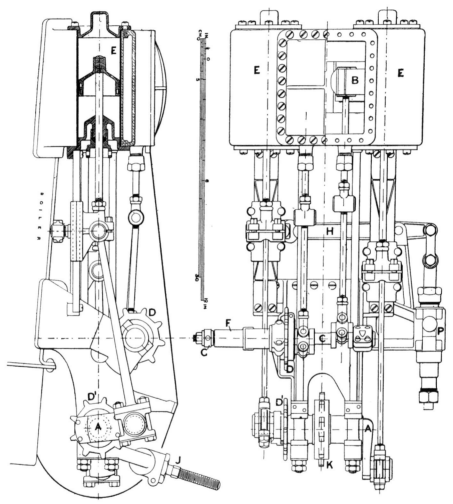

FIGS. 353 & 354.—WHITNEY STEAM CAR: ELEVATION OF ENGINE AND CONNECTIONS.

the tube a and of the tube c. This makes a very satisfactory burner for the vapour of gasolene or petrol, and with a really effective vaporiser it might be useful for burning kerosene.

Engine   The little double-cylinder engine is shown to $\frac{1}{5}$th full size in Figs. 353 and 354, which need very little description. The two cylinders E E,

474

each of 2-in. diameter and 4-in. stroke, are coupled in the usual way to the crankshaft A, which has two overhanging crank dips at right angles to each other. The slide valves B B are driven by a little double

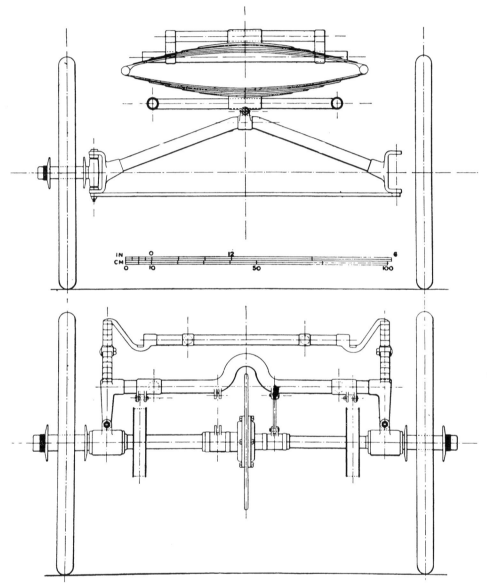

FIGS. 355 & 356.—FRONT AND REAR ELEVATIONS OF THE UNDERFRAME AND SPRINGS OF WHITNEY STEAM CAR.

crankshaft C, driven by a chain on the sprocket wheels D D′. At the end F of the little crankshaft is a sleeve fitted with a collar in a groove G, and upon this sleeve the boss of the sprocket wheel D fits, connected to it

Reversing gear

475

by a spiral feather, the sleeve before mentioned being connected to the end of the little crankshaft by a straight feather. By pulling the sleeve F outward, the wheel D is made to take a very different relative position with regard to the crankshaft to that which it holds when the sleeve is pushed in, and hence the direction of rotation of the engine is reversible. This operation is carried out by means of the connections 9, 10, 11, 12, and pedal 13, Fig. 348.

Feed pump The feed pump P is shown in Fig. 354, worked by means of a rocking lever H, connected by a sliding pivoted block to the crosshead of one of the engines. The sprocket wheel for the transmission of the power of the engine to the rear axle is seen at K in the centre of the crankshaft, the connection of the adjustable stay rod of the engine is shown at J, Fig. 353. The supply of steam to the engine by the pipe B, from the combined stop and safety valve W , is controlled by the rod X and lever Y, Fig. 347, which are connected by levers, not shown, to the rod Z operated by the pedal U.

Steering gear and suspension From Figs. 355 and 356, which are respectively front and rear views of the axles and underframes, and from the sectional elevation, Fig. 347, and plan, Fig. 348, the construction of the underframe, steerage, and suspension connections will be readily understood.

Whitney car with differential gear crank In August last, the author made a special trial run of the Whitney car, with the differential crank transmission mechanism, for the purpose of advising Messrs. Brown Bros., upon the car, its working and its patents. Since that time several improving modifications have been made, and the cars now to be put before the public have leaf springs in the front, the rotative engines shown by Figs. 353, 354, chain transmission, and the air and water pumps have also been improved. The car ran very quietly and mounted all the hills with ease, and during the greater part of the run the exhaust steam was very little seen. The distance travelled was about 108 miles, and water was taken in at Crawley on the out journey and at Horley on the return, five gallons of petrol being taken in at Brighton.

The cars as now made are known as the Whitney-Stanley cars, and much of the general design is the same as that of the Stanley car described, but details herein described are different, and better forms of gauge-glass fittings are used. Messrs. Brown Bros. ran one in the Automobile Club 1,000 mile trial recently.

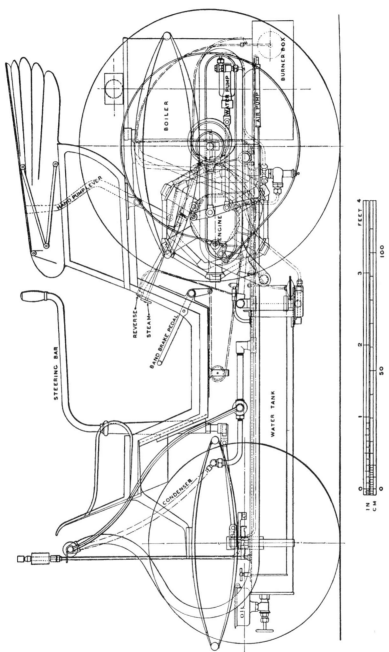

HAND PUMP LEVER

STEERING BAR

BOILER

BURNER BOX

WATER PUMP

AIR PUMP

ENGINE

REVERSE

STEAM

BAND BRAKE PEDAL

WATER TANK

CONDENSER

OIL

IN
CM

1
50

2
100

3

FEET 4

Fig. 357.—Elevation of Clarkson & Capel's Steam Victoria, 1897.

477

MOTOR VEHICLES AND MOTORS

THE CLARKSON & CAPEL STEAM VICTORIA.

General description

As of interest among the steam vehicles of recent construction, the elevation and plan of the Clarkson & Capel steam Victoria, as exhibited at Richmond, are given in Figs. 357 and 358. The names of the main elements of the carriage are printed in their positions in these engravings, thus making any lengthy descriptions unnecessary, the operation of steam **Six-cylinder** mechanism being commonly known. The motor employed is a six-cylinder, **motor** single-crank, boxed-in engine, driving one of the hind wheels of the carriage by single reduction gear, consisting of a large internal toothed ring fixed on the driver, and pinion upon the end of the internal shaft which passes through the crank case extension. The engine is provided with

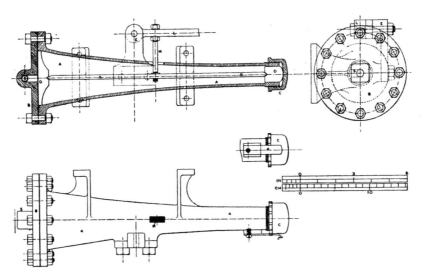

FIG. 359.—CLARKSON'S AUTOMATIC OIL-FEED REGULATOR.

piston valves, which are successively operated by a single cam on the crankshaft, the angular position of which can be changed by the movement of a spiral feather and collar. The cylinders are 2-in. in diameter, 6-in. stroke, and the engine is said to be capable of giving 8 B.HP. The number and disposition of the cylinders made a flywheel unnecessary, so that the weight of the cylinders was balanced by the reduction in weight in this respect. Steam was supplied by a pair of flash type boilers, heated by kerosene burners of the same type as those which will be described hereafter with reference to the heavy vehicles. The oil in the tank, seen between the two water tanks in the plan, was kept under air pressure by a small air pump, worked off a continuation of the lever **Oil burners** which worked the feed water, as seen in the elevation, Fig. 357. Under the influence of this pressure, the oil was fed to the vaporiser part of the oil burners, controlled by the automatic regulator shown by Fig. 359, and

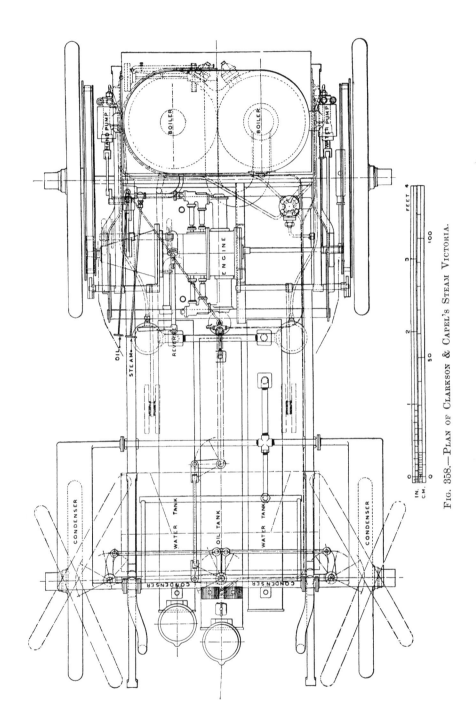

FIG. 358.—PLAN OF CLARKSON & CAPEL'S STEAM VICTORIA.

479

in position in plan by Fig. 358. The oil connections may be traced between the boilers and the oil tank and distributor valve.

Oil feed regulator

The regulator, Fig. 359, consists of a casing A, containing a steel rod R, pointed at its ends to a chisel edge, and supported between two abutment pieces D D. About the middle of its length there is freely attached to this bar an adjustable bolt M, by means of which a lever L, pivoted at E, may be operated. At the end B of the case is a cover containing, between itself and the flange of the case, a diaphragm, against which the circular piece inside the case forming one of the abutments rests. At the small end of the regulator case is a screwed cap C, the inside of which bears against one end of the slightly bent steel rod R. It will be seen that with steam entering at S, and pressing against the diaphragm between the cover B and abutment D, it tends to increase the bend or initial curvature given to the rod R, and by a very small movement of D, to impart to M a very considerable movement. By these means the growth or fall of steam pressure could with very small diaphragm movement regulate the oil supply to the burner through the medium of the lever L, and a suitable form of throttling valve not in the oil supply but in the oil vapour orifice opening into the burner.

Clarkson condenser

From the steam motor steam passed into a condenser formed in part by the tubular framing and by the wheel mudguards which were made up of small tubes and also of a large tubular grid the position of which is shown in Figs. 357 and 358, made of Clarkson's wire-covered condenser tubes, like that shown in Fig. 257. This form of carriage motor and mechanism is not being repeated by the designers.[1]

[1] Of this carriage and its details numerous engravings were given in *Industries and Iron*, Nov. 25th, 1898.

Chapter XXVIII

HEAVY MOTOR VEHICLES

French heavy vehicles and trials

PREVIOUS to the trials organized by the Automobile Club of France in 1897, for testing the capacities of the vehicles then made for heavy work, including lorries, wagons and omnibuses, several firms had made vehicles of heavier types not yet described, some of which were, however, present at the *Petit Journal* Competition of 1894, including a steam tractor, already referred to, by de Dion & Bouton, and a steam wagonette by M. Scotte.

The vehicles to which special reference need be made are those which first appeared in the Automobile Club's "Poids Lourds" trials above mentioned. The vehicles which accomplished the most notable performances were those of de Dion & Bouton and Scotte, which were propelled by steam, and the petrol motor omnibus of Panhard & Levassor, and the petrol lorry of de Dietrich. The most important figures obtained during these trials will be found in Table XX.

It will be noticed from this that the dead load of most of the vehicles was very heavy in comparison with the paying load, and that the petrol motor lorry was in this respect very much more efficient than any of the others.

Results of French trials

So far as Table XX. relates to French vehicles it has been compiled from the very full reports published by the Automobile Club of France, two of which have already been published. The various items given in the table are believed to be those which are most wanted, and space need not be here occupied in again dissecting it.

THE DE DION & BOUTON STEAM OMNIBUS.

The de Dion & Bouton tractor and omnibus

In the 1897 trials Messrs. de Dion & Bouton ran two different vehicles, namely, the tractor, hauling a long char-à-banc called a Pauline, and the steam omnibus, which is illustrated, so far as its exterior is concerned, by Fig. 360, and in some of its details by Figs. 361 and 365.

General

This vehicle is capable of carrying sixteen passengers, twelve inside and four on the back platform; it weighs complete 6 tons, 1 cwt., and without passengers 4 tons, 10·8 cwt.; when fully loaded 4·15 tons are carried by the rear wheels, and 1·9 tons on the front wheels. The length over all is 21 ft.; of this, the boiler, coke box, and seats for driver and engineer occupy the front 6½ feet; the space for inside passengers is 10¾ ft., leaving 3¾ ft. for the platform at the back; the breadth over all is 6½ ft.

The de Dion boiler
The boiler, Fig. 361, is of the de Dion & Bouton type. It consists of two annular cylindrical vessels A, F, which form the water and steam spaces and are connected by 500 steel tubes D D, which are slightly inclined so as to assist the circulation of the water. A diaphragm E causes the steam from the inner chamber, as well as that from the outer, to pass through the upper tubes, so that it is partially dried before passing from the boiler. The steam pipe G is connected to the upper part of the inner

FIG. 360.—DE DION & BOUTON STEAM OMNIBUS.

chamber by means of the casting which carries the stop valve and the safety valve H. The covers of the annular spaces A, F are held together by long bolts as seen in Fig. 361. The products of combustion after passing up between the tubes escape by the chimney C. Fuel is supplied **Grate and heating surface** through the aperture B at the top of the boiler, and the ashes from the furnace are removed through door L. One of the two water gauges is seen at I. There are also two pressure gauges attached to the boiler. Water is supplied by means of a feed pump worked off an excentric on the rear axle and may also be supplied by an injector. Before entering the boiler the feed water is heated by passing round the coil of pipe J J in the fire box. The working pressure is 200 lb. per sq. in. The grate area is 1·95 sq. ft. and the heating surface about 60 sq. ft. About 6 lb. of water are evaporated per 1 lb. of coke, and steam can be raised in thirty minutes. The weight of the boiler empty is 900 lb., and with water and fuel 1,070 lb.

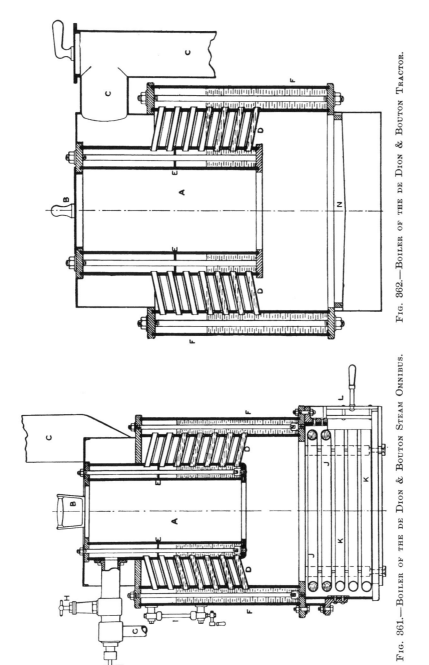

Fig. 362.—Boiler of the de Dion & Bouton Tractor.

Fig. 361.—Boiler of the de Dion & Bouton Steam Omnibus.

483

The de Dion engine

The engine, which is shown diagrammatically by Figs. 363 and 364, is carried below the underframe and is of the horizontal compound type, with cranks at 90°. The cylinders A, B are 3·95 in., and 7·5 in. diameter, respectively, and 6·7-in. stroke. The cut off in each cylinder is at ¾ stroke. The pistons and connecting rods are coupled to disc cranks D D on the crankshaft E. In case of necessity a special valve enables the full

Live steam to low-pressure cylinder

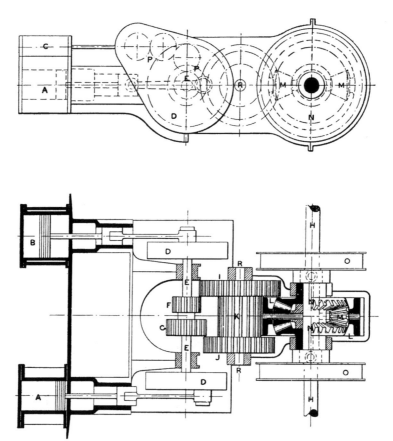

FIGS. 363 & 364.—DIAGRAM PLAN AND ELEVATION OF ENGINE AND DRIVING GEAR OF DE DION & BOUTON STEAM OMNIBUS.

pressure of steam to be admitted to both cylinders. The slide valves C are worked by excentrics on a shaft which is geared to the engine shaft through a train of pinions P, reversal being effected by moving a lever, which alters the number of pinions in the train, in the manner already described with reference to Fig. 161. The exhaust is superheated by passing through the coils K K in the fire box, and then escapes from the chimney in a less visible

Power

condition. The engine develops 24½ HP. at a speed of 600 revolutions per minute. The speed of the vehicle may be varied from 8·7 to 12·4 miles per

HEAVY MOTOR VEHICLES

hour by putting one or other of two pinions F G on the engine shaft E into gear with wheels I J on a counter shaft R, which carries a pinion K, gearing with the driving wheel L of the differential gear M N. The rear wheels are driven direct through the de Dion system of jointed rods H H described hereafter. All the gearing and working parts of the engine are enclosed in a cast-iron casing and run in an oil bath. The total weight of engine, gearing and case is 1,780 lb. There are two pairs of band brakes, the one acting on the naves of the driving wheels and the other on drivers O O mounted on the outside of the universal joints near the differential gear. The coke box is in the extreme front of the vehicle surrounding the boiler, and contains 270 lb. The lubricating oil tank is at the side of the boiler. The water tanks are under the seats of the passengers' compartment, and together contain 100 gallons. The vehicle is steered by means of the front wheels, which are mounted at the ends of the fixed front axle on the Ackerman system. They are $31\frac{1}{2}$ in. diameter, with tyres $3\frac{1}{2}$ in. wide, the driving wheels are 4 feet in diameter, with tyres 4 in. wide. At an average speed of 8·85 miles per hour, the coke consumption was, on trial, 6·45 lb. per mile = 1·08 lb. per ton mile, and water 40 lb. per mile, so that a run of about 3 hours can be made without taking in supplies.

Gearing enclosed

Brakes

Coke bunkers and water tanks

Steering

Fuel consumption

Fig. 362 illustrates a form of the de Dion boiler as it was used in the tractor. It will be seen that the chimney C has a downtake instead of an uptake and that the tie bolts between the top and bottom plates have their lower ends accessible for tightening with nuts, instead of formed as the requirements of the feed-heating tube arrangement below made necessary. The boiler being of a water-jacket, fire-box, and water-jacketed chimney and water connecting-tube type, should be of considerable efficiency, and probably would be if it were not as a rule, or during so much of its working time, steaming under forced conditions. The result is, as shown in Table XX., the evaporative efficiency is rather low.

The tractor boiler

Efficiency

The de Dion system (Patent 14698 of 1894) of transmitting the power from the motor, or a second motion shaft driven by it, to the driving wheels, is illustrated by Fig. 365. From the diagram plan of the engine, Fig. 364, it will have been seen that the shaft which carries the differential gear has upon its ends coupling boxes in which the ends of the rods H are freely pivoted ; these rods H are seen in Fig. 365 as connected to the short driving axles by means of the flexible joints within the collars K. The inner ends of the short axles L, and the outer ends C of the short sleeves C on the differential motion spindles, are formed as shown in the small detail in which the cheese-shaped head is cut away, so as to leave the four pieces L L. Between these the brasses on the pins G are free to move with the angular movement of the rod H, and the eye E at either end of this rod works between these lugs L, with freedom for moving through a sufficient angular distance in all directions, so that under the action of the springs the axles L may fall or rise, while the driving spindle C remains at the fixed distance imposed by the brackets D carrying it. After the

The de Dion axle drive

485

cheese-heads at the ends of the sleeves c and the inner ends of the axles L are slotted, so as to leave the four lugs shown by the small detail view, a ring K is fixed upon them as shown. The axles L run in bushes carried in a steel socket, fastened at Q to the bent fixed axle R, and at their ends they carry the drivers M, in which are fixed the spring driving arms, N, freely moving in the small housings o fixed to the feloes P. The naves or hubs of the wheels are on what may be called the Hancock system, and are fitted with large bushes, which run upon the turned exterior of the horizontal part of the socket Q. It will be noticed that the driving pins G are both in the same plane, and not placed at right angles to each other as they are in ordinary universal joints. This arrangement is adopted to prevent the slight variation in speed at different parts of a

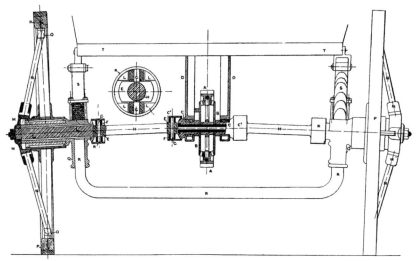

FIG. 365.—DE DION & BOUTON'S ARTICULATED DRIVING AXLE.

revolution of the thing driven when the pins are at right angles. In the engraving a slight mistake has been made, the pin c′ being at right angles to the driving spindle instead of at right angles to the spindle H.

This arrangement of driving tackle avoids the difficulties connected with either spur gear or chain gear on the driving wheels.

Remarks   During the "Poids Lourds" trials of 1897, 1898, the de Dion & Bouton vehicles were considered to be highly satisfactory in their performance. The conditions of the trial-running, under the eyes of the designers, and with some of their best workmen, were undoubtedly not those which would obtain in commercial running; but it will be noticed that the highest form of flattery has been accorded to de Dion & Bouton in that their general arrangement of vehicles with boiler in front, surrounded by its coal bunkers, with enclosed engine and second motion,

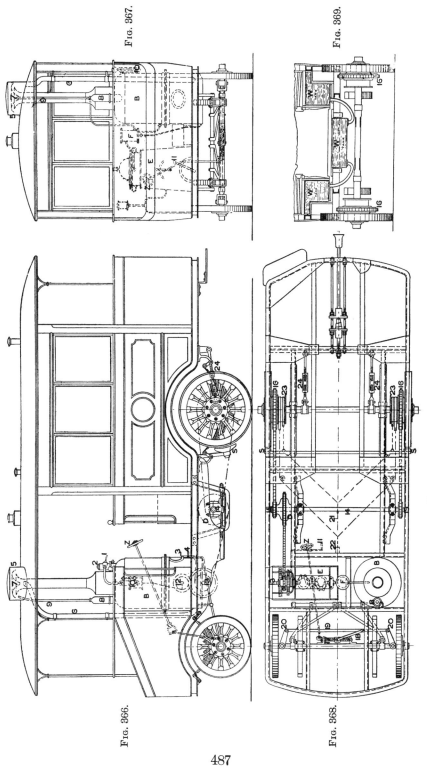

FIG. 367.

FIG. 369.

FIG. 366.

FIG. 368.

FIGS. 366–369.—THE SCOTTE OMNIBUS AND TRACTOR.

487

MOTOR VEHICLES AND MOTORS

and differential gear shafts at the rear, and spring driving device for the
main wheels, have been largely imitated, and some of our best vehicles
driven by means of flexible spindle connections on systems similar to that
illustrated by Fig. 365.

THE SCOTTE STEAM OMNIBUS AND ROAD TRAINS.

Results of trials
The results of the trials in the "Poids Lourds" competition, and the
chief statistics and features relating to the well-known and somewhat
extensively used Scotte vehicles, will be found in Table XX., the general
arrangement of the steam omnibus and tractor is shown by Figs. 366–
369.

General description
The Scotte road train for passengers consists of a *tracteur porteur*, as in
Fig. 366, capable of carrying eight inside passengers and three outside, and
hauling a trailer carrying fifteen persons and their baggage. The total
weight of the train with passengers is 9 tons; the vehicles themselves weigh
6 tons 6 cwt., and the load carried 2·5 tons. The tractor is fitted with two
speed gears, the one giving about 7½ miles, and the other 3¼ miles per hour.
The tractor is 18 ft. long over all, and the trailer 15½ ft., the breadth of
each being 5 ft. 9 in. The underframe of the tractor is built up chiefly
of steel channel and T bars, and is sprung from the axles on four plate
springs. The engine and boiler are bolted to the front cross bracings, and
with the sloping-fronted coke box occupy a space of about 5 ft. 4 in. The
seats for driver and engineer 2 ft. 6 in., leaving 7 ft. 6 in. for inside
passengers, and 2 ft. 8 in. for the rear platform.

Boiler
The boiler B is of the vertical Field tube type. The working pressure
is 170 lb. per sq. in., and steam can be raised from all cold in about
Grate and heating surface
35 minutes. The grate area is 1·6 sq. ft.; it burned, on trial, 72 lb. of coke
per square foot per hour, and evaporated 4·7 lb. of water per 1 lb. of coke.
The weight of the boiler empty is 1,100 lb., and with water 1,250 lb.
It is provided with a feed pump, an injector, and a water-circulator, which
consists of a pipe passing through the centre of the fire box and uptake,
where it is protected from the direct action of the furnace by a cast-iron
baffle plate, which also serves to regulate the draught, and connecting the
steam space with the water space around the lower portion of the fire box
Steam supply
by means of outside connections. The steam supplied to the engine is
controlled by the handle 1, which works a slide in the steam regulator 2,
and is superheated by passing through a coil in the chimney base at the
top of the uptake. The grate has fixed as well as jointed bars, which
can be moved by the lever 3, and clinkering prevented. It is also possible
to raise or lower the whole grate by means of the handle 4, so that the fire
Closed ashbox
may be kept low when the engine is not working. The ashes fall into a
closed ashbox, and are damped by means of condensed steam from the feed-
Chimney top
water heater F. The chimney is bell-mouthed at the top, and covered by
a hood 5, the bottom of which slopes down to a pipe 6, through which
cinders and water fall into the ashbox; the deflector 7 prevents the emission

488

of sparks. The exhaust steam from the engine, after passing through the feed heater is led into the chimney, where it escapes in an almost invisible condition. The blow-off from the safety valve 8 passes into the hood through the pipe 9. The gauge glass and pressure gauge are attached to a vessel 10 in connection with the boiler.

Exhaust steam

The engine E is vertical, with two cylinders, each $4\frac{1}{2}$ in. diameter × $4\frac{3}{4}$-in. stroke, it developes 16 HP. at 400 revolutions per minute, and weighs 670 lb. Reversal is effected by means of an ordinary link motion, and an expansion valve makes it possible to vary the cut off from $\frac{3}{8}$ stroke for ordinary running to $\frac{3}{4}$ stroke on hills or bad parts of the road. By means of the lever 11 one or other of two pinions on the engine shaft 12 can be put into gear with wheels on the countershaft 13 corresponding to speeds of $7\frac{1}{2}$ or $3\frac{1}{4}$ miles per hour of the vehicle. A pinion on shaft 13 drives the outer wheel of the differential gear D by means of an ordinary pitch chain. The differential gear is carried on a countershaft 14, which drives each of the rear wheels independently by means of pitch chains 16, 16.

Engine

Transmission gear

The wheels are of wood with iron tyres, the rear wheels are 3 ft. diameter, with tyres $4\frac{3}{8}$ in. wide, and the front wheels 1 ft. 6 in. diameter, tyre 3 in. wide. The weight of the vehicle is nearly equally distributed, so that with passengers the load on the rear wheels is about 3·8 tons, and 2·5 tons on the front wheels: total 6·3 tons. The trailer weighs 1·7 tons, and carries 1·3 tons.

Wheels

The steering is effected by the hand wheel, which works the screw 17 by means of a jointed rod and bevel gear, the vertical spindle to which is free to allow of vertical play of front axle. A nut is thus worked backwards or forwards on the screw, and by means of the bars 18, 19 and arms 20, 20 moves the front wheels, which are pivoted on the ends of the fixed axle.

Steering gear

There are two brakes: a screw brake worked from 21, which presses the shoes s s on the tyres of the driving wheels, and a foot brake worked from 22, tightening the Lemoine brakes 23, which are wound twice round the drums. The driving chains 15, 16 can be tightened by the screw arrangements 24, which vary the distance between the shaft 14 and the main axle.

Brakes

Driving chain

The coke box is in the extreme front of the vehicle, and is capable of containing 270 lb. There are three water tanks, one under each seat in the passengers' compartment, and one under the footboard, as seen in Fig. 369. They will hold 149 gallons. The consumption of coke, when the train is fully loaded, is about 17 lb. per mile for an average commercial speed of about $6\frac{1}{2}$ miles per hour, and of water 80 lb. per mile. It therefore can travel about 16 miles, i.e. $2\frac{1}{2}$ hours, without taking in supplies. The coke consumption works out to about 1·83 lb. per ton mile total.

Water tanks

Speed

Although the Scotte vehicles have been made in considerable numbers, the use of the double set of chain driving gear, the heavy weight, and

Remarks

the high tension at which everything seemed to work when ascending the hilly parts of the trial roads, did not favourably impress the English engineers. The design, as shown by the engravings, is succeptible to some improvements in the mechanical arrangements, but it is, on the whole, better than some of the vehicles which have been made in England of more recent date.

THE WEIDKNECHT STEAM OMNIBUS.

General description   The Weidknecht is mother of those which entered the "Poids Lourds" competition of 1898. As, however, it did not complete the trial runs, its performance is not recorded in Table XX. It is illustrated by Figs. 370–372. It is constructed to carry sixteen passengers and half a ton of luggage. It weighs, complete, nearly 7 tons, the useful load being 1·6 ton, and the weight of the vehicle, with coke, water, and attendants, 5·3 tons. The over all length is 18 ft.; of this the front compartment, containing the driving mechanism, occupies 7 ft., and the passengers' compartment 11 ft., the luggage being carried on the roof. The distance between the axle centres is 7 ft. 10 in., and the breadth from centre to centre of the driving wheels, 6 ft.

Underframe   The underframe is built up of light channel-iron girders, the boiler, engine, and driving-gear being bolted to the cross bars of the front portion, which is of rectangular outline. It is sprung from the front axle on two leaf springs. On this axle are the driving wheels, which are 4½ ft. diameter, with wooden spokes and rims, and iron tyres 3¾ in. wide. The rear axle is rigidly attached to the underframe, which is considerably reduced in breadth as it approaches the axle, giving it a slight flexibility in order to reduce the shocks transmitted to the driving mechanism.

The passengers' compartment is entirely separated from that containing the machinery, and is sprung from the rear axle on three plate springs, and from the frame by means of an arrangement of volute springs, one set of which is seen at A in Fig. 370. The object sought in this arrangement being to isolate the passengers from the vibration and jolting of the heavy working parts, and also to keep the load on the driving-wheels constant.

Boiler   The boiler B, Fig. 373, is vertical, and contains both water and smoke tubes. It is of rectangular section, the sides of the fire box and shell being strongly stayed together. The water tubes, 87 in number, cross the upper half of the fire box in a slightly inclined direction. They are 1·18 in. diameter. There is a cleaning door for these tubes at each side of the boiler. The uptake is divided between 16 smoke tubes, which also serve **Grate and heating surface** the purpose of staying the top of the fire box and shell. The total heating surface is 64 sq. ft., and the grate area 3 sq. ft. The boiler is stated to evaporate 572 lb. of water per hour, at an average working pressure of 150 lb. per sq. in. The fuel burnt is coke, the furnace being fed by **Automatic stoker** means of an automatic stoker, which requires replenishing about every

490

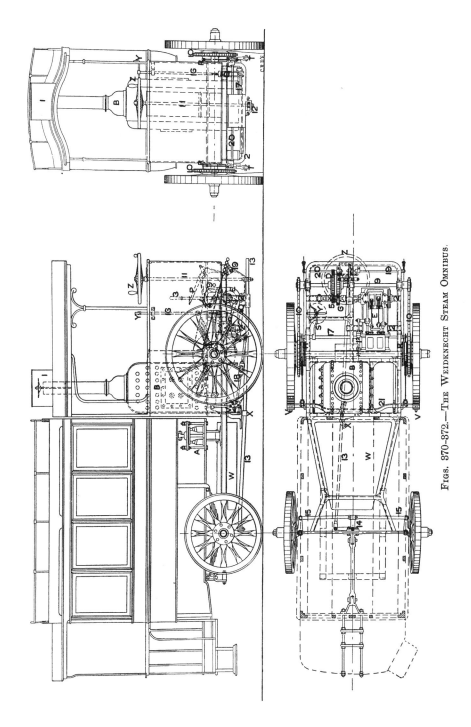

FIGS. 370-372.—THE WEIDKNECHT STEAM OMNIBUS.

491

Grate difficult of access

2½ miles. The grate bars are considerably inclined and are made in two parts, the lower back portions being fixed, and the front parts are all carried from a bar which can be turned by a lever, thus opening sufficient of the grate for cleaning purposes without completely drawing the fire.

Exhaust

The bars are, however, difficult of access for cleaning. The exhaust steam

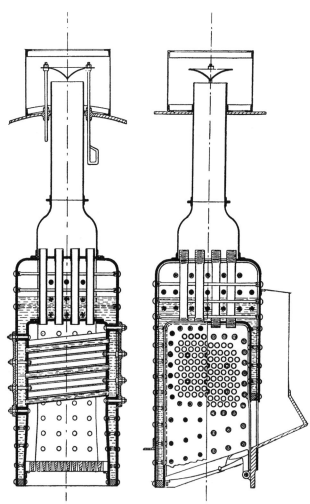

FIG. 373.—SECTIONS OF THE WEIDKNECHT VERTICAL BOILER.

from the engine passes, with the products of combustion, up the vertical chimney, which is covered by the hood I, and fitted with a silencer, which sufficiently prevents the noise of exhaust.

Feed-pump

A feed-water heater is attached to the boiler, and water is supplied either by a pump P, worked off one of the cross-heads of the engine, or by a Sellar's injector.

HEAVY MOTOR VEHICLES

The horizontal engine E is carried on the cross channel bars 22 of the The engine underframe. It has two cylinders, each 4·92 in diameter and 4·92 in. stroke, the cranks being at 90°. It is fitted with a Solms' variable expansion and reversing gear, which is worked from the lever 3, moving between two toothed sectors 4. This gear allows of a variation of cut off from 0·1 to 0·83 of the stroke. At a speed of 350 revolutions per minute the engine develops 19·7 HP. under ordinary working conditions. The working parts are protected by a cast-iron casing, open only at the top.

Running loose on an extension of the engine shaft are two pinions Transmission gear 5, 6, both of which are continuously in gear with wheels on the exterior of the differential gear D by means of claw clutches. Either of the pinions may be locked to the shaft and the speed of the vehicle varied from 9·3 to 4·65 miles per hour. By using the variable cut off in connection with this gear it is possible to give the vehicle any speed between $2\frac{1}{2}$ and $12\frac{1}{2}$ miles per hour. The power is transmitted to the driving wheels from the ends of the countershaft 9, carrying the differential gear, by means of the sprocket wheels and pitch chains 10 10.

The vehicle is steered by means of the rear wheels, which are pivoted Steering gear on the ends of the fixed axle. They are moved by the hand wheel z on the vertical shaft 11, at the lower end of which is a pinion 12 gearing with a rack on the end of the rod 13, which is guided at x. The motion of this bar is transmitted to the pivoted axles by the steering arm 14 and links 15.

There are two brakes: a screw hand brake, worked from the handle Y Brakes on the vertical shaft 16, screwed at its lower end and working in a nut on the link Q, the motion of which is transmitted to the shaft 17 through the link S; at each end of this shaft links T T work in the slotted ends of the adjustable rods 18, and press the shoes V on the tyres of the driving wheels. The other is a Lemoine coiled wire hand brake acting on drums attached to the driving wheels. It is worked from the pedal P. By depressing this, motion is imparted to the shafts 19, 20, upon the ends of which are links connected by wire ropes to the free ends of the brake bands R R. The other ends of the bands are attached to the bar 21, which carries the shoes V.

The water tank w will hold 62 gallons, and the coke box by the side Water tank and coke bunker of the boiler 132 lb. The coke consumption on trial was 11 lb. per mile, equal to 1·6 lb. per ton mile total, and the water used, 57 lb. per mile. The vehicle can therefore make a run of about 11 miles without taking Consumption in supplies.

One of these vehicles was brought to this country, but neither its performance nor its appearance was calculated to extract opinions which would be confirmed by orders. The omnibus is not, however, without points of interest, though it will be more interesting after it has been re-designed.

Fig. 374.—The "Lifu" Steam Omnibus, 1890.

494

Chapter XXIX

THE LIQUID FUEL ENGINEERING COMPANY'S "LIFU" VEHICLES

THREE different forms of vehicles made by this company from the designs of Mr. H. A. House are illustrated as to exterior appearance and general arrangement by Figs. 374–376, and a steam van by the same builders is illustrated in outline by the side, front, and back elevations, and in plan by Figs. 377–380. With certain exceptions the arrangement of the running gear of the whole of these vehicles is of the same design. The machinery will be described in detail hereafter with reference to the recent forms of 3-ton steam lorries, but the general arrangement and main features may now be described with reference to the engravings of the steam van.

It is generally known that the name "Lifu" is the trade name of the Liquid Fuel Engineering Company, the word being the first syllables of the two words "liquid" and "fuel." With this fuel the company had, at least up to a certain date, been more successful, both for marine and land road propulsion purposes, than any other makers.

Much of this success was due to the designs of engines and boilers, but the successful evaporiser burner was really a most important element in the special combination. All the vehicles now to be described are fitted with, and steam raised by the combustion of, kerosene, or still heavier oils, by means of these burners.

The underframe of these vehicles is constructed of light channel steel F, but in all cases, and as a special feature of the "Lifu" design, a pair of strong tubular longitudinal stays H H connect the front and rear axles, adding very much to the strength and rigidity of the whole structure and relieving the frame F of many severe stresses. In all but the first vehicles of "Lifu" manufacture the boiler B is placed centrally in the front of the vehicle, where it is carried by the steering wheels. The engine E is centrally situated within, and attached to, the upper frame, which also carries the bevel transmission gear in an oil box at C, this gear transmitting motion by a spindle within the tube J on a sleeve on the end of

495

FIG. 375.—THE "LIFU" STEAM LORRY, 1898-99.

496

which at D is a bevel pinion, running in the oil box shown, and driving the spindle carried by the bearings K K. The bevel wheel, with which the pinion at D gears, forms part of the differential gear. The spindle driven by this gear is, of course, in two parts. Each part of this spindle has, at its outer extremity, a spur pinion, having teeth at a suitable angle for gearing with the internal toothed rings s s, fixed to the dished wheels which are upon axles x downwardly inclined in the ordinary way.

From the front view, Fig. 378, it will be seen that the fixed axle A carries the pivoted axles at P, at an angle of a few degrees from the

"Lifu" steam-van

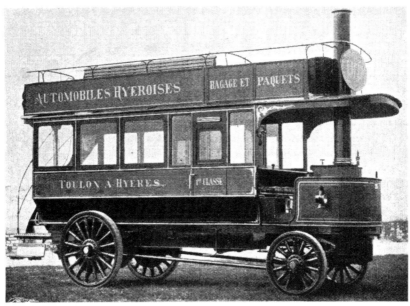

FIG. 376.—THE "LIFU" 1ST & 2ND CLASS STEAM OMNIBUS FOR TOULON.

vertical, the object of which is to obtain approximate coincidence in the point of contact of the wheels on the road, and of the axis of the pivot P produced to the road, so that the effect of meeting with any obstructions on the road is very little felt at the steering handle.

Turning now to Figs. 381 and 382, which are respectively elevation and plan of the underframe and machinery of the "Lifu" 3-ton steam **"Lifu" lorry** lorry. In these engravings all the parts shown in Fig. 380 are included, but the same letters of reference are not employed. The frame is made as before of light steel channel bar, the width remaining the same throughout. The longitudinal tubular stays are employed in the same manner and for the same purposes, including the support of the bearings for the main transmission shaft and the strong **T** iron brace by which they are connected as at x in Fig. 379.

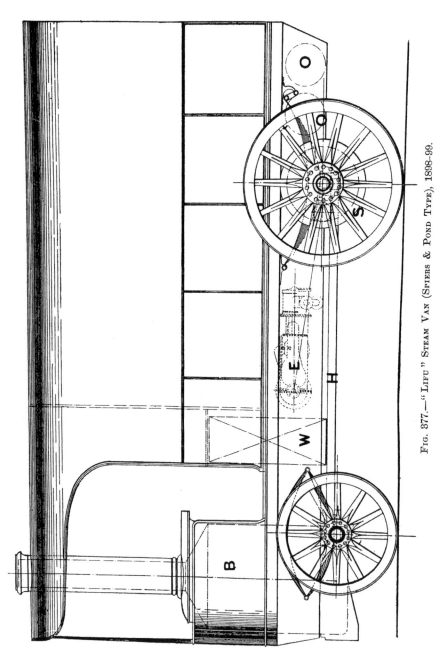

FIG. 377.—"LIFU" STEAM VAN (SPIERS & POND TYPE), 1898-99.

498

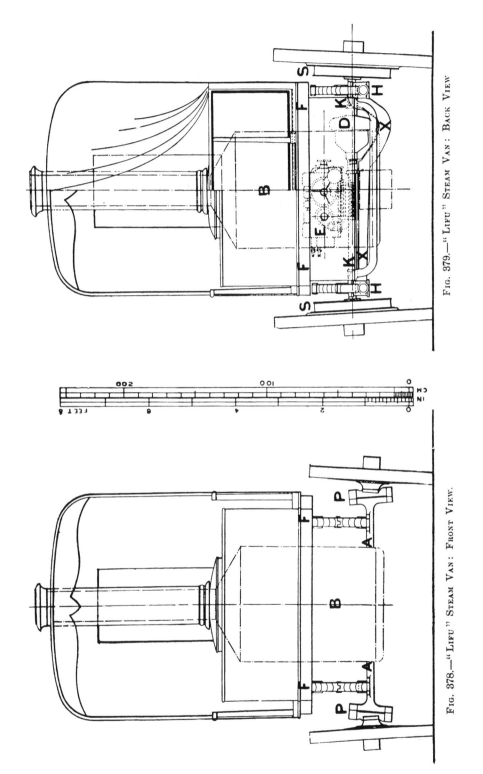

FIG. 378.—"LIFU" STEAM VAN: FRONT VIEW.

FIG. 379.—"LIFU" STEAM VAN: BACK VIEW

499

Boilers   The boiler used in these vehicles is illustrated by Figs. 382A and 383. It is of the water-tube type, the lower ends of the tubes, which are $\frac{1}{2}$ in. internal diameter, being connected each one by a gun-metal union nut to the circular trunk tube C, and by similar union nuts to the central pot boiler A, which provides the necessary steam space except in so far as this is added to by the dome G. The dimensions of the pot are, externally, diameter $14\frac{1}{4}$ in., length 30 in. The lower circular trunk is of gun-metal, formed in one piece, with the bridge water connecting arch B, to which the copper drum A is connected by a screwed joint. The boiler is encased with light iron plate F, lined with thick asbestos sheet, the casing resting upon a flange cast round the trunk ring C to receive it. At D about one-sixth of the ring trunk is raised for the purpose of providing suitable door covered space for access to the burner for lighting up and cleaning.

Boiler tubes   Below the boiler and supporting the burner J is a casting L, with arms V attached to lugs K, which also carry the coned plate M, and the annular partly coned plate R. As will be seen from Fig. 382A, the tubes E, after a vertical start of about 9 inches, except at the raised part D, are all given a spiral bend in alternate directions. The character and centre line of this bend is shown by the dotted line N, Fig. 383, in which the letters P P represent the water gauge connections, Q the steam gauge, O the main steam connection, and S the main feed connection. The water gauge, as shown in Fig. 382A, is of the double asbestos packed and protected glass pattern, both gauges being connected at P P, see Fig. 383, to one trunk pipe, the upper part of which carries the steam gauge. The feed pipe which is screwed in at S has two accessible check valves, as shown in Fig. 382A, p. 510.

The "Lifu" burner   For the variable work which characterises the operation of all kinds of motor vehicles, and introduces so many difficulties, the atomiser burner, in any of its forms, as used for locomotive, marine, and fixed boiler purposes in Russia and elsewhere, successful as they may be for these purposes, are quite useless. They will give perfect combustion, or sufficiently so, under the nearly fixed conditions that determine their adjustment of oil, steam, and air, but that combustion would be rendered immediately imperfect if the calorific requirements varied from minute to minute, as they do so frequently with motor vehicles. Hence, although much time and ingenuity have been spent on this problem, the high temperature vaporiser form of burner is the only one which satisfies the exigencies of very variable steam demands. One of the best known forms of these burners is that illustrated by Figs. 384 and 385, which are a vertical section of the "Lifu" burner and a horizontal section of the cast-iron vaporiser or generator. Oil enters this at the inlet shown, and passes over the backwards and forwards path indicated by the arrows in the plan, and thence passes downward, also as indicated by arrows in the vertical section, to the burner which is fixed on the central piece L, shown in Fig. 383. When only a small burner is required, the pressure in the oil tank is only sufficient to cause the central valve with its small

500

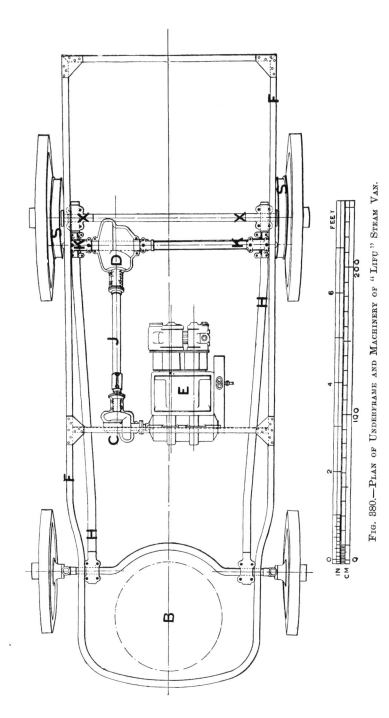

FIG. 380.—PLAN OF UNDERFRAME AND MACHINERY OF "LIFU" STEAM VAN.

501

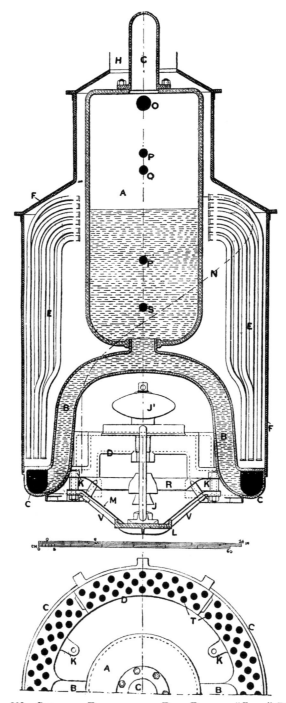

FIG. 383.—SECTIONAL ELEVATION AND PART PLAN OF "LIFU" BOILER.

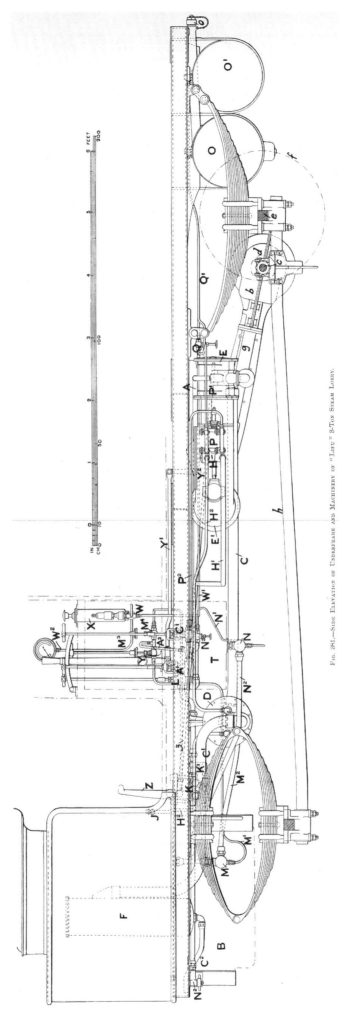

Fig. 381.—Side Elevation of Underframe and Machinery of "Lifu" 8-Ton Steam Lorry.

The material originally positioned here is too large for reproduction in this reissue. A PDF can be downloaded from the web address given on page iv of this book, by clicking on 'Resources Available'.

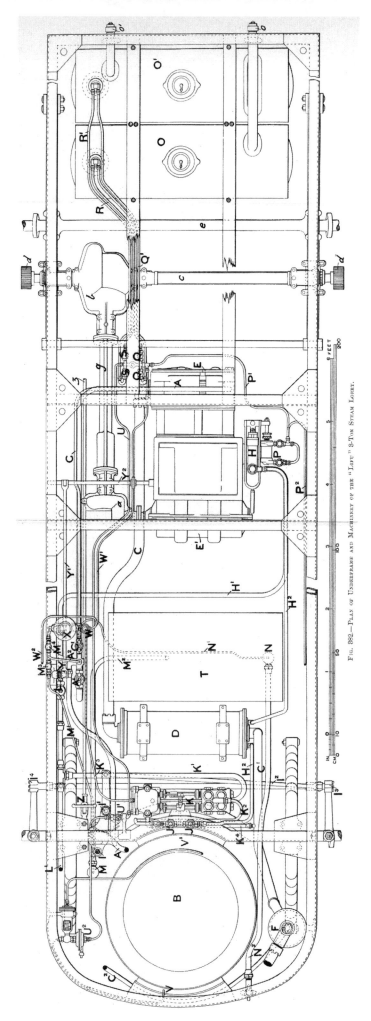

FIG. 382.—PLAN OF UNDERFRAME AND MACHINERY OF THE "LIFU" 8-TON STEAM LORRY.

The material originally positioned here is too large for reproduction in this reissue. A PDF can be downloaded from the web address given on page iv of this book, by clicking on 'Resources Available'.

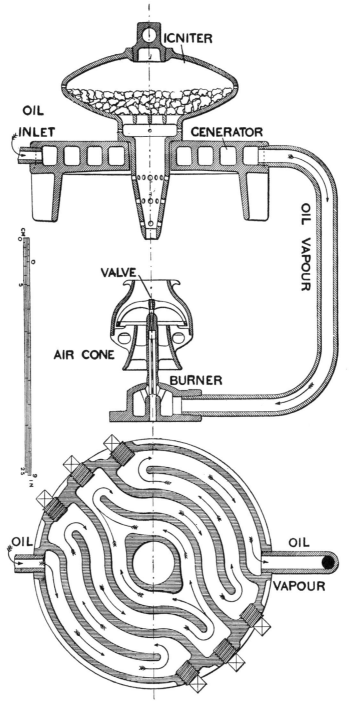

ICNITER

OIL INLET

CENERATOR

OIL VAPOUR

VALVE

AIR CONE

BURNER

OIL

OIL VAPOUR

FIGS. 384 & 385.—THE "LIFU" HEAVY-OIL BURNER.

503

pointed spindle to lift. When the full size powerful flame is required, the pressure is increased, and this valve and the larger one at the bottom of the hollow stem near the word "burner" also lifts, and then an enormous flame, impinging upon and surrounding the oil gas generator, is formed, filling the inside of the fire box, and maintaining the generator and the

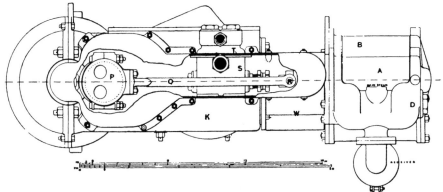

Fig. 386.—Side Elevation of "Lifu" Compound Engine.

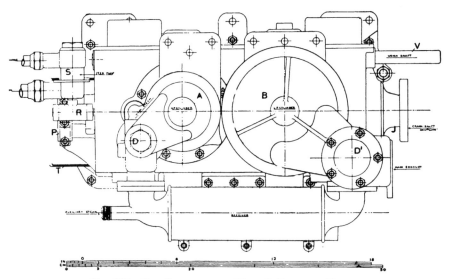

Fig. 387.—End Elevation of "Lifu" Compound Engine.

igniter at a high temperature. The igniter is simply a cast iron vessel partly filled with lumps of fire brick or other refractory material, its purpose being to retain heat long enough to re-ignite the oil vapour from the burner should it be blown out by a very strong gust of wind, or a too sudden release of the air pressure on the oil supply tank.

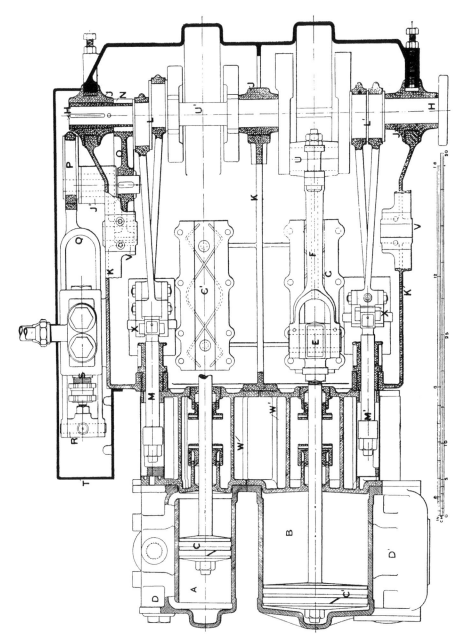

FIG. 388.—PLAN OF "LIFU" COMPOUND ENGINE.

505

The compound engines

The engines, seen in position at E E', Fig. 382, in side elevation in Fig. 386, in end elevation in Fig. 387, and in sectional plan in Fig. 388, are of the compound horizontal, slipper guide, reversing-link type, with pistons of 3 in. and 6 in. diameter, and 5 in. stroke. Between the cylinders A, B and the guide bars c c' are distance pieces within which are the glands, those on the cylinders for the usual purpose of preventing steam escape, and those on the outer parts of the distance castings nearest the crossheads E being to prevent the carriage of water of condensation from the piston rods into the oil-containing crank box. The slide-valve rods M M' are also carried by tubular bearings carried by the crank box. The valves themselves are placed at a considerable angle below the centres of the cylinders, as seen by the positions of the valve boxes D D'. The valves are driven by dog links, which are seen in Fig. 387. Piston valves

Feed pump of simple form are employed. The main feed pump is driven by an excentric P (seen also in Fig. 385), and forked connecting rod Q, attached to

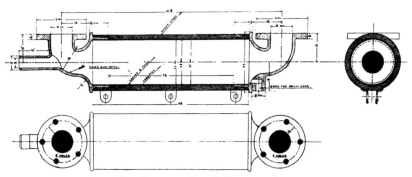

FIG. 389.—PLAN OF "LIFU" RECEIVER SECTIONS.

Crank box a crosshead R at the rear of the pump. The excentric P is on the outer end of a short spindle driven by the gear wheels N, O. The crank box K is of bronze, except the front part and the pump covering part shown in black in Fig. 387, which is of aluminium. The receiver, or connecting piece, and chamber, between the two cylinders, seen in elevation in Fig.

Live steam to L.P. cylinder 386, is shown in section and plan in Fig. 389. At the high pressure end of this receiver will be seen the connection for admitting high-pressure steam from the boiler to the low-pressure cylinder, for occasionally overcoming difficulties.

Steam and water connections From the elevation and plan, Figs. 381 and 382, it will be seen that the engine E[1], is supplied with steam through the pipe A from the boiler at B, and that the exhaust passes from the low-pressure cylinder by the pipe C, to the feed heater D, and thence by the pipe c[1] to the silencer F, whence it passes into the uptake by a pipe not shown, the water of con-

Feed pump connections densation passing away by the pipe c[2] to the tank from which the feed pumps draw by pipe L[1] on the off side of the boiler. The latter are two in number, one operated by the engine at H (see plan), and the other a direct-

506

acting independent steam pump K below the foot plate. Both draw from the front tank by way of the pipe L^1. By means of union connections, the suction pipe K^1 of the direct-acting feed pump is connected to the pipe L^1, which is continued to L, and along to H^1, and forms the suction pipe of the pump H. The delivery from the latter is by way of pipe H^2 to the feed heater D, and by another small pipe marked H^2 to the boiler by way of the check valves J J^1. The delivery from the steam pump K is by way of the pipe K^2 to the check valves. From the pipe A steam passes to a valve A^1, which controls the steam admission to a pipe which bends down to a lower level, and is continued rearwards and appears again at A, where it joins the high-pressure side of the engine at G^1. In line with the steam valve A^1 is a smaller valve which controls steam admission to the pipe G, which delivers into the low-pressure cylinder by way of the junction seen on the left hand of the receiver, shown in detail in Fig. 389. At A^2 is a small steam-pipe connection, from which steam may be sent by a valve M^3 and pipe M^1 to the injector M, by means of which water from the tank T is delivered into the higher level tank above M; or steam may be sent by valve M^4 and the pipe bending round under x to K^3, and thence to the steam cylinders of the little direct-acting feed pump K. The exhaust from this pump passes by pipe K^4 into the main exhaust pipe from the heater D to the silencer F.

At P is a small air pump worked off the plunger head of the feed pump H, the inlet being gauze covered at the forward end of pipe P^2. It supplies air under a pressure of about 15 lb. by way of pipe P^1 to distributing valves Q Q, and by means of the two pipes at Q^1 delivers air into the two oil tanks o o^1. By a T-piece on the left of the valves Q, air passes by way of pipe W^1 to a maximum pressure valve at W, and to the presure gauge W^2. By means of the pressure in the tanks o o^1, oil is sent through the pipes R R^1 by way of the valves s s, and thence by the pipe U to the valve at U^1, and then by a continuation of this pipe to the filter U^2, and from this by the pipe shown by dotted lines within the front part of the channel-iron framing to the burner at V. The quantity of oil being regulated by the steam diaphragm near V^2, which receives steam from the boiler by pipe V^1. At o o^1 are the capped oil-filling pipes to the tanks o o^1, and at N^2 is a pipe which at the front end is, when necessary, connected with a hose for filling the tank T, by the steam injector lifter at N, supplied with steam by pipe N^1 from the steam connections at G^1.

The crankshaft of the engine has upon its end a mitre wheel a, Fig. 382, which transmits motion to the driving gear by a mitre wheel on the end of a spindle in the tube g to the gearing in the box l, upon the transverse differential gear shaft c, which carries at its end the pinions d, which gear with the internal gear wheels, which are similar to those shown in Fig. 379, clip-bolted to the spokes of the driving wheels. As these wheels are dished, the axles incline downwards, and the teeth of the pinions d have consequently to be arranged at an angle from the shaft.

Telescopic transmission shaft
The gearing in the boxes *a, b* is shown in detail by Figs. 390 and 391. In these the upper figure shows the mitre gear A, B connected to the engine shaft by the continuation spindle and coupling E. The wheel B is fixed on the upper end of a spindle F, which like the spindle E runs in white metal bearings in the enclosing case C C. The continuation of the spindle F, shown in black section, is tubular, and runs within an exterior

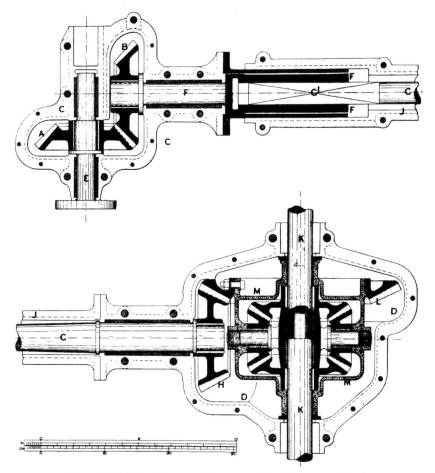

FIGS. 390 & 391.—SECTIONAL PLANS OF "LIFU" DRIVING GEAR.

tube, also shown in black section. Surrounding this outer tube is the upper end of the strong exterior tube J, which, as seen by the lower engraving, Fig. 391, is firmly connected to the gear box D. Within the tubular part of the spindle F, is a spindle G, square at the part G′ in the tube. The tube J is free to slide through a limited range on the tube which is flange-connected to the box C, the range being shown by the space at F F. The spindle G has corresponding freedom of longitudinal movement within the square tube by which it is driven. By this means the vertical movement (with the main axle) of the differential gear spindle ᴋ, marked c in Figs. 381 and 382, is accommodated.

This spindle runs in bearings which are clamped to the longitudinal frame tubes *h*, Fig. 381, and are transversely connected by a **T** section transome, similar to that shown at x in Fig. 379. As this, and all that it carries, is independent of the deflections of the rear springs, the freedom given by the telescopic arrangement shown by Fig. 390 is essential. The lower ends of the spindle ɢ, Figs. 390 and 391, carry a bevel pinion ʜ, gearing with a bevel ring ʟ on the differential gear box ᴍ. The construction of the differential gear is clearly shown in Fig. 391.

Fɪɢ. 392.—Hᴏᴜꜱᴇ's "Lɪꜰᴜ" Pʀᴏᴛᴇᴄᴛᴇᴅ Sᴇᴄᴛɪᴏɴᴀʟ Rᴜʙʙᴇʀ-Cᴜꜱʜɪᴏɴ Wʜᴇᴇʟ.

The change of direction of motion of the engine is effected by ordinary **Reversing** link motion and double excentrics, seen in Fig. 388, operated by the lever pivoted at ʏ, Fig. 381, and the connecting rod ʏ¹ and arm ʏ².

The steering and driving wheels are of what may be called the Han- **Wheels and tyres** cock type, and for the steam brakes and omnibuses, which have done such splendid services, they are fitted with solid rubber tyres; but for some of the heavier vehicles, on which iron tyres have been found objectionable, wheels of the design shown in Fig. 392 have been made by Mr. House, and are in promising experimental use. The outer ends of the spokes are not tanged as in ordinary construction, but are continued full size to their ends, where they are tightly inserted into flanged ferrules riveted to a flat steel rim. Surrounding this rim is a series of rubber and canvas cushion pieces, protected by cast steel shields, one outside every spoke. At their

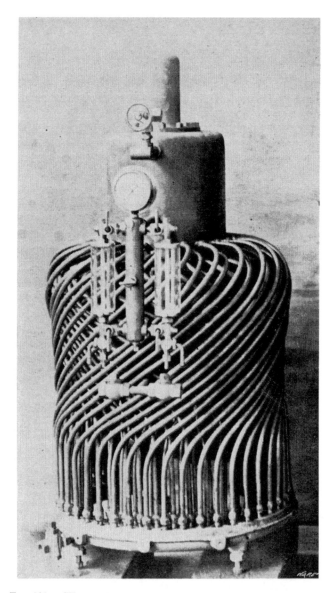

FIG. 382A.—WATER TUBE BOILER OF THE "LIFU" STEAM VEHICLES.

ends these shields are extended inwardly by lugs, through which pass bolts upon which are strong canvas rubber tubes; the shields are prevented from creeping round the wheel by cross ribs, one between every spoke, and they

510

are free to move radially in correspondence with whatever compression occurs in the rubber canvas cushion pieces under the load.

The steering gear is of the Ackerman type. On the lower end of a vertical spindle 1 (see plan, Fig. 382) is a sector gearing with one of the same radius on a short spindle 1^1, carrying a backwardly extending arm, on the end of which is pivoted, near κ^3, the rod 1^2 at the end 1^3, of which on one of the axle steering arms is a ball and socket joint. Pivoted at this joint is also a rod extending to another ball and socket joint on the end 1^4 of the off steering axle arm; these arms, it will be seen, are at right angles to the axles, but the pivots are inclined as already described.

A considerable number of different forms of the "Lifu" vehicles have been made, including small and large omnibuses and brakes, vans, and heavy lorries. Most of these have proved the goodness of the design of the vehicles and their machinery, and the highest praise has been universally bestowed upon the fourteen-passenger brake type. One of these brakes has been very largely used by the Automobile Club on its English tours, and it has invariably behaved splendidly as a steady, very fast, perfectly controllable vehicle, and good hill climber. The only objection that has been made to the vehicle during its running has been based upon the imperfection of the combustion by the burner when the variation of the demands upon it has been very great, the incomplete combustion almost always resulting upon a change from full load to small or no load as at stops, and from occasional extinction of the flame. The latter very seldom happens; but when it does, the small quantity of oil remaining in the vaporiser passes away as an objectionable unburned heavy oil fog. When working on tolerably uniform loads, anywhere between half and full load, the burner gives satisfaction, although the combined efficiency of the burner and boiler is low. The "Lifu" vehicles have only entered one of the competitive trials, and this was at Liverpool in 1898 (see Table XX.), and the awards then made afford little or no indication of the merits, from a practical point of view, of the vehicles exhibited.

Chapter XXX

HEAVY STEAM VEHICLES (*continued*)

THORNYCROFT STEAM VEHICLES.

AMONG the first of established English engineering firms to take in hand the construction of the heavier class of steam vehicles of various forms was that of The Steam Carriage and Waggon Company, of Chiswick, now the Thornycroft Steam Carriage and Waggon Company, of Chiswick and Basingstoke.

Steam van, 1896

One of the first of the vehicles of their construction was that shown at the Crystal Palace Exhibition of 1896, which was a steam van, fitted with a little vertical engine such as might be used in small launches, and with a small Thornycroft water-tube launch boiler. Previously to this, however, Mr. Thornycroft had designed and made an experimental passenger carriage for carrying eight persons.[1] The first of the heavy vehicles made at Chiswick were the steel dust-tip waggons shown by Figs. 393 and 394. Several of these have been made, two of them for the Chiswick Vestry, and are in successful use, having a capacity of 6 cubic yards, which may mean the carriage of anything between 1 ton and 4

Steam dust cart

tons of town refuse. Particulars of the trials of one of these vehicles with the body replaced by a lorry top, as tested by the Royal Agricultural Society in 1898, will be found in Table XX., and also the leading dimensions of the machinery. The extreme length of the dust waggon is 15 ft., the breadth 6½ ft., and height about 10 ft. The wheel base is 8 ft. by 6 ft., the rear wheels 3 ft. diameter, and the tyres 4½ in. wide. The engine

Compound engine

is of the horizontal compound type, with cranks at 180°, and cylinders placed on opposite sides of the crankshaft. All the main parts of the engine, including the frame, guides, valves, piston-rods and guides were of gun-metal or phosphor bronze, partly for lightness and partly for avoiding damage by rust when the vehicle is laid by. The engine drives a shaft, which is a continuation of its own, through the medium of a friction-

Friction-clutch fly-wheel

clutch of the multiple disc or Weston type which serves also as flywheel. Upon the shaft is a pinion which drives a chain pinion shaft by running

[1] *Fielden's Magazine*, Nov. 1899, p. 432.

in gear with the outside of a differential gear box. The Renolds saw-tooth or silent chain was at first used. The front wheels run on pivoted axles, and are steered by simple form of gear and hand-steering wheel.

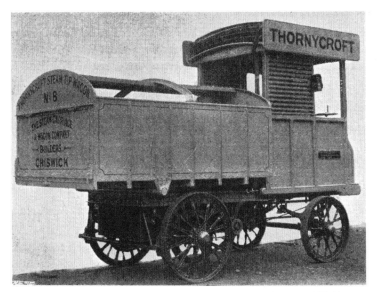

FIG. 393.—THE THORNYCROFT STEAM DUST TIP WAGGON, 1897–98.

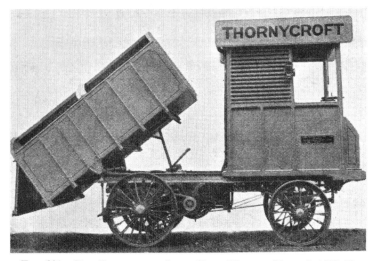

FIG. 394.—THE THORNYCROFT STEAM DUST WAGGON (TIPPED), 1897–98.

The boilers at first used in these vehicles were of the Thornycroft water-tube form,[1] similar to the launch boilers, but the subsequently used

[1] See Paper by Author, *Soc. Arts Journ.*, Nov. 27th, 1896.

boilers, as on the vehicle tested by the Royal Agricultural Society, and at Liverpool in 1898, see Table XX., are of the circular form which is illustrated by Fig. 408. These are employed on all the recent Thornycroft steam waggons and lorries. On the top of the dust vans a simple form

FIG. 395.—THORNYCROFT'S 3-TON LORRY, 1898–99.

FIG. 396.—THORNYCROFT'S 3-TON LORRY, 1898–99.

of condenser was carried, consisting of $\frac{1}{2}$-inch copper tubes of 24-gauge in thickness, connected at their ends to trunk pipes, and having a total cooling surface of 145 sq. ft. Reference to the cost and cost of working of these vehicles will be given hereafter in the chapter on "Cost of Transport by Motor Vehicles."

Fig. 397.—Thornycroft Standard 3-Ton Waggon and Tractor, 1899.

515

Steam drays and lorries

Figs. 395 and 396 are photographs of the two sides of the Thornycroft 3-ton lorries, several of which are in daily use in different trades. Fig. 397 is from a photograph of a 3-ton steam waggon, similar to a dray, built for Messrs. Guinness & Co., capable of carrying at least 3 tons, or of weighty materials of small bulk 4 tons, and when carrying 4 tons to haul at least $2\frac{1}{2}$ tons. The general arrangement in these two forms of lorry is very much the same, the main difference being in the form of the engine, which,

The engines

as used in Fig. 395, is illustrated in Figs. 400–402. The leading particulars as to dimensions of this engine will be found in Table XX. It is a compound engine in which the steam from the high-pressure cylinder F′, admitted to it at R, passes through the pipe s to the steam chest H of the low-pressure cylinder F, to which it is distributed by the valve V, and finally escapes through the pipe T. The construction of the engine may be said to be obvious from the engravings, but it may be pointed out that

Reversing gear

the reversing is done by means of single excentrics M M′, which, mounted on a very quick pitch screw on the centre of the crankshaft, are moved by the collar and fork at P, by the arms Q Q on the shaft O. As shown in the engraving, the excentrics M M′, which are double the width of the guided excentric straps, are in the position for running out on the under stroke. By throwing the collar at P over towards N the excentrics are moved from a position of about 120° from the cranks in one direction to the same ahead of the crank in the other direction. It will be noticed that torpedo-boat experience has not been neglected in designing these engines. They are completely boxed in, and their design makes them practically independent of more than occasional attention. The wearing parts are large where necessary, and lock nuts and pins are used, and spring catches w to hold the screw glands in the position to which they are turned. The engines are easily accessible by the covers x^1, x^2, x^3.

Heavy lorry 1899

The latest pattern of lorry, Fig. 397, has an engine which differs slightly in its arrangement, as will be gathered from the photographic view and from Figs. 398 and 399. The dimensions of its cylinders and its stroke are given in Table XX., with the results of the trials carried out at and near Liverpool.

Underframe

The underframe v v of this vehicle consists of light angle steel, well braced by cross-pieces and gusset plates, which serve also to carry the machinery. Below the underframe, but acting as part of the under-gear, are two strong channel-steel bars Y, extending from front to rear axle. The mounting of the underframe upon the axles will be gathered from Figs. 398 and 399, and also from Figs. 403 and 405.

Engine

The engine E E is carried by brackets attached to the underframe v, and to the cross angle irons above the cylinders and above the crank case. The crankshaft outside the case is squared, and carries two pinions F,

Transmission gear

G, which may alternately be put respectively into gear with the wheels I, H on a short shaft carried in bearings J J attached to the main underframe. At the inner end of this short shaft is a shaft K, the ends

516

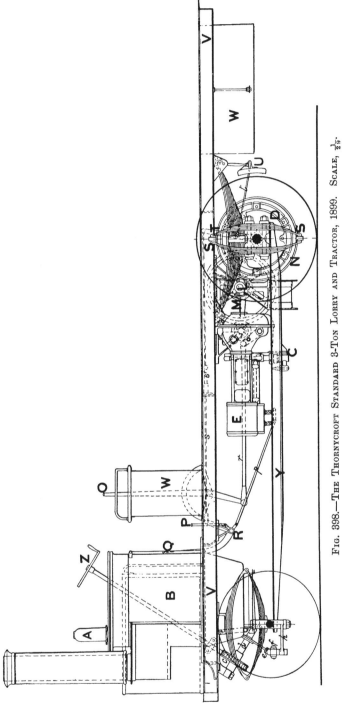

FIG. 398.—THE THORNYCROFT STANDARD 3-TON LORRY AND TRACTOR, 1899. SCALE, $\frac{1}{29}$.

517

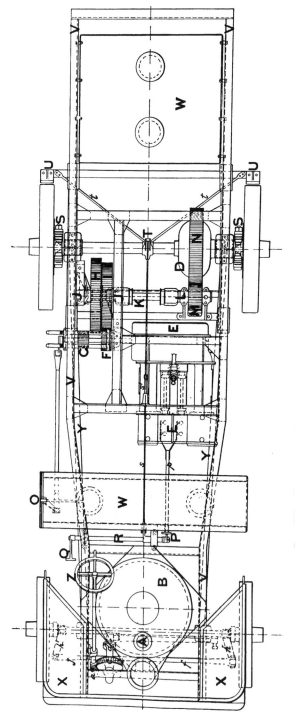

Fig. 399.—Thornycroft 3 to 4-Ton Lorry and Tractor: Plan. Scale, $\frac{1}{20}$.

518

of which are of the type of universal joint shown by Fig. 404, in which κ is
the articulated shaft, driven by pins 5 5 in brasses 4 4, free to move in their

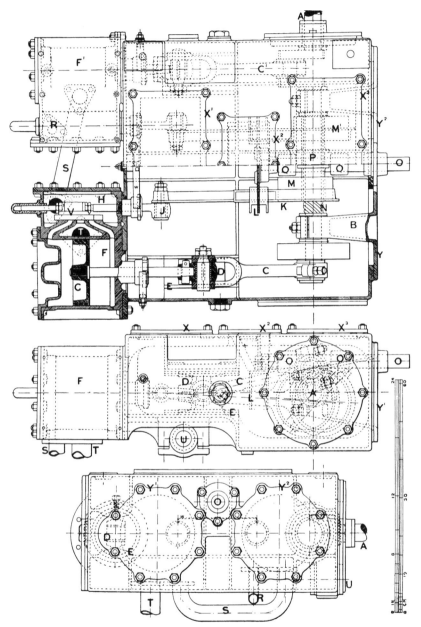

FIGS. 400–402.—SECTIONAL PLAN, ELEVATION, AND END ELEVATION OF ENGINE,
THORNYCROFT 3-TON LORRY.

respective angular directions within the joint boxes 2 2, at the ends respectively of the shafts which run in the bearings J J, L, Fig. 399, the

latter carrying the pinion M. The shaft carrying the pinion M is a short one, carried by bearings fixed to the "reach" or subsidiary underframe Y Y, and a heavy cross-tie, seen in Fig. 405. The bearings L move with the main axle unaffected by the deflection of the main springs, and the method of construction just described maintains constant distance between the pinion M and the large wheel N on the outside of the differential gear which it drives, the relative movement of the frames V, Y being allowed by the spindle K above mentioned. The differential gear D is of the three-pinion kind, and drives two hollow shafts x, z, Fig. 405, through which project the ends of a solid axle, upon which the main driving-wheels run freely through whatever range is enforced by the turning of corners.

FIG. 397A.—COMPOUND ENGINE, THORNYCROFT 3 TO 4-TON STEAM LORRY.

Spring driving connection

Upon the ends of the hollow axles are square bosses, to which are attached the strong double-leaf springs s s, by which motion is imparted to the driving wheels at their felloes. The arrangement of this spring drive is shown by Figs. 406, 407, in which the square boss on the end of the hollow shaft (which should have been shown in section, see Fig. 405) is seen at c, through which and the springs A pass heavy bolts holding all together. The ends of the inner and longest leaves of the spring, it will be seen, are turned over and given suitable form for bearing against the drivers B, of cast malleable iron, fastened each by four bolts through two segments of the wood felloes. This is about the most satisfactory job that has yet been made of a spring-driving arrangement with wheels which represent a combination of the suggestions of Hancock and of Gurney.

Boiler

The boiler used in this vehicle is shown in some detail by Figs. 408

520

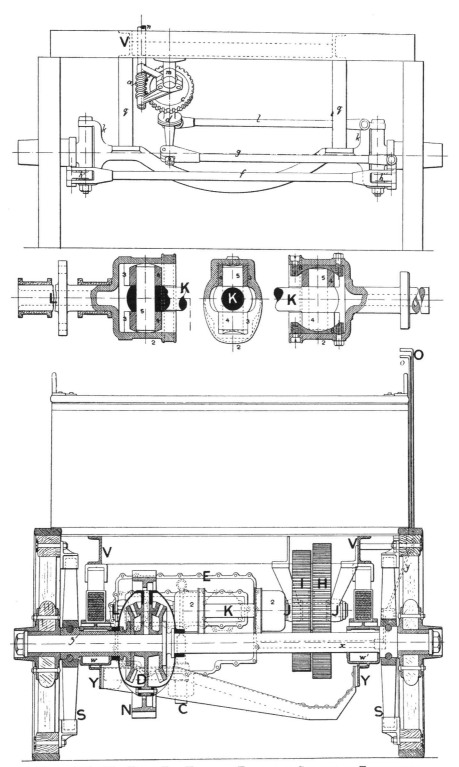

FIGS. 403 & 405.—FRONT END VIEW AND TRANSVERSE SECTIONS OF THORNYCROFT
3 TO 4-TON STEAM LORRY. SCALE ¾ INCH=1 FOOT.
FIG. 404.—DETAILS OF ARTICULATED DRIVING SHAFT (K OF FIGS. 399 AND 405).

521

and 409, in which the straight tubes T, of which there are two rings, are fixed at their bottom and top ends in the tube plates of two annular vessels U W, the upper one W of which forms in part a steam space. It will be seen that both the tube plates are deep pressings, the conical part of the lower one being drawn down in thickness in the part that forms the fire box. The upper stamping is extended in depth by angle-steel rings, one limb of the angle being used for the insertion of studs by which the annular cover plate is held on. The upper edges of the annular chamber so constructed are turned, and corresponding grooves or chases are turned

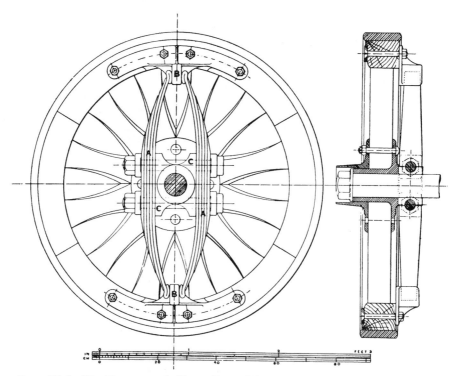

FIGS. 406 & 407.—THORNYCROFT'S HEAVY LORRY WHEELS AND SPRING DRIVING CONNECTION.

in the cover plates. These grooves are partly filled with asbestos tape for making the joint.

Above the upper end of the coil s s is fixed a small steam dome, marked A in Figs. 398 and 399, from which steam is taken, passed through the superheating coil s, across the steam space in w, and out at the off-side of the boiler, where is situated the regulator valve placed under command of the right hand of the driver, as seen in Fig. 396. At A, Fig. 408, is the closed ashbox, accessible by the damper door K. The firebars B are arranged in two forms, as shown in the plan. At D, in front of the fire box, is a door for access to the fire for occasional clinkering. The water-

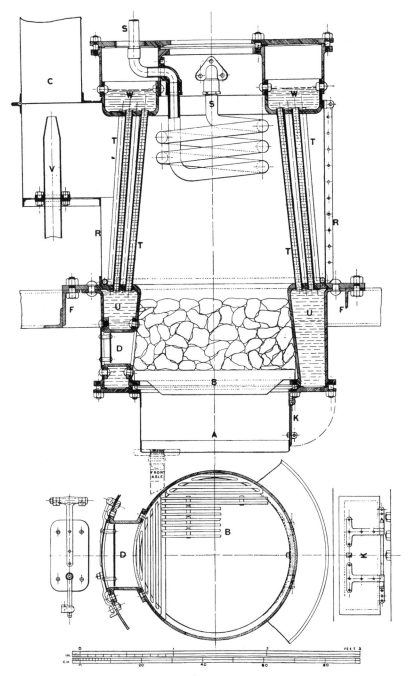

FIGS. 408 & 409.—WATER-TUBE SUPERHEATER BOILER OF THORNYCROFT LORRIES.

523

tube part of the boiler is surrounded by a sheet-iron casing and asbestos sheeting, the casing being extended at the front, so as to form an uptake into the chimney c, into the lower part of which the exhaust steam is delivered by a pipe v, after passing through a puffer-box, as Hancock called it, or silencer and feed-water heater, not shown in the drawings.

In the plan, Fig. 409, the doors at κ, D are separately shown. Two water tanks w w, Figs. 398 and 399, carry about 160 gallons, sufficient for about 20 miles journey. Coal bunkers at the sides of the boiler carry about 5¾ cwt. of coal.

Feed pumps The feed is ordinarily supplied to the boiler by a feed pump at c, but a self-starting injector is also fitted for feeding the boiler when standing.

Reversing The engine of this standard pattern vehicle is seen in side view in Fig. 398, in which o is the reversing lever, which is coupled by the connecting rod R to a way shaft in the bottom of the engine case, by arms on which link motions are operated. See also Fig. 397A.

Steering gear The steering gear is worked by the hand wheel z, on an inclined shaft, carrying at its lower end a worm a gearing with the worm wheel c on the inclined spindle d (see Figs. 398, 399 and 403). The upper end of the spindle d is carried by a bearing at m, and near the lower end, namely at d, it carries an arm jointed to the rod g, which in its turn is connected to the steering arm on the pivoted axle at h, this arm being connected by the rod F to the corresponding arm of the off axle. The proper distance between the lower end of the spindle d and the pivoted axles is maintained by the rod l.

Brakes The vehicle is provided with a pair of shoe brakes u u, Figs. 398 and 399, operated by the driver's foot on the pedal at the end of the spindle R, which gives motion to the rod s, supported by a boss at s', and connected at T to the two rods t t, which pull upon the break shoe bar; this is suspended, as is seen in Fig. 398, from the main frame. The engine is relied upon as a further brake by means of its reversing gear. The cylinder drain cocks are opened by the handle P and connecting rod p.

As shown in the engravings, the heaviest spur gear M, N has ordinary straight cogs, but some of these vehicles are now fitted with double helical gearing.

Wheels The steering wheels of the standard 3-ton vehicles are 2 ft. 10 in. diameter, with 4½-in. tyres, and the driving wheels 3 ft. 3 in. diameter, with 5¼-in. tyres.

From what has been said in the preceding notice of the Thornycroft vehicles, it will be seen that the makers have in the five years of working passed through a number of styles of invention and design of arrangement, and that they are now in a position to supply thoroughly well-made vehicles having about 35 HP. on a tare weight of 3 tons. In the five years the position of the boiler has been changed from the rear to a little in front of the centre, and lastly to the extreme front. The engine has changed in form and position. The gear has changed from spur gear and chain to spur gearing only, with jointed driving shaft, and the wheels have changed from iron spoked, iron rimmed, rather light wheels, to heavy close-spoked wood wheels.

Of the success of the vehicles there can now be no doubt, and the accumulating experience will in increasing ratio rapidly add to the number of trading requirements for vehicles of various sizes.

Already a commercial speed of from six to seven miles per hour is undertaken by the builders. This increase of speed over that common with horses may be successfully followed on good roads, but it rapidly destroys the wheels and other parts of the heavy-load lorries, which may last several years in good condition at horse speeds. This is especially the case when much of the travelling is over very badly paved stone sett roadways, like many of those about Liverpool.

Chapter XXXI

HEAVY STEAM VEHICLES *(continued)*

The Clarkson & Capel 3-Ton Lorry.

General description

A 3-TON lorry, designed and constructed by Messrs. Clarkson & Capel, London, is illustrated by Figs. 410–418. From the elevation, Fig. 410, it will be seen that this lorry is arranged with the boiler B in the extreme front, with a cab behind it covered by a Clarkson condenser C, and enclosing a vertical engine E, which drives, through a two-speed gear at D, a chain, on a sprocket wheel 5, conveying motion to the sprocket wheel 6 on the outside of the differential gear G. The differential gear is on a transverse shaft, which is in two parts, in the usual way, and at its ends 7 7 carries sprocket pinions, upon which run chains imparting motion to sprocket wheels on the naves of the rear driving wheels.

Engine

The engine is a small compound, with link reversing gear and piston valves, and with cylinders 2·75 in. and 6 in. diameter, with a piston stroke of 4 in.; it is enclosed in an oil-tight case, and at 600 revolutions per minute is said to give 14 B.HP. The steam supply to the engine is taken from the stop valve 1, and thence by way of pipe 2 to a regulator valve s by the side of the engine, operated by the lever 3.

Boiler

The boiler is of the cross-tube, vertical, fire-engine type, and is illustrated by Figs. 413 and 414, in which B is the water jacketed fire box, crossed by small inclined tubes. It is provided with a large steam space. Particulars as to heating surface will be found in Table XX. At the water level is placed a feed-water supply regulator, consisting of a float x, balanced by a weight w, carried by arms Y Y, with a stud spindle at s', and a longer spindle at s, passing through a gland-packed bearing J.

Automatic feed-water regulator

Outside this bearing is a ring H, running within a groove in a collar on the spindle, and operated by a lever Q, so as to give a reciprocating movement of small range and small speed for the purpose of ensuring the full and free action of the float x, when it varies in level due to the variation in the water. The spindle s is in fact given this reciprocating movement, so that when the float tries to rotate the spindle s, a compound motion is imparted, like that employed by Wicksteed for the pump plungers of testing machines, or like that given

526

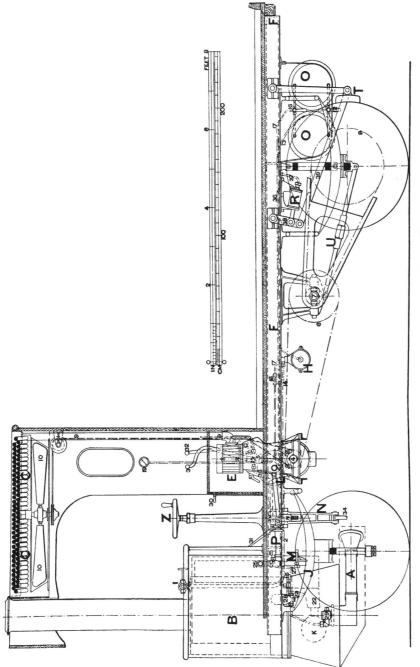

FIG. 410.—CLARKSON & CAPEL'S 3-TON STEAM LORRY.

527

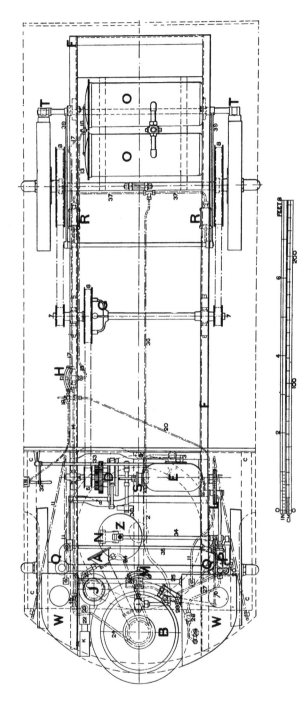

Fig. 411.—The Clarkson-Capel 3-Ton Lorry, 1898.

528

by ordinary mortals anxious to remove a cork from a bottle having desirable contents. The lever Q is pivoted at L, and at its outer end receives its motion from a connection with the feed pump crosshead near P, by means of the rod p.

Within the boiler, to limit the range of movement of the arms carrying the float and balance weight, projections z z are employed. A short lever T on the spindle s, Fig. 415, controls by its angular movement

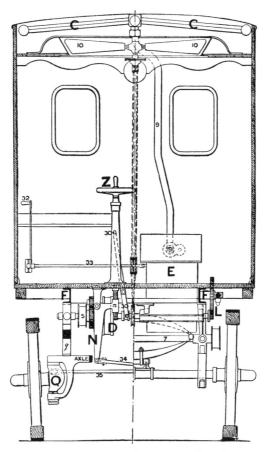

FIG. 412.—TRANSVERSE SECTION OF CLARKSON & CAPEL'S STEAM LORRY.

a suction cock at 26, a pipe from which is shown by dotted lines to the pump P, Fig 411. When the water in the boiler reaches the top level, the suction valve is so nearly closed by the lever T, that only the minimum supply necessary at any time can be taken by the pump.

The pump delivers by the pipe 27 and valve 29, on which is the air **Feed pump** vessel 28, seen also in Fig. 410. The water is taken from the two tanks w w, which are connected by a cross pipe 25, the water on its way from

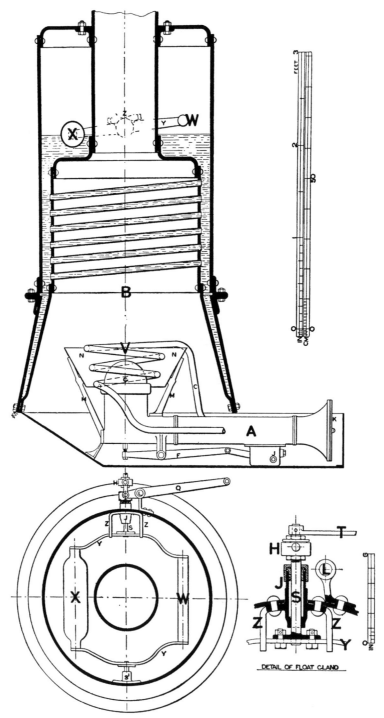

FIGS. 413 & 414.—VERTICAL AND HORIZONTAL SECTIONS OF THE CLARKSON & CAPEL BOILER.
FIG. 415.—DETAIL OF FEED-WATER SUPPLY REGULATOR.

Feed
strainer

these passing through a strainer at M. The feed water is not heated, but as most of it consists of the water of condensation from the condenser c at the top of the cab, this is perhaps not necessary. The steam from the engine passes by the pipe 9, Figs 411 and 412, to the condenser, and the water from the collecting troughs of the latter, c, Fig. 411, passes down to the tanks w by the pipes 11, 11.

Condenser

The condenser consists of three longitudinal trunk pipes to which are connected about 380 ft. of thin copper tube of $\frac{1}{2}$ in. exterior diameter covered with the metallically connected spiral radiating wire, shown by Fig. 256, p. 341. The total surface of this condenser, not including the spiral-wire coating, is about 50 sq. ft. This surface would be very small for 14 B.HP. of steam; but aided by the fan 10, 10, Figs. 410 and 412, it appears

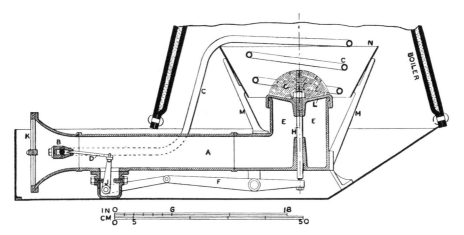

FIG. 416.—SECTION OF THE CLARKSON-CAPEL BURNER: VERTICAL TYPE.

to have been sufficiently powerful to have condensed about 60 per cent. of all the steam which passed through the engine, inasmuch as the boiler is credited with evaporating 3·7 lb. of water per 1 lb. of oil fuel used. The fan 10 is driven by means of gut or round leather belt running on a grooved pulley in the crankshaft, and guided by idle pulleys, as seen in Figs. 410 and 412.

Condenser
fan

The burner used for heating the boiler of the lorry is of the vertical Clarkson & Capel form. It is shown in position in Fig. 413, and in section in Fig. 416, and also in the general plan, Fig. 411, where at its air inlet it is marked A. It will be convenient here to explain the oil pipe connections between the burner, the oil tanks o o, and the starting burner and vaporiser respectively at 22 and J. Worked off the pump crosshead at P is a small air pump of bicycle type, which sends air under pressure by pipe 20 to the four-cross union 18, and thence by pipe 17 to the cock 16, where it joins the air vessels. A branch from the air supply at 18 leads to the air pressure gauge in the cab at 19. From below the cock 16, oil is taken

Oil-fuel
burner

from the tanks o, by the pipe 13, to a strainer at H, and from this by pipe
14 to a cock at 15, and thence on by a pipe 24 to the coil in the burner,
from which it passes in the form of vapour by pipe 24, by a bend in which
it enters the main air inlet trunk A of the burner. The construction of
the burner is clearly shown in the section, Fig. 416, and in the exterior view,
Fig. 413, in which only the inverted cone N is in section. Turning to
Fig. 413, the oil admission pipe is shown broken off near the letter A;
through this pipe, bent into a coil v, the oil passes, emerging at the top

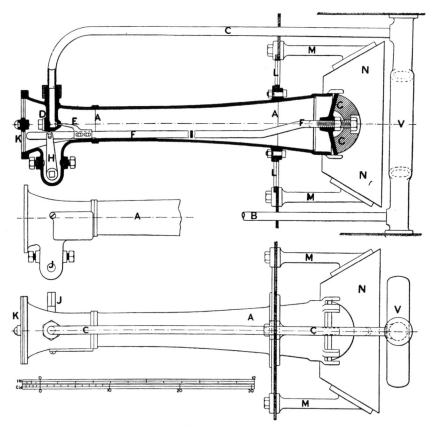

Fig. 417.—Sectional Elevation of Clarkson & Capel's Burner: Horizontal Form.

and through the part c, Fig. 413, as a high-temperature vapour, and thence
as seen by dotted lines, Fig. 416, to the jet piece B. Into the orifice of this
jet is fitted a taper rod D', acting as a valve to control the emission of this
vapour. The rod D' is carried by an arm J on a spindle, which also
carries an arm connected with a pivoted lever F. The latter at the end of
its shorter arm supports the spindle H of the large circular valve L, which
carries a fireclay or other refractory or hemispherical lamp G. The spindle
and arm J are controlled as to position by an external hand lever, according

532

to the volume of the flame required, and it will be seen that as the taper rod D is moved further from the jet orifice, the valve L is lifted higher from its seat. At K is a large air-admission disc, perforated like an air ventilator, by which the quantity of air drawn by the action of the jet B, and carried away by the combustion at the burner, is regulated.

For starting this burner a fan K, worked by hand, delivers an air blast **Starting the burner** into the chamber 22 and conical vessel 23, in which is placed a piece of waste soaked in petrol, the powerful bunsen flame from which is sent into the vessel J containing a coil into which oil passes from the three-way cock 15. This oil is vaporised in the coil, and passes away by pipe 21 into the jet D of the burner. When the burning of the vapour thus obtained has heated the main burner coil, the fan is stopped, the cock 15 is turned so as to divert the oil supply from J to the pipe 24, so as to enable the burner to provide its own vapour.

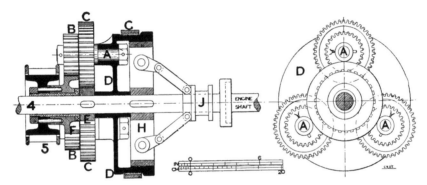

FIG. 418.—CLARKSON & CAPEL'S CHANGE-SPEED GEAR.

Fig. 417 shows another form of this burner arranged to project a **Horizontal flame burner** horizontal flame. In it B is the oil admission pipe to the vaporiser V, which in this case is a cast annular chamber of the section shown by dotted lines. The oil vapour passes out by pipe C to the jet piece D in the trunk A of the burner. The needle valve E is mounted on the bar F, which at its front end carries the flame-regulating valve G, similar to that of the vertical type of burner. The bar F and valves E, G are adjusted from a hand lever on the spindle J, a second lever H on which engages with a pin on the bar F. The air supply to the trunk case A is regulated by apertures at the back end, closed more or less by the adjustable disc K. A second adjustable disc at L regulates the quantity of air passing into the fire-box of the boiler, which for the burner illustrated would be of the horizontal flue or of the locomotive type. The sheet nickel cone shield N, which directs the flame from the mouth of the burner trunk, is supported by arms M from the fire-box end.

Variable steam pressure is not relied upon as the only means of vary- **Speed change gear** ing speed or propelling power. A two-speed change gear is employed, as

533

MOTOR VEHICLES AND MOTORS

illustrated by Fig. 418, in which the crankshaft extension 4 carries a pinion
E, which drives the wheels C, C on stud spindles A, carried by a drum D.
By fixing the drum D to the shaft 4, by means of the internal clutch H,
operated by the collar J and its connected toggle levers, the chain pinion 5
and the whole of the gear are carried with the shaft 4 at the speed of the
engine. To obtain a slower speed, as for hill climbing, the clutch H is freed
and the hand clutch G put into action, so that the drum D becomes a non-
rotative fixture. The pinion E then drives the wheels C, and thereby the
pinions B, which give motion to the central wheel F and chain pinion 5 at
the lower speed. The collar J of the clutch H is operated by the lever 30.

Brakes Three brakes are provided, beside that which the reversing gear
supplies. The first of these is the pair of brake shoes T, T acting on the tyres
of the driving wheels, and actuated by steam in the cylinders R, R through
the levers 38 and rods 39. Steam is supplied to the little brake cylinders by
the pipe 36, and the branches 37 from it. A supplementary brake is
formed by the change speed gear, Fig. 418. By fixing the drum D to the
shaft 4 by the clutch H, the clutch band G may be used as a brake band.

Steering gear The steering is effected by a hand wheel Z, on the top of a spindle
which is screwed at its lower end, and upon which runs a nut n, pivoted
between the horizontal arms of a double bell crank lever N, seen in Fig.
410 and in Fig. 412. In the latter figure the pivot of the nut n is not
shown. To the lower end of the vertical part of the bell crank lever is
attached the connecting rod 34, by which motion is given to the rod 35
connecting the steering arms on the stud axles.

There are, it need hardly be said, numerous points in the arrangement
of the machinery of this car which are ingenious and interesting, but there
are others, such as the form of the fixed forked part of the front axle, the
method of support of the whole fore carriage, and the somewhat unstable
mounting of the body over the rear axles, and the double-chain driving
gear, which are not good. It is also very questionable whether the
arrangements adopted for automatic maintenance of the water level in
the boiler are likely to remain so effective and trustworthy as to be
reliable. If not uniformly effective, the more usual hand-controlled feed
would be preferable.

Chapter XXXII

HEAVY STEAM VEHICLES (*continued*)

The Coulthard, the Leyland, and the Simpson & Bodman Steam Lorries.

F
IG. 419 is from a photograph of the 2-ton steam lorry by Messrs. T. Coulthard & Co., Preston, of which some particulars are given in Table XX. Several of these vehicles have been made, some of which have been sent abroad and some put into use in this country.

The Coulthard Lorry

They are of the chain-gear type, but are fitted with three-speed spur wheel and pinion change gear, and with spur wheel reversing gear. The photographic view, Fig. 419, represents one of the type which entered the Liverpool trials of 1899.

General description

The main frame is a parallel structure of channel steel, and it carries in the front a cab surrounding the boiler and engine, arranged side by side with a condenser, consisting of Rowe tubes, on the dashboard in front.

The boiler is of the vertical smoke-tube type of the simplest form, as shown to a scale of $\frac{1}{8}$ full size in the engraving, Fig. 420, which also shows in section the oil burner by which the boiler is heated. An inspection of the engraving will show that the shell is constructed so that, by undoing the ring of nuts in the upper tube plate and those in the angle rings in the bottom, the shell complete may be lifted from and leave the fire-box, tubes, and top tube plate, with the steam door and separator s, all intact. The heating surface and evaporative power are given in Table XX.

Boiler

The burner is of the vaporiser kind, and is carried by an extension on the stay bolt B. Oil enters the tubular ring v, as indicated by dotted lines, in which it is partly evaporated, and passes off, in the direction shown by the arrows, into a small flat round chamber c, where vaporisation is completed and from which the vapour passes by the tube D to the jet E, where, mingled with air which enters at the bottom of the tube F, it burns as a powerful bunsen burner all round the narrow chamber c, except where the pipe D and the supporting leg opposite it (see section on A A) form slight obstructions. The flame spreads well over the bottom tube

Oil-fuel burner

535

plate, and the tube v is maintained at a high temperature. The burner is heated for starting by putting petrol into the shallow asbestos-bottomed dish, seen on the door at the bottom of the fire-box. The steam from the boiler to the separator s passes through three small holes, and steam is drawn off from the box shown after passing over the top of the vertical diaphragm shown in section.

Triple expansion engine The engine used in this car is of triple expansion type, with piston valves arranged along the side of the cylinders, so that there shall be no space occupied between the cylinders, which are separated only by the thickness of a cylinder wall. By these means the three crank pins are brought sufficiently near together to make it possible with a good stiff

FIG. 419.—THE COULTHARD 2-TON OIL-FUEL STEAM LORRY.

crank to do without an intermediate bearing, and the use of the triple cylinder and suitably divided crank circle makes balancing easier. The engines are shown in vertical sections and plan to a scale of ⅛, by Figs. 421-423. The pistons in the three cylinders A, B, C are connected to slipper guide crossheads. The connecting rods are of marine type, forked at the little end and tubular for oil supply throughout. The **Lubricating** oil for all these parts is fed into a cup caught under the piston rod nut. From this it runs to the hollow crosshead pin and thence by a small pipe drips into the cup in the fork of the connecting rod and thus reaches the crank pin. The piston valves in the valve cylinders D, E, F are worked by excentrics G, having sheaves connected to tubular rods as seen in Fig. 421. These tubular rods take the oil put into the oil cups at the top of the slide valve joint pin and deliver it to the excentric straps. The position of the excentrics K, L, M is seen in the sectional elevation, Fig. 422, on an excentric

536

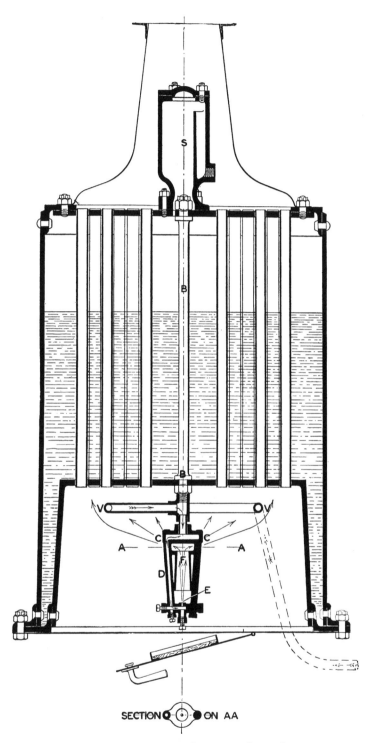

SECTION ON AA

FIG. 420.—BOILER OF THE COULTHARD STEAM LORRY.

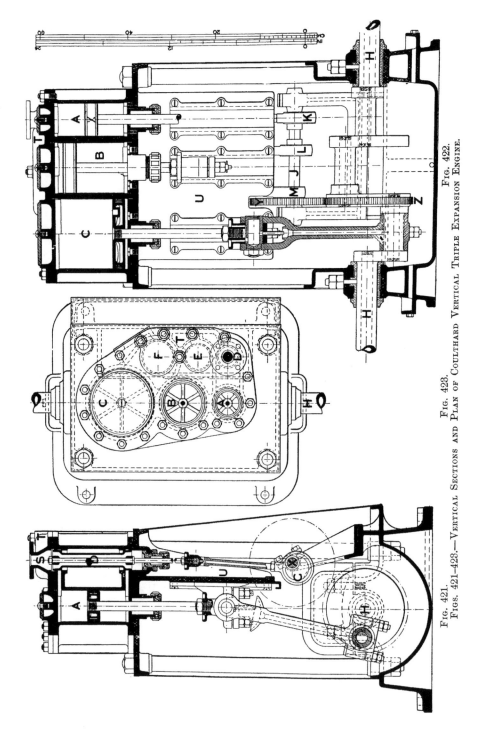

Fig. 422.

Fig. 423.

Fig. 421.

Figs. 421-423.—Vertical Sections and Plan of Coulthard Vertical Triple Expansion Engine.

538

shaft J, driven by the spur wheels Z, Y, the pitch lines of which are seen in Fig. 421. As here shown, no reversing gear is provided, the reversing being effected by spur gearing, to be hereafter described. An arrangement has been designed, however, for reversing the direction of rotation of the engine. This is shown by Fig. 424, in which X is the shaft driven by the spur wheel Y upon it, which carries the excentric sheaves K, L, M on a sleeve J. This sleeve receives motion through the pinion P, which is on the spindle Q loose upon the shaft X, the pinion P being driven by the pinion N near X, shown as fixed by a set screw. Now it will be seen that with the pinion P on its spindle Q in the position shown by the lower figure in Fig. 424, the relative positions of the two pinions

Reversing

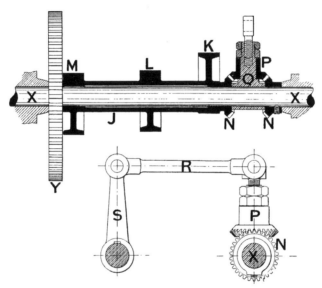

FIG. 424.—COULTHARD ENGINE REVERSING GEAR.

N, N, with reference to any points on the shaft X, will remain the same for any number of revolutions of the shaft X; but if the position of the spindle be moved by angular movement of the reversing arm S and connecting rod R their relative positions are affected by the amount to which P has been forced to roll upon N, and thus move the bevel pinion on the end of the sleeve J and the excentrics through about 120° from their previous position. The objection to the arrangement is that the three excentric sheaves are driven through the one constantly working set of mitre pinions P, N, N.

Fig. 425 illustrates in section the spur transmission and change speed gear already referred to. In this C is the engine crank shaft fitted with a sleeve D, carrying spur pinions E, F, G. These gear selectively, by means of the two clutches M, N, with the wheels I, J, K on

Change
speed and
reversing
gear

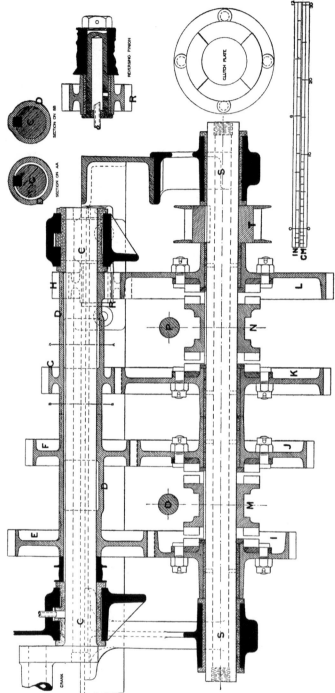

FIG. 425.—SPUR DRIVING AND CHANGE-SPEED GEAR OF COULTHARD STEAM LORRY.

the bottom shaft s s. By forked striking levers on the two spindles, seen in section at o, p, the clutches m, n may be thrown into gear with the renewable clutch faces on either side of the last-mentioned wheels. Reversing is effected by means of the pinion h gearing with a pinion r on a stud shaft, which in turn gears with the wheel l, thus giving to the shaft s a slow speed in the opposite direction of rotation.

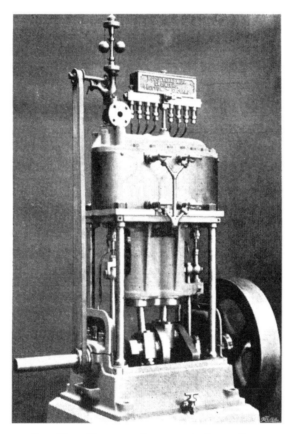

FIG. 425A.—ENGINE OF COULTHARD'S STEAM CART.

It will be seen that the whole of these spur wheels and pinions are always in gear, though only one pair of the three pairs of change wheels, or the reversing set, can be at work at any one time. Motion from the shaft s is conveyed by a Renold's silent chain, running on the flanged spur pinion t to the differential gear on a third motion shaft in the centre of the lorry, as seen in Fig. 419. On the ends of this shaft are flanged spur pinions upon which run two more Renold's silent chains. **Chain driving gear**

The gearing is arranged for speeds of 2·5, 4·5 and 7·5 miles per hour, with a two miles per hour reverse speed. These speeds correspond with ratios of 19·5, 11·5 and 7 of engine revolutions to one of driving wheels. The use of speed change gear has the advantage that the boiler pressure and the rate of evaporation may be kept more nearly constant than when only one or no changes of speed by gear are adopted, and this gives the burner more regular work.

Steering gear   Steering is effected by a hand wheel on the top of a vertical spindle, on the lower end of which is an arm connected by links and steering arms, to the pivoted axles of the front wheels.

Oil and water supply   A cylindrical oil tank, to hold 20 gallons, is carried at the end of the lorry, and two water tanks, holding together 50 gallons.

FIG. 426.—THE LEYLAND 4-TON OIL-FUEL STEAM LORRY.

The condenser and a feed-water heater which is provided both drain into the feed-water tanks.

Messrs. Coulthard & Co. have also made a form of four-wheeled steam cart driven by machinery somewhat similar to that now described, but with a smaller two-cylinder engine of their stationary type, of which a photographic view is given in Fig. 425A.

Past experience and the recent experience of other makers does not lead me to expect any very satisfactory results from the use of the kind of iron spoke, and combined rim and tyre wheels, used by Messrs. Coulthard, nor can continued satisfaction be expected from the chain gear employed.

THE LEYLAND STEAM LORRY.

The Leyland Lorry   Fig. 426 is from a photograph of a 4-ton oil-fuel steam lorry made by the Lancashire Steam Motor Company at Leyland, by whom several vehicles have been constructed. Particulars of the performance in experimental trial runs and of the main features of the vehicle, shown by the photograph here reproduced, will be found in Table XX. It is one of those which depend on long chain gear transmission, which cannot be satisfactory for heavy and continuous work when light chains with thin links and small pins are used.

The tensile strength of a chain as a whole and the shearing strength of the pins may be many times greater than the normal pull upon them due to ordinary working. This, however, is but a poor guide to the behaviour of material in a chain running over a comparatively small sprocket wheel, and subject to the heavy inertia stresses, aggravated by backlash due to necessary slackness of a comfortable working chain drive. Wear under adverse conditions, heavy impact stresses and elongation, effect of burring over of thin links and ribbing of small pins, and other things come into play, and make necessary what would appear, from mere tensile strength considerations, excessive strength and dimensions.

Strength and gear of driving chains

THE SIMPSON & BODMAN STEAM LORRY.

A steam lorry containing many features which differentiate it from those yet described has been made by Messrs. Simpson & Bodman, of Cornbrook, Manchester. Its general appearance and arrangement are shown by Fig. 427, which is from their experimental vehicle intended to carry from 3 to 4 tons. From this it will be seen that the position selected for the boiler and the use of horizontal bars connecting the front and rear axles, and in a measure supplementing the main underframe, is similar to that adopted in two of the vehicles already described. Here the similarity ends. The underframe is stiffened by strong transomes above the ends of the driving wheel springs, these projecting beyond the longitudinal members of the frame and there carrying guide housings for the spring ends. Similar housings are attached to these cross bars for accommodating the ends of the inner springs, there being two to each wheel.

Simpson & Bodman's Lorry

General description

The driving wheels run loose upon a through axle, and each wheel is separately driven by an independent three-cylinder single-acting engine, which drives through single reduction spur gear and a short Renold's silent chain.

The boiler is of the Serpollet instantaneous generator type, steam from which is collected in a steam drum from which the engines are supplied.

The outlines of the general arrangement of the lorry shown by Fig. 427 are given in Figs. 428–430, which are respectively elevation, front elevation, and plan. From these the main features of the underframe may be gathered. It is constructed of wood, strengthened by thin bar iron, as will be seen in Fig. 431. Other points in the arrangements of the framework may be gathered from Figs. 432 and 434.

Underframe

The generator is shown by these last mentioned illustrations and by Fig. 439, which is from a photograph of the generator and steam drum. The generator consists of twelve two-legged elements of thick steel tube

The steam generator

indented after the manner of the Rowe patent tube, and connected to each
other in series by bands and Haythorn connectors. The boiler may be

Fig. 427.—Simpson & Bodman's 3-Ton Steam Lorry, 1899.

described by the following extract from particulars supplied by the makers.
They say :

"We take plain welded steel tubing of tolerable thickness of shell,
swage it down at the ends and in the centre, and indent it with a peculiar

544

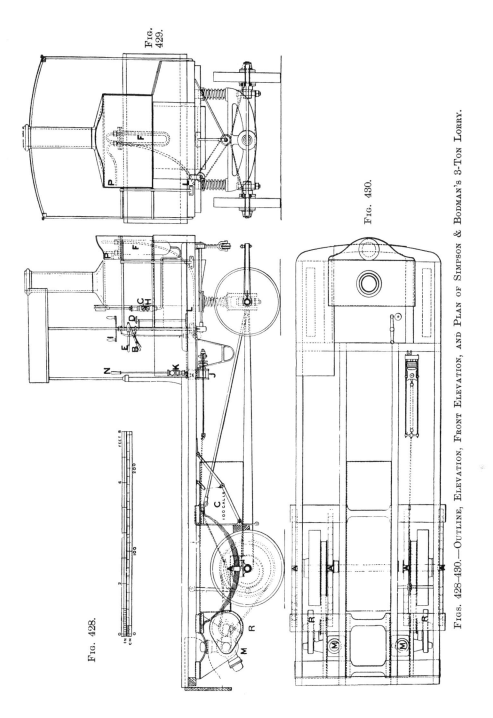

F<small>IG.</small>
429.

F<small>IG.</small> 430.

F<small>IG.</small> 428.

FEET 8

200

100

IN
CM 0

F<small>IGS.</small> 428-430.—O<small>UTLINE</small>, E<small>LEVATION</small>, F<small>RONT</small> E<small>LEVATION</small>, <small>AND</small> P<small>LAN OF</small> S<small>IMPSON</small> & B<small>ODMAN'S</small> 3-T<small>ON</small> L<small>ORRY</small>.

545   35

indent patented by O. M. Row, and then bend the tube in the centre to bring the two ends together to the required pitch. These ends are screwed 11 threads per inch. The tubes are capable of standing approximately one ton per square inch in pressure as received by us, the effect of indenting by its shape certainly does not weaken the tube, it probably strengthens it, and the process is so drastic that any flaw must at once show itself."

The tube thus prepared is, after screwing at the ends, bent at the centre, as is done by Serpollet, to form the double leg elements E E.

The boiler tube connections "To connect these members in series, nipples made from bar steel are fitted on expanded steel bends, the nipples being screwed 9-thread pitch. From hexagon bar steel, nuts c, Fig. 438, are made, screwed correspondingly 11-thread and 9-thread pitch. To make the joint the nut is run right down on the tube end, the bend put to the face, and the nut un-

FIG. 431.—END VIEW OF SIMPSON & BODMAN'S LORRY.

screwed from the tube end; it engages with the nipple and, gaining one-ninth whilst losing one-eleventh, it gives a final tightening equivalent to a 50-thread screw, whilst retaining the stripping strength of 11 threads. This joint is the property of the Haythorn Tubulous Syndicate. We test this joint and generator to four or five times the working pressure and never have the slightest trouble to make it, and heating it nearly to **Feed pumps** redness does not affect its soundness. The starting and feeding of this boiler is effected by means of two pumps, one worked by hand to send the water through, to raise the initial pressure for starting the lorry, and the other is driven by a cam on the road wheel, and comes into operation when **Feed over-flow** the car starts. The pumps have power in excess of the work to be done, and this excess of water pumped is returned from the delivery side of the pump back to the suction by a wheel valve regulated by the driver and

HEAVY STEAM VEHICLES

called the bye-pass valve. The sectional drawings show somewhat diagrammatically the kind of casing and fire box used.

" An essential feature of this generator is the weldless steel cylinder shown externally to the casing in Figs. 434, 437 and 439. Prior to its use we found the generator very delicate in working. The temperature at which this generator will work well is from 600° F. to under 1,000° F., and maintained at that it is practically indestructible, provided with a strength enormously above the pressure it will have to stand, and supplying steam of such quality that it seems a totally different fluid to the ordinary steam of a water tube or pot boiler. It will work steadily at from 100 to 250 lb. pressure, and supply higher pressures for temporary use in a few seconds; but this temperature of 600° to

Steam drum and feed heater and steam temperator

FIG. 432.—SIDE VIEW OF DRIVING WHEEL AND GEAR OF SIMPSON & BODMAN'S LORRY.

1,000° F. is too great for lubrication, and especially is it high when starting work after perhaps 30 or 40 minutes firing up with an empty boiler. We therefore attach the steam drum F to the generator, placing it outside the casing away from the furnace heat. In this drum all the steam from the generator is collected, and from the drum passes to the stop valves on its way to the engines.

Superheat

" In this drum depends a **U** tube F (see Fig. 434), and through this tube passes the feed water, either direct from the pumps or after it has passed through one or two rows of tubes below the funnel to take heat from the flue gases. Assuming now that the generator is in a very hot state and about to start work, water is pumped by hand through the top

547

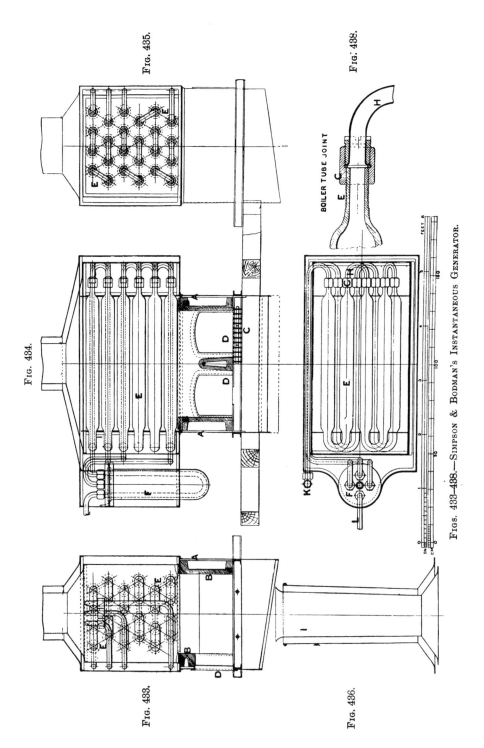

FIG. 435.

FIG. 434.

FIG. 433.

BOILER TUBE JOINT

FIG. 438.

FIG. 436.

FIGS. 433-438.—SIMPSON & BODMAN'S INSTANTANEOUS GENERATOR.

548

rows of tubes, and probably passes through the **U** tube in the drum in the form of steam, thence to the bottom and hottest row of tubes, issuing from the middle row as highly superheated steam, to collect in the drum. Here it comes in contact with the **U** tube and imparts some of its excessive heat to the contents of the tube, cooling itself and assisting to vaporise the contents of the tube. By the time sufficient steam is collected to start the engines, the steam, although highly superheated, does not give the engines the bad two or three minutes they get immediately after the start of a flash boiler not provided with this drum. A further device that enables this drum to materially improve the convenience of the generator, is a back-pressure valve placed on the feed pipe between the bye-pass valve and the generator. The driver is thus prevented from easing the pressure

Necessity for steam drum

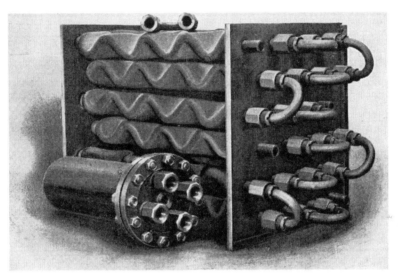

FIG. 439.– SIMPSON & BODMAN'S GENERATOR AND STEAM DRUM.

in the generator as is usually done by allowing hot water to blow back through the bye-pass valve on the water pipe, which we found very uneconomical if a series of sudden checks occurred, and also very troublesome to the pumps. The steam is allowed to collect in the drum and pressure to be relieved by blowing off steam through an ordinary safety valve loaded to a fairly high pressure. We thus get a good steady volume of steam for restarting without continual recourse to the hand pump. The questions usually asked us in reference to this generator are : (a) Will the tubes be affected by deposit? (b) What is the action on them of a sudden check with a full fire on, or a stand of some hours with the fire kept ready to start at any moment? (c) How does the driver know whether his boiler is ready for starting? (d) How does he know the speed to feed his water

Tests of the boiler

549

TEST OF SIMPSON & BODMAN GENERATOR.

Duration, 3 hours; Fuel used, furnace coke. Starting.—Fire lit and boiler blown out for one hour, the fire then drawn and thrown aside. Temperature of boiler taken, 380° F., fresh firing applied and test proceeded with.

	1st hour.			2nd hour.			3rd hour.		
	H. M.	Deg. Fah.	lb.	H. M.	Deg. Fah.	lb.	H. M.	Deg. Fah.	lb.
Times of stoking and temperature at each stoking taken before fire door is opened . .	11.25	380°	—	12.25	500°	—	1.25	500°	—
	11.35	600°	—	12.35	600°	—	1.37	600°	—
	11.45	730°	—	12.45	730°	—	1.45	650°	—
	11.55	500°	—	12.55	500°	—	1.55	700°	—
	12.5	500°	—	1.5	500°	—	2.5	730°	—
	12.15	500°	—	1.15	550°	—	2.15	600°	—
Times of clinkering, cleaning, etc. . . .	—	—	—	1.25	—	—		—	—
Fuel weighed out at start	—	—	57	—	—	64		—	78
Fuel weighed in at finish	—	—	—	—	—	—		—	—
Fuel used	—	—	57	—	—	64		—	78
Gross weight, water and tank at start . .	—	—	500	—	—	488		—	412
Gross weight, water and tank at finish . .	—	—	234	—	—	136		—	36
Weight of water evaporated	—	—	266	—	—	352	Water pumped through at 2·25 to reduced temperature to 380°	—	376
Pressure worked at per sq. in.	—	—	100	—	—	100		—	100
Temperature of feed water	—	70°	—	—	70°	—		70°	—
Evaporation per lb. fuel used	—	—	4·66	—	—	5·5		—	4·82
Evaporation per sq. ft. heating surface . .	—	—	5·78	—	—	7·65		—	8·17
Combustion per sq. foot grate . . .	—	—	20·7	—	—	23·27		—	28·36
Average temperature .	—	535°	—	—	563°	—		630°	—
Degrees of superheat above saturation . .	—	197°	—	—	225°	—		292°	—

Remarks.—Weight of ash not taken, being very slight.

Nozzle in blast pipe (exhaust), 1¼ in. diameter.

Evaporation per lb. fuel is taken at pressure and temperature shown.

Second Test.

Duration, 3 hours; Fuel, anthracite. Starting as in Test I. Temperature, 550° F.

	1st hour.			2nd hour.			3rd hour.		
	H. M.	Deg. Fah.	lb.	H. M.	Deg. Fah.	lb.	H. M.	Deg. Fah.	lb.
Times of stoking and temperature	9.40	550°	—	10.40	500°	—	11.55	450°	—
	9.50	780°	—	10.45	450°	—	12.0	550°	—
	10.0	550°	—	10.50	450°	—	12.5	700°	—
	10.5	480°	—	10.55	500°	—	12.15	700°	—
	10.10	560°	—	11.0	510°	—	12.25	550°	—
	10.15	560°	—	11.5	375°	—	12.35	530°	—
	10.20	460°	—	11.10	400°	—	12.45	530°	—
	10.25	460°	—	11.15	500°	—	—	—	—
	10.30	460°	—	11.20	550°	—	—	—	—
	10.35	460°	—	11.25	550°	—	—	—	—
	—	—	—	11.30	550°	—	—	—	—
Times of clinkering and cleaning	10·40	—	—	{11.5 / 11.40}	—	—	—	—	—
Fuel weighed out at start		—	131	—	—	136	—	—	69
Fuel weighed in at finish		—	53		—	69	—	—	—
Fuel used		—	78		—	67		—	69
Weight of ash		—	30		—	30		—	30
Nett fuel consumed		—	48		—	37		—	39
Gross weight of water and tank	Hour finished at 10.40	—	468	[Hour finished at 11.40 5 mins. stokes.	—	500	Finished at 12.55 with a temperature 380°	—	500
Gross weight of water at finish		—	126		—	189		—	196
Weight of water evaporated		—	342		—	311		—	304
Pressure worked at per sq. in.		—	100		—	100		—	100
Evaporation per nett lb. fuel used		—	7·12		—	8·2		—	7·97
Evaporation per sq. ft. heating surface		—	7·0		—	6·7		—	6·6
Combustion per sq. ft. grate		—	17·5		—	13·3		—	14·2
Average temperature		532°	—		485°	—		572°	—
Degrees of super heat above saturation		194°	—		147°	—		234°	—

Remarks.—The firebars were evidently not suitable for burning anthracite, as the amount of ash was excessive.

The bars also choked up badly.

Nozzle in flue was ¼ in. diameter, full open to the steam.

at, as it is evident that he may run his car at a very high speed at starting, pump through all the water his pump can supply, and speedily cool his boiler? M. Serpollet has effectually demonstrated that generators of this class are absolutely unaffected by deposit. There is no still place where deposit can lie. The temperature is much too high for most deposits, and mud, which might collect at the base of the drum, is easily blown out periodically. (e) We have made a number of tests with heavy coke fires to ascertain the rise in temperature and pressure occasioned by suddenly stopping the draw-off steam, as would happen in practice. The ashpit is close shut down, and the fire door, which hangs about 4 in. above the fire bars, open. The temperature will rise to about 650° F., if below that, and pressure gradually to about 250 lb. as the water present is evaporated, falling slightly as the fire blackens down, which it very speedily does. The fire has been left for two or three hours to die away gradually, and there is no scaling of the tubes.

Tests of boiler

"C. and D. are questions of some experience; no motor car boilers can be worked without a little practice, this experience in fact constituting the driver. The appended tests relate to the generator shown in Figs. 433 to 439. The temperature is of the issuing steam, not of the generator which is probably 100° and 200° higher."

Particulars of generator and tests (see pp. 550, 551):—

"Weight of generator as photographed, 645 lb.; weight of regulating drum, 98 lb.; number of elements, 12; tubes per element, 1 with 2 legs; length indented per tube, 2 ft. 6 in.; pitch of indent, 3½ in.; heating surface indented, 46 sq. ft.; size of tubes before indenting, 1st, 2nd, and 3rd rows, 3 in. × 2½ in bore, 4th, 5th and 6th rows, 3 in. × 2¼ in. bore; hydraulic (cold) tests for joints, 750 lb. sq. in.; thermometer used, Mercury with nitrozine chamber reading to 900° F., bulb depending 1 ft. 6 in. into drum; grate area 2¾ sq. ft.; fire bars, ½ in. wide; air spaces, ½ in. wide. Lighting up test:—Fire made up from plain kindling wood and 7 ft. seam coal; fire lit at 3 hr. 30 min.; steam for blower at 3 hr. 40 min.; 400° F. in drum at 3 hr. 55 min.; ready for work at 4 hr."

Control levers

Referring again to Figs. 428–430, it will be seen that the steering and reversing handles I and B, respectively, are all brought up to the right hand of the driver. Steam passes from the boiler at P to the drum F, and from F by the pipe L to the engines. At D is the bye-pass handle, previously referred to, and E is the throttle-valve lever, N the hand-pump lever, K the hand pump, G the blast-pipe valve, and H the two-way valve for steam-water lifter and for the brake. At J is the main steam pump which is actuated by the large excentric shown on the side elevation, Fig. 428, surrounding the nave of the off driving wheel. Against this excentric rests a friction roller carried by the end of a pendent lever, on the forward side of which is a projecting arm, to which a chain is attached running over a pulley above the letter G on the 100-gallon water tank. The forward end of this light chain is coupled up to a frame attached to the

Main feed pump

552

pump crosshead, and between the rods of this frame is a spiral spring, shown by dotted lines. The pump plunger is pulled into the pump against the spring pressure by the excentric and chain, and returned by the spring.

The brakes consist of two shoe brakes on the tyres of the driving wheels, as seen in Fig. 432. For further brake power the engines are relied upon.

Brakes

Fig. 440.—Simpson & Bodman's Three-Cylinder Engine.

The main springs, one on each side of the hind wheels, are supplemented by spiral springs, and spiral springs only are used on the front axle, as seen in Fig. 429. They rest upon the two ends of a plate axletree pivoted on the centre of the fore axle, which lies within the plate sides.

Double springs and bearings on axle

Spiral springs

The steering of the axles pivoted on the end of this front axle, is effected by means of a large inwardly turned forging, seen in Fig. 417, fixed to the outer ends of the pivoted axles and connected across the front of the vehicle by a connecting rod (see Figs. 428 and 429). This connecting rod is moved in either direction transversely by means of chain gear, a chain being wound on and off a central pulley, as seen in the figures just referred to.

Steering gear

Three-
cylinder
engine

The engine used in this lorry is illustrated by the photographic reproduction, Fig. 440, and by the longitudinal and transverse sections, Figs.
441 and 442. As these figures show, the engine has a central crank box A,

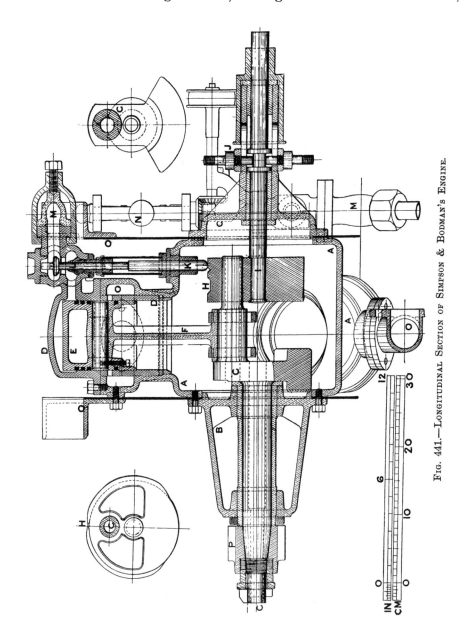

FIG. 441.—LONGITUDINAL SECTION OF SIMPSON & BODMAN'S ENGINE.

in which runs a crank c, to which the pistons E in the two-part cylinder
D are connected by means of the rod F, which are kept in their place upon
the crank pin by a ring embracing the two sides of the three-part boss

554

HEAVY STEAM VEHICLES

made by the three connecting rod ends. The part of the cylinders within which the pistons work is in one piece, flanged and fixed upon the cylinders projecting from the crank case, in which the cylinders proper are inserted

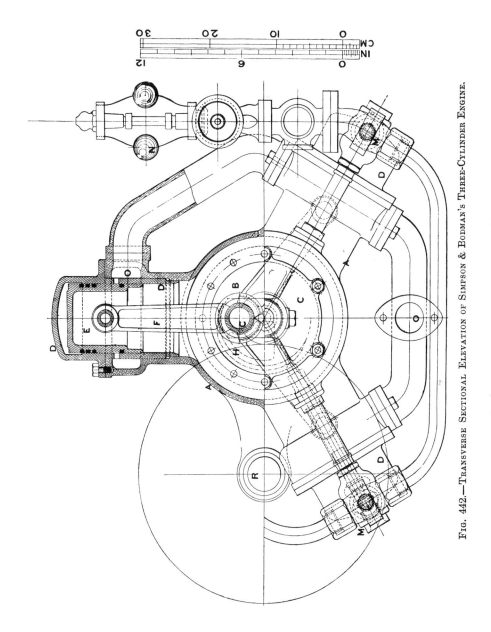

FIG. 442.—TRANSVERSE SECTIONAL ELEVATION OF SIMPSON & BODMAN'S THREE-CYLINDER ENGINE.

and joint-seated at their inner ends. Exhaust takes place through a number of holes round the cylinder at the end of the piston stroke into an annular exhaust port o, and thence by an exhaust pipe common to the three

No exhaust valves

555

cylinders, and finally away at the bottom of the engine at o. This system involves considerable compression. The admission of steam is effected by the valve on the stem L, like a gas engine exhaust valve, moved by the rod K, which receives its motion from the large cam H, on the inner end of the spindle I, and loose upon the end of the crank pin. The cam has three different faces or grooves, one giving the engine full steam ahead, one giving steam through a part only of the stroke and the other giving steam for the reverse motion. These three positions are obtained by connections with the lever B, Fig. 428, and the crosshead and pins at J, Fig. 441, and in perspective in Fig. 440. The cam H is shown in side view to the left of Fig. 441. The engine is governed by a Pickering governor N, which controls the admission of steam to the pipes M M. The framework of the engine or that by which it is connected to the vehicle frame consists simply of a pair of thin triangular steel plates and angle bars as shown in Figs. 440 and 441. The cylinders are 4 in. diameter, the stroke 3 in., and, at 500 revolutions per minute, each engine is said to give off 8·5 B. HP.

Cam lifted steam valves

Valve motion cam

Reversing

Governor

Transmission gear

At the outer end of the crank shaft of the engines is a steel pinion P, which gears with a bronze spur wheel, seen in Fig. 432, on a spindle which runs in a bearing at R, Fig. 442, and indicated also in Fig. 430, and presumably in an interior bearing not shown. Inside the bearing R, on this spindle, is a chain pinion carrying a Renold's saw tooth silent chain, which runs on a toothed ring fixed by bolts to the inner felloes of the compound driving wheels seen in Fig. 432. Variations in speed are effected only by variations in steam pressure, no clutches or speed change gear being employed. It is intended, however, to provide each car with three sets of change wheels and pinions to give speed ratios between engines and road wheels of 8·5, 14·5, and 24·5 to one. It is arranged that the lock nuts which hold these wheels and pinions in their place shall draw them off when they are unscrewed, so that the change of gear can be quickly made.

Change wheels for change speeds

Although I am not aware that any notable practical work has been done by Messrs. Simpson & Bodman with the vehicle illustrated, and would not expect satisfaction from it as it is, it must be admitted that the numerous points of departure from ordinary practice may be considered a sufficient reason for giving up the space which has been allotted to it.

Toward & Co.'s Vehicle and Boiler.

Toward's boiler

Messrs. Toward & Co., of Newcastle, have made several vehicles, including a tractor to act as steam horse to an ordinary lorry with its front wheels removed, and locking plate resting on the tail of the tractor, and a steam waggon, the boiler of which is illustrated by Figs. 443–445, of which Fig. 444 is a longitudinal section through near the centre of the boiler, Fig. 443 a vertical section through the funnel and water and tube space shown in Fig. 444, and Fig. 445 a front elevation. It will be seen that

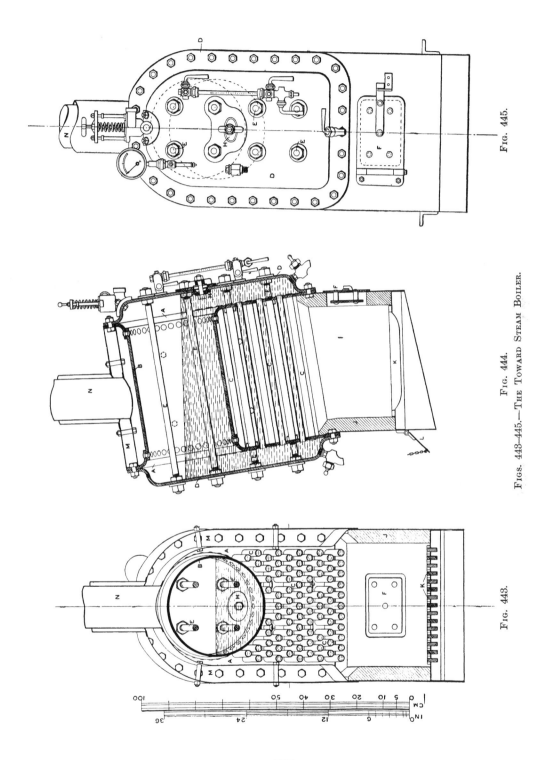

FIG. 445.

FIG. 444.

FIGS. 443–445.—THE TOWARD STEAM BOILER.

FIG. 443.

557

the boiler consists of a firebrick-lined fire box supporting the lower part of a front and rear pressed plate water chamber D D, the inside plates of which form tube plates for inclined tubes C, and by outward flanging form the means of connection to a cylindrical steam and water vessel connecting the two water chambers. The pressed outer plates of the latter are stayed by strong cross bolts E E through the steam chamber and also through the lower parts of the water chambers. By taking nuts off the stay bolts and from the bolts by which the flanged plates D are held upon the tube plates, the whole boiler becomes accessible. Surrounding the upper part of the tube space and steam and water chamber is a double-skin sheet-iron shell, into the lower part of which air can freely enter and then pass up the space M and between the funnel N and the surrounding petticoat pipe. The scale by the side of Fig. 443 gives the dimensions, of which the length from plate to plate D is 2 ft. 2 in., and the heating surface about 40 sq. ft. It is stated that with a boiler of this type 4 ft. in height, as compared with 3 ft. of the boiler illustrated, and having about 65 sq. ft. of heating surface with a total of weight of 810 lb., the evaporation was found to reach 615 lb. of water, from cold to the temperature of steam at 190 lb. pressure, in one hour.

Chapter XXXIII

ON BRAKES

THE frequent necessity for powerful brakes, especially on high-speed cars, confers an importance on these appliances not exceeded by any other parts.   The usefulness of high-power brakes depends upon and implies strong wheels and well-fixed tyres.   Wheels which will stand the heavy but more gradually or uniformly applied stresses, due to climbing stiff gradients, may be severely strained or destroyed by the sudden application of brakes sufficiently powerful to meet the demands made by occasional emergencies of high-speed road travelling, especially if the wheels be fitted with well-fixed solid-rubber tyres.

Importance of brakes

In considering brake requirements, the sufficiency of wheel as well as of tyre strength must be assumed ; but it may be noted that the high co-efficient of adhesion of rubber tyres, as compared with iron tyres, will make the application of greater effective brake power possible, and therefore throw greater stresses upon the wheels than have been common with smooth iron tyres.

Stresses thrown on wheels

Band brakes will be most generally employed, although spoon or block brakes will always be a desirable addition when iron and solid rubber tyres are used, because they are effective and throw less stress upon the wheels than band brakes.   They will, however, probably be but little used on pneumatic tyres.

Type most favoured

There is perhaps no essential feature which has so commonly been badly designed and badly made as the brake and its gear.   In many cars the band brakes and their fittings have been so deficient in strength, proportions and arrangement that nothing but good fortune and no emergency can have saved the passengers from the worst of accidents.

Bad forms of construction

It is necessary to determine the brake power required to stop a vehicle of given weight running at a given speed in a given distance, and by this means to arrive at something like due comprehension of the necessary dimensions of the parts brought into play to effect this stop.

Brake power required

It must first be pointed out as a reminder to those who have overlooked the fact, that the stress put upon a brake to effect a stop in a given distance increases as the square of the increase of speed, so that to stop a

559

car running at 20 miles per hour requires four times the power required to stop it in the same distance at 10 miles per hour.

Great increase of power required with high speeds

Commonly all calculations relating to the acceleration of masses at high speed are calculated on the basis of distance covered in feet per second, and hence the work or energy K lodged in a mass having a weight W moving at a velocity v in feet per second is given by the expression

$$K = \frac{W \, v^2}{2g} \qquad (^1)$$

g representing acceleration due to gravity or 32·2 ft. per second. It is convenient, however, in dealing with motor vehicles, the speed of which is always expressed in miles per hour, to modify this expression to suit that unit.

There being 5,280 ft. in one mile, and 3,600 seconds in an hour :—

$$1 \text{ mile per hour} = \frac{5,280}{3,600} = 1·466 \text{ ft. per second.}$$

Hence for 1 mile per hour

$$\frac{W \, v^2}{2g} = \frac{W \times 1·466^2}{64·4} = \frac{W \; 2·15}{64·4} = W \; 0·0334,$$

and putting V. = miles per hour the expression (^{1}) becomes

$$K = W \, V^2 \times 0·334. \qquad (^2)$$

Hence the energy K stored in a vehicle which, with its load, weighs 1 ton, at speeds of 10 miles per hour and at 20 miles, excluding road and internal resistance, will be $K = W \, V^2 \times 0·334 = 2,240 \times 100 \times 0·0334 = 7,480$ ft. lbs. for 10 miles per hour, and $2,240 \times 400 \times 0·0334 = 29,920$ ft. lbs. for 20 miles per hour.

This has to be absorbed by the brakes during the time which elapses between their application and the stopping of the car within an observed distance or a predetermined distance.

Tyre adhesion coefficient

Coefficients of friction k between tyre and road must be assumed. These may be taken as 0·30 for iron tyres on roads, and 0·60 for rubber tyres on roads. It is desirable to underestimate rather than overestimate this coefficient.

As brakes are only applied to two of the four wheels of a car, the whole of the work lodged in the car and its load has to be dissipated by the brakes on these two. Hence it is only the frictional or adhesive resistance of the two tyres, and the load upon them, that can be called upon for brake power, and the demand should always be less than the tyre adhesion.

For the purposes of an example the proportion w of the whole weight, which may be assumed to be carried by the wheels to which the brakes are applied, may be taken as 0·6 of the whole, or 12 cwt. = 1,344 lbs.

We then have

$$\frac{W \, V^2 \times 0\,0334}{k \, w} = l, \qquad (^3)$$

l being the minimum distance within which the vehicle can be stopped with brakes capable of causing the wheels to skid, or nearly skid, on the

560

level, the ordinary road or draught resistance being disregarded, except so far as it is included in the frictional coefficient adopted.

In the two cases assumed we have then

$$\text{For iron tyres at } 10 \text{ miles, } l = \frac{7,480}{0.3 \times 1,344} = 18.6 \text{ ft.}$$

$$\text{,, \quad ,, \quad ,, \quad } 20 \text{ ,, \quad } l = \frac{29,920}{0.3 \times 1,344} = 74.2 \text{ ,,}$$

$$\text{For rubber tyres at } 10 \text{ miles, } l = \frac{7,480}{0.6 \times 1,344} = 9.3 \text{ ,,}$$

$$\text{,, \quad ,, \quad ,, \quad } 20 \text{ ,, \quad } l = \frac{29,920}{0.6 \times 1,344} = 37.1 \text{ ,,}$$

If the road resistance be regarded, it may be taken as a quantity to be added to the product of $k\,w$.

For structural purposes we want the pull p on the brake band and its connections.

Pull on band brake and connections

In this case we have firstly

$$\frac{W \cdot V^2 \times .0334}{l} = k\,w = p, \qquad (4)$$

and this is the mean resisting effort at the tyre throughout the distance l.

For the rubber-tyred vehicle we have for 20 miles per hour,

$$k\,w = \frac{29,920}{37.1} = 806 \text{ lb.} = p.$$

This pull of 806 lb. is the holding resistance or frictional grip required in a pair of spoon or shoe brakes on the tyres, or 403 lb. on each, but the pull upon the ends of the brake bands, of the smaller brake wheels or rings, will be greater in inverse proportion to the diameter of road wheel and brake rim or

$$p = k\,w\,\frac{D}{d} \qquad (5)$$

D being diameter of road wheel and $d =$ diameter of brake rim.

Hence if $D = 36$ in., $d = 18$ in., then in the above case $p = 806 \times \dfrac{36}{18} = 1,612$, or

806 lb. mean pull on each of two brake bands, and this will be exceeded considerably in any attempts at such sudden stops as will skid the wheels.

The pull p, which may possibly arise, will vary with the road resistance, because the conditions which will increase rolling resistance will generally increase adhesion, and this will on some road surfaces be much larger than on others. With iron wheels, too, the variation will be great, and the firm hold of the brake band may on the one hand cause the wheel, even before it skids, to dig into the road, when its resisting power will be increased, or on the other hand the road pavement may be greasy when the power will be minimised.

The pressure on the two spoon brakes applied to the tyres to make the stop in the distance l will, with a coefficient k as before, $= \dfrac{403}{0.6} = 671.7$ lb.

Pressure on brake blocks

MOTOR VEHICLES AND MOTORS

Brake Power Required on Hills.

Increased brake power on hills

This will increase the distance l or the pull p in proportion to the angle of inclination of the hill. We may take it as a quantity which reduces the efficiency of the frictional or adhesive resistance $k\,w$, and expressing the gradient as a fraction, we have, taking G to represent the gradient as a fraction,

$$l = \frac{\text{W V}^2\, 0{\cdot}0334}{(k\,w) - (\text{G} \div \text{W})} \qquad (^6)$$

In the case of the rubber tyres and 20 miles per hour, this becomes, on a gradient of 1 in 15,

$$l = \frac{2{,}240 \times 400 \times {\cdot}334}{(0{\cdot}6 \times 1{,}344) - (2{,}240 \div 15)} = \frac{29{,}920}{806 - 149} = 45{\cdot}5 \text{ ft.}$$

or if the gradient be 1 in 10 then

$$l = \frac{29{,}920}{806 - 224} = 52{\cdot}4 \text{ ft.}$$

Having now the pull p on the fixed head or most heavily pulled end of the brake band, it is necessary to ascertain the pull or tension p^1 on the tail end of the band, namely that which is pulled by pedal or handle by the operator to put the brake in action.

Pull on tail end of brake and at pedal

Now p^1 will be less than p as the number of degrees or part of the whole circumference of the brake wheel embraced is greater (but not in simple proportion thereto). Hence, in order that the effort at the pedal or the handle of a brake lever may be as small as possible, it is desirable that the part of the brake rim embraced should be as large as possible—the whole circumference whenever convenient.

Proportion of brake pulley embraced

It is not necessary here to enter into the theory of the relation of the friction between say leather and the total friction on more or less of the whole surface of a cylinder such as a brake pulley, and thereby the difference in tension on the two ends of a brake band. It is the same as that relating to the theory of tension on a belt.

Putting $k =$ the coefficient of friction of the leather or similar lining of a brake band on the brake wheel or rim as $0{\cdot}4$, and $a =$ the angle of brake rim embraced, then

$$\text{hyp log } \frac{p}{p^1} = k\,a,$$

or in common logarithms $(^6)$

$$\log \frac{p}{p^1} = 0{\cdot}434\, k\, a. \qquad (^7)$$

From this the following table for simplifying calculations on this question has been calculated:—

ON BRAKES

TENSIONS ON TAIL ENDS OF BRAKE BANDS.

Part of brake wheel embraced.		$\dfrac{p}{p^1}$	p being 1 $p^1=$
Degrees.	Fraction of circumference.		
180	$\frac{1}{2}=0\cdot500$	3·51	0·28
240	$\frac{2}{3}=0\cdot667$	5·34	0·19
270	$\frac{3}{4}=0\cdot750$	6·59	0·15
300	$\frac{5}{6}=0\cdot833$	8·11	0·12
315	$\frac{7}{8}=0\cdot875$	9·02	0·11
360	$1=1\cdot000$	12·85	0·08

From this table it will be seen that with the assumed coefficient of friction of brake band on brake wheel, which probably obtains in most cases, the pull on the tail end of the brake band is always small as compared with that on the head end, and for 300°, or $\frac{5}{6}$ of the circumference embraced, it is only $\frac{1}{8}$ of the pull p, and only $\frac{1}{12}$ when the whole circumference is embraced, which is often the case. The tail may thus be easily worked by a pedal when a complete band brake is used.

Strength of connections of both ends of band

Both ends of the band should be capable of withstanding the pull p, as although the stopping effort for backward running is not as great as that brought about by forward speed, it may approach it sometimes by sudden application.

Having the pull p, the necessary sectional area of the band may be assigned, a very large factor of safety being employed.

Dimensions of connections

The pull of 806 lb. above arrived at should to begin with be doubled, in view of shock due to sudden application, making it 1,612 lb. on each brake band head. Taking 28 tons as the ultimate strength of the mild steel used for the brake band, or say 62,000 lb., and a factor of safety of 6, then we have as the necessary effective sectional area of the brake band or its connections at the weakest part, at rivet holes or screw threads,

$$\frac{1{,}612 \times 6}{62{,}000} = 0\cdot155 \text{ sq. in.,}$$

or say a diameter at the bottom of the threads, if any, of nearly half an inch.

Chapter XXXIV

CONCERNING VIBRATION, BALANCING, AND BALANCED MOTORS

INVENTORS have during the past two years been very busy with what, in their minds, is a most important thing, not as they believe yet obtainable for a motor vehicle, namely, a perfectly balanced motor. They see a petrol motor carriage standing with its motor running, and they notice the trembling of the steering handle and other parts, which are so held as to amplify by their flexure the vibratory impulse of very small range set up by the motor. At the time the motor is perhaps cutting out twenty charges for every one it uses. It is doing nothing, but the reaction of the occasional impulse sets up the vibration which is magnified by the flexible nature of the support of the motor, and vastly more magnified by the uninitiated and destructive critic of the street pavement, or by his pen, in some of the journals. He does not notice that this vibration is inconsiderable in the better class cars, but condemns all alike. He perhaps notices it in the most expensive cars, namely, those built for very high speeds; cars for the use of experts, and which have perhaps 12 HP. on a total weight of about 21 cwt. In these cars the vibration when standing, with a 12 HP. engine running at high speed on a light frame, is not considered by those for whom these cars are made and **Vibration** by whom they are used. The engines are almost perfectly balanced for **not due to** their speed of running in actual work, and the vibration ceases when the **want of** race-horse of iron and steel is doing its work. These cars are not built for **balance** standing about or loitering in the parks. The slower speed cars of much smaller power are built for this purpose and for touring purposes, and in these modern cars of good make the vibration when standing is small and even negligible, and it is nothing when the car is running. The motors are balanced, and what is more important in this connection, well regulated, and the critic condemns because he does not know that when the motor is put to work and the unbalanced torque ceases, the trembling ceases also.

Much of the invention and of the writing concerning balanced motors arises from a misconception of the conditions. An internal combustion motor may be, and often is, extremely well balanced, so far as most or all of

564

the disturbing forces are concerned ; and a steam engine, double-acting especially, may be almost perfectly balanced for the speed at which it is to run. But there is a great distinction between these two motors in all that affects vibration when running light.

Difference between steam and oil engines

With the steam engine, running light is done by reducing the impulses in power and not in number, either by throttling or early cut off, or both. The steam may be looked upon as giving the pistons very gentle uniform pushes, at least only just sufficient in force to effect a quarter of a revolution.

On the other hand the running light of an oil engine is done by reducing the number of impulses and not the force of the impulse. The push upon the pistons in most petrol motors is but little less in suddenness and intensity when running light than when running full load. Hence, as the governing is done by cutting out all but the one occasional charge, and as there may be from at least 10 to 16 misses to one charge, the steam engine impulses may be from 40 to 64 times more numerous than those of the internal combustion motor, and each impulse of only $\frac{1}{40}$ to $\frac{1}{64}$ the intensity.

This secures for the steam engine a uniform torque of very small effort during any one revolution, while the petrol motor running light has an extremely variable torque, ranging from full intensity during part of one stroke to a minimum at the end of about forty strokes, before another impulse is received. That is to say then, that the rate of increase of acceleration is enormous at widely separated intervals, and that the disturbing effort, which must be dissipated as it cannot be utilised, is expended in setting up vibration.

For this reason the effective regulation of an internal combustion engine is of more importance than minute accuracy in balancing. By effective regulation is meant governing by decreasing the intensity of the impulses instead of their number, and this, in motors as at present made, means the use of a form of carburettor and method of mixture and of ignition which will secure the uniform formation of a slow combustion mixture and its certain ignition. Governing by throttling provides an approach to the requirements, a powerful electric ignition being necessary, or at least best, for the purpose.

Most of the attempts to produce a vibrationless engine have proceeded on the assumption that there is some virtue in opposite cylinders and equally and oppositely acting pistons and connecting rods. In so far as this construction makes balance weights on the cranks unnecessary, it simplifies the balancing of the moving parts. The inertia effects of reciprocation, and the effect of connecting-rod angularity are almost eliminated. But the balance of these parts and of the cranks is now sufficiently understood to enable manufacturers generally to approximate nearer and nearer to correct balancing, so far as disturbances by these parts is concerned. The fact is, however, that the simultaneous action of

565

MOTOR VEHICLES AND MOTORS

two opposite pistons, both connected to a single crank, does nothing to
neutralise the main cause of the vibration. That main cause is the
irregular doing of work on a crankshaft and fly-wheel; the irregularly
recurrent explosive impulse which for its dissipation has to act through a
fly-wheel, whose necessary inertia is only less than that of the engine body
and the parts of the car to which it is rigidly attached. If the fly-wheel
were rigidly held and the engine body not fixed to the car frame, the
engine would rotate while the crank stood still. This is what the engine
is always trying to do, and does do through a small range which is propor
tional to the moments of inertia of the fly-wheel and of the engine body,
the engine running light. In accelerating the piston, and thereby the fly-
wheel, it imparts motion to the cylinder and all connected to it. This is
the small range motion of which the evidence is vibration of parts of the
car when standing with the engine running. The relation between the
instantaneous range of motion of the engine and fly-wheel masses considered with reference to their centres of inertia, is very simple, and is
directly proportional to their relative masses. The direction of this small
but strong impulse movement is, however, difficult of determination.

It would be simplest in an engine such as the Daimler inside crank
disc fly-wheels on either side of the crankpin, and worst with a single
fly-wheel at one end of the crankshaft. It would be least with arrangements such as that adopted in the Lanchester car motor, in which two
fly-wheels on two separate cranks running in opposite directions connected by two cog wheels and impelled by four connecting rods on two
pistons, are used. It is doubtful, however, whether the advantage gained
is equal to the disadvantage of multiplication of parts.

So long, then, as the engine does irregular internal or external work, or
explosive gases do work on the piston of an engine at irregular rates, there
will be vibration, and this vibration will be worst when symmetry and
kinetic equality of rotative and reciprocating masses is not observed.
Governing by variation of energy of piston impulse, instead of by cutting-
out impulses, is the most effective mode of reducing this vibration; and
increasing the number of cylinders rather than increase their diameter
beyond certain expedient limits is also a most important aid in this and
other respects.

When the car is running, the motor pistons receive, say, ten impulses
to one when the motor is running light. Hence the repetition of the
vibratory effort with a two-cylinder motor, running at say 800 per minute
is about 13 per second instead of 1·3, and the vibratory transmission is
nullified by interference. On the other hand, when the motor is running
light, and only 1 or 1·3 impulses occur per second, the vibration of parts of
the car is increased in amplitude by the coincidence, constant or periodic, of
the period of elastic flexure of the car parts and of the motor impulses.

Chief cause of vibration

566

Chapter XXXV

PIVOTED STEERING AXLES

THE invention of pivoted front or steering axles by Lankensperger, of Munich, patented for him in this country by R. Ackermann, as described in the sixty-page pamphlet published in 1819, has proved to be of more importance for motor vehicles than the makers of horse-drawn vehicles appear to have thought it for their purposes. As claimed by Ackermann, the main advantages were the use of higher front wheels, which the system made possible, the small radius of the curve which could be traversed, and the greater stability of the vehicle. The system was, however, very unfavourably criticised by many of the coachbuilders of the time, and, as every one knows, it was very little used from 1819 down to 1894. It must be admitted that for horse-drawn vehicles the arrangement of the axles involved the use of a number of parts, most of which required good workmanship not easily then obtained at moderate cost, and that the advantages were not really of sufficient importance to justify any complication or weakness in so important a part of a horse-drawn coach.

Lankensperger thoroughly understood the possibilities of the system and the arrangement of the steering arms so as to obtain differential angularity of the two wheels with a single centre of radiation for both axles.

Where the design of a carriage frame and body leaves room for the locking round of an ordinary straight axle with locking plates of considerable diameter (or fifth wheel, as the pair of plates is called in America), it cannot be urged that the vehicle with pivoted axles is notably superior in stability. For very sharp turning the pivoted axle is no doubt the more stable, as the wheels are not moved so far under the body; but it must be remembered that when the turning is so very sharp as to bring this advantage into play, the speed must in any case be very slow, or if the attempt be made to turn at high speed the result would be the same in either case. For such turns as a motor carriage can with impunity take at the higher speeds, the straight axle, as used for instance by MM. Panhard & Levassor in their voiturette, illustrated on p. 249, is quite as good as the pivoted axle, especially when transverse stiffness of support

567

of the front of the carriage is lessened by the use of a longitudinal pivot in the centre of what we may call the fixed part of the axle, as is the case with the Columbia Electrical Stanhope, the Peugeot car, illustrated on p. 119, and other vehicles.

With either system of steering axle, four tracks of more or less distance apart, according to the radius of the curve described, are traversed by the four wheels, and it will be seen from Fig. 446 of the following diagrams that whatever the radius of the curve covered, the four wheels describe curves having a common centre when the old form of steering axle is used. That centre is always upon the produced axis of the hind wheels, but at a distance from the carriage which decreases with the increasing sharpness of the turn made. There is therefore not necessarily any more grinding of the tyres by lateral rubbing than with those of wheels on pivoted axles.

One of the first questions in connection with the use of the pivoted axle relates to the determination of the proper angular position of the steering arms connected to these axles.

The object is to secure an angle that will, with simple connecting rods between the two arms, give an increasingly greater angularity to the wheel on the inner side of the curve than to the outer wheel, especially for the curves of smaller radii.

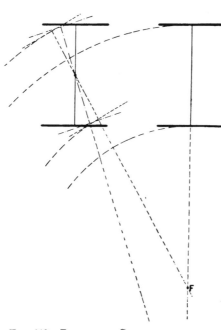

FIG. 446.—DIAGRAM OF CURVES DESCRIBED BY VEHICLE WITH STRAIGHT-STEERING AXLE ON CENTRAL PIVOT.

In order to accomplish this, the arms are set so that with a given movement of the rod, one of the arms describes a greater number of degrees than the other, just in the same way as the crank of a steam engine moves over a greater number of degrees of its path at the ends of the piston stroke per inch of rectilinear travel than it does during the middle of its travel. Various ingenious devices have been worked out for obtaining the exact differentiation of angular position of the steering wheels for truly following curves of any variations of radius. These however, in practice, are seldom carried out, and are not necessary, as an easy, practical approximation gives results which under the conditions of running of vehicle wheels on roads is in every way sufficient, and probably quite as accurate.

In diagram Fig. 447 the front wheels are shown in position for

Setting out angle of steering arms

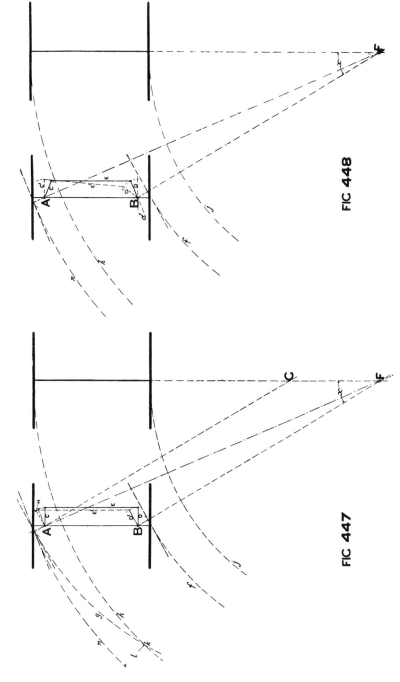

FIC 447

FIC 448

FIGS. 447, 448.—DIAGRAMS OF CURVES DESCRIBED BY WHEELS OF VEHICLES WITH PIVOTED STEERING AXLES
WITH DIFFERENT ANGLES OF STEERING ARMS.

569

straight running upon their axles pivoted at A, B. For the purposes of explanation the steering arms C, D are placed at right angles to the axles, and are connected by a rod E, which is assumed to receive its motion by one of the usual methods. The gauge of the wheels being say 4 ft., and the distance from the inner hind wheel to the centre F of the curve about to be traversed is 8 ft., that is to say the conditions which would probably obtain in passing from a road say 25 ft. wide into a road of similar or narrower width at right angles to it. To describe so short a curve the angularity to be given the front wheels is very considerable, and is indeed greater than can be given by the wheels of most cars.

Now assuming the inner front wheel to be moved to the position, shown by the dotted lines, for describing a curve having its centre at F, the arm D will have taken the position D', and arm C will have described exactly the same number of degrees of arc, giving to the outer wheel the same angularity as the inner wheel. The wheels will be in fact parallel, and as a result, instead of having a common centre for their curved path, they will have two centres, namely F and G, and the curves traversed by them will correspondingly be *f* and *g*, the curves traversed by the hind wheels being *h* and *j*.

Side rub of wheels on road
It will be noticed that the curve *g* crosses the curve *h* at a point *k* which has a distance *l* of 1 ft. from the curve *n* which would have been traversed by the outer wheel if it had radiated from the common centre F; that is to say, that in travelling about 6 ft. the tyre of the wheel has been forced to slide or grub over the road surface a distance of 12 in. It is obvious then that the outer wheel has received too much angular movement and that the right-angled position of the steering arms C, D is wrong, and further that the length of the connecting rod E is too great. Now from this diagram we can see how much too far the arm C has been pushed, and how much the rod E is too long. With the arm D moved to the position D', by a length of rod E, which is the necessary length to couple the arms when at right angles, the arm C has been moved too far by the distance H which is the number of degrees by which the proper angle of the outer wheel, namely tangent to the curve *n*, differs from the angle to which it has been pushed by the rod E when in obedience to the angle of the steering arms it took the position E'.

Having now obtained the angle of error H, we can alter the position of the right-angled arms to one which will give very approximately the correct radiation for the axles for all ordinary curvatures. The new angle will be greater than a right angle by 3 times H; that is, they will depart from the right angle position and lean towards each other by very nearly 3 times the angle H, Fig. 447. This may be shown by reference to diagram Fig. 448, in which the same scale is adopted and the arms C, D are placed in accordance with what has been found by Fig. 447. The result is that when D moves to D', and E to E', C moves the lesser distance to C', and both the steering wheels remain normal to radii originating at F. C has

570

been moving *to* the right angle position before it can move *from* it, while the *whole* of the movement of D has been *from* the right angle position to a greater angle and towards the fixed axle, when for several degrees its movement would have no effect on c. In other words, part of the angular movement of D to D' has been used in diverting the rod E from a rectilinear movement to the position E'. Hence a given range of motion of the rod gives more angular movement to D than to c, or if the turn be made to the right instead of the left, c would on the contrary be moved through the greater angle.

For any curves greater in radius than that of the diagrams, the position of the arms c, D will remain practically true for the curves to be traversed. The angle will be theoretically slightly too great. For curves of a smaller radius, which in most cases cannot be turned because the wheels touch the springs, the angle is too small; but this is of no importance. The curves *n h* and *f j* remain practically parallel as thus set out for any radius greater than the 8 ft. assumed in the diagrams, so that the wheels do not rub over the road surface laterally.

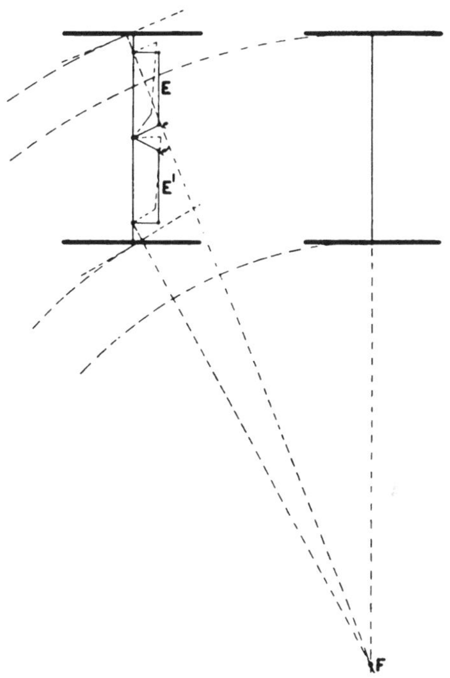

Some makers prefer to use steering arms which are at right angles to the pivoted axles, and this is done by having two connecting rods E, E', Fig. 449, connected to a central steering arm with pivots *e e'* separated by a distance

Divided steering connecting rod

FIG. 449.—DIAGRAM OF PIVOTED STEERING AXLE ARMS, WITH DIVIDED CONNECTING ROD.

which includes double the angle shown for c, D, Fig. 448, or includes 6 times the angle H, Fig. 447.

In all these diagrams the steering arms are shown as at the back of the axles which is sometimes objected to as putting the rod E under compressive stress during ordinary running. This objection is of little importance as rod E may be tubular and easily made capable of resisting the varying compressive stresses. When, however, it is found for any structural reason necessary that the rod should be in front and under tension, all that is necessary is that the angle made by the steering arms and the axles shall be as much less than a right angle as they were

Connecting rod in front and at back of axle

571

greater for the arrangement shown in Fig. 448; that is to say, they should be in the position shown at *d*. This arrangement, however, brings the ends of the steering arms very close to the spokes of the wheels.

It will be found that the necessary angularity of the steering arms increases slightly with the length of the wheel base of the car.

For explanatory purposes two diagrams, Figs. 447 and 448, have been used; but in practice one drawing would suffice, because the new position found from Fig 447, and drawn upon Fig. 448, can be shown on the same drawing.

Chapter XXXVI

THE POPULAR AND COMMERCIAL USES AND COST OF WORKING MOTOR VEHICLES

SUCCESSFUL employment of heavy motor vehicles depends upon many questions of trading and industrial expediency apart from the simple question of cost per ton-mile of working the vehicle.

The choice of a motor vehicle for what may be called popular purposes, namely, pleasure and health, depends firstly upon the purse, and secondly upon the number of persons it is desired generally to carry.

Omitting from consideration those reasons for the purchase and use of a vehicle, such as the pleasure of riding or driving, sight-seeing and health-giving, reasons which may obtain for the use of any vehicle, then motor tricycles will be chosen by those who enjoy the exercise of skill in driving, who enjoy high speed, who want to travel long distances in a day, and in some cases because they are economical in first cost and in maintenance, and require little space for housing.

The quadricycle will be chosen by those who want to carry two persons, for one of whom a more comfortable seat is required than the ordinary motor tricycle saddle gives, though some tricycles are fitted with seats.

The voiturette and the lighter carriages will be chosen by those who require at least two comfortable seats and perhaps room for some luggage. With some purchasers high possible speed will not be looked upon as necessary if hills can be mounted at a speed a little in excess of that of horses in carriages. Vehicles of this class will also be purchased by those who, like doctors, have business purposes in view. There are, or soon will be, several different good British cars in this class. When purchased for pleasure purposes the question of cost in comparison with other modes of travelling does not come into consideration. When, however, business reasons induce the purchase, the questions of comparative cost have to be considered, and these are not completely answered by reference to first cost and cost of working, even when the user intends to perform only the same journeys as he previously travelled by pony and trap or horse and carriage.

Comparative cost by horse and motor

573

Generally it is only the first cost which is greater. The working expenses will always be less with a good motor vehicle than with horses for the same distance covered per year if the owner gives anything like as much attention to the maintenance of his motor vehicle as he would to the care of his horse. If he fails to do this, either by negligence or incapacity, or by the employment of an incompetent attendant or amateur repairers, the cost per year, including interest on capital outlay and depreciation, may reach or even exceed that of horses. The distance travelled by a motor vehicle owner, however, soon exceeds that previously travelled by his one or two horses, and the value of this extra mileage is generally forgotten in reckoning cost, though it might not have been possible without an extra horse. Sometimes the distance covered in one day is greater than could have been covered at all by horses. This also represents money value.

The cost may be considered with reference to two cases, assuming the price of the voiturette to be £200.

Cost by horse haulage

In the first instance I will assume the requirements of one whose purposes are fulfilled by a two-wheeled vehicle and cob or strong pony assumed to be capable of doing a round of 15 miles per day during 4 days per week, this distance being all that is required.

TABLE XXI.—COST OF TWO-SEAT CARRIAGE ACCOMMODATION BY HORSES.

	Cost per year for 15 miles per day and 200 days per year.			Cost per year for 30 miles per day and 200 days per year.		
	£	s.	d.	£	s.	d.
Capital outlay—						
Two-wheel vehicle	40	0	0	—		
Four-wheel vehicle	—			75	0	0
Cob or pony	40	0	0	—		
Horses	—			100	0	0
Stable and coach-house requisites	8	0	0	10	0	0
Harness	8	10	0	25	0	0
Total capital outlay	96	10	0	210	0	0
Interest on capital at 4 per cent.	3	17	0	8	4	0
Depreciation on horses and harness, at 20 per cent..	9	14	0	20	0	0
Depreciation on vehicle and repairs, at 10 per cent..	4	16	0	7	10	0
Horse food and bedding, at 15s. each per week . .	39	0	0	—		
" " " " 16s. " " . .	—			83	0	0
Shoeing	3	5	0	7	5	0
Veterinary, insurance, axle oil, harness paste, etc. . .	3	10	0	10	10	0
Rent and rates	10	0	0	20	0	0
Licence (tax)	0	15	0	1	1	0
Extra labour for 2 horses, at 12s. per week . .	—			31	4	0
	74	17	0	188	14	0
Total cost per mile (driver not included)	5·98d.			7·55d.		

574

USES AND COST OF WORKING MOTOR VEHICLES

Secondly, I will take as an illustration the case of a proprietor whose requirements are a round or a journey of 30 miles per day during 4 days of the week.

The average cost in these two cases will generally be represented by the expenditures in Table XXI. There are in this, no doubt, some quantities which would not be met with by those whose circumstances place them in possession of stables and horse-keeping facilities, but the figures represent the expenditure which must be provided for by any one proposing to do either the 15 miles per day or the 30 miles per day, which, for 4 days a week or 200 days per year, represents respectively 3,000 and 6,000 miles per year.

Inasmuch as 30 miles per day could not be done by one horse for 4 days per week, the cost of keep of two horses is assumed. For 15 miles a day one horse only is considered in the estimate but it would be a very lucky proprietor who got through the year with only one horse, and an addition to this estimate of probably £5 for hire would make it more correct. Without this, however, it will be seen that when everything is taken into consideration, as it is in these estimates, the cost for 3,000 miles per year by single horse and trap is £74 17s., while the cost for 30 miles per day, or 6,000 miles per year, in which case a more commodious and comfortable carriage has been allowed for, reaches £188 14s. Even if the same form and cost of vehicle be assumed in the second case as in the first, so that interest and depreciation be reduced by £5 4s., the cost will still be £183 per year.

Now we may compare these figures with those of the estimate in the following table of the cost of performing precisely the same service by

Cost by motor vehicles on solid tyres

TABLE XXII.—COST OF TWO-SEAT ACCOMMODATION BY MOTOR CARRIAGE.

				Cost per year running 15 miles per day on 200 days per year, or 3,000 miles.			Cost per year running 30 miles per day on 200 days, or 6,000 miles.		
	£	s.	d.	£	s.	d.	£	s.	d.
Capital outlay—									
Car	200	0	0						
Sundry apparatus	10	0	0						
Interest on capital, at 4 per cent.				10	8	0	10	8	0
Depreciation, at 15 per cent., on car				30	0	0	30	0	0
Depreciation on sundries, at 10 per cent.				1	0	0	1	0	0
Repairs and renewals				3	10	0	6	0	0
Tyre renewals				3	3	0	6	6	0
Petroleum spirit, at 1s. 3d. per gal.				15	0	0	26	0	0
Lubricating oil, at 3s. per gal.				0	12	0	1	0	0
Sundries, grease, waste, etc.				0	8	0	0	12	0
Insurance				5	0	0	5	0	0
Licence (tax)				2	2	0	2	2	0
Rent and rates				10	0	0	10	0	0
Total cost				81	3	0	98	8	0
Total cost per mile				6·5d.			3·95d.		

575

means of motor cars. Care has been taken to include every item of cost, and the same rates for rent, etc., have been assumed as in the preceding cases, and the cost of, on the one hand, an ostler, and on the other a mechanical driver, have been excluded, because in each case the owner might perform these duties, except so far as cost of occasional assistance is concerned, which might by either method be about the same in amount.

These estimates show that for 15 miles per day the total cost by motor car may be greater than with the horse car. On the other hand, however, if the distance covered be only 40 miles per week instead of 60 miles the motor car will cost less than the horse and trap.

For the greater distance of 120 miles per week the saving resulting from the employment of the motor car reaches no less a sum than £89 16s. per year, or 48 per cent. in favour of the motor car.

In these calculations solid tyres have been assumed, and a life of 15,000 miles. It is true that with very careful use a greater mileage than this can be obtained from the tyres considered, such as the Connolly and the North British, but so much of the tyre has gone after 15,000 miles of running that the elastic cushion value remaining is but slight.

The cost of tyres per mile based on these prices and mileage is 0·51d., a cost which could only be lessened at the cost of extra engine power if the tyres be run after they have been much worn down ; and inasmuch as the cost of petrol may be taken as from 0·5d. to 0·6d. per mile for vehicles of the size considered, weighing say 8 to 12 cwt., it would be false economy to use the old inelastic tyres.

It will be seen that the cost per vehicle mile carrying two passengers is 6·5d. when 60 miles are run per week, but only 3·95d. per mile when 120 miles are run per week. The cost, on the other hand, of travelling by horse and trap is practically 6d. (5·98d.) per mile when only 15 miles per day is done, and no less than 7·55d., or 7½d., per mile when 30 miles per day are done.

Cost by motor vehicle on pneumatic tyres

We may now turn our attention to the cost when pneumatic tyres are used. It is unnecessary to repeat the items in the foregoing estimates because most of them would remain the same. It is only necessary to consider the very much greater first cost of the tyres and their very much shorter life. For the purposes of arriving at a comparative estimate the cost of a set of covers and inner tubes has been taken, the cost of rims being excluded as not being a repeating charge. In addition, the cost of road bands for prolonging the life of the tyres has been included, together with the cost of putting them on, as a fair representation of a set of wheels, 32 in. tyres, have been assumed. The total cost of this set would be about £30. Taking the life of such a set of tyres with the road bands properly put on as 4,500 miles, then the cost per mile would be 1·517d. instead of the 0·51d. which has been adopted in the preceding estimates, that is to say an extra cost of 1d. per mile. From this, however, may be deducted a small amount to represent the reduced cost of petrol for the

576

smaller power required with pneumatic tyres, and further a slight reduction may be made on the cost of renewals, owing to the superior tyre cushioning. These two quantities may be taken as 0·15d. and 0·05d. respectively, making a total saving of 0·2d. per mile, so that we have an additional cost for the luxury of pneumatic tyres of 0·8d. per mile.

Turning for the moment to the heavier carriages, weighing say 17 to 19 cwt., and carrying 4 to 6 passengers, we have the following as an estimate of the annual cost of such a car fitted with 2-in. solid tyres and as fitted with 3·5-in. pneumatic tyres.

Cost by four-seat motor vehicles

TABLE XXIII.—Cost of Four-seat Accommodation by Motor Carriage.

	£	s.	d.	Cost per year running 30 miles per day on 200 days, or 6,000 miles. 2-in. solid tyres.			Cost per year running 30 miles per day on 200 days, or 6,000 miles. 3½-in. pneumatic tyres.		
				£	s.	d.	£	s.	d.
Capital outlay—									
Car	400	0	0						
Sundry apparatus	10	0	0						
Interest on capital, at 4 per cent.				16	8	0	16	8	0
Depreciation, at 15 per cent., on car				61	10	0	61	10	0
Depreciation on sundries, at 10 per cent.				1	0	0	1	0	0
Repairs and renewals				6	0	0	5	0	0
Tyre renewals				12	18	0	59	15	0
Petroleum spirit, at 1s. 3d. per gal.				26	0	0	22	5	0
Lubricating oil, at 3s. per gal.				1	0	0	1	0	0
Sundries, grease, waste, etc.				0	15	0	0	15	0
Insurance				5	0	0	5	0	0
Licence (tax)				2	2	0	2	2	0
Rent and rates				10	0	0	10	0	0
Total cost				142	13	0	184	15	0
Total cost per mile (driver not included)				5·71d.			7·39d.		

This must be compared with the cost of similar accommodation by a horse-hauled carriage, but it must be first noted that the 7·39d. may easily be enormously increased by high speeds.

Speed raises cost

To work a vehicle of the capacity assumed 30 miles per day on 4 days a week, even assuming it to be in country over which one horse could haul the vehicle, it would be necessary to have at command three horses, and these would be absolutely necessary in country which occasionally needed two horses to the vehicle. Good horses suitable for the work could not be obtained for less than £55, the items of expenditure will then be approximately as given in the following estimate :—

Comparison with cost by horses

37

TABLE XXIV —COST OF FOUR-SEAT ACCOMMODATION BY HORSE VEHICLES.

	Cost per year running 30 miles per day on 200 days, or 6,000 miles.		
	£	s.	d.
Capital outlay—			
Vehicle.	120	0	0
Horses .	165	0	0
Stable and coach-house requisites	10	0	0
Harness	45	0	0
Total capital outlay	340	0	0
Interest on capital, at 4 per cent. .	13	12	0
Depreciation on horses and harness, at 20 per cent.	42	0	0
Depreciation on carriage, at 10 per cent., and repairs .	12	0	0
Horse food and bedding, at 16s. each per week .	124	15	0
Shoeing .	10	0	0
Veterinary, insurance, axle oil, harness paste, etc.	15	0	0
Rent and rates	26	0	0
Licence .	1	1	0
Labour for 3 horses, at 22s. per week .	57	3	0
Total cost	301	11	0
Total cost per mile (driver not included) .	12·06d.		

Saving by use of motors

From this then it follows that the cost by horse haulage reaches 12·06d. per mile, as compared with 5·71d. on solid rubber tyres, or as compared with 7·39d. on large pneumatic tyres. Thus assuming that the £120 included in the estimate for the horse-drawn carriage for the accommodation of six persons, including the driver, is the price with iron tyres, then we have by motor vehicle the luxury of riding on pneumatic tyres at a cost of 7·39d. per mile, as against 12·06d. on iron tyres, a saving of 39 per cent., or of 38½ per cent. if a carriage costing only £80 be used, even when we put no money value on the extra comfort. Moreover these comparisons place no money value on the advantage, often a very great one, of being able to run any distance, up to say 100 miles in a day, with the motor carriage, as compared with 30 miles by the horses. That 100 miles may be safely reckoned upon is shown by the results of the periodic trials con-

Recent 100 mile tests

ducted by the Automobile Club. Only recently (April 11th) three vehicles, typically representing three classes, namely a Peugeot three-seated vehicle, as on p. 167, a de Dion voiturette, as on p. 181, and an Ariel quadricycle, as on p. 301, ran the 100-mile trial between 11.30 a.m. and 7 p.m.

On the 23rd of April 58 vehicles reached Bristol, having left London at 7 a.m. in the morning, 50 of them completing the journey before 8 p.m., having stopped for breakfast, lunch, and tea on the road.

Following up these estimates of cost, which are based on actual

USES AND COST OF WORKING MOTOR VEHICLES

experience, the cost of running an electrical carriage may be dealt with, although cost must be a secondary consideration, or indeed a matter of very small importance to those who adopt this system of propulsion for private carriages. This is obvious from the following figures :—

TABLE XXV —COST OF TWO-SEAT ELECTRICAL CARRIAGE ACCOMMODATION.

	Cost per year for 25 miles per day, and 200 days per year, or 5,000 miles.
	£ s. d.
Capital outlay—	
Electrical vehicle with solid rubber tyres.	350 0 0
Carriage house requisites .	5 0 0
	355 0 0
Interest on capital, at 4 per cent. .	14 4 0
Depreciation on vehicle, at 10 per cent.	30 0 0
Depreciation on battery, at 50 per cent.	25 0 0
Battery renewals, at 15 per cent. .	7 10 0
Repairs and renewals	3 0 0
Tyre renewals	12 6 0
Charging current, at 5d. per unit .	37 10 0
Lubricating oil, grease, waste, etc.	1 0 0
Insurance.	5 0 0
Licence (tax) .	2 2 0
Rent and rates	10 0 0
Total cost	147 12 0
Total cost per mile (driver not included) .	7·1d.

The conditions here assumed are somewhat favourable to this electrical vehicle, namely a 25-mile charge used each day, leaving the whole night for re-charging. Possibly the current might under some circumstances cost less than 5d., but in many cases it would cost more, and other costs would under any circumstances be greater if a higher speed than has been here assumed for an average, namely 8 miles per hour, were adopted. In determining the quantity of current necessary for charging, a discharge efficiency of 66·6 per cent. has been assumed.

Working 4 days a week, and thus giving 3 days rest, may be considered extremely favourable for any of the vehicles herein under survey. The result, so far as concerns the electrical vehicle, is that it may possibly cost less than the same accommodation by horses, but that it costs more per mile than when similar, but not the same, accommodation is provided by a small horse and trap for 15 miles per day. It does not, however, at all follow that the electrical vehicle would prove satisfactory as to cost of working for hire purposes, working 6 days per week, paying establish-

579

ment charges, paying rather high wages to a driver, and bound to run over roads and streets regardless of either gradient or condition of the surface. In comparison with the mineral spirit motor vehicle the electrical vehicle, even if it could work the mileage, is inferior to the extent of over 40 per cent.

At present there is not sufficient data to enable me to form an estimate of the cost of running of light steam vehicles. The Whitney and the Stanley steam vehicles, both being fired with mineral spirit, will cost for fuel about double as much as internal combustion motors, but the wear on the tyres will not be quite as great. The gain in this respect, however, will not modify perceptibly the increase in petrol consumption. The Serpollet car, which uses kerosene oil, might be said to escape this objection; but of the latest forms of the Serpollet car no systematic fuel consumption trials have been made.

The de Dion voiturette weighs, when ready for the road, 7·25 cwt. without passengers, and the Stanley car weighs about 6·75 cwt. The former ran 100 miles, including the Dashwood and Ashton hills, on the road to near Oxford and back, on the 11th April last, on 3 gallons of petrol, or 33·3 miles per gallon, carrying driver and one passenger. The Stanley or the Whitney would not, as far as has at present been shown, run half this distance on the same supply.

With regard to the cost of running the heavier steam vehicles, reference must again be made to Table XX. It will, however, have been noticed from the foregoing comparisons that the cost of transport per mile may be a misleading guide to the cost of working under different circumstances, and in fact the tabulated results of mere experimental trial runs, such as those of Liverpool, have often proved misleading. The total cost of running includes a number of fixed charges which remain the same for a short haul as for a long haul.

Economic length of journey The maximum economy is only reached when the maximum distance is covered in the proper working day. For a given kind of vehicle and given load, there is, therefore, a distance or length of haul which is more economical than any other. That is best which enables the vehicle to start early with its load, deliver it and receive return load, if any, and be back again at the end of a day which does not exceed in length the customary hours. If the haul be shorter than this the fixed charges become greater in proportion to the mere cost of transport. If longer the labour expenses increase. With heavy-load vehicles a distance of more than about 50 miles per day is not economical, because either overtime occurs or out expenses are incurred. Twenty-five miles out and home, at an average of 6 miles per hour, may be carried on continuously. Any distance between about 25 miles, and approaching 50, involves long hours if only a few over 25 miles, or special arrangements for exchange of drivers at suitable depôts *en route*, or conducting the service by men who sleep alternately at one end of the journey and the other, if approaching 50 miles.

USES AND COST OF WORKING MOTOR VEHICLES

A proof that an out and home journey of say 48 to 50 miles can be performed at a total cost of say 6*d*. per ton-mile is far from being a sufficient guide to the cost of any other length of haul a few miles shorter or a few miles longer. The exigencies or customs of trades based on expediency will also have effect on the best length of haul and best load.

Ton-mile cost on experimental runs not a sufficient guide

As an example of the variety of matters that sometimes enter into a consideration of the question of transport of goods by the different methods available, I am enabled by the kind permission of Messrs. John Broadwood & Sons, to quote from a report which I was called upon to prepare for them in connection with their contemplated building of a new manufactory out of London. A large proportion of the pianofortes referred to are heavy " grands." The extracts from the report explain themselves.

Effect of manufacturing and trading exigencies

Service required. This is :—

1. Transport from works at say Bedford to warehouse in Great Pulteney Street of from six to eight pianofortes per day, or say forty per week, and of about six pianofortes per week from London to Bedford.

Assumed required service

2. The transport of about 700 tons of timber per year from London Docks to, say, Bedford. This would be in irregular quantities at irregular intervals.

3. An alternative service to and from Blank line, say 13 miles out and 13 back.

This service must be provided either by rail and cartage at terminals, by motor vehicles, or by horse haulage the whole distance. Canal service has also been considered.

Several arrangements or systems for the conduct of this transport present themselves for consideration, namely :—

A. Loading the pianofortes at the works into horse vans of the kinds now employed in London, horsing the vans to the railway station at Bedford, and there putting the vans on to railway trucks to go by passenger train to St. Pancras, where the vans would be taken by your horses to Great Pulteney Street. This involves keeping three horses at each end and having five 1-ton vans, two going to London each day, two returning each day, and one as a stand by for repairs, and it supposes that regular service by the Railway Company can be assured. The railway charge to be 16*s*. per van each way, a loaded van not to exceed 2½ tons total. Consigners risk.

Cost of transport by rail and horsing to station

B. Same as *A*, except that the vans would each carry 2 tons, and one each way would be put on trucks of luggage trains instead of on passenger train trucks, the railway charge to be 31*s*. 6*d*. per ton; so that the charge would be less for the return than for the incoming loaded van.

C. Pianos packed in cases collected and delivered by the Railway Company at 31*s*. 6*d*. per ton. Empty cases to be returned at 8*d*. per cwt.

D. Railway carriage Bedford to London, without collection, at 24*s*. 10*d*. per ton, and 30*s*. per ton London to Bedford.

These systems involve the cost of packing and unpacking in cases, and for system *D* horses and vans would have to be kept at Bedford and London as in system *A*.

E. The second available system is the conduct of the transport by motor vans, in which case the pianos would be loaded into the vans, without packing cases, at Bedford, and delivered at Great Pulteney Street.

Cost by motor 2-ton vans

This involves the purchase and working of:—

(*a*) Three suitable vans for carrying loads of 2 tons, and as such vans cannot be reckoned to make a speed of more than an average of 5½ to 6 miles per hour the whole way, thus occupying about 8½ to 9½ hours on the journey, the drivers would have to sleep alternately in London and Bedford. The present legal speed under Local Government Board regulations for such heavy vans is only 5 miles per hour; but this will probably be extended to say 7 miles ere long, and vehicles of this weight run at that speed now on clear roads.

Cost by 1-ton motor vans

(*b*) Or alternatively the purchase and working of lighter vans, each capable of carrying 1 ton and running at an average speed of 7 to 8 miles per hour, and thus accomplishing the journey from Bedford to London in from 6½ to 7½ hours, and the return journey light at a speed of from 9 to 10 miles per hour, or in from 5 to 5½ hours. If the driver took his dinner while his van was being unloaded he would with an hour stop make about a 14½- to 16-hour day. This speed may, in the near future, be increased so that the double journey in the day would be possible, but men could not stand it as a regular daily job.

Cost by 2-ton motor vans and exchange of drivers

(*c*) Motor vans to carry 2 tons, and stopping about half way, say Luton, St. Albans, or Hatfield, and there exchanging drivers, so that each driver would do 50 miles per day and would each start and return from and to Bedford or London, so as to avoid sleeping alternate nights at Bedford and London. The whole journey occupying about 10 hours.

Cost by 1-ton motor vans and exchange of drivers

(*d*) Motor vans to carry 1 ton each, and stopping about half way and there changing drivers, each driver doing 50 miles per day, and the whole journey occupying about 8 hours, and employing four drivers. This would appear to be one of the best systems, as each car would work under better conditions and the drivers would have time to do their own cleaning, and the whole system would be much more flexible both as to cars and drivers.

Cost of shorter journey by 1-ton vans

(*e*) Transport by a 1-ton motor van between two places, 13 miles apart, the van making two double journeys, say 52 miles per day Two vans required to maintain service.

Cost of same by 2-ton vans

(*f*) Transport by a 2-ton motor van between two places, 13 miles apart, the van making one double journey, or 26 miles per day. Assume one van sufficient to maintain service, and that chief repairs could be done at slack part of year, an addition being made of say £13 per year for hire of horses, using ordinary delivery vans for say twice per year when the one motor van is under repair taking more than slack time allows.

582

USES AND COST OF WORKING MOTOR VEHICLES

F. Conduct of the service by means of horses. The horses could do say 25 miles per day, and stables would have to be established at about half way between Bedford and London. Horses and driver would leave Bedford with full load and be met 12½ miles out by horses and man from the half-way stables with the nearly empty vans. The horses would exchange vans and return to Bedford and to half-way stables respectively.

Similarly horses would take the full van from half-way stables to 12½ miles towards London, and there exchange vans with horses from London, the horses returning respectively, one set to London with full load and the other set to the half-way stables. Each van would thus be two days on the road for each up and down journey. Thus five vans would be required and at least twelve horses.

Four drivers would be required, and a man and boy, or two men, to look after the six horses at the half-way stables.

Stables would have to be procured and maintained.

(*a*) Transport by horses between two places, 13 miles apart, with 2-ton van, making 26 miles per day. Two vans required to maintain service.

G. Traction engine service for timber transport.

The following estimates for the different methods are given in sums per week :—

Cost by Rail. By Passenger Train and Cartage.

A. By 2 loaded vans, each carrying 1 ton, placed on railway truck.
(1) By passenger train.

	£ s. d.	£ s. d.	
Cost of 6 horses, at £60	360 0 0		Capital outlay
Cost of 5 vans, at £90	450 0 0		
Cost of 6 sets of harness, at £6 10s. . . .	39 0 0		
Cost of chaff cutters and stable sundries . .	50 0 0		
Total capital outlay	£849 0 0		£849
Interest on capital outlay, at 4 per cent		0 13 0	
Depreciation of horses, at 20 per cent.		1 7 9	
Depreciation on 4 working vans, at 10 per cent. . . .		0 13 11	
Repairs, etc.		0 5 3	
Depreciation on 4 sets of harness, at 20 per cent. . . .		0 3 6	
Cost of man's wages, at 28s.; and fodder for 3 horses, at 16s., at Bedford, to take van to station		3 16 0	
Cost of same at Great Pulteney Street		3 16 0	
Horse shoeing, at 6s. per month, per horse		0 9 0	
Insurance, veterinary, taxes, and rent, at £5 per horse . .		0 10 6	
Total cost of keep of 4 horses, and 2 men's wages for working 2 vans, per week		11 14 11	
Railway Company's charge for conveying 2 vans, each 6 journeys each way to and from St. Pancras and Bedford, at 16s. per journey		19 4 0	
Total cost by vans sent on passenger trains . . .		£30 18 11	

Total cost per load ton-mile, 11·75*d.*

Note.—If 2 horses can be depended on at each end, this estimate would be reduced to £29 6s. 6*d.*

583

MOTOR VEHICLES AND MOTORS

Cost by Rail and Cartage.

B. By 2-ton loaded vans on railway trucks, by goods train, delivered to and from station by own horses.

		£	s.	d.	£	s.	d.
Capital outlay	Four horses at £60	240	0	0			
	Three vans, at £140	420	0	0			
	Four sets harness, at £6 10s.	26	0	0			
	Chaffcutter and sundries	50	0	0			
£736	Total	£736	0	0			
	Interest on capital, at 4 per cent., per week				0	11	4
	Depreciation on horses, at 20 per cent.				0	18	0
	Depreciation on 2 vans				0	10	9
	Depreciation on harness, at 20 per cent.				2	0	0
	Repairs to vans and harness				0	2	8
	Horse shoeing				0	6	0
	Wages : two drivers, at 28s.				2	16	0
	Food and bedding, 4 horses, at 16s.				3	4	0
	Insurance, veterinary, rates, etc.				0	4	0
	Total cost of men, horses, vans, etc., per week				10	12	9
	Railway Company's charge for 6 loaded vans, Bedford to London, at 31s. 6d. per ton ; load, 2 tons ; van, 1½ ton. Total weight, 3½ tons × 6, at 31s. 6d.				33	1	6
	Ditto for 6 nearly empty return vans, say 1 ton 15 cwt. each . .				15	10	9
	Total cost by vans on trucks on goods trains per week . .				£59	5	0

Total cost per load ton-mile, 20·2d.

Note.—If rebates are allowed off the 31s. 6d. rate, when carried in vans, the total cost will be £51 10s.

C. Cost by railway. Pianos packed in cases and collected and delivered by Railway Company, at 31s. 6d. per ton, Bedford to London. Packing cases returned at 8d. per cwt. Pianos in cases, London to Bedford, at 31s. 6d. per ton.

	£	s.	d.
Pianos, say 12 tons.			
Forty-eight cases, at average of 1½ cwt.=3 tons 12 cwt.			
Total weight, 15 tons 12 cwt.			
Carriage, 15 tons 6 cwt., at 31s. 6d.	24	11	5
Packing 48 pianos in cases, at 4s. 6d.	10	16	0
Interest on cost of 24 special packing cases, at 30s. . . .	0	0	7
Depreciation on cases, at 100 per cent.	0	14	0
Return of 42 empty packing cases, at 8d. per cwt. . . .	2	2	0
Say 1¼ ton of pianos, London to Bedford, at 31s. 6d. per ton .	1	19	4½
Packing 6 pianos for London to Bedford	1	7	0
Unpacking, say, 54 pianos	3	10	0
Total cost of transport of pianos packed in cases and sent by goods train	£45	0	4½

Total cost per load ton-mile, 15·85d.

Note.—This above includes delivery and collection by Railway Company, and no rebate has been assumed. If rebates be deducted, the total cost would be £39 15s.

USES AND COST OF WORKING MOTOR VEHICLES

D. Cost by railway. Pianos packed in cases, and delivered and fetched in own vans to and from railway stations.

	£	s.	d.	
Horses, men, and vans for collection and delivery—all costs, per week, as in Case B	8	14	9	**Capital outlay as in Case B £736**
Railway charges, pianos to London for 15·6 tons, at 24s. 10d. . .	19	7	5	
,, ,, ,, say 1·35 tons to Bedford, at 30s. . .	1	17	6	
Packing 48 cases at Bedford and 6 in London	12	3	0	
Unpacking 54 pianos	3	10	0	
Interest and depreciation on cases	0	14	7	
Return of 42 empty packing cases to Bedford	2	2	0	
Total cost	£48	8	9	

Total cost per load ton-mile, 19·4d.

Only 4 horses are considered above, but probably more would have to be kept.

The rebates off railway charges are taken off in the above, namely, 1s. 6d. per ton at Bedford, and 6s. 8d. per ton in London. Full rate, 31s. 6d.

E. Transport by two motor vans from and to works at Bedford and warehouse in Great Pulteney Street.

(a) Three vans (two working) to carry 8 pianos, corresponding to a load of 2 tons, at £600 each.

	£	s.	d.	
Interest on capital, at 4 per cent.	1	8	6	**Capital outlay Vans, £1,800 Sundries, £50 £1,850**
Depreciation, at 10 per cent.	3	9	2½	
Wages: 2 drivers, at 33s.	3	6	0	
Fuel: coal, 1·5 lb. per gross ton-mile, 1·5×5×300, or for each van 2250 lb., at 20s.	2	0	0	
Repairs and adjustments	2	10	0	
Oil and stores, at 4s. per week each	0	8	0	
Out money for drivers loading in London, at 3s. 6d. per night .	1	1	0	
Water, say 7¼ gals. per mile, say 8 gals., at 1s. 6d. per 1000 .	0	7	3	
Licence, at £3 3s. per year	0	2	6	
Insurance, at £5 per year per van	0	6	0	
Total cost by motor vans	£14	18	5½	

Total cost per mile, 5·95d.

Total cost per load ton-mile, 5·95d.

If the return load of only 6 pianos per week be considered, the cost per load ton-mile will be 5·21d. Possibly this return load could be increased by sending in this way various goods that would have to be conveyed to Bedford, and thus further reduce the cost per load ton-mile.

585

MOTOR VEHICLES AND MOTORS

(*b*) Three 1-ton motor vans. Two to run 100 miles per day each, to carry 4 pianos, or a load of 1 ton ; to run at high speed, at £500 each, and sundry appliances, £50.

		£	s.	d.
Capital outlay £1,500 £50	Interest, at 4 per cent., on £1,550	1	3	10
	Depreciation, at 15 per cent.	4	6	6
	Wages : 2 drivers, at 35*s*.	3	10	0
	Feul : petroleum spirit, at 1*s*. per gal., at 0·3d. per ton-mile, allowing for light return load	3	0	0
	Repairs, adjustments, and renewals	3	0	0
	Licence for 2 working vans, at £2 2*s*.	0	1	10
	Insurance, etc.	0	10	0
	Total cost per week by high-speed motor vans	£15	12	2

Total cost per mile, 3·12*d*.
Total cost per load ton-mile, 6·24*d*.

At present it does not appear that this system could be followed, as the drivers could not stand the long heavy day as a regular thing.

(*c*) Transport by 2-ton motor vans, stopping half way and exchanging drivers. The cost would be the same as in E (*a*), except that there would be no night-out money to pay drivers, but some expense would be incurred at the halting-place, beyond the cost attending the stop for water.

(*d*) Transport by four 1-ton motor vans with 4 drivers, each doing 50 miles per day.

		£	s.	d.
Capital outlay £2,500	Five motor vans, at £500 each, and £50 for sundry appliances.			
	Interest on capital, at 4 per cent.	1	18	5½
	Depreciation, at 12 per cent.	5	15	4½
	Wages of 4 drivers, at 32*s*. 6*d*.	6	10	0
	Fuel : petroleum spirit	3	10	0
	Repairs and adjustments	2	10	0
	Licence for 4 vans, at £2 2*s*.	0	3	5½
	Insurance, etc.	0	15	0
	Total cost by four 1-ton vans, and 4 drivers . . .	£21	2	3½

Total cost per mile, 4·225*d*.
Total cost per load ton-mile, 8·45*d*.

Query.—Whether any addition should be made for halting-place charges, nothing being required by drivers, except sometimes a bucket of water.

586

USES AND COST OF WORKING MOTOR VEHICLES

(e) Transport by a 1-ton motor van between two places 13 miles apart, the van making two double journeys, or 52 miles per day. Two vans required to maintain service.

	£	s.	d.	
Two 1-ton motor vans, at £500, sundry appliances included.				
Interest on capital, at 4 per cent., per week	0	15	5	**Capital**
Depreciation, at 12½ per cent.	2	8	1	**outlay**
Wages of drivers, at 35s.	1	15	0	**£1,000**
Fuel	0	17	6	
Repairs .	0	15	0	
Insurance, etc.	0	2	0	
Licence for 2 vans, at £2 2s.	0	1	10	
Oil, waste, etc.	0	2	0	
Total cost per week	£6	16	10	

Total cost per mile, 5·26d.
Total cost per load ton-mile, 10·52d.

(f) Transport by a 2-ton motor van between two places 13 miles apart, the van making one double journey, or 26 miles per day.

	£	s.	d.	
One 2-ton motor van, at £600.				
Sundry depôt appliances, £50.				
Interest on capital, at 4 per cent., per week	0	10	6	**Capital**
Depreciation, at 15 per cent.	1	17	6	**outlay**
Wages for driver, at 35s.	1	15	0	**£650**
Fuel : coal, at 20s., including lighting up .	0	13	0	
Oil, waste, etc.	0	4	0	
Water	0	2	0	
Insurance, etc.	0	2	0	
Licences, at £3 3s.	0	1	3	
Repairs .	1	0	0	
Total cost per week	£6	5	3	
If two motor vans be purchased having one as stand by and for extra duties, the total cost would be	£7	11	0	

Total cost per mile, 9·63d.
Total cost per load ton-mile, 9·63d.

F. Transport by horses and 2-ton vans between London and Bedford.

	£	s.	d.	
Five vans, at £120 .	600	0	0	**Capital**
Twelve horses, at £60	720	0	0	**outlay**
Twelve sets of harness, at £6 10s..	78	0	0	**£1,798**
Three sets stable appliances, say £100 .	100	0	0	
Half-way stable, say £300 .	300	0	0	
Total	£1,798	0	0	

MOTOR VEHICLES AND MOTORS

	£	s.	d.
Interest on capital, at 4 per cent.	1	7	9
Depreciation on vans, at 10 per cent.	1	3	1
Repairs, etc., of vans	1	0	0
Depreciation of horses, at 20 per cent.	2	16	4
Depreciation of stables, at 5 per cent.	0	6	0
Depreciation of harness, at 20 per cent.	0	6	0
Repairs to harness	0	10	0
Wages of drivers, at 28s.	5	12	0
Stableman and boy	1	15	0
Fodder, at 16s. per horse	9	12	0
Shoeing, etc.	1	2	0
Insurance, veterinary, taxes, etc.	1	4	0
Total cost of transport by horses	£26	14	2

Total cost per mile, 10·28d.

Total cost per load ton-mile, 10·28d.

Perhaps something should be added to this for charges for convenience at two halting places, say 13 miles out of Bedford and of London.

(*a*) Transport by horses between two places, 13 miles apart, with 2-ton van, making 26 miles per day.   One van assumed to be sufficient to maintain service.

Capital outlay £450

	£	s.	d.
One van, at £120	120	0	0
Four horses, at £60	240	0	0
Four sets harness, at £10	40	0	0
Stable, etc., appliances, £50	50	0	0
Total	£450	0	0

	£	s.	d.
Interest on capital, at 4 per cent.	0	6	11
Depreciation on van, at 10 per cent.	0	4	8
Depreciation on horses, at 20 per cent.	0	18	6
Depreciation on harness, etc., at 20 per cent.	0	4	0
Man's wages, at 30s.	1	10	0
Fodder for 4 horses, at 16s.	3	4	0
Repairs to van, harness, and stable requisites	0	6	6
Insurance, veterinary charges, taxes, etc.	0	7	6
Stable assistance	0	18	0
Shoeing	0	6	0
Total cost by horses per week	£8	6	1

Total cost per mile, 12·78d.

Total cost per load ton-mile, 12·78d.

G. TIMBER.   FROM DOCKS TO BEDFORD.

700 tons per year.   Occasionally about 50 tons in a week.

The distance would probably be at least 54 miles.

The work would require traction engines, each performing 27 to 30 miles per day.

Each engine would be four days on the double journey, except under pressing

588

circumstances, when extra work and hours were arranged for, and then each engine could do the double journey in 3 days. This would give 2 loads per week. Engines of 8 HP. nominal would haul 16 tons over the roads to be considered.

To deliver 50 tons per week would therefore require more than one engine, say two engines and four traction waggons, each engine of 8 HP.

If the tonnage were equalised throughout the year, or taking 50 weeks, and say 14 tons per week, then one 8 HP. engine and 2 waggons would perform the work, a third waggon being obtained for reserve during repairs.

If the road locomotive system be adopted it would probably be best to have the plant for this regular weekly service, and hire from traction engine and waggon owners for the occasional extra help when dock deliveries make more transport power facilities temporarily necessary.

Cost of service by one 8 HP. traction engine and three 8-ton waggons.

	£	s.	d.
One 8 HP. engine	600	0	0
Three waggons	270	0	0
Drawbars (spring)	15	0	0
Sundries	50	0	0
Total	£935	0	0

	£	s.	d.
Interest on capital outlay, at 4 per cent., per week	0	14	6
Depreciation, at 10 per cent., per week	1	18	0
Repairs and renewals, per week	2	0	0
One driver	1	12	0
One steersman	1	5	0
One assistant on engine (Act 1899)	0	18	0
One assistant on waggons	0	18	0
Oil, waste, and water	0	5	0
Coal	1	2	6
Out money for men two nights per week, at 3s. 6d. each	1	1	0
Cost of stopping-place for engines, waggons, and load, say	0	13	0
Insurance, etc.	0	5	0

Total cost per week of engine, waggons, and men for delivery of say 14 tons of timber per week, or at a pinch by making 2 double journeys per week, 28 tons, occasionally only ... £11 12 0

This is about the same as the cost by carting and rail as per the Railway Company's quotation, and it does not include anything for superintendence.

Questions may arise as to difficulties imposed by the different hours during which different local authorities permit traction engines to work through their towns. These may add somewhat to the number of hours a journey may occupy, and for times of exceptional work may add the cost of labour of an extra man or two.

The worst town would perhaps be London, but as the London County Council Bye Laws are still under consideration, and cannot be accepted by the Local Government Board, nothing definite can be settled. It is not yet known which streets and roads will be closed against traction engines, nor are the hours of exclusion settled. The hours of arrival within the limits of other towns and the effect of their hours of exclusion cannot yet be determined.

In this estimate I have not included the cost of two engines, because for the greater part of the year there would be two days per week when the one engine would be at home, so that arrangements could always be made for at least four days off for repairs when required.

From these several estimates for the performance of work, under circumstances which may be quite common, it will be seen that the cost per vehicle-mile may vary from equality with the cost per ton-mile to only half that amount. The annexed is a summary of the cost per load ton-mile, by the different systems.

Transport of 2 tons, 50 miles per day, for 6 days per week :
1. By two 1-ton horse vans placed on railway truck, by passenger train   11·75*d.*
2. By two 1-ton horse vans placed on railway truck by goods train .   20·2*d.*
3. By railway, collected and delivered . . . . .   15·85*d.*
4. By railway and own horse vans . . . . .   19·4*d.*
5. By three 2-ton motor vans (two working) . . . .   5·93*d.*
6. By three 1-ton motor vans (two working) . . . .   6·24*d.*
7. By five 1-ton motor vans (four working) . . . .   8·45*d.*
8. By five 2-ton horse vans (four working) . . . .   10·28*d.*

Transport of two tons, 13 miles per day, for 6 days per week
9. By two 1-ton motor vans . . . .   10·52*d.*
10. By one 2-ton motor van . . . . . .   9·63*d.*
11. By one 2-ton horse van . . . . . .   12·78*d.*

In each of the above cases the cost per load ton-mile, would be less if the load were 3 tons instead of 2, nearly in proportion to the increased load. Thus in Case 6 the cost might fall to a little over 4*d.*

Steam Motor Tip Waggons.

Some figures have been obtained from the working of the Thornycroft steam dust collecting tip waggons, several of which have been tested, and are at work for different vestries in or near London. Reports from several sources indicate that the work of collection, carrying, and tipping by these steam vehicles may, without doubt, be conducted at a considerable profit as compared with the smaller horse dust vans. One of these reports was by Mr. T. W. Higgens, Assoc. M. Inst. C.E., who is surveyor to the Vestry at Chelsea. The invitation for tenders for dust vans, issued in October, 1899, by this Vestry, together with the various tenders received from the different makers, and the remarks upon them for the information of the Vestry by Mr. Higgens, may be usefully reproduced here with slight omissions. They are as follows :—

Cost of Steam Tip Vans or Dust Carts.

"The tenders received by the Vestry for the supply of 3 motor vans were from the undermentioned firms, and were as follows :—

	Each van. £	For 3 vans. £
1. Messrs. Coulthard & Co., of Preston	475	1,425
2. The Lancashire Steam Motor Co., of Leyland :—		
For coke-fired motor	490	1,470
For oil-fired motor	510	1,530
3. Messrs. E. H. Bayley & Co., of London, S.E. . . .	650	1,950
4. The Steam Carriage and Waggon Co., of Chiswick (Thornycroft's)	700	2,100[1]

[1] Less 2½ per cent. with partial prepayment.

USES AND COST OF WORKING MOTOR VEHICLES

In order to enable them to tender, each firm was supplied with the following specifications, in accordance with which they were asked to frame their estimates.

SPECIFICATION FOR THE SUPPLY AND DELIVERY OF THREE MOTOR VANS FOR THE VESTRY OF THE PARISH OF CHELSEA.

1. The motor vans must come under the definition of light locomotives, as defined in the Locomotives on Highways Act, 1896,[1] and must comply with the regulations of the Local Government Board framed under that Act.

2. The vans are to be four-wheeled vehicles, provided with easily removable covers, and are to be of sufficient capacity to carry 6 cubic yards of sand or other material, of a weight not exceeding 4 tons. The extreme width of the vehicles is not to exceed 6 ft. 6 in., they are to be capable of going anywhere where a horse-drawn vehicle carrying the same load is ordinarily required to go, each vehicle is to be completely under the control of one man, both for driving and steering, and is to be capable of being tipped by one man over a baulk of timber 14 in. x 14 in. and held at any angle.

3. The vans will not be required to go at a greater speed than 6 miles an hour, and must be able to carry a load up an incline of 1 in 20 for 100 yards at a speed of 4 miles an hour.

4. Makers are not restricted to any kind of motive power, but must state which they intend to use, and must be prepared to guarantee that the cost of working, apart from the driver's wages, shall not exceed a certain sum per mile, to be named in the tender.

5. The tare weight of each vehicle must be given, both exclusive of water or fuel and also in full working order.

6. The body is to be of suitable seasoned wood, but the frame may be of iron or steel, and the top side-boards are to be hinged.

7. The inside of the platform is to be not more than 3 ft. 4 in. above the ground level when unloaded.

8. All working parts are to be properly encased.

9. The Surveyor to the Vestry shall be at liberty to inspect the vans at any time when being built at the makers' yard.

10. The vans are to be painted as directed, and lettered Vestry of Chelsea in bold letters, and are to be delivered to the Vestry Wharf, Lots Road, Chelsea; one, within four months of receipt of order, the other two within six months of receipt of order.

11. The vans are to be driven for one week after being delivered, by the makers' drivers, free of expense to the Vestry.

12. Payment will be made on the certificate of the surveyor, as follows: 90 per cent. within one month of approval, and the balance within three months from the payment of the first instalment.

13. The makers are to enter into an undertaking to maintain the vans against fair wear and tear for a period of two years from the date of approval.

14. Tenders are to be delivered in a sealed envelope, endorsed "Tender for Motor Vans," before 4 p.m. on Tuesday, November 21st, 1899.

15. A drawing or photograph of the vans is to be submitted with the tender.

16. The Vestry do not bind themselves to accept the lowest or any tender.

---

[1] To comply with the requirements of the Acts the motor vans must be constructed :—
  (1) To weigh less than 3 tons when unladen, without taking into consideration the weight of water, fuel or accumulators.
  (2) To emit no smoke or visible vapour.
  (3) To measure less than 6 ft. 6 in. in width between their extreme projecting points.
  (4) To have flat tyres to each wheel, and if the weight exceeds 1 ton the width of the tyre is not to be less than 3 in., and if 2 tons not less than 4 in.
  (5) To be capable of travelling either forwards or backwards.
  (6) To have two independent brakes.

MOTOR VEHICLES AND MOTORS

The various makers call attention to the following, amongst other particulars, in respect of their vans :—

1.—Messrs. T. COULTHARD & Co., of Cooper Road, Preston.

Engine   The engine is to be of the triple expansion type, the engine bed to be so arranged that all the advantages of an open-fronted engine are obtained, whilst being fully protected from dust. All moving parts are lubricated by means of an oil bath; they are completely covered in, but such covers can be easily detached.

Boiler   The boiler is to be of the liquid fuel fire-tube type, and to have 110 square feet of heating surface. All the tubes are straight, readily disconnected, and easily accessible. The suitable working pressure is from 200 to 225 lb. per square inch, tested to 450 lb. per square inch.

The oil tank holds 25 gallons of ordinary petroleum, which is more than sufficient for a day's work. Liquid fuel is said to be more suitable for this motor.

The water tank holds 60 gallons of water, which is sufficient for a run of 15 miles.

Wheels   The power is transmitted to the drawing wheels by means of Hans Renold's silent chains.

The wheels to be of gun-carriage pattern, the felloes and spokes to be best selected oak, the hubs to be of cast steel, the tyres to be of charcoal iron.

Front wheels to be 2 ft. 6 in. diameter by 5 in. face.

Hind wheels to be 2 ft. 9 in. diameter by 6 in. face.

The front and hind axles to be the best mild steel.

Weight   The weight of the vehicle, exclusive of fuel and water, is 2 tons 19 cwt.; and, in full working order, 3 tons 8 cwt.

Body   The tipping van is to be of best selected oak framing with yellow pine boarding. The covers to be of sheet steel.

The tipping gear is by means of a rack to which is geared a spur wheel, operated by means of a worm and wheel.

Price   The price for three vehicles is £475 each ; the makers' terms are 90 per cent. within one month of approval at their works, the balance within three months.

Maintenance   The makers will maintain the vehicles against fair wear and tear for two years for the sum of £75 per vehicle per year, and will provide a man to be accessible for repairs during that period.

Cost of working   They cannot give *guaranteed* costs of working, but submit the following *estimate of annual expenditure.*

Prime cost, £475.

	£	s.	d.
Interest, 5 per cent.	23	15	0
Depreciation, 15 per cent.	71	5	0
Fuel: 6 gallons American or Russian petroleum per vehicle per mile, at 25 miles per day, at 4d. per gallon × 280 days	70	0	0
Lubricating oil and waste	12	0	0
Repairs	75	0	0
	£252	0	0

Work done : 3 tons average load × 25 miles per day × 280 days = 21,000 ton-miles.
Cost £252 ÷ 21,000 = 2·88d. per nett ton-mile.

2.—THE LANCASHIRE STEAM MOTOR Co., of Leyland.

Engine   The engine to be of the triple expansion type, to be entirely cased in, and to run in an oil bath.

Boiler   The boiler can be either coke or oil fired, and a large feed water heater is fitted to raise the temperature of the feed water to about 180° F.

The water tank holds 60 gallons, which is sufficient for a run of from 12 to 15 miles.

USES AND COST OF WORKING MOTOR VEHICLES

The framing is to be of oak, the floor and sides to be of well-seasoned pine, the whole well supported with iron work. The underframe to be of channel steel, and the covers are to be of sheet steel, arranged so that they can all be removed or opened separately. **Body**

The tipping gear is by means of double cut-steel racks and pinions.

The price for three coke-fired motor vans is £1,470, and, if arranged for oil firing, £1,530. The makers strongly recommend the coke-fired motor. **Price**

The makers will maintain the vehicles in working order against fair wear and tear for two years for 12 per cent. on the cost price. **Maintenance**

The makers give two estimates of the cost of working for one week. **Cost of working**

(*a*) A coke-fired machine :—

	£	s.	d.
Depreciation, 15 per cent. on £490	1	8	0
Coke firing, 3s. 6d. per day	1	1	0
Repairs, £55 per annum	1	2	0
Oil, grease, and waste	0	8	0
	£3	19	0

Running 30 miles per day, 6 days per week=180 miles, at a cost of 5·26d. per vehicle-mile.

(*b*) An oil-fired machine :—

	£	s.	d.
Depreciation, 15 per cent. on £510	1	9	0
Oil fuel, 6s. per day (oil at 4d. per gallon)	1	16	0
Repairs, per week	1	0	0
Oil, grease, and waste	0	8	0
	£4	13	0

Running 30 miles per day, loaded with 4 tons=180 miles per week, at a cost of 6·2d. per vehicle-mile.

3.—Messrs. E. H. BAYLEY & Co., Newington Causeway, S.E.

The engine is compound, of the closed-in type. **Engine**

The boiler is of the water-tube type, and the fuel is coke. The water tank is carried under the body, and water is conveyed from it to the boiler by a pump on the engine and an injector. The working steam pressure is from 175 to 250 lb. per square inch. **Boiler**

Power is transmitted from the engine through steel-toothed gearing to the hind wheels.

The wheels are of oak spokes and ash felloes, and they are properly "dished." **Wheels**

The tare weight is not less than 3 tons. **Weight**

The price of the vehicles would be £650 each, but the makers state that it is to only be in accordance with clauses 2, 3, 6, 7, 8, 10 of our specification. **Price**

The makers estimate that £2 per week per vehicle would be ample to allow for maintenance, but add that sufficient experience has not yet been gained upon which to base figures. **Maintenance**

The cost of the fuel will be 7 lb. of coke per mile; this, with coke at 10s. per ton, will amount to slightly under ½d. per mile. The consumption of oil will be about 2 gallons per week. **Cost of working**

MOTOR VEHICLES AND MOTORS

The following estimate is submitted :—

Estimated annual cost of Steam Motor Van.

50-mile run per diem.

	£	s.	d.
Fuel : 6·15 lb. coke per vehicle + 50 (miles) × 312 (days), at 10s. per ton	21	8	4
28 lb. coke per day for raising steam ; 28 (lb.) × 312 (days) at 10s. per ton	1	19	0
Wood for lighting fire.	0	15	0
Water : 4·9 gallons per vehicle-mile ; 4·9 (gallons) × 50 (miles) × 312 (days), at 1s. per 1,000 gallons	3	16	5
Wages : Driver, 35s. per week	91	0	0
Lad, 10s. per week, to fire boiler and assist driver. . .	26	0	0
Repairs : Material and labour	52	0	0
Lubricating oil and waste	12	0	0
	£208	18	9

Work done = 4 tons load × 50 miles a day × 312 days = 62,400 nett ton-miles per annum.

Cost = £208 18s. 9d. ÷ 62,400 = ·8d. per nett ton-mile.

4.—The Steam Carriage and Waggon Company, Ltd.,

Homefield, Chiswick (Thornycroft's).

Engine   The engine is to be compound, reversing, and entirely enclosed in a dust-proof and oil-tight casing.

Boiler   The boiler will be of the Thornycroft water-tube type, so constructed that all its tubes can be thoroughly cleansed and it can be retubed in position.   Welsh coal or coke is used as fuel.

Wheels   The gearing is to be of the silent chainless type, and the wheels of military type with metal naves, oak spokes and felloes, and steel tyres, 5 in. and 4¼ in. respectively.

Body   The body is to be of galvanized steel or of timber, the frames to be of steel throughout.

Price   The price is to be £2,100 for the three vans, less 2½ per cent., and is payable, one-third with the order, and the remaining two-thirds on delivery.

Maintenance   The makers are not at once prepared to enter into an express undertaking as to up-keep, but would be willing to go into the question with us later on.

Cost of working   The following estimate is submitted :—

Estimated up-keep cost of one steam tip waggon for collection of dustbin refuse.

Capital outlay :—

One steam tip waggon complete £700

Annual cost :—

	£	s.	d.
Interest on capital, at 4 per cent.	28	0	0
Depreciation (12 years life)	58	7	0
Wages : 1 driver at 30s.	78	0	0
Do. 3 fillers at 21s.	163	16	0
Coke, 30 tons at 18s.	27	0	0
Adjustment and repairs	52	0	0
Oil and stores	10	0	0
	£417	3	0

USES AND COST OF WORKING MOTOR VEHICLES

Annual Performance :—

Working days per annum, 5 per week	260
Cubic yards dealt with per working day	30

Total cost of running :— £ *s. d.*

Per working day	1 12 1
Per week	8 0 5
Per cubic yard collected and tipped	12·8*d.*

On comparing the estimates with my Specification it will at once be noticed that no one has strictly complied with it, and it will, perhaps, help the Vestry to justly estimate the value of these tenders if I draw attention to their departures from my Specification in any essential points, the following paragraphs being numbered to correspond with the clauses of the Specification set out on page 2.

1. *Compliance with the Locomotives on Highways Act.*

Messrs. Coulthard's motor van is 2 tons 19 cwt. or just one hundredweight under the weight allowed by the Act.

The Lancashire Motor Company do not mention the weight of their vehicle.

Messrs. Bayley & Co. say all motor vans at present in use exceed the statutory weight, it being impossible to construct one, to carry a load of 4 tons, of a tare weight less than 3 tons. They add that the Local Government Board are alive to this fact and do not, therefore, enforce the strict letter of the law.

Messrs. Thornycroft state that their standard van (for which, apparently, they tender) does not of itself carry more than 3 tons, but would haul another 2 or 3 tons on a trailer.

There is no doubt that the motor van industry is sorely crippled by the 3-ton limit There is a consensus of opinion amongst engineers and manufacturers that this limit should not be insisted upon, and, I believe, Messrs. Bayley correctly interpret the attitude of the Local Government Board thereon. But, as regards the present tenders, Messrs. Coulthard & Co. undertake to comply with the 3-ton limit.

As regards conditions (2) *The types of vans*, and (3) *Speed*, there does not appear to be anything to call attention to, except that Messrs. Bayley, in their estimate of the cost of working, provide for a lad to assist the driver, but this is probably only for long-distance runs.

4. *The motive power and cost of working.*—All the firms tender for steam motors. Messrs. Coulthard prefer oil fuel, the other three prefer coke; but, as regards the guaranteed costs of working, no one gives any. They give estimates. Messrs. Coulthard, 2·88*d.* per nett ton-mile; the Lancashire Motor Company, 5·26*d.* per vehicle-mile. Messrs. Bayley say that the cost of fuel on their motor has worked out at one-twelfth of a penny per mile, and the cost complete at ·8*d.* per nett ton-mile, and Thornycroft's submit an estimate based on dust-removal figures. Probably the meaning of all these differences is that no motor vans have been long enough in work for the makers to give guarantees upon past results.

5. *The weight.*—I have dealt with this under No. 1. I should, however, add that Messrs. Coulthard's van, in full working order, weighs 3 tons 8 cwt.

6. *The body.*—All are satisfactory.

7. *The height of platform.*—All comply with this, except Messrs. Thornycroft, who say that the platform of their standard van slightly exceeds 3 ft. 4 in. above the ground level.

8. *Encasing machinery.*—All are satisfactory.

9. *Inspection.*—Messrs. Bayley do not include this in the list of conditions to which they tender, but this is possibly an oversight.

10. *Delivery.*—All agree to this.

11. *Week's trial.*—Messrs. Coulthard and the Lancashire Motor Company do not mention this, but they probably imply it. Messrs. Thornycroft agree to it, but Messrs. Bayley have again not included this clause as one to which they tender.

MOTOR VEHICLES AND MOTORS

12. *Payment.*—Messrs. Coulthard ask for 90 per cent. within one month of approval *at their works.* The Vestry, however, would require the delivery in Chelsea, and after delivery here I would approve them, if satisfactory. Neither the Lancashire Motor Company nor Messrs. Bayley mention any different terms to those of the Vestry, but Messrs. Thornycroft say one-third to be paid with the order, and two-thirds on delivery. Of course, the Vestry could not consent to this.

13. *Maintenance for two years.*—Messrs. Coulthard ask £75 per van per year, but as they undertake for this to have a man continually available for the period of two years it might be possible to come to some terms whereby the man might be employed by the Vestry as a driver. The Lancashire Motor Company's price, 12 per cent., would amount to £58 16s. 0d. per van per year. Both these firms are prepared to guarantee these prices, but Messrs. Bayley estimate this cost at about £104 a year per van, and Messrs. Thornycroft are not prepared to give a definite undertaking at present, but their manager has since informed me that they would maintain the vans free for six months.

The other conditions do not call for any comment.

Taking the specification as a whole, it is seen that Messrs. Coulthard & Co.'s tender shows the nearest compliance with the conditions under which the various firms were asked to submit estimates; but in considering any variation from the Vestry's conditions it should be borne in mind that there are two classes of variations: first, those which are caused by the van being constructed according to a maker's own pattern, towards which the Vestry might, if they think fit, allow a little latitude; secondly, variations from the specification as to payment and approval, which should not be permitted, any tender being only accepted on condition that the Vestry's conditions in these respects should be strictly complied with.

It might be asked why some of the firms do not adhere more closely to the Vestry's conditions. The reason, I think, is that some firms have special types of vehicle—Messrs. Thornycroft, in particular, quote for their standard type of van—and it is much more satisfactory to a large firm, and in the end to those who buy their vehicles, that they should make all their vehicles of standard sizes and so be enabled to make all similar parts in all machines from the same patterns.

In considering the price of a motor van it must be borne in mind that each vehicle does the work of at least 2 or 2½ horses and carts (possibly much more), so that the 3 vans advertised for would represent more than seven horses and vans and would not wear out the roads, nor make them require cleansing, as the seven horses do. When the matter is looked at from this point of view, and when it is remembered that the machinery of a motor can always be repaired, while horses are but mortal, in spite of the characters which horse dealers give them when they sell them, a large capital expenditure may be a wise outlay and in the end be the least costly policy for a local authority to adopt.

In conclusion, I should add that I believe any of the makers who have tendered are capable of supplying the Vestry with suitable motor vehicles. Messrs. Coulthard have made several motor vehicles which are in use in the North of England; the Lancashire Steam Motor Company have, amongst other vehicles, made a motor van for the Corporation of Liverpool, which is said to be working most satisfactorily: Messrs. Bayley & Co. are the well-known cart and van builders whose motor dust van gave very satisfactory results at the Richmond Show this summer; and the Steam Carriage and Waggon Co., of Chiswick (Thornycroft's), are the makers of the three motor vans in use at Chiswick, and also of the motor van which I found so satisfactory when it worked for a month in Chelsea."

Chapter XXXVII

THE 1896 ACT AND THE LOCAL GOVERNMENT BOARD REGULATIONS

IT is probably incomprehensible to any one but an Englishman that the first steps taken by a British Parliament towards what should be encouragement of an industry of a most important kind, such as the production of the means of cheap transport, is the imposition of restrictions and taxes not imposed on the older and more costly methods.

Some years ago, before the passing of the 1865 or 1878 Acts, the road The 1865 and 1878 Acts locomotive was permitted a speed of 10 miles an hour in the country and 5 miles an hour in towns. The 1865 Act gave town councils powers to make orders as to speeds which should not in any case exceed 2 miles an hour through towns and 4 miles an hour on the turnpike roads, and amongst other mediæval prohibitions it introduced the famous red flag 60 yards ahead of the " locomotive, on foot." The 1878 Act introduced a large number of technical points, such as dimensions of wheels, forms of their surface, and weight of locomotives; and the famous red flag was repealed, and the attendant, shorn of his badge of office, was allowed to precede the locomotive on foot by only 20 yards, instead of 60 yards.

This Act, however, instead of insisting that every locomotive " shall be constructed on the principle of consuming and so as to consume its own smoke " on pain of £5 penalty for every offence, graciously gave the locomotive owner the advantage of the introduction of the words " so far as practicable."

Until the passage of the 1896 Act it has, unfortunately for English The 1896 Act engineering, been the practice of Parliament to waste its time, and that of everybody else concerned, in discussing, and subsequently inserting into Acts, technical details which may very soon, and which have often become obsolete, restrictive, and even repressive. This policy was changed to some extent when the 1896 Act was under discussion, and these detail matters placed under the control of a technical department whose officers would be able to modify regulations in accordance with the necessities and the progress of invention.

It is also the fate of English people to have to submit to laws made

Advice of Select Committees of 1831 and 1873

by the biased, or prejudiced and uninformed part of the legislature, in total disregard of the reports and advice of Select Committees of men chosen for their knowledge and capacity for dealing with the subject under consideration, and weighing the evidence which is placed before them. Thus, in 1831, a Committee of intelligent men went thoroughly into all that could be said by the leading men of the day concerning steam on common roads. The report of this Committee urged great improvements in the laws relating to roads, and the running upon them of locomotive vehicles. The result, however, was as usual, where a question is finally decided by the mere vote of prejudiced and ill-informed men.

Again, in 1873, the Select Committee on Locomotives on Roads went thoroughly into the grievances as to danger to horses, damage to roads and bridges; as to the requirements of agriculturists; as to licences, and as to regulations. In their report, amongst other things, after having done that which satisfied certain heavy traction low-speed interests, they recommended "that self-contained locomotive carriages (or engines), not exceeding 6 tons in weight, making no sound from the blast, and consuming their own smoke, be classed as *light*, and that they be permitted to travel at the ordinary speed of vehicles, and only subject to the same restrictions as such vehicles." If this recommendation of the Select Committee of 1873 had been adopted by Parliament, as it should have been, the whole of the trouble, worry, and cost of the 1896 Act would have been avoided, English engineers would have been free to have pursued the construction, trial, and use of light locomotives, and general industries would for several years have been reaping the advantage of cheap high-road transport. A further advantage which would have accrued if this recommendation had been adopted, is the establishment of an engineering industry which is only beginning now, after our restrictive Acts of Parliament have given the lead to France.

Failure of rigid enactments

The lesson to be learned from the many years of experience of this class of legislation, is that Parliament should legislate only on broad lines, and leave all technical details to properly constituted departments for the drawing up and occasional revision of regulations which should be enforced according to the terms of the general act. If this had been done it would not now be necessary that Parliament should be again approached, as it must be, for the purpose of obtaining modifications in the Locomotives Act of 1896 The whole of the restrictions with regard to tare, weight per wheel, and width of wheels, as to reversing, and as to speed, should be removed; but it is desirable that a suitably constituted board should be empowered to make, and modify, from time to time, as may be found expedient through experience, regulations relating to all these things, but the

Necessity for discretionary powers

speed should be no more limited than it is with regard to horse vehicles, and punishment for excessive speed on the part of the user of the mechanical vehicle should be based on experience, not on opinion as to what is at present possible with due regard to public safety.

LOCAL GOVERNMENT BOARD REGULATIONS

As to weight of vehicle and weight per wheel, these are questions **Excessive weights and "extra-ordinary wear"** which should be settled with reference to the wear resistance powers of such road surfaces as it is expedient should be adopted and maintained at public expense for general traffic purposes. Excessive weights and consequent extraordinary wear should be judged and dealt with by present knowledge and under present laws for recovery against offenders. Heavy weights on narrow wheels will crush road materials like edge runners in a mortar-mill pan. Excessive wear of this kind can be detected and proved.

Notes on the Locomotives on Highways Act of 1896, and Regulations thereunder which affect Construction.

This Act, passed under the advice and consent of the Lords Spiritual **The Act of 1896 and regulations under it** and Temporal, and Commons, provides and encourages by restrictions as follows, among other things:—

Weight.—The weight of the locomotive shall not exceed 3 tons **Weight** without fuel or water. The weight of locomotive and trailer shall not exceed 4 tons without fuel or water. If electrical, the batteries count as fuel and water. Thus a 3-ton electrical vehicle may weigh 6 or 7 tons, and still be counted a light locomotive, though with its fuel on board, and with its load of passengers or merchandise, it may weigh 10 tons or more. This shows the absurdity of any attempt to deal with technical details in Parliament. Moveover, if accumulator vehicles were successful, it would give electrical vehicles an extremely unfair legal advantage over all others. It would enable heavy electrical vehicles to run with tyres so narrow, in comparison with the total weight carried, that no roads could withstand their destructive action.

A good deal has been said by those interested in the heavier classes **Tare** of vehicles in adverse criticism of the 3-ton tare limit. It is urged that there is a demand for vehicles which will carry from 5 to 8 tons, and that these cannot be made on a tare of 3 tons. Already, however, vehicles have been made which will carry 5 tons on 3-ton tare, and there is reason to believe that future improvements will enable them to do this successfully. It may be questioned, however, whether it is desirable, in the interests of those who have to make and pay for the roads these vehicles run upon, that anything more than 8 tons, or 2 tons per wheel (on the four wheels which are the only heavy vehicles dreamed of in these rules), should be carried. It may not do much harm in an occasional journey, but it would certainly be exceedingly destructive if a continuous service of such vehicles were kept up, as has been proposed between Liverpool and Manchester, unless very wide wheels were used. If heavy loads are to be carried on common roads, more than four wheels will have to be used, and then the tare may be increased. To wheel width, reference will be made hereafter.

Speed.—Maximum speed under the Act, 14 miles per hour, or less than **Speed** this, as the Local Government Board may decide. *Fourteen* miles an hour;

and our grandfathers ran with perfect safety in steam coaches at 20 to 30 miles an hour *sixty years ago* ! To-day 40 miles an hour is a common speed in France, and we are at the present tame and deceitfully acquiescent under the Local Government Board regulation which makes 12 miles the maximum, whether the vehicle be a 1-cwt. motor bicycle, a 2-cwt. tricycle, or a 29-cwt. carriage with a ton of people on board.

Clause 1 of Article IV. of the Regulations says: "He shall not drive the light locomotive at any speed greater than is reasonable and proper, having regard to the traffic on the highway, or so as to endanger the life or limb of any person, or to the common danger of passengers." This is all that is necessary. It is similar to the law under which horses and teams of horses are allowed to be driven. Penalty and recovery might also be the same.

Many miles have recently been run in this country with large pneumatic-tyred vehicles at speeds which are consistent with modern developments and well established practice, and only inconsistent with bucolic limits of vision. Mile after mile at 30 and 35 miles an hour has been run by vehicles which are as safe as a 6-mile coach, and more easily controlled than a perambulator. Only two things happen. One is that the motor carriage passengers reach their destination with a celerity which has due regard to the value of time. The other is that some of the excessive numbers of policemen, now wastefully employed, experience a conviction that the speed is dangerous because only runaway horses, that have not been detected and shot soon enough, have ever in their experience appeared to go as fast.

The speeds, then, at present made legal are :—

Weight of vehicle unladen	Under 1½ ton.	Over 1½ ton.	Over 2 tons.
Maximum speed in miles per hour.	12	8	5

Licences

LICENCES.—For some inscrutable reason the mechanical vehicle industry is further encouraged by the imposition of, as a tax, an *additional* duty of Excise as follows :—

	£	s.	d.
If the weight unladen exceeds 1 ton, but is under 2 tons	2	2	0
If the weight exceeds 2 tons	3	3	0

Motor vehicles of lesser weights pay the same tax as ordinary horse carriages under the Revenue Act, that is to say :—

	£	s.	d.
Hackney carriages, any sort	0	15	0
Four-wheel carriages drawn, or adapted to be *drawn*, by two or more horses, or by mechanical power	2	2	0
Four-wheel carriages drawn, or adapted to be drawn, by one horse only	1	1	0
Two or three-wheeled carriages	0	15	0

Extra tax to encourage motor industry

The Act has, it will be seen, imposed a duty or tax on motor carriages weighing over 1 ton of £2 2s., in addition to the £2 2s. or £1 1s. tax as an ordinary carriage adapted to be drawn by two or by one horse respectively. (Motor carriages, it may be remarked, are not drawn, or adapted to be drawn, by horses or by mechanical power.) Why the poor man owning a motor carriage should be called upon to pay twice the tax required from the rich man with a carriage and four remains to be explained— and reversed.

REVERSING GEAR.—Any motor vehicle exceeding 5 cwt. unladen, "must be capable of being worked so that it can travel either forwards or backwards." This should be increased to 8 cwt. at least. **Reversing gear**

WIDTH.—Extreme width not to exceed 6 ft. 6 in. Many horse vehicles reach 7 ft. Why not a motor vehicle?

TYRES.—Must all be smooth, and where they touch the ground (it does not matter about the other part) must be flat and of the following width :— **Tyres**

Weight of vehicle unladen.	Minimum width of tyre.
Exceeding 15 cwt. and not exceeding 1 ton . . .	2½ inches.
,, 1 ton ,, ,, 2 tons . . .	3 ,,
,, 2 tons	4 ,,

We are graciously permitted to use pneumatic tyres, or tyres of a soft material, and they may be round or curved (where they touch the ground as well as at the other part), and the width is the width when not subject to pressure. With some pneumatic tyres the width would be somewhat problematical, if not enigmatical.

This making of a regulation as to the width which makers of motor cars shall adopt for rubber tyres is one of the instances of what is nothing short of absurdity in connection with restrictions in engineering details, a useless interference, like several of the other rules. This particular rule is one to which makers will pay absolutely no attention. Rubber tyres will not hurt the roads anyhow. If they are too narrow they will wear too quickly and users will not have them, and they are not likely to be too wide. It is one of the many structural details which will rapidly settle themselves by experience and expediency, and that which is best for the user will be best for the community. One cannot help repeating an old question to our legislators and the departments under Parliament, and asking "Why can't you let things alone?" Just stand by for two or three years and "keep on saying nothing," and at the end of that time check any abuse, if abuse there be, and punish those who err by common law. One might also say "Let things develop, help industries on and upward, and do not everlastingly sit on things. Leave off checking industries by restrictive enactments, and then trying to make up for the depression that follows by voting money for some nice sounding hollow futility like Technical Education." **Useless official interference in technical matters** **"Why can't you let it alone?"**

BRAKES.—Two independent brakes are required, either of which shall be capable of holding two of the wheels "on the same axle." Why any two wheels should be stipulated does not appear, but to require ample **Brakes** brake power is a good thing. Half the brakes at present provided are of inferior construction and action.

Trailers must be fitted with brakes and carry a person to apply them if the speed is over 4 miles per hour.

LAMP.—The Act and the Regulations require a lamp on the extreme off side. It must show a sufficient white light ahead and a red light to the rear. This regulation should be made to apply to all vehicles on roads, more especially the slower vehicles, in country as well as town. It is hardly necessary with respect to motor vehicles, inasmuch as for safe working all vehicles carry two lamps, and many of them three. It does not apply to motor cycles.

Chapter XXXVIII

TYRES

CUSTOM in the cycle world has altered the spelling of the word, which used to signify the tire or band which encircles a wheel and withstands the wear and tear of running on the road surface, to "tyre." Why this should be is now a matter of no importance, for it seems this spelling must be accepted.

"Tire" or "tyre"

From the figures which have been given in previous chapters it will have been seen that the cost of tyres per mile run varies with different tyres and different vehicles, from somewhat below the cost of the fuel used to propel the vehicle, to about three times as much when pneumatic tyres are used.

Cost of tyres

The greatest mileage is obtained with good solid tyres on vehicles, such as the small Benz cars, which are not heavy and seldom run at more than what may be called moderate speeds. The mileage somewhat rapidly decreases with the heavier cars of the same or other types when driven at a high speed. The smallest mileage is obtained with the most expensive pneumatic tyres on the heavier high-speed vehicles. High speed alone does not cause rapid destruction, as is shown by the great mileage which has been obtained from some motor tricycle tyres. On the other hand great weight with even slow speeds is attended by rapid wear. This is, assuming good quality for any of the tyres under consideration, generally a result of insufficient size of tyre, and mainly of insufficient width, and consequently too great a pressure per inch of width. When under great pressure even rubber is easily cut or broken by rolling upon the succession of imperfect cutters which an ordinary road surface provides. Under a heavy load the tyre of a wheel compresses to a small extent, but the deformation and extruding action is so severe that the crudest cutters in the form of stone edges are sufficient to start the cut, which enables the outer part of the rubber at the sides near the rim edges to free itself of the excessive tension. These cuts once begun soon extend towards the centre, and then the heavy compression of the central surfaces carries the surface-cut into the body of the rubber.

Mileage of tyres

Effect of pressure and speed

Tyres should, then, for long life, be of greater width than is generally

Width of solids

thought sufficient; and they should be bevel-edged, not square-edged, and should have a slightly rounding transverse profile. They should not be of lesser thickness than is commonly used with the narrow section, and generally it must be remembered that the greatest wear only occurs with the combination of heavy load per unit width and high speed. In fact weight may be looked upon as the most important determining factor.

Estimates of mileage and cost

In the estimates given in Chapter XXXVI., sets of tyres of a common size have been assumed, and the wear upon them is based upon a careful weighing of experience of numerous users of vehicles of different kinds with various tyres. It must, however, be remarked that experience is sufficiently dissimilar to make exact estimates impossible. Experience is rapidly accumulating as to the wear of different kinds of tyres made with different quantities or proportions of rubber and different qualities, but as the speeds of travelling adopted by different users varies so much, the mere mileage recorded by any one is of little value. The experience is required from a number of those who may be called "similar users of similar tyres at similar speeds on similar roads."

The real knowledge of the probable life of a set of tyres is at present not much, but the life assumed in the estimates given fairly represents it such as it is, and those who constantly travel at high speeds will find that the life of a set of tyres is less than that assumed.

Pneumatic tyres—conditions of wear and tear

The gradual reduction of weight of cars will extend the life of pneumatic tyres until their life may approach that of tricycle tyres; but if their quality as obstruction absorbers is to be retained, they must ever be subject to a great range of deformation, which means internal work as well as external wear upon materials which have a limited work-resisting or punishment-bearing existence. To resist the deformation, heavier air pressures must be adopted, which means the introduction of more and more weight of materials of inferior elastic flexure, giving less and less of the air cushion effect. Tyres have grown enormously in size and in cost, and in the demand for very careful attention as to the pressure maintenance. Given the necessary strength, these large tyres will no doubt wear longer because of their larger surface and will be better than the smaller tyres as obstruction absorbers for carrying the same weight. Makers have, however, a problem of considerable difficulty before them if they are to respond to all the requirements of large pneumatic tyres for considerable weights. It is actually on the tread that the obstacle-absorbing or deforming capability is required. This cannot be fully provided, because some considerable thickness of wearing material is there required. Most of the free deformation must therefore take place elsewhere, and this relegates the bending to the thinner sides near the rim and concentrates it there. Only by adopting very high pressures and greater thickness of textile material can this be avoided and this means hard tyres. Except for those users to whom cost is of no importance, this process may go on until the choice between pneumatic and solid or "Compound" tyres is a narrow one. It will,

604

however, always be in favour of the pneumatic where the extra cost per mile run is not a first consideration.

It must also be remembered that greater comfort to the rider is due to lessened severity of vibration and shock, and this is a relief in which everything above the tyres participates. Now this means a reduction in the wear and tear of every part of the car and motor which can easily be underestimated. The experience of the London cab-owners, whose records of every cost are carefully kept, is a proof of this ; and they find that rubber-tyred wheels suffer very much less than the iron-tyred, every part that could be loosened or broken by constant severe dither or hard vibration remains tight very much longer, the breakages of lamp brackets, hangers and other parts does not occur, and that even the varnish, which being hard and breakable, lasts a great deal longer. The same immunity of the high speed car is obtained by pneumatics as compared with solids, and its value is greater in proportion to the greater value of the vehicle.

Reduced cost of car maintenance with pneumatic tyres

The numerous attempts that have been made to combine the qualities of a solid tyre and a pneumatic tyre have hitherto been so much waste time. The solid tyre element loses the small-obstruction-absorbing power of the pneumatic tread, and gives the effect of a very hard tyre, with all the rapidity of wear of pneumatic tyre sides, the whole of the deformation in which takes place near the seating in the rim.

For high-speed running with comfort over street crossings and level railway crossings, the expensive pneumatic is necessary, but it is a high price to pay for this luxury, and it will only be paid by the few who will pay anything for speed. After a while, when automobile travel settles down to the moderate speeds of the majority, and to the requirements of business, the better forms of solid or nearly solid tyre, in which a comparatively small amount of internal movement of the rubber takes place, will probably be most used. A hard pneumatic tyre is superior to this for ease at the bad places in roads and over crossings, but greater strength of material suitable for the purpose than is yet available is required to meet all the conditions.

One of the difficulties connected with tyres results from the almost necessary use of wheels of small diameter. A run along the Thames Embankment and other of the shamefully bad roads in London may be made with but small discomfort in a hansom with its 4 ft. 10 in. or 5 ft. wheels, the peripheries of which bridge all the smaller hollows. When, however, the wheel is reduced to about half this, it sinks from two to three times the depth into hollows of the size that are quite common, with the result that the blow given to the tyre and the axle are delivered at an angle which is at least twice that from the vertical which the hansom cab wheel and axle experience. The buffer qualities of the tyre need thus to be enormously greater for the small motor car wheel, or the motor must go at half the speed of the cab.

The use of small wheels

This may be well shown by reference to the diagram, Fig. 450, in

MOTOR VEHICLES AND MOTORS

Advantages of high wheels

which the peripheries of two wheels, one 60 in. and the other 30 in. diameter, are shown crossing a depression in the road which is 12 in. from edge to edge. It will be seen that the 60-in. wheel bridges this depression (which is a small one for the Thames Embankment and the edges of crossings) by sinking only half an inch. The 30-in. wheel, on the other hand,

Destructive impact on small wheels

sinks two and a half times this, or to a depth of 1¼ in. The blow given to the loaded tyre by this fall is thus enormously greater than with the 60-in. wheel, and correspondingly so is the push that has to be given to the tyre and axle to lift the car by the wheel, up and over the forward edge. The effect may be judged moreover by the difference between the angle made with the ground by the lines A D and B C connecting the centres of the wheels and the edge of the hollow. The one is more than double the other and the component along the line B C of the horizontal thrust is sufficiently obvious from the diagram to make numerical proof unnecessary.

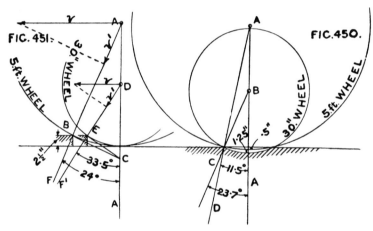

FIGS. 450–451.—DIAGRAMS SHOWING EFFECT OF USE OF SMALL WHEELS ON TYRES.

In Fig. 450, the 30-in. wheel has sunk only 1¼ in., or the obstruction in front of it after its fall is only 1¼ in. high. A sharp-edged obstruction 2½ in. high, is not, however, uncommon, and Fig. 451, will give some idea of what it means to a tyre, its rubber, its fabric, and all above it, when it hits such an obstruction at say 20 miles an hour. It also shows the advantage which the 60-in. wheel has in this case.

The angle of incidence D, E of the small wheel is 33·5°, while that of the larger wheel A, B is only 24°. The larger wheel has to lift the same height as the smaller wheel but rolls much farther in doing it, and therefore does it much more gently and more easily in proportion to the greater relation of A, B to B, C than of D, E to E, C. Further, it will be seen from the length of the horizontal line v, representing the velocity of movement of the large wheel and the resolved velocity of impact v′, that the latter is 0·41 of the velocity of horizontal translation, while for the small wheel it

606

is 0·58. The damage done to a tyre is, however, the kinetic value of the impact, which is proportional to the square of the velocity, and hence the damage done to the tyre of the small wheel will be greater than that to the larger wheel in proportion to $\dfrac{\sqrt{58}}{\sqrt{41}} = 2$.

In addition to the causes for the rapid destruction of pneumatic tyres already dwelt upon, there is the heavy stress due to tangential traction effort, aggravated by the reversal of this stress when the brakes are used and this latter stress is often most severe. The use of the Palmer fabric with all the threads at a considerable angle from a radius, or approaching the tangential, has, apart from its other great advantages, done much to meet these stresses in the best way. The size of the wheel has little practical effect on this stress. Tangential pull on driving tyres

The difficulties in the way of using large wheels with the modern light high-speed motor, however, are considerable. There are, however, no real difficulties in the making and maintenance of good roads, and this is what will have to be done in the future, to the great comfort and great economical advantage of everybody.

There is one comforting thought in this pneumatic tyre question, and that is that motor-vehicle travelling is very fascinating to many who are quite willing to pay the high price necessary to cover the cost of the development stage.

The pneumatic tyres most used in the larger vehicles are the Michelin, which are a French make of good tyre on the lines of the Clincher cycle tyre, but held in place not only by the ribbed edge in the side grooves of the rim, but by chaplet head E of screws F, with fly nuts G on the inner side of the rim D, as shown by Fig. 452. This Fig. shows a light tyre as used for tricycles, but a similar tyre, with thicker cover B, with several more layers of embedded fabric in the body of the cover and worked round on the inside and outside of the thickened edge C, is made by the Michelin Co., the North British, and other firms. The whole tyre is very much thickened and is made with or without a thick road band layer, as seen in Fig. 453. When worn, after say 2,500 or 3,000 miles, this band can, with the proper appliances at the works, be stripped off and renewed. As shown in Fig. 452, the cover is thickened where it bears against the rim, and the bending that must take place is distributed over a considerable distance from above the letter A to near the rim. This distance decreases with the increase of thickness of this or of any of the other tyres. Fig. 452, with the differences mentioned as to thickness and number of layers of fabric, may be taken as representing the Clipper as well as the Clincher and and Michelin types. A similar tyre is also made by the Dunlop Company Some of the tyres in use Michelin—clincher and clipper

Fig. 453 represents the Dunlop wire-edged tyre for heavy cars. It is held in place and got into place by the same means as in the Dunlop-Welch cycle tyre. The part of the tyre first got on is dropped or pushed the distance E to the bottom of the deep rim. The diametrically opposite part Dunlop of the tyre will then by some persuasion lift over the outer edge of the rim. The drop E is then divided, and becomes drop F when the tyre is in place, and the inextensibility of the wires c prevents the tyre from leaving the rim so long as pressure in the inner tube is maintained. Difference of opinion exists as to the effect of the greater depth of the part F held in the rim, as compared with the lesser depth of the enlarged edge c, but there does not appear to be much reason for this difference, and at present this tyre is not as favourably received as the others. The round edge of the cover at the wires leaves an undesirable corner into which the inner tube is forced under the high air pressure necessary.

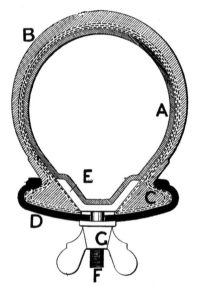

FIG. 452.—LIGHT CLINCHER CLIPPER OR MICHELIN TYRE.

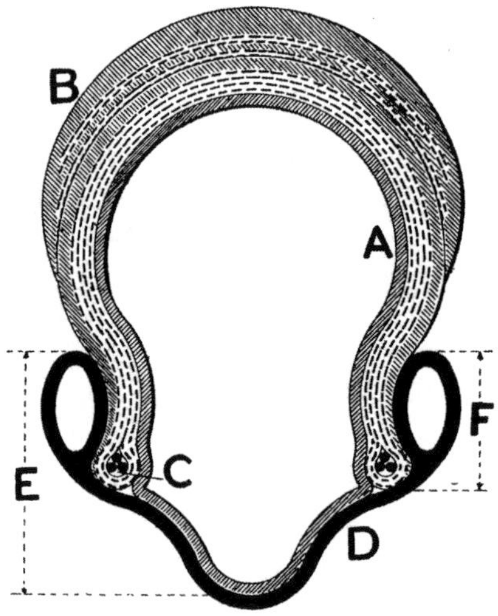

FIG. 453.—HEAVY DUNLOP TYRE.

Fig. 454 shows a representative section of the American single tube motor carriage tyres, such as those used on the "Locomobile" of the Stanley Company, still thicker tyres being used on the heavier Whitney cars now being introduced. With the great thickness of fabric and of rubber it would seem that punctures should be very improbable. In practice, however, this does not appear to be the case in this country, whatever it may be on the dirt roads in America. The great strength of a tyre made with a Palmer type of fabric enables it to receive the driving and brake stresses without harm, but cutting, splitting, and even breaking of the rubber will occur with treads under tension, as it did during the Automobile Club 1,000-mile trials. Like the Palmer cycle tyre, the Michelin tread is under compression when inflated, and at present it is the best type, though British makers may soon equal them.

It seems rather remarkable that, for motor vehicle purposes, a good detachable single tube, tubeless, or cover and tube tyre, or a solid tyre, has not been made on the lines of the Thomas patent of 1889.

With regard to solid tyres, the experience of the London hansom cabs is of much interest. A pair of $1\frac{5}{8}$ or $1\frac{3}{4}$ in. tyres will last from a little over six months to at most nine months. The most rapid wear is on those cabs which have the best and fastest horses, if we except those cabs that have constantly to run in districts where the road surfaces are destroyed by the prevalence of tramways, those expensive metallic admissions of the badness of the ordinary roads, and of the incompetence and penny-wise policy of most of the road authorities.

Suggested detachable tyres for vehicles over 5 cwt.
American single tube tyre
Wear of solid cab tyres

If 30 miles per day for the hansom driven by men who are, as most are, allowed two horses per day, and assuming 300 days per year, then a year's mileage would be 9,000. They run, however, not more than eight months at best before tyre renewal, so that the mileage is probably not more than about 5,500 to 6,000. This will make the cost about $0{\cdot}4d.$ per mile run.

The mileage of the tyres on the four-wheel cabs is much greater, as

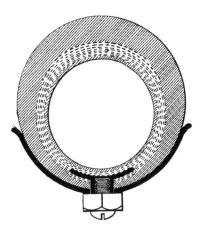

FIG. 454.—AMERICAN SINGLE-TUBE MOTOR CAR TYRE.

would be expected, from the smaller weight each wheel carries and the lower speed. The miles travelled per month will also be less.

What the tyres cost on the electrical cabs, weighing nearly 2 tons, it is perhaps needless to enquire.

The mileage of the privately owned and more carefully driven broughams is again very much greater, the life being as much as two, three, and even four years, the miles run per month varying very much.

The best solid tyre yet available for the heavier vehicles is the tyre which is vulcanised on the rim, although one left its wheel on the recent run to Bristol. Assuming the proper quality of rubber for the tread and the proper section for effective cushion and least wear of tyre this should be the best system of fixing; but it must be well done, and it is not always

Tyres vulcanised on rims

convenient to send the rims to the rubber works to have the tyres put on, and generally it is quite inconvenient.

Pneumatic tyre valves

The valves for pneumatic tyres are various, and are good and bad from both the convenience and holding-power points of view.

Sangster valve

Figs. 455 to 457 show three forms of valves, each employing a different arrangement for securing air-tightness. The form of valve which from

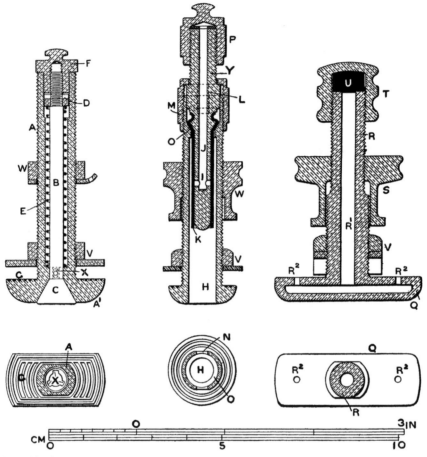

Fig. 455.—Sangster's Valve.   Fig. 456.—Wood's Valve.   Fig. 457.—Welch's Valve

structural and convenience points of view is best, is that shown by Fig 455, and known as Sangster's valve. In this c is a conical hard rubber valve, the seat for which is formed in the head A′ of the valve tube A, which passes through the inner tube of the tyre and the rim of the wheel. The joint between the valve and the inner tube is made by the roughened face G of the head A′ of the valve and the washer at v, the valve being held to the wheel rim by nut v and the washer inside, and the lock nut w and

610

TYRES

washer outside. The valve c has a long stem b, which passes through the valve tube, and is guided by the projections of the three-notch passage at x near the valve seat, and by the circular nut d at the other end. Between x and d is a light spring e, which tends always to keep the valve c upon its seat. The end of the valve tube is closed by a cover f, which screws on the top of the valve stem b, and pulls the valve tight on its seat. To inflate the tyre the cover f is removed and the pump nozzle screwed on. The air under pressure from the pump then passes through notches in the edge of the nut d, and through the valve tube and the passages at x to the valve c, which is thus lifted against the internal air pressure and the pressure of spring e. The tyre may be deflated by removing the cap f, and pressing down the valve stem b, so as to lift the valve off its seat.

Fig. 456 shows Wood's valve, which is similar to the ordinary Dunlop **Wood's valve** bicycle tyre valve. In it k is the rubber valve tube which expands when pressure is raised in the pipe j, and admits air to the tyre through the pipe h. The pump nozzle is screwed on to the end of tube y, which is closed by the cap p when the tyre has been inflated. To deflate the tyre the cap m is loosened. This allows the internal air pressure to lift the valve o on the tube y, and the air escapes through holes in the exterior valve tube.

Fig. 457 shows Welch's valve. In this the air-valve seat is formed by **Welch valve** the inner tube of the tyre itself, which by means of the nut and washer v is pressed against the face r^2 of the head q of the valve tube, and thus closes the holes at $r^2 r^2$. It will be understood that the wheel rim is gripped between the washer and nut v, which is locked by the cap s. To inflate the tyre, the cap t, with its rubber cover u, are removed, and the pump nozzle screwed on the end of the tube r. Air is then pumped through the tube r', and by lifting the inner tube at $r^2 r^2$ gains access to the interior of the tyre tube. To deflate the tyre the nut v and cap s are loosened and the cap t removed.

Chapter XXXIX

MISCELLANEOUS

Motor Vehicles for War Purposes

A GOOD deal has been of late said on this question, much of it in consequence of Col. Temple's successful use of traction engines and Major R. E. B. Crompton's departure to South Africa with more traction engines with searchlight and other tackle. My own experience with traction engines, steam ploughing engines, and later with motor vehicles, forces me to admit that much that has been said of the usefulness of the latter in South Africa has been urged in ignorance of the work and the circumstances under which it has to be done. Traction engines and their winding gear would undoubtedly be very valuable, and special forms of lighter tractions could be made which would have the necessary robust health that would suit them for the work in a new country if in the hands of well-trained men. In the same new country most of the motor vehicles would be of very little use, though in the future steam and heavy oil engine tractors with suitable waggons will be forthcoming that will meet undoubted requirements. They have, however, yet to be designed, and traction engine knowledge and experience will be most required.

In old countries, well provided with good roads, the same remarks apply, except that the motor tractor waggon will be more widely useful, and the fast motor carriage will be able to perform invaluable services. With these remarks, and such exceptions as they involve, I would refer those interested in the general question to a very suggestive paper recently read before the Automobile Club by the Right Hon. J. H. A. Macdonald, C.B., F.R.S.

The Best Motor Vehicle.

There is no one kind of motor vehicle which is the best, any more than there is a best locomotive, or best boot, or best plough. There are many which may be grouped as the worst. All the others are best according to the requirements, country, and user. These variants make it undesirable to generalise either with regard to the carriage or light vehicle classes, or the heavy vehicles.

612

MISCELLANEOUS

It may be useful, however, to remind some of the most enthusiastic that there is a lot of work which will always have to be done by horses; and others that the high roads and motor vehicles are not the proper places, or the means of transport for heavy continuous loads, such as, say, 10,000 tons of goods a week between Liverpool and Manchester. When it comes to this a special tramway or railway is wanted.

Motor Vehicle Development—a Road Question.

It has been amply shown in these pages that great ingenuity and skill have been devoted to the increase of the power that can be got into a vehicle of given weight. The greater part of all this power is required for hill climbing, nearly all of which is absolute waste. Gradients much greater on common roads than are possible on ordinary railways are permissible, but we in this country, like those in others, make our roads follow, with few exceptions, the rise and fall of the land as nature left it, just as slavishly as though we were so many ants. We thus tax ourselves with double and treble the work and power requirements that would otherwise be necessary. We make good roads in places, and having spent large sums of money upon them, allow our councils and vestries to dictate to the road surveyors and engineers a policy of the most indefensible and wasteful disregard of all effective maintenance. A road is made and then abandoned to all and everything that may happen. No organised attention to defects as they show themselves, and repairs as they are required, is made. All the stitches are allowed to go until the whole nine are required—the whole road to be made anew. Before this is done, the roadway, like the Thames Embankment, reaches a condition which is ruinous to everything that passes over it. The road costs double per decade what it would if proper maintenance and repair were organised and carried out. The case is even worse on many of our expensive wood-paved streets, where a defective block here and there is allowed to remain until they become the centres of areas of destruction. The road, after this has gone on some time, is remade, and half at least of what might have been good blocks for years are thrown away.

In another place I have shown, on the basis of the saving in horse keep and horse haulage alone, that about £125,000,000 might be expended, with a return to the country at 3 per cent., by the improvement and proper maintenance of the roads all over the country, including the revision of many of the hill and hollow roads and the diversion of the worst.[1]

The importance of this equally applies with motor vehicle transport. This may be gathered from the tables herein given of road resistances, and of the power required on good and bad roads and on stiff hills as compared with moderate hills and levels. It will, however, probably take this country, businesslike as it is in most things, a long time to apply these

---

[1] Address as President of the Society of Engineers, February, 1898.

613

facts, although everybody sees daily a locomotive hauling hundreds instead of the tens it could haul if its rails followed even the most moderate hills hardly observed by motor vehicles.

If for nothing else but a reduction in the annual cost, every one ought to turn their attention to the disgracefully wasteful system adopted by our road authorities. The vast saving, however, which would result in radical road improvement, should equally attract and receive the attention of motor car owners and engineers. The whole country would benefit by the greatly reduced cost of road transport and goods delivery which means cheapened goods. Mention should be made here of the Viagraph, an instrument, designed by Mr. John Brown, of Dunmurray, Belfast, which should be at the service of every county and borough road surveyor.

THE AUTOMOBILE CLUB.

The Automobile Club of Great Britain and Ireland (with which the previously existing Self-Propelled Traffic Association has been incorporated) is the centre of motor vehicle representation and development. This club, largely due to the judiciously employed initiative and energy of its secretary, Mr. C. Johnson, has encouraged and has carried out numerous trials of motor vehicles under practical conditions, the last effort in this direction being the 1,000 miles endurance trial from London to Edinburgh by a western route, and Edinburgh to London by an eastern route, which will have been completed before this page makes its appearance. This club is doing work of inestimable value to the future industry of motor vehicle manufacture and transport in this country, and work which will hasten the dawn of the day when horses will no longer be made to endure the sufferings that mark the three to five years that it now takes to "knock them out" in the hack work of London and other big towns in England and abroad. The club has at present over six hundred members, is independent of any interest, and its premises are at 4, Whitehall Court, S.W.

OTHER IMPORTANT MATTERS.

These are many, but this book must go to press some time or other, because it has been promised these last two years, so these other matters must be held over. Some readers may have expected a separate chapter on the "whole art of driving a motor vehicle." This, however, has been avoided because a careful perusal of the descriptions of and remarks upon the forms, purposes, construction, and working of the motors and their mechanism and other parts, makes everything else a matter of personal handling. A man cannot be taught skilfulness or care in driving by a book, any more than he could to ride a bicycle, or drive a pig or a plough.

Appendix

RESULTS OF THE 1,000-MILE TOUR OF THE AUTOMOBILE CLUB OF GREAT BRITAIN AND IRELAND, APRIL–MAY, 1900

THE severest tests to which motor vehicles have been put in this country were those of the 1,000 miles, including the numerous and long hill tests. The hills climbed were Taddington Hill, between Matlock Bath and Buxton; the Shap Fell, between Kendal and Borrow Bridge; the Dunmail Raise, between Rydal and Wythburn; and the Birkhill, between Capplegill and St. Mary's, Lough. Of some of the salient results I am now enabled to include a short account and examination in this appendix.

Of 83 vehicles entered for the trial, 65 started from Hyde Park Corner, but several of these could hardly have been expected to get through the tour and trials; and one at least has shown that even the best organization may be perverted to misleading business uses. Of this number, 23 are recognised officially as having gone through the 1,000 miles with complete records, and climbed all the hills except the Shap Fell, which was optional. Twenty-six others, with more or less incomplete records, arrived, and were exhibited at the Crystal Palace Exhibition, which opened on the 14th May, making a total of 49. Of this number also 14 entered and ran through the speed trials in Welbeck Park, the results of which will be mentioned hereafter. **Number of vehicles entered and run**

The hill-climbing trials form an important part of the whole of what may be looked upon as a durability test, and it is by means of the figures obtained on these hill trials that we may best derive some indications as to the power and mechanical efficiency of the cars generally. Hill-climbing trials alone would not of course be sufficient as a test of the wearing power or durability of a car, and for this reason the long distance under trial conditions was necessary as an endurance test. There can be no doubt that several of the cars which dropped out of the trial in consequence of breakages or excessive wear of some parts would have got through the 1,000 miles if running under ordinary users' circumstances, and not under the spur of trial conditions and time tables, and under the **The hill-climbing trials** **Trials more severe than ordinary working**

615

APPENDIX

incentive of competition to run at speeds for which some vehicles were not suited. As to general quality, the accompanying Table XXVI., which gives an analysis of the hill-climbing trials by 53 vehicles, will afford a good indication. Generally it may be taken that a car which performs well on a succession of stiff hills may be expected to do well in ordinary and long-distance running; but this is not necessarily always the case, as will be shown by reference to the table, which contains the names and particulars of 53 cars, of which 4 failed to complete the whole trial.

Classification of vehicles
The vehicles were divided into two sections, namely, of vehicles entered by manufacturers, and of vehicles privately owned and driven by the owners or by their substitutes or servants. These sections were divided into classes as follows :—

(a) Sections 1 and 2 divided into four classes, viz.—
    Class A. Cars declared at a selling price of £200 or less.
    Class B. Cars declared at a selling price of more than £200, but not more than £300.
    Class C. Cars declared at a selling price of more than £300, and not more than £500.
    Class D. Cars declared at a selling price of more than £500.
    Class E. Motor Cycles.
        (a) Tricycles.
        (b) Tricycles or Quadricycles for two persons or more.
    Class F. Vehicles for public service.

There was a further section for parts of cars, but under this section only one entry was made, namely, of tyres. These several classifications and sub-classifications will be found in the first column of the table, the whole of the cars in the privately owned or amateur class being given in the latter part of the table, with the letter A prefixed to all the numbers. Amongst those entered and started in section 1 by the manufacturers, the names of several do not appear in the table. These include,—

Vehicles not in the table
Section 1, class A, No. 18, the "Endurance" car ran to Bristol only; No. 19, "Orient Express," retired, after several mishaps, between Kendal and Carlisle; Nos. 29, 30, "Eureka" cars, retired between Carlisle and Edinburgh after various mishaps; No. 42, Voiturette, Briére system, retired first day after breaking flywheel shaft. Class B, No. 11, Motor Manufacturing Co., "Princess" car, withdrawn soon after start; No. 45, S.S. Motor Co.'s carriage, retired in Derbyshire. Class C, No. 43, London Motor Van and Waggon Co.'s Daimler Phaeton, withdrawn after starting. Class D, No 21, Lanchester carriage, breakage of cast-iron bearer and other mishaps, retired at Manchester. On the other hand, the table includes the names, numbers, and some figures relating to vehicles which did not complete the 1,000-mile trial, although they were in the hill-climbing trials, as far as shown, some of which were optional, namely, the two Shap Fell climbs. These cars were, taking them in the order in which they appear in the table :—

616

APPENDIX

Sec. 1, Class A, No. 52 ; Class B, No. 24 ; Class D, No. 22 ; Class E (*a*), Nos. 12, 20. Sec. 2, Class C, No. A23 ; Class E (*a*), No. A16. (No. 12, Class E (*a*), was exhibited at the Palace in error, as its record excluded it.)

Vehicles in table which did not complete

In Sec. 1, Class F, No. 38, the Daimler "Public Service" vehicle, completed the tour after various mishaps and repairs. Sec. 2, Class D, A19, 12 HP. Daimler, Mr. Hargreaves, ran unofficially ; another 12 HP Daimler belonging to Mr. Hargreaves was entered, and ran a great part of the tour unofficially.

One of the most important of all the deductions to be made from the hill-climbing trials is the relation which the power exerted at the peripheries of the road wheels bears to the power given off at the end of the crank shaft of the motor—in other words, the mechanical efficiency of the whole of the transmission gear. As a means of arriving at this we have the gradient and total rise of five hills on which the speed of ascent of the vehicles of known total weight was taken. We have also the B.HP. of the engine as stated by the makers, as confirmed in some cases from other sources of information, or as corrected thereby. We have also, as a check upon the power of the motors, their dimensions and revolutions from which the power has in some cases been calculated as was done for Table XVI., p. 387.

Mechanical efficiency of the cars

The mechanical efficiency of 53 of the 54 vehicles in Table XXVI. is based on these figures and observations ; but some examination and criticism is necessary because it is known that the power exerted by the motor was in some cases more, and in some less, than that given in the entries.

Power of the motors

Dealing with the results found, in the order in which they are placed in the table, we first come to the Benz Ideal cars with engines nominally of 3 HP. at 400 revolutions per minute. This, on the hill climbing, gives to the whole car, No. 1, an efficiency of only 51 per cent. It is quite clear that this motor, at the revolutions per minute which it would make on the hill, does not give 3 HP. ; and this is shown by reference to Table XVI., which shows that the stated B.HP. is greater per 100 revolutions and per 100 cub. in. capacity than the mean of all the motors by 23 per cent., or as shown by Table XVI., the power of the motor at that speed is more nearly 2·3 HP. than 3 HP. The No. 2 Benz, fitted with a motor of the same size, but with a single belt and gear transmission, gave an efficiency of 61·5 per cent. In this car the single belt runs on pulleys of better ratio with an open, instead of a crossed belt, and the gear runs in oil. If the motor be taken as 2·3 B.HP., based upon its dimensions and speed, then the mechanical efficiency reaches about 80 per cent. So high an efficiency does not seem probable, so that we are bound to conclude that the motor with the lowest speed in gear was running considerably above the normal speed.

Analysis of power and efficiency figures

The next car on the list is the "Stanley" steam carriage, which has

617

APPENDIX

an apparent mechanical efficiency of 61 per cent. The 2 HP. stated by the makers is of course merely nominal, inasmuch as with the steam pressure available and the variation of speed of the motor with the speed of the car, the power of the motor may be anything between 1 HP. and, for short periods, about 4 HP. At an average speed of 9·5 miles per hour the motor itself, with a mechanical efficiency of 85 per cent. and allowance for fall in pressure between boiler and steam chest, would be about 3·1 B.HP., at which the efficiency of the car is that given in the table. Considering, however, the simplicity of the transmission, it would seem that the car was not working at its best; and there is reason to believe that part of the time at least the boiler was priming, through the introduction into the water tank of something other than water.

With reference to No. 16, the "Gladiator" voiturette, which generally ran exceedingly well, the efficiency appears to be only 32·6 per cent. This is a case in which the makers do themselves harm by overstating the power of their motor. This car, it may be mentioned, should be fitted with better brakes before it is sold for use in this country.

Nos. 27, 28, "New Orleans" cars, made by Messrs. Burford & Van Toll, Twickenham. These are driven by air-cooled motors, and generally ran extremely well. The power of the motor is given as 3 HP., but it is obvious that it does not give this continuously on a long heavy pull, as the mechanical efficiency of the car, which has very simple transmission gear, works out as only 35·7 per cent. For short periods, as in average travelling, while the motor remains cool it may occasionally approach 3 HP., but it obviously was not doing it on the hill trials. There are several meritorious points in this car as well as simplicity, and a means of varying the belt tension, and one of interest is the belt-driving pulley on the motor shaft, which is made of alternate ribs of cast iron and soft metal, and seems to give an excellent belt hold, which reduces the otherwise necessary tightness of the belt. Nos. 33 and 34, " Decauville " cars, have motors the power of which is obviously overstated.

No. 52, Roots & Venables' petroleum motor car, ultimately ran the whole 1,000 miles, but not officially. It broke a pivot axle at Bolton, when their man, instead of repairing it at once, put it on the railway without instructions, and thus lost four days. It was afterwards returned to Bolton, and ran the whole distance driven by inexperienced men. The speed was taken on the Taddington hill, and the efficiency of the car works out at 50 per cent. It is obvious that the driver did not obtain full duty out of the motor. For general arrangement of the car, see p. 333.

Nos. 31, 32, Renault cars, were entered by the Motor Car Co. as the "M.C.C. Triumph" cars, and fitted with 3·5 HP. motors. As, however, these motors are the same as those in Nos. 14, 15, the "de Dion and Bouton" voiturettes entered as 3 HP., that power has been taken in calculating the mechanical efficiency. These cars ran very well indeed—one of

618

APPENDIX

them especially—and often at very high speed. The general arrangement of the transmission is shown on page 237.

No. 40, the " Wolseley " voiturette, designed by Mr. H. Austin, is a new car, comprising several very meritorious features. It ran exceedingly well throughout the trial, and did well on the hills. The motor is entered as of 3 B.HP., but has in the works given 3·3 HP. on the brake when new, and was probably giving at normal speed considerably more than this after it had been running two or three days. The normal speed is from 700–750 revolutions per minute, with a cylinder 4·5 in. diameter and 5-in. stroke. It was, however, run without a governor, and if allowed to make, say, 900 revolutions per minute, the power, instead of being 3 HP., as in the official entry, would be 3·6 HP. This alone would materially reduce the very high apparent mechanical efficiency based on the official entries. To begin with, the total weight on the greater part of the journey was 16 cwt. 1 qr., and on the Taddington hill, when three passengers were carried, it was more than this. This gives the mean effective HP. at the road wheels as 2·83. The motor itself was probably giving 3·5 HP. at normal speed ; and if the low gear was in use and the engine allowed to run at 800 revolutions per minute, the power would be 3·75 B.HP., and the mechanical efficiency of the whole car would then reach 75 per cent., which is extremely good.

No. 22, the " Lanchester " carriage. Here the mechanical efficiency works out to only 31·9 per cent., which is again in agreement with the efficiency derived from the Petersham hill trials, as given in Table XV., page 384. There can be no doubt that, firstly, the motor which is air-cooled is less than 8 HP. on a long heavy pull ; that a good deal of power is absorbed by the fan ; and that the efficiency of the transmission is not what was hoped from it.

There are two other vehicles to which reference must be made, namely, No. A10, the 8 HP. Napier, driven by Mr. S. F. Edge ; and No. A17, the 12 HP. Panhard, of the Hon. C. S. Rolls, by whom the car was driven. Both these cars have an apparent efficiency of 82 per cent. This is based on the total weights given in the table and on the B.HP. entered by the owners. There cannot, however, be any doubt that both these motors were practically run without any governor, and that their normal speed was very considerably exceeded. It will certainly be no exaggeration if a speed of 950 revolutions per minute be assumed as that which obtained on these hill trials ; and it is possible that it was more than this if the gear which could be used was a low one. Even allowing, however, for Mr. Rolls' remarkable skill in driving, and the perfection with which he manipulates his speeds and mixture adjustments, and, above all, if full credit be given him for the care with which he tends every point of lubrication, it cannot be for one moment supposed that this skill and attention raises the mechanical efficiency of the transmission gear of his Panhard car to anything like 82 per cent. Nor can it be admitted that this has been reached with the 8 HP. Napier Daimler motor car. Indeed, the considerations of the speed trials confirm the supposition that the actual power of the motor was largely increased by increase in the number of revolutions.

Speed trials

The measured mile on which these trials were run was by no means level, the gradient reaching in one place 1 in 26·1, and the total rise being 70·37 ft., or 1 in 72·5. The surface of the road was excellent, and there was no noticeable wind. Fourteen cars took part in the trials with the results given in the following table, which is reproduced from the *Autocar*.[1]

Analysis of speed trial figures

THE SPEED TRIALS.

TIME AND SPEED RECORDS.

	No. of car.	Description.	Time uphill.	Time down-hill.	Mean time.	Average per hour of mean time.	Average per hour of down-hill time.
			M. S.	M. S.	M. S.		
1	A17	Hon. C. S. Rolls's 12 HP. Panhard . .	1 $46\frac{4}{5}$	1 $24\frac{3}{5}$	1 $35\frac{3}{5}$	37·63	42·55
2	A 4	Mr. Mark Mayhew's 8 HP. Panhard .	2 $17\frac{3}{5}$	1 $45\frac{4}{5}$	2 $1\frac{3}{5}$	29·60	34·02
	A10	Mr. Kennard's 8 HP. Napier	2 16	1 $47\frac{2}{5}$	2 $1\frac{3}{5}$	29·60	33·51
3	4	Ariel tricycle with Whippet trailer .	[2]2 $11\frac{1}{5}$	1 $53\frac{1}{5}$	2 $2\frac{1}{5}$	29·45	31·80
4	A22	Mr. J. A. Holder's 12 HP. Daimler . .	2 34	2 $0\frac{2}{5}$	2 $17\frac{1}{5}$	26·23	29·90
5	A11	Hon. J. Scott Montagu's 12 HP. Daimler	2 $38\frac{1}{5}$	1 $57\frac{4}{5}$	2 18	26·08	30·56
6	39	[3]Century tandem tricycle.	2 $34\frac{3}{5}$	2 $24\frac{2}{5}$	2 $29\frac{2}{5}$	24·09	24·93
7	16	$3\frac{1}{4}$ HP. Gladiator voiturette	2 $48\frac{3}{5}$	2 $22\frac{1}{5}$	2 $35\frac{2}{5}$	23·16	25·31
8	A31	Mr. Johnson's 6 HP. Parisien Daimler	3 $1\frac{3}{5}$	2 $10\frac{4}{5}$	2 36	23·07	27·52
9	40	3 HP. Wolseley voiturette	3 $1\frac{1}{5}$	2 $14\frac{3}{5}$	2 $37\frac{4}{5}$	22·81	26·74
10	14	3 HP. de Dion voiturette	2 $54\frac{1}{5}$	2 $33\frac{4}{5}$	2 44	21·70	26·00
11	A 3	Mr. T. B. Brown's 6 HP. Panhard . .	3 26	2 $12\frac{2}{5}$	2 $49\frac{1}{5}$	21·27	27·19
	A 2	Mr. Butler's 6 HP. Panhard	3 $10\frac{1}{5}$	[4]2 $28\frac{1}{5}$	2 $49\frac{1}{5}$	21·27	24·13
12	15	3 HP. de Dion voiturette	3 $22\frac{4}{5}$	2 $47\frac{1}{5}$	3 5	19·45	21·53
—	—	[5]Mr. Lord's 7 HP. Peugeot	2 45	1 $58\frac{1}{5}$	2 $21\frac{3}{5}$	26·97	30·45

All the above trials, with the exception of those noted, were made with driver only on board.

From this table it will be seen that four vehicles ran at considerably over 31 miles per hour on the down grade, and nearly 30 miles per hour mean speed up and down grade; while Mr. Rolls' car reached 42·55 miles per hour down hill, and 37·63 as the mean of the up and down grade.

Now the actual work done in propelling this car at the mean speed of 37·63 miles per hour is only 4·46 HP.; while the air resistance at that speed equals 8·06 HP., making a total of 12·52 HP. In estimating the air resistance, the expressions given on page 58 have been used, and head resistance only considered, the head wind area being taken as 15 sq. ft., which is as near the area as can be ascertained. Assuming an efficiency of 75 per cent. for the transmission gear, the actual power of the motor was 16·7 HP. This being the case, the motor speed must have been increased from the normal of 750 revs. per minute to at least 1,040 per minute. On the hill trials, therefore, Mr. Rolls may have been using anything between 12 HP. and 16·7 HP., depending partly upon the particular pair of speed

[1] May 19th, 1900, p. 476. [2] Both riders pedalling. [3] Two riders.
[4] With brake slightly on. [5] Not officially competing.

gear wheels he was able to use. Thus, assuming that he was able to use a low gear and a high speed of motor, and thus get, say, 15 HP. out of the latter, the true mechanical efficiency of the transmission gear would be 65·5. It is thus obvious that, unless the hill-climbing trials were conducted with a known speed of motor, the true mechanical efficiency cannot be obtained, and that given in the last column of Table XXVI. must only be taken as comparative. The same remarks hold good with regard to several of the cars which appear to show very high mechanical efficiency of transmission ; while on the other hand some of the cars, such as Mr. Butler's 6 HP. Panhard, which appear to have a low mechanical efficiency in this respect, show that the motors were not in a condition to give the high speed and power obtainable from them if in thoroughly good condition and in the hands of exceptionally skilful drivers. The table does, however, show by the last column what may be looked upon as the capacity of the car under the conditions of its trial; and thus, although it is not accurately the mechanical efficiency of the transmission which is presented, it does give an index of the mechanical sufficiency of the car as a whole.

It will be noticed that the 8 HP. Napier car of Mr. Kennard, which is fitted with a Daimler motor with slight modifications, and was driven by Mr. S. F. Edge, had an apparent mechanical efficiency which is exactly the same as that of Mr. Rolls' car, namely, 82 per cent ; and the speed trial gives an efficiency of 87·8 per cent. if the motor be taken as only of the power stated. Of the total power accurately used, 3·58 HP. represents the net work of propelling the car at the speed observed, and 3·43 HP. the work done against air resistance, or a total of 7·01 HP., the head area being in this case taken as 14 sq. ft., instead of 15 as in the case of Mr. Rolls' car. The motor was probably running at about 1,000 revolutions per minute and giving about 10 HP., the real of transmission being correspondingly lower.

Considering the general excellence of the machinery of Mr. Kennard's car and its performance on the hills, it might have been expected that it would have given a higher speed on the Welbeck Park road ; but as the motor was driven throughout the tour without a governor, it may be said to have been " accelerated " during the whole of the running, and there was little to call upon when the higher speed of rotation was required.

The same remarks obtain with regard to the Wolseley car, which was run during a great part of the tour on the throttle, and not with a governor. On the speed trial the tractive work was 1·75 HP. and the air resistance work 1·18 HP., with a head area of 10 sq. ft., giving a total of 2·95 HP. Thus, assuming a mechanical efficiency of 0·75, the motor was obviously capable of giving 3·92 HP.; and this would give for the last column of Table XXVI. a true mechanical efficiency of 68·5 per cent. This is high, and is about what might be expected from the car, the motor of which drives by one belt to spur gearing in a box pivoted upon a shaft, by the movement of which the tightness of the belt is adjustable to any requirement, and by which either of the pairs of spur wheels can be put into gear, the belt being automatically freed at the instant the change of speed is made. It is unnecessary to follow out this examination to more of the cars; but the generally excellent behaviour of some of the 6 HP. Daimler Parisien cars makes it desirable to refer to them. Taking one of them, Mr. Johnson's car, which made a mean speed of 23·07 miles per hour, and a speed of 27·52 miles per hour on the down grade, it will be noticed that the difference in the uphill and downhill speeds was greater than with the previously mentioned cars. By reference to Table XXVI. it will be seen that this car, having a 6 HP. motor, weighed with passengers 24·5 cwt.; while the Panhard, with its 12 HP. motor, weighed 26·25 cwt. Hence the weight of the car with the 6 HP. motor tells heavily against it on the uphill, and gives an apparent disproportion between the uphill and downhill speeds, which would not of course be found with the 12 HP. car. This is apparently not in agreement with the figures obtained with the 12 HP. English Daimler cars; but in one at least of these cars, namely Mr. Holder's, the weight is excessive and the motor was only giving on the speed trial 8·4 HP., assuming 75 per cent. efficiency of transmission. On the other hand, with Mr. Johnson's 6 HP. car, the tractive work was 2·54 HP., air resistance work 1·68 HP., or a total of 4·22 HP. With similar transmission efficiency, namely 0·75, Mr. Holder's motor should have given 11·2 HP., Mr. Johnson's motor giving on the same basis 5·62 HP.

Among the smaller comfortable two-seat cars was Mr. Phillips' "Mors Petit Duc," supplied by the Automobile Association. This was quite new from the works when it started on the tour, and Mr. Phillips but little experienced in driving it. It was run on its governor always, and its performance was very satisfactory throughout, the only stoppage of any note from the time of leaving London to the time of reaching the Crystal Palace being due to an electrical defect which could have been remedied in one minute with more exact knowledge of the arrangement of the circuits. As in many other cars, chains gave some trouble, and there are details which may easily be very considerably improved.

Air resistance has not been taken into consideration in the calculations in Table XXVI., firstly, because the speeds in most cases were not sufficient to make it important; and, secondly, because on some of the smaller vehicles the riders left the vehicles part of the time. With two of the vehicles, however, the air resistance was important, namely, with the 8 HP. Napier Daimler on Shap Fell, and the 12 HP. Panhard of Mr. Rolls on the same hill. In these cases the total power given out at the road wheels was, respectively, 6·61 HP. and 11·78 HP. Taking a speed of 950 revolutions of the motor of each vehicle, the power given off by the motors would be, with the Napier Daimler car 9·06 HP., and with the Panhard car 15·2 HP. The mechanical efficiency in the two cases would then be, respectively, 73 per cent. and 77·5 per cent.

On the whole, although a number of the cars which went through the

622

tour and trials had to be repaired or to receive a good deal of careful attention either on the road or at the stopping-places, they have shown that there are now to be had vehicles of several types capable of enduring a great deal of very heavy work under abnormal conditions. The trials on the whole may be taken as gratifying to the Automobile Club and to the motor vehicle industry, and as of great public value. They taught many lessons which will be fully utilised by those manufacturers who are susceptible of instruction and capable of putting it into practice.

Index

INDEX

INDEX

INDEX

INDEX

INDEX

INDEX

INDEX TO TABLES

INDEX TO TABLES.

Butler & Tanner, The Selwood Printing Works, Frome, and London.

TABLE XX.

Results of Organised Trials of Heavy, Steam, Oil and Electrical Vehicles during the Years 1897, 1898 and 1899.

1	2	3	4	5	6	7	8	9	10	11	12	13	14	15	16	17	18	19	20	21	22	23	24	25
Name of Maker and type of vehicle.	Trials.	Total weight of vehicle and load.	Mean tare of vehicle with fuel and water, but not attendants.	Load carried.	Seating capacity of the passenger vehicles.	Ratio of mean tare to load.	Mean tare per B.H.P. declared.	Total moving weight per declared B.H.P.	Declared boiler pressure per sq. in.	Total heating surface of boiler.	Fuel.	Lb. of water evaporated per lb. of fuel, but not including lighting up.	Diameter of cylinders of engine.	Stroke.	Speed of engine, revolutions per minute.	Declared B.H.P.	Distance travelled on trial.	Average speed per hour.	Fuel used per loaded vehicle mile, including lighting-up.	Water used per loaded vehicle mile.	Cost of fuel and water per loaded vehicle mile.	Cost of fuel and water per loaded gross ton mile.	Estimated cost of running the loaded vehicle per mile.	Name.
		tons cwt.	tons cwt.	tons cwt.			cwt.	cwt.	lbs.	sq. ft.			inches.	inches.			miles.	miles.	lbs.	gallons.	d.	d.	d.	
Scotte Steam Omnibus	Poids Lourds, 1897	8 19	5 19	1 3/6	12	3.67	8.7	9.2	170		Coke	5.5	4.35	4.55	400	13.7	192	6.7	13.5	7.3	2.76	0.465	a 7.3	Scotte.
Scotte Road Train (for Passengers)	"	8 19	6 6	2 3/2	26	2.56	8.62	11.4			"	4.7	4.55	4.75	400	15.7	194.2	6.38	17.1	7.95	4.38	0.383	10.0	Scotte.
Scotte Road Train	"	11 11	6 6	5 3/2		1.63	8.6	14.7			"	5.5			400		192	1.19	21.2	11.4	4.38	0.378	13.9	Scotte.
De Dion & Bouton Steam Omnibus	"	6 11	4 10.8	1 2/2	16	4.13	3.64	4.84	200		De Dion & Bouton	6.2	3.96 7.5	6 7/8	400	24.5	192	8.96	6.15	4.9	1.37	0.226	5.5	De Dion & Bouton.
De Dion & Bouton Tractor and Trailer	"	9 15	6 15.5	3 3/2	35	2.75	3.95	5.7			"	6.0	4.55 7.7		400	34.3	193.8	6.6	12.6	6.78	2.6	0.267	9.45	De Dion & Bouton.
Panhard & Levassor Omnibus	"	3 7	2 3/6	0 19.8	14	2.2	3.7	5.7			Petrol		3.55 3.55	5.3	750	11.75	195.7	6.36	pints 1.35	0.84	2.12	0.634	b 7.2	Panhard & Levassor.
De Dietrich Lorry	"	2 9.2	1 3/3	1 3/6		0.99	3.66	7.2			"		3.75	6.3	690	6.96	193.4	5.28	pints 6.785	7.87	1.84	0.75	4.46	De Dietrich.
De Dion & Bouton Omnibus	Poids Lourds, 1898	8 5	6 1	1 3/4	20	3.06	4.12	5.62	200	424	Coke	1.87	3.95	6.3	400	29.4	150	8.96	9.3	4.5	1.88	0.228	a 5.66	De Dion & Bouton.
De Dion & Bouton Char-à-bancs	"	8 9.2	5 17.2	2 7/2	21	2.48	4.0	5.76			"	4.98			400		189.2	8.57	10.3	4.97	2.08	0.246	6.14	De Dion & Bouton.
De Dion & Bouton Lorry	"	8 14	5 4.2	3 3/2		1.6	3.55	5.92			"	4.5			400		192	6.83	13.1	5.8	2.50	0.298	7.31	De Dion & Bouton.
Serpollet Omnibus	"	6 15	5 2.6	1 6/6	11	3.85	6.97	9.2		758	Oil	c 11.0	4.75	3.95	415	14.7	187.5	7.77	pints 1.55	4.27	4.33	0.642	c 7.8	Serpollet.
Leyland Van	"	2 10.7	1 16	0 15	7	2.4	6.0	8.45	190	500	"	10.6			750	6.0	183	5.88	pints 2.1	d 0.675	2.5	0.985	e 5.9	Leyland.
Panhard & Levassor Van	"	3 4	1 17.2	0 19.75		1.89	4.77	8.2			Petrol		3.15 3.15	4.75	750	7.8	192	8.8	pints 0.985	insignificant	2.43	0.76	g 4.48	Panhard & Levassor.
De Dietrich Brake	"	3 6	1 12	1 9/6		1.06	3.64	7.3			"		4.35	6.3	700	7.8	191	6.7	pints 1.34		3.8	1.18	6.5	De Dietrich.
Roser Mazurier Omnibus	"	3 0	1 18.5	0 19.75	?	1.98	4.38	6.82			"			6.3	700	8.9	191	7.15	pints 1.33		2.8	0.93	5.5	De Dietrich.
Compagnie Française des Voitures Electromobiles (Berse) Van	"	3 12	2 10.8	0 18.7	14	2.72	5.47	7.75			"	8.8	4.75/6.3	5.9	500	9.8	95.8	5.87	0.78		1.45	0.40	h 5.75	Roser Mazurier.
Krieger Van	S.P.T.A., Liverpool Branch, 1898	2 16	1 19.5	0 15		2.44	11.3	16.0			Electricity					3.35	150.5	6.2	k.w.-hours 0.68		0.153	0.055	i 4.95	Cie Française des Electromobiles.
Thornycroft Lorry	"	2 0	1 8.7	0 10		2.88	3.65	5.1	175		Coal	4.91	4.0	5.0	500	7.85	97.5	6.4	k.w.-hours 0.305		0.08	0.04	4.26	Krieger.
Thornycroft 6-wheel Lorry	"	5 7.2	3 5	2 10.6		1.29	3.18	6.85		65	"	4.78	4.0	5.0	500	18.9	148	5.22	pints 9.2	3.69	0.75	0.12	j 118	Thornycroft.
Lifu Lorry	"	9 11	4 8.2	4 14.6		0.93	4.9	10.6			"		4.0	5.0	500	18.0	71.8	2.79	pints 19.07	9.11	1.55	0.16	146	Thornycroft.
Leyland Lorry	R.A.S., Birmingham, 1898	6 8	2 18.4	2 4/4		1.21	2.67	5.35	210	80	Oil	8.66	3.0	5.0	630	29.0	143.5	7.02	pints 5.24	4.54	2.62	0.49	l 111.4	Lifu.
Thornycroft Lorry	"	7 8.6	3 2	4 12		0.78	4.5	10.4		110	Coal	8.81	3.0	5.0	500	14.0	143	5.22	pints 4.22	3.41	2.11	0.28	9.7	Leyland.
Leyland Van	"	6 17.5	3 2	3 0/5		1.13	3.78	7.15	175	65	Oil	9.23	4.0	5.0	500	14.0	142.5	6.2	pints 19.8	4.55	2.09	0.30		Thornycroft.
English Daimler Van	(Automobile Club, Liverpool Branch)	2 9.8	2	1 5/5		1.33	4.42	9.85	200	110	Petrol	8.8	3.0	6.0	500	14.0	468	6.48	pints 4.02	3.53	2.38	0.364		Leyland.
Thornycroft Lorry	Automobile Club, 1899			3 0		1.2	3.65	9.1		65	"	9.1	3.62	6.87	700	5.5	468	7.82	pints 0.48	0.28	o 0.45	0.18		Daimler.
Bayley Lorry	"	3 12.6	3 12.6	3 5		1.08			175		Coal	6.9	4.0	5.0	440		20	5.45	9.85	6.8	m 1.13	0.16		Thornycroft.
Cannstatt-Daimler Lorry	"	3 9.5	3 9.5	3 5		1.08					Coke	7.97	4.0	5.0	500		20	5.2	6.15	4.9	q 0.588	0.055		Bayley.
Cannstatt-Daimler Lorry	"	4 8	2 2	2 4		0.92	5.4	11.3		77	Petrol		3.81	5.875	600	7.8	20	5.85	pints 1.1	insignificant	1.6	0.364		Cannstatt-Daimler.
English Daimler Postal Van	"	8 10	3 1	5 5		0.56	5.2	14.4			"		4.94 4.94	6.125	540	11.8	20	3.88	pints 1.6		2.4	0.282		Cannstatt-Daimler.
Thornycroft Lorry	"	4 1	2 2	1 16		1.17	3.65	7.0			"		3.56 3.56	4.75	800	11.3	20	50	pints 1.2		1.8	0.44		English Daimler.
Thornycroft Lorry and Trailer	(Automobile Club, Liverpool Branch, 1899)	7 9.3	3 11.6	3 14.5		0.96	2.05	4.27	175	83	Coal	8.13	4.0	5.0	770	35	71.5	5.31	8.98	7.15	m 1.02	r 0.135	9.55	Thornycroft.
Coulthard Lorry	"	11 12	4 15	6 13		0.71	2.97	5.8	200		"	7.86	4.0	5.0	770	40	71.5	5.67	12.46	9.81	1.45	0.125	11.9	Thornycroft.
Leyland Lorry	"	5 0	2 11.4	2 6/3		1.00	7.15	7.15	212	77	Oil		2.75 4.18	6.0	500	14	35.9	4.78			2.75	r 0.255	10.8	Coulthard.
Clarkson & Capel Lorry	"	7 15	3 2.8	4 8/8		0.71	4.5	11.08	200	110	"	9.2	2.75	5.0	400	14	66.8	5.02	4.0		3.63	0.522	11.7	Leyland.
Bayley Lorry	"	6 16.3	3 6.7	3 3/4		1.0	4.76	9.45	200	80	"	u 4.87	2.75	4.0	600	14	71.5	4.94	2.13		5.76	0.522	s 10.45	Clarkson & Capel.
	"	7 5.6	3 6.95	3 13.4		0.95	8.04	6.61		70	Coke	u 5.76	4.0		500	22	71.5	4.98	4.9		0.45	0.082		Bayley.

a Coke, at 21s. 6d. per ton.
b Petrol, at 2½d. per litre.
c Petroleum of 900/91 sp. gr., at 1½d. per litre.
d Condenser employed.
e Kerosene, at 2d. per litre.
f Compound engine.
g Petrol, at 4d. per litre.
h Kerosene, at 3d. per litre.
i Kerosene, at 1¾d. per litre.
j All below this calculated from cols. 20 and 21.
k Coal, at 15s. 6d. per ton.
l Kerosene, at 1¾d. per gallon.
m Coal, at 20s. per ton.
n Kerosene, at 4 7/10d. per gallon.
o Benzoline, at 7¾d. per gallon.
p These weights are approximate.
q Coke, at 1s. per ton.
r Coke, at 1s. per 1,000 gallons.
s Coke, at 11s. 6d. per ton.
t Cost of water, 1s. per ton.
u Condenser in use and water re-evaporated.

The material originally positioned here is too large for reproduction in this reissue. A PDF can be downloaded from the web address given on page iv of this book, by clicking on 'Resources Available'.

TABLE XXVI.

RESULTS OF THE AUTOMOBILE CLUB 1,000 MILE TOUR, APRIL–MAY, 1900; HILL-CLIMBING TRIALS. DEDUCED MECHANICAL EFFICIENCY.

Number. Class.	Name.	Total weight, incl. passengers. (cwt. qrs.)	Horse-power of Engine stated by maker.	Transmission Gear.	Taddington Hill: Av. speed m.p.h. on rise of 651 ft. in 13,320 ft.	Taddington Hill: Av. actual H.P. at driving road wheels.	Shap Fell (1): Av. speed m.p.h. on rise of 840 ft. in 38,300 ft.	Shap Fell (1): Av. actual H.P. at driving road wheels.	Shap Fell (2): Av. speed m.p.h. on rise of 500 ft. in 7,388 ft.	Shap Fell (2): Av. actual H.P. at driving road wheels.	Dunmail Raise: Av. speed m.p.h. on rise of 450 ft. in 9,040 ft.	Dunmail Raise: Av. actual H.P. at driving road wheels.	Birkhill: Av. speed m.p.h. on rise of 460 ft. in 10,560 ft.	Birkhill: Av. actual H.P. at road wheels.	Approximate Mechanical Efficiency of Vehicle. (per cent.)
A. 1	Benz Ideal. Messrs. Hewetson's	13 1	3	Belt, chain	5·89	1·56	11·5	2·07	4·8	1·76	5·01	1·47	6·0 d	1·57	51
9	Benz Ideal. Messrs. Hewetson's	13 0	3	Belt, tooth, chain	7·18	2·06	6·0	—	6·64	1·61	6·64	1·92	6·6 d	1·65	61·5
5	Locomobile Steam Carriage	9 1	2 k	Chain	9·76	1·86	—	—	—	—	9·79	1·89	10·9	1·98	61 k
16	Gladiator Voiturette	8 0	3·25	Chain, tooth	8·17	1·24	—	—	6·64	—	7·67 e	1·185	7·5 d	0·995	32·6
27	New Orleans Car. Messrs. Burford & Van Toll.	8 2	3	Belt, tooth	6·3	1·125	—	—	4·42	0·88	6·32 d	0·985	7·7	1·28	35·7
28	New Orleans Car. Messrs. Burford & Van Toll.	8 2	3	"	6·05	1·10	—	—	—	—	4·74	0·80	—	—	31·6
33	Decauville Car. The Motor Car Co.	8 3	3·5	"Tooth"	6·7	1·20	—	—	—	—	5·13	0·84	—	—	34·0
34	Decauville Car. The Motor Car Co.	8 3	3·5	"	6·3	1·10	—	—	—	—	6·84	1·21	8·6 e	1·23	39·4
41	International Victoria	11 0a	3	Belt, chain	5·29	1·26	—	—	—	—	6·94	1·35	—	—	48·5
44	International Victoria	10 2a	3	" "	6·17	1·43	—	—	—	—	5·55	1·33	4·1	—	46
51	Star Voiturette	11 2a	3·5	" "	9·15	2·30	—	—	—	—	6·04 e	—	8·0	1·76	54·4
52	Roots & Vembles Car (Kerosene oil)	11 2a	2·87	" Chain	5·8	1·205	—	—	—	—	—	—	—	—	50
B. 14	De Dion Voiturette. Mr. R. Fuller	9 2	3	Tooth	10·08	1·975	13·5	1·612	7·24	1·79	9·79	1·94	10·9	2·04	61·5
15	De Dion Voiturette. The Motor Power Co.	9 2	3	"	—	—	14·5	1·735	7·58	1·885	8·0 e	—	8·0 e	1·41	58·3
24	Marshall Carriage	15 1	5·0	Belt, chain	9·45	1·76	11·5	2·20	4·42	1·81	9·33	1·75	9·6	1·70	41
31	Renault Car. The Motor Car Co.	9 0	3·5 c	Tooth	7·56	1·41	—	—	—	—	7·33	1·37	—	—	58
32	Renault Car. The Motor Car Co.	15 2a	3·5 c	"	10·08	3·21	13·0	2·52	6·37	2·59	7·9	2·55	8·6	2·62	46·3
40	Wolseley Voiturette	15 3	5	Belt, tooth, chain	—	—	10·5	1·94	4·54	1·75	6·04	—	6·3	1·83	89·5 l
49	Marshall Carriage	14 3	5	Belt, chain	4·94	1·49	10·5	1·94	4·54	1·85	6·04	1·85	6·3	1·83	35·5
C. 8	Daimler Phaeton. Motor Manufacturing Co.	25 3	6	Tooth, chain	5·29	2·785	10·5	3·37	5·14	3·47	5·86	3·13	6·5	3·26	58·5
9	Daimler Phaeton. Motor Manufacturing Co.	25 3	6	"	5·49	3·09	10·5	3·75	4·42	2·8	5·13	2·93	6·3 e	3·20	52·8
23	Whitney Steam Car. Messrs. Brown Bros.	12 2a	3·8 b	Chain	4·56	—	—	—	—	—	7·67	—	8·9	—	62
26	Peugeot Carriage. Mr. Friswell	19 3	6	Tooth, chain	9·45	3·83	13·5	3·98	6·37	3·72	9·79	4·08	10·9	4·23	55·0
35	Daimler Car. The Daimler Co.	21 0	6	"	6·3	2·94	13·5	3·98	4·42	3·05	7·9	3·70	6·8	3·08	55·7
36	Daimler Car. The Daimler Co.	26 1	6	"	6·55	3·58	16	4·27	—	—	6·64 e	3·72	6·6	3·43	61·6
37	Daimler Parisian. The Daimler Co.	22 0	6	"	9·15	4·12	—	—	—	—	8·92	4·06	8·0	3·45	66·2
46	Georges Richard Car	21 0	7	Belt, tooth, chain	6·43	2·78	—	—	—	—	7·67	8·41	7·5	3·09 e	46·5
47	Georges Richard Car	21 0	7	" "	—	—	—	—	—	—	5·86 f	1·86	5·0 g	1·98	31·6
D. 22	Lanchester Car. Mr. Millership	16 0a	8	Worm	8·62	3·01	11·5	2·54	4·98	2·22	6·84	2·41	7·7	2·57	31·9
E (a). 12	De Dion Tricycle	3 0a	2·25	Tooth	9·45	0·87	—	—	—	—	7·9	0·495	12·6	0·747	27·6
20	Simms' Motor Wheel	4 0a	2·75	"	—	—	—	—	—	—	2·1 d	—	—	—	31·6
E (b). 39	Century Tandem Tricycle	5 2a	3	Chain	15·13 h	2·03	20·5	1·67	6·64 e h	0·98	11·41 d	1·42	10·9 d	1·08	36·2
3	Ariel Quadricycle. Mr. J. Stocks	6 0	3·15	Tooth	14·4	1·62	20·5	1·41	6·64 d j	—	8·22 d j	—	13·3 e	1·51	48·3
4	Ariel Tricycle with Whippet Trailer	5 2	2·25	"	—	—	—	—	—	—	—	—	12·6 d j	1·26 d j	67·5
A. A25	Benz Ideal. Mrs. Bazalgette	12 2	3	Belt, chain	6·7	1·81	—	—	—	—	4·66	1·25	7·5	—	51
B. A24	Mors Voiturette. Mr. Phillips	13 2	4	Tooth, chain	8·39	2·33	—	—	—	—	7·67	2·15	7·5	1·98	58·7
C. A2	6 H.P. Panhard. Mr. Butler	21 3a	6	Tooth, chain	5·69	2·54	10·5	2·86	4·8	2·74	6·22	2·81	5·7	2·43	44·6
A3	6 H.P. Panhard. Mr. Browne	20 3	6	"	8·17	3·49	18·0	3·20	5·68	2·90	7·9	3·06	8·6	3·49	54
A7	6 H.P. Daimler. Mr. Harmsworth	18 2a	6	"	13·74	7·39	19·0	5·71	—	—	13·69	7·24	6·6 e	2·62	43·6
A10	8 H.P. Napier. Mr. Kennard	26 0	8	"	6·43	2·89	11·0	3·67	—	—	7·08	3·1	11·5	5·86	82
A12	6 H.P. Daimler. Mr. Edmunds	20 2a	6	"	—	—	—	—	—	—	6·22	3·33	4·8	—	49·8
A21	6 H.P. Daimler. Mr. Pitman	24 2a	6·25	"	6·3	3·26	13·0	3·37	4·42	3·81	6·84	3·9	4·1 e	—	60·3
A23	6 H.P. Daimler. Mr. Cordingly	28 3a	6	"	5·8	2·54	—	—	—	—	6·84	3·9	4·1 e	—	59·7
A26	6 H.P. Daimler. Mr. Gregson	25 3	6	"	8·39	3·26	11·5	2·61	4·82	—	6·92	2·95	7·0	2·84	47·6
A30	6 H.P. Daimler. Mr. Siddeley	20 2	6	"	8·17	2·54	13·0	3·35	5·68	2·61	6·92	2·95	7·0	2·84	59·7
A31	6 H.P. Daimler. Mr. Johnson	24 2	6	"	—	3·98	11·5	3·35	5·68	3·48	7·9	3·92	8·2	3·93	59·7
D. A4	8 H.P. Panhard. Mr. Mark Mayhew	20 1a	8	Tooth, chain	10·08	4·28	12·0	—	5·48	—	5·26 f	—	10·4	4·64	53·5
A11	12 H.P. Daimler. Hon. J. Scott Montagu	23 3a	12	"	11·19	5·46	27·5	8·78	17·71	10·9	20·54 f	11·81	16·0	8·21	42
A17	12 H.P. Panhard. Hon. C. S. Rolls	26 1	12	"	17·77	9·55	—	—	7·24	6·4	10·27	6·44	16·0	6·46	82
A22	12 H.P. Daimler. Mr. J. A. Holder	30 1	12	"	14·4	8·94	—	—	—	—	9·79	—	10·0	—	57·6
A29	7 H.P. Peugeot. Mr. Mark Mayhew	19 2a	7	"	7·74	3·13	15·5	3·8	7·97	4·1	9·79	3·99	10·0	3·82	53·7
E (a). A16	Ariel Tricycle. Mr. A. J. Wilson	3 2a	2·25	Tooth	18·91	1·35	18·5	0·815	15·94	1·46	5·13	1·25	6·8 j	1·83	48·3
A20	Empress Tricycle. Mr. H. Ashby	3 2b	2·75	"	—	—	20·0	0·88	—	—	17·06	1·06	7·5	—	43·2
E (b). A28	Enfield Quadricycle. Mr. E. M. Iliffe	6 2b	2·25	"	9·15 e	1·11	—	—	—	—	9·33 e	1·13	12·0 e h	1·83	60·5

a Weight based on maker's statement.
b Estimated.
c The de Dion water-cooled engine used in this car is generally stated to give 3 H.P., and this figure has been taken in calculating the efficiency.
d Both passengers off temporarily.
e One passenger off temporarily.
f Two passengers off temporarily.
g All off temporarily.
h Pedalling.
j Had to push.
k Brake H.P. of engine here understated. See remarks, p. 618.
l Brake H.P. of engine taken as 3·1.
Air resistance is not included in these calculations. See remarks, p. 632.

The material originally positioned here is too large for reproduction in this reissue. A PDF can be downloaded from the web address given on page iv of this book, by clicking on 'Resources Available'.

Printed in the United States
By Bookmasters